DATE DUE

DEMCO 38-296

WILEY'S ENGLISH-SPANISH
SPANISH-ENGLISH
CHEMISTRY DICTIONARY

DICCIONARIO DE QUÍMICA
INGLÉS-ESPAÑOL
ESPAÑOL-INGLÉS WILEY

By the same author

English-Spanish, Spanish-English Legal Dictionary,
Second Edition published 1997.

Wiley's English-Spanish, Spanish-English Business Dictionary.
Published 1996.

*English-Spanish, Spanish-English Electrical and Computer
Engineering Dictionary.* Published 1996.

*Wiley's English-Spanish, Spanish-English Dictionary of
Psychology and Psychiatry.* Published 1995.

Por el mismo autor

Diccionario jurídico inglés-español, español-inglés,
Segunda Edición publicada en 1997.

Diccionario de negocios inglés-español, español-inglés Wiley.
Publicado en 1996.

*Diccionario de ingeniería eléctrica y de computadoras
inglés-español, español-inglés.* Publicada en 1996.

*Diccionario de psicología y psiquiatría inglés-español,
español-inglés Wiley.* Publicada en 1995.

WILEY'S ENGLISH-SPANISH SPANISH-ENGLISH CHEMISTRY DICTIONARY

DICCIONARIO DE QUÍMICA INGLÉS-ESPAÑOL ESPAÑOL-INGLÉS WILEY

Steven M. Kaplan
Lexicographer

Editorial Advisors
Dr. Rolf Altschul
Dr. Sylvia Márquez-Pirazzi

A WILEY-INTERSCIENCE PUBLICATION
JOHN WILEY & SONS, INC.
New York • Chichester • Weinheim • Brisbane • Singapore • Toronto

Library of Congress Cataloging in Publication Data:

Kaplan, Steven M.
    Wiley's English-Spanish, Spanish-English chemistry dictionary =
Diccionario de química inglés-español, español-inglés Wiley / Steven
M. Kaplan, lexicographer ; editorial advisory board, Rolf Altschul,
Sylvia Márquez-Pirazzi.
        p. cm
    "A Wiley-Interscience publication."
    Includes index.
    ISBN 0-471-19288-0 (cloth : alk. paper). — ISBN 0-471-24923-8
(paper : alk. paper)
    1. Chemistry—Dictionaries. 2. English language—Dictionaries—
Spanish. 3. Chemistry—Dictionaries—Spanish. 4. Spanish
language—Dictionaries—English. I. Title.
QD5.K292 1998
540'.3—dc21   97-34936
                                                  CIP

Printed in the United States of America

10 9 8 7 6 5 4 3 2 1

PREFACE

Throughout the world, over 350 million people speak Spanish and over 450 million speak English. Most of them live in America. Over the Internet, dozens of countries are providing a massive amount of information concerning chemistry. This has further increased the need for precise communication between users of chemical terminology in the two languages.

The primary objective of this dictionary is to provide accurate equivalents for the words and phrases most commonly encountered in the chemistry literature. There are over 40,000 total entries, approximately 20,000 in each section of the dictionary.

Prefixes such as cis, trans, ortho, meta, para, 1, 2, 3, D, L, alpha, beta, N, O, and sec, among many others, are not placed in front of compound names in the dictionary. There are three main reasons for this. First, this eliminates the possible confusion concerning alphabetization. It also avoids wasting space with certain groups of entries which would have almost identical equivalents. For example, to have three entries for

1,2,3-trimethylbenzene
1,2,4-trimethylbenzene, and
1,3,5-trimethylbenzene,

instead of just **trimethylbenzene,** would add to the bulk of the dictionary without increasing usefulness. Finally, when users of the dictionary encounter compounds of this nature, simply placing the prefix in the target language in the same spot it was found in the source language is all that is needed for transliteration. So **1,3,5-trimethylbenzene** would be **1,3,5-trimetilbenceno.** Another example will illustrate how this works with another prefix in a different location: **m-sulfobenzoic acid** has **ácido m-sulfobenzoico** as its equivalent.

Many thousands of chemical compounds are included in this dictionary, allowing for quick and simple reference. Unfortunately, many chemical compounds have names that are either too long or are not encountered frequently enough in the literature to warrant inclusion. To help users in these cases, most every prefix and suffix that can be an integral part of a chemical name is included. For example, cyclo, di, ethyl, and ate, among over one hundred others, will help piece together even the most involved name.

I am very grateful to John Wiley Editor Carla A. Fjerstad who, after recognizing the merits of this project, enlisted the prominent advisors and supervised the work. The editorial advisors, Dr. Rolf Altschul and Dr. Sylvia Márquez-Pirazzi, made significant contributions to the quality of the dictionary, which is greatly appreciated. Special thanks go to

John Wiley Senior Editor George J. Telecki, who kindly insured that the idea for this book was communicated to the appropriate person. I also wish to thank Wiley Editorial Assistant Jennifer Campaniolo who helped things march smoothly throughout the project.

Steven M. Kaplan

Miami, Florida

PREFACIO

A través del mundo, más de 350 millones de personas hablan español y otras 450 millones hablan inglés. La mayoría de estas viven en América. A través de la Interred, docenas de países están proveyendo una cantidad masiva de información concerniente a la química. Esto ha aumentado aun más la necesidad para la comunicación precisa entre los usuarios de la terminología química en los dos idiomas.

El objetivo principal de este diccionario es proveer equivalentes precisos para las palabras y frases más comúnmente encontradas en la literatura de la química. Hay más de 40,000 voces de entradas en total, aproximadamente 20,000 en cada sección del diccionario.

Prefijos tales como cis, trans, orto, meta, para, 1, 2, 3, D, L, alfa, beta, N, O, y sec, entre muchos otros, no están colocados delante de los nombres de compuestos en el diccionario. Hay tres razones principales para esto. Primero, esto elimina la posible confusión referente a la alfabetización. Además evita desperdiciar espacio con ciertos grupos de voces de entradas las cuales tendrían equivalentes casi idénticos. Por ejemplo, el tener tres voces de entrada para

1,2,3-trimethylbenzene
1,2,4-trimethylbenzene, y
1,3,5-trimethylbenzene,

en vez de sólo **trimethylbenzene,** añadiría tamaño al diccionario sin aumentar la utilidad. Finalmente, cuando los usuarios del diccionario encuentren compuestos de esta naturaleza, el sencillamente colocar el prefijo en el idioma objeto en el mismo lugar en que se encontró en el idioma de origen, es todo lo que se necesita para la transliteración. Así que, **1,3,5-trimethylbenzene** sería **1,3,5-trimetilbenceno.** Otro ejemplo ilustrará cómo esto funciona con otro prefijo en otro lugar: **m-sulfobenzoic acid** tiene **ácido m-sulfobenzoico** como su equivalente.

Muchos miles de compuestos químicos están incluidos en este diccionario, permitiendo referencia sencilla y rápida. Desafortunadamente, muchos compuestos químicos tienen nombres los cuales son o muy largos, o que no se encuentran suficientemente frecuentemente como para justificar su inclusión. Para ayudar a los usuarios en estos casos, casi cualquier prefijo y sufijo el cual podría ser una parte integral de un nombre químico está incluido. Por ejemplo, ciclo, di, etil, y ato, entre más de cien otros, ayudarán a formar hasta el nombre más envuelto.

Le estoy muy agradecido a la editora de John Wiley Carla A. Fjerstad, quien tras reconocer los méritos de este proyecto, reclutó los asesores prominentes, y supervisó el trabajo. Los asesores editoriales, el Dr. Rolf Altschul y la Dra. Sylvia Márquez-Pirazzi, hicieron contribuciones significativas a la calidad del diccionario, lo cual se aprecia grandemente. Gracias especiales van al editor de John Wiley George J. Telecki, quien amablemente se aseguró de que la idea para este diccionario fuera comunicada a la persona apropiada. También quiero dar mis gracias a la asistente de editor de Wiley Jennifer Campaniolo quien ayudó a que todo marchara sobre ruedas a través del proyecto.

Steven M. Kaplan

Miami, Florida

ENGLISH-SPANISH
INGLÉS-ESPAÑOL

A

abaca abacá
Abbegg rule regla de Abbegg
Abel reagent reactivo de Abel
aberration aberración
abherent antiadherente
abieninic acid ácido abienínico
abietadienoic acid ácido abietadienoico
abietic acid ácido abiético
abietic anhydride anhídrido abiético
ablation ablación
ablative material material ablativo
abrasion abrasión
abrasive abrasivo
abscisic acid ácido abscísico
absinthe absintio
absinthium absintio, ajenjo
absolute absoluto
absolute alcohol alcohol absoluto
absolute boiling point punto de
    ebullición absoluto
absolute configuration configuración
    absoluta
absolute density densidad absoluta
absolute error error absoluto
absolute humidity humedad absoluta
absolute stereochemistry estereoquímica
    absoluta
absolute system sistema absoluto
absolute temperature temperatura
    absoluta
absolute temperature scale escala de
    temperatura absoluta
absolute unit unidad absoluta
absolute zero cero absoluto
absolute zero temperature temperatura
    de cero absoluto
absorbance absorbancia
absorbancy absorbancia
absorbent absorbente
absorptiometer absorciómetro
absorptiometric analysis análisis
    absorciométrico
absorption absorción

absorption apparatus aparato de
    absorción
absorption band banda de absorción
absorption cell célula de absorción
absorption coefficient coeficiente de
    absorción
absorption constant constante de
    absorción
absorption edge borde de absorción
absorption indicator indicador de
    absorción
absorption limit límite de absorción
absorption line línea de absorción
absorption of light absorción de luz
absorption paper papel de absorción
absorption peak pico de absorción
absorption spectroscopy espectroscopia
    de absorción
absorption spectrum espectro de
    absorción
absorption tower torre de absorción
absorption tube tubo de absorción
absorptive power poder de absorción
absorptivity absortividad
abundance abundancia
abundance ratio razón de abundancia
acacetin acacetina
acacia acacia
acacia gum goma de acacia
acadesine acadesina
acanthite acantita
accelerating anode ánodo acelerador
accelerating electrode electrodo
    acelerador
acceleration aceleración
accelerator acelerador
accelofilter filtro acelerador
acceptable explosive explosivo aceptable
acceptable risk riesgo aceptable
acceptor aceptor
accessory accesorio
accumulator acumulador
accuracy precisión, exactitud

acenaphthene acenafteno
acenaphthenequinone acenaftenoquinona
acenaphthylene acenaftileno
acetal acetal
acetal resin resina acetálica
acetaldehyde acetaldehído
acetaldehyde cyanohydrin cianhidrina
  de acetaldehído
acetaldol acetaldol
acetamide acetamida
acetamide chloride cloruro de acetamida
acetamide hydrochloride clorhidrato de
  acetamida
acetamide nitrate nitrato de acetamida
acetamidine acetamidina
acetamidine hydrochloride clorhidrato
  de acetamidina
acetaminophen acetaminofeno,
  acetaminofén
acetaminosalol acetaminosalol
acetanilide acetanilida
acetate acetato
acetate film película de acetato
acetazolamide acetazolamida
acetiamine acetiamina
acetic acid ácido acético
acetic acid amide amida de ácido acético
acetic acid amine amina de ácido acético
acetic acid glacial ácido acético glacial
acetic aldehyde aldehído acético
acetic anhydride anhídrido acético
acetic ester éster acético
acetic ether éter acético
acetic oxide óxido acético
acetin acetina
acetoacetanilide acetoacetanilida
acetoacetate acetoacetato
acetoacetic acid ácido acetoacético
acetoacetic ester éster acetoacético
acetoin acetoína
acetol acetol
acetolysis acetólisis
acetone acetona
acetone bromoform bromoformo de
  acetona
acetone chloroform cloroformo de
  acetona
acetone cyanohydrin cianhidrina de
  acetona

acetone dicarboxylic acid ácido
  acetonadicarboxílico
acetone monocarboxylic acid ácido
  acetonamonocarboxílico
acetone number número de acetona
acetone semicarbazone semicarbazona
  de acetona
acetone sodium bisulfate bisulfato
  sódico de acetona
acetonitrile acetonitrilo
acetonylacetone acetonilacetona
acetonylidene acetonilideno
acetophenetidine acetofenetinina
acetophenone acetofenona
acetophenone acetone acetona de
  acetofenona
acetophenone oxime oxima de
  acetofenona
acetostearin acetoestearina
acetotoluide acetotoluida
acetotoluidide acetotoluidida
acetoxime acetoxima
acetoxylation acetoxilación
acetozone acetozona
acetyl acetilo
acetyl bromide bromuro de acetilo
acetyl chloride cloruro de acetilo
acetyl coenzyme acetilcoenzima
acetyl fluoride fluoruro de acetilo
acetyl iodide yoduro de acetilo
acetyl ketene acetilcetena
acetyl nitrate nitrato de acetilo
acetyl oxide óxido de acetilo
acetyl peroxide peróxido de acetilo
acetyl value valor de acetilo
acetylacetic acid ácido acetilacético
acetylacetonate acetilacetonato
acetylacetone acetilacetona
acetylamino acetilamino
acetylating agent agente acetilante
acetylation acetilación
acetylbenzene acetilbenceno
acetylcarbromal acetilcarbromal
acetylcholine acetilcolina
acetylcholinesterase acetilcolinesterasa
acetylene acetileno
acetylene black negro de acetileno
acetylene complex complejo de acetileno

acetylene dibromide dibromuro de acetileno
acetylene dichloride dicloruro de acetileno
acetylene hydrocarbon hidrocarburo de acetileno
acetylene tetrabromide tetrabromuro de acetileno
acetylenogen acetilenógeno
acetylenyl acetilenilo
acetylferrocene acetilferroceno
acetylformic acid ácido acetilfórmico
acetylide acetiluro
acetylmethylcarbinol acetilmetilcarbinol
acetylphenol acetilfenol
acetylpropionic acid ácido acetilpropiónico
acetylsalicylic acid ácido acetilsalicílico
acetylurea acetilurea
Acheson process proceso de Acheson
achiral aquiral
achromatic acromático
achromatic indicator indicador acromático
acicular acicular
acid ácido
acid acceptor aceptor de ácido
acid alcohol alcohol ácido
acid amide amida de ácido
acid ammonium sulfate sulfato amónico ácido
acid ammonium tartrate tartrato amónico ácido
acid anhydride anhídrido de ácido
acid-base catalysis catálisis ácido-base
acid-base equilibrium equilibrio ácido-base
acid-base indicator indicador ácido-base
acid-base pair par ácido-base
acid-base titration titulación ácido-base, valoración ácido-base
acid chloride cloruro de ácido
acid content contenido ácido
acid dissociation disociación de ácido
acid dye colorante ácido
acid fuchsin fucsina ácida
acid group grupo ácido
acid halide haluro ácido
acid magnesium citrate citrato ácido magnésico
acid methyl sulfate sulfato ácido de metilo
acid number número de ácido
acid oxide óxido ácido
acid phosphatase fosfatasa ácida
acid phosphate fosfato ácido
acid potassium sulfate sulfato ácido potásico
acid precipitation precipitación ácida
acid radical radical de ácido
acid rain lluvia ácida
acid reaction reacción ácida
acid-resistant resistente al ácido
acid salt sal ácida
acid solution solución ácida
acid solvent solvente ácido
acid value valor de ácido
acidic acídico
acidic group grupo acídico
acidic hydrogen hidrógeno acídico
acidification acidificación
acidify acidificar
acidimeter acidímetro
acidimetry acidimetría
acidity acidez
acidity coefficient coeficiente de acidez
acidity constant constante de acidez
acidolysis acidólisis
acidproof a prueba de ácido
acidulant acidulante
Acker process proceso de Acker
aconic acid ácido acónico
aconite acónito
aconitic acid ácido aconítico
aconitine aconitina
Acree-Rosenheim reaction reacción de Acree-Rosenheim
acrid acre
acridine acridina
acridine orange anaranjado de acridina
acridinic acid ácido acridínico
acriflavine acriflavina
acrimonious acrimonioso
acroleic acid ácido acroleico
acrolein acroleína
acrolein dimer dímero de acroleína
acrolein polymer polímero de acroleína
acrolein test prueba de acroleína

acrometer acrómetro
acrylaldehyde acrilaldehído
acrylamide acrilamida
acrylamide copolymer copolímero de acrilamida
acrylamide polymer polímero de acrilamida
acrylate acrilato
acrylate plastic plástico de acrilato
acrylate resin resina de acrilato
acrylic acrílico
acrylic acid ácido acrílico
acrylic acid polymer polímero de ácido acrílico
acrylic aldehyde aldehído acrílico
acrylic ester éster acrílico
acrylic polymer polímero acrílico
acrylic resin resina acrílica
acrylonitrile acrilonitrilo
acrylonitrile copolymer copolímero de acrilonitrilo
acrylonitrile dimer dímero de acrilonitrilo
acrylonitrile polymer polímero de acrilonitrilo
acryloyl chloride cloruro de acriloilo
actinic actínico
actinic radiation radiación actínica
actinide actínido
actinide series serie actínida
actinism actinismo
actinium actinio
actinium series serie de actinio
actinochemistry actinoquímica
actinoid actinoide
actinoid contraction contracción de actinoides
actinometer actinómetro
actinometry actinometría
actinomycin actinomicina
actinon actinón
actinoquinol actinquinol
action acción
action spectrum espectro de acción
activated activado
activated adsorption adsorción activada
activated alumina alúmina activada
activated carbon carbón activado
activated charcoal carbón activado

activated complex complejo activado
activated molecule molécula activada
activated state estado activado
activation activación
activation analysis análisis por activación
activation energy energía de activación
activator activador
active activo
active carbon carbón activo
active center centro activo
active component componente activo
active element elemento activo
active group grupo activo
active mass masa activa
active site lugar activo
active transport transporte activo
activity actividad
activity coefficient coeficiente de actividad
activity curve curva de actividad
activity series serie de actividad
actuator actuador
acyclic acíclico
acyclic compound compuesto acíclico
acyl acilo
acyl group grupo acilo
acyl halide haluro de acilo
acylation acilación
acyloin aciloína
acyloin condensation condensación aciloínica
adamantane adamantano
Adamkiewicz reaction reacción de Adamkiewicz
Adams catalyst catalizador de Adams
adamsite adamsita
adapter adaptador
adatom adátomo
addition adición
addition agent agente de adición
addition complex complejo de adición
addition polymer polímero de adición
addition reaction reacción de adición
addition resin resina de adición
additive aditivo
adduct aducto
adenine adenina
adenosine adenosina

adenosine diphosphate difosfato de adenosina

adenosine monophosphate monofosfato de adenosina

adenosine phosphate fosfato de adenosina

adenosine triphosphate trifosfato de adenosina

adenylic acid ácido adenílico

adherence adherencia

adhesion adhesión

adhesion agent agente de adhesión

adhesive adhesivo

adiabatic adiabático

adiabatic calorimeter calorímetro adiabático

adiabatic change cambio adiabático

adiabatic compression compresión adiabática

adiabatic cooling enfriamiento adiabático

adiabatic expansion expansión adiabática

adiabatic heating calentamiento adiabático

adiabatic process proceso adiabático

adipate adipato

adipic acid ácido adípico

adipinketone adipincetona

adiponitrile adiponitrilo

adjacent position posición adyacente

adjuvant adyuvante

Adkins-Peterson reaction reacción de Adkins-Peterson

adrenaline adrenalina

adsorbate adsorbato

adsorbent adsorbente

adsorption adsorción

adsorption analysis análisis de adsorción

adsorption chromatography cromatografía por adsorción

adsorption coefficient coeficiente de adsorción

adsorption indicator indicador de adsorción

adulterant adulterante

adulterated adulterado

adulteration adulteración

advection advección

aerate airear

aerated water agua aireada

aerobic aerobio, aeróbico

aerogel aerogel

aerometer aerómetro

aerometry aerometría

aerosol aerosol

affinity afinidad

affinity chromatography cromatografía por afinidad

affinity curve curva de afinidad

aflatoxin aflatoxina

afterglow luminiscencia persistente, luminiscencia residual

agar agar

agaric acid ácido agárico

agaricin agaricina

agate ágata

agent agente

agglomeration aglomeración

agglutinate aglutinar

agglutination aglutinación

aggregate agregado

aggregation agregación

aging envejecimiento

agitator agitador

aglucone aglucona

aglycone aglicona

agmatine agmatina

agricultural chemical producto químico agrícola

agricultural waste desperdicios agrícolas

Agulhon reagent reactivo de Agulhon

Aich metal metal de Aich

air aire

air bath baño de aire

air compressor compresor de aire

air filter filtro de aire

air oxidation oxidación por aire

air pollution contaminación de aire

air pump bomba de aire

air regulator regulador de aire

alabandite alabandina

alabaster alabastro

alacreatine alacreatina

alanine alanina

alantic acid ácido alántico

alanyl alanilo

alazopeptin alazopeptina

albaspidin albaspidina

albata albata

albedo albedo
albertite albertita
albite albita
albomycin albomicina
albumen albumen
albumin albúmina
albuminoid albuminoide
albumitate albuminato
albutoin albutoína
alclophenac alclofenac
alcogel alcogel
alcohol alcohol
alcoholate alcoholato
alcoholic alcohólico
alcoholic fermentation fermentación
　　alcohólica
alcoholysis alcohólisis
alcosol alcosol
alcyl alcilo
aldehyde aldehído
aldehyde condensation condensación de
　　aldehído
aldehyde ketone cetona de aldehído
aldehyde polymer polímero de aldehído
aldehydine aldehidina
Alder-Rickert rule regla de Alder-
　　Rickert
Alder-Stein rules reglas de Alder-Stein
aldicarb aldicarb
aldohexose aldohexosa
aldoketene aldocetena
aldol aldol
aldol condensation condensación aldólica
aldolase aldolasa
aldonic acid ácido aldónico
aldopentose aldopentosa
aldose aldosa
aldosterone aldosterona
aldoxime aldoxima
aldrin aldrina
aleuritic acid ácido aleurítico
alexandrite alexandrita
alfin alfin
alfin catalyst catalizador alfin
algae algas
algarose algarosa
algicide algicida
algin algina
alginate alginato

alginic acid ácido algínico
algulose algulosa
alicyclic alicíclico
alicyclic compound compuesto alicíclico
alignment alineación
aliphatic alifático
aliphatic acid ácido alifático
aliphatic compound compuesto alifático
aliphatic group grupo alifático
aliphatic hydrocarbon hidrocarburo
　　alifático
aliphatic polycyclic hydrocarbon
　　hidrocarburo policíclico alifático
aliphatic series serie alifática
alipite alipita
aliquot alícuota
alite alita
alizaramide alizaramida
alizarate alizarato
alizaric acid ácido alizárico
alizarin alizarina
alizarin black negro de alizarina
alizarin blue azul de alizarina
alizarin cyanine cianina de alizarina
alizarin dye colorante de alizarina
alizarin red rojo de alizarina
alizarin sulfonate sulfonato de alizarina
alizarin yellow amarillo de alizarina
alkadiene alcadieno
alkalamide alcalamida
alkalescence alcalescencia
alkalescent alcalescente
alkali álcali
alkali alcoholate alcoholato alcalino
alkali blue azul alcalino
alkali metal metal alcalino
alkali reaction reacción alcalina
alkalide alcalida
alkalimeter alcalímetro
alkalimetry alcalimetría
alkaline alcalino
alkaline earth tierra alcalina
alkaline-earth metal metal alcalinotérreo
alkaline-earth oxide óxido alcalinotérreo
alkaline reaction reacción alcalina
alkaline solution solución alcalina
alkalinity alcalinidad
alkalize alcalizar
alkaloid alcaloide

alkaloidal alcaloidal
alkalometry alcalometría
alkamine alcamina
alkanal alcanal
alkane alcano
alkanesulfonic acid ácido
    alcanosulfónico
alkanization alcanización
alkannin alcanina
alkanol alcanol
alkanolamine alcanolamina
alkasal alcasal
alkene alqueno
alkene series serie de alquenos
alkoxide alcóxido
alkoxy alcoxi
alkoxylate alcoxilato
alkyd resin resina alquídica
alkyl alquilo
alkyl halide haluro de alquilo
alkyl nitrite nitrito de alquilo
alkyl oxide óxido de alquilo
alkylamine alquilamina
alkylate alquilato
alkylation alquilación
alkylbenzene alquilbenceno
alkylbenzene sulfonate sulfonato de
    alquilbenceno
alkyle alquilo
alkylene alquileno
alkylide alquiluro
alkylidene alquilideno
alkylolamide alquilolamida
alkylolamine alquilolamina
alkylphenol alquilfenol
alkyne alquino
Allan-Robinson reaction reacción de
    Allan-Robinson
allanic acid ácido alánico
allanite alanita
allantoic acid ácido alantoico
allantoin alantoína
allantoxanic acid ácido alantoxánico
allanturic acid ácido alantúrico
allelochemical aleloquímico
allelomorph alelomorfo
allelopathic chemical producto químico
    alelopático
allelotrope alelótropo

allelotropism alelotropismo
allemontite alemontita
allene aleno
allethrin aletrina
allicin alicina
alligation aligación
allo alo
allochromatic alocromático
allocinnamic acid ácido alocinámico
alloclasite aloclasita
alloisomerism aloisomerismo
allomaleic acid ácido alomaleico
allomeric alomérico
allomerism alomerismo
allomerization alomerización
allomorphism alomorfismo
allophanamide alofanamida
allophanate alofanato
allophane alófana
allophanic alofánico
allophanic acid ácido alofánico
allose alosa
allosteric effect efecto alostérico
allosteric enzyme enzima alostérica
allotoxin alotoxina
allotriomorphism alotriomorfismo
allotrope alótropo
allotropic alotrópico
allotropism alotropismo
allotropy alotropía
allowable permisible
allowed permitido
allowed band banda permitida
allowed transition transición permitida
alloxan aloxana
alloxanic acid ácido aloxánico
alloxanthin aloxantina
alloy aleación
alloy steel acero aleado
allulose alulosa
allyl alilo
allyl acetate acetato de alilo
allyl acrylate acrilato de alilo
allyl alcohol alcohol alílico
allyl aldehyde aldehído alílico
allyl benzoate benzoato de alilo
allyl bromide bromuro de alilo
allyl butyrate butirato de alilo
allyl carbamate carbamato de alilo

allyl chloride cloruro de alilo
allyl cinnamate cinamato de alilo
allyl cyanide cianuro de alilo
allyl ether éter alílico
allyl fluoride floruro de alilo
allyl group grupo alilo
allyl iodide yoduro de alilo
allyl isocyanate isocianato de alilo
allyl isosulfocyanate isosulfocianato de
    alilo
allyl isothiocyanate isotiocinato de alilo
allyl oxalate oxalato de alilo
allyl plastic plástico alílico
allyl polymer polímero de alilo
allyl resin resina alílica
allyl sulfide sulfuro de alilo
allyl thioether tioéter alílico
allyl tribromide tribromuro de alilo
allylacetone alilacetona
allylacetonitrile alilacetonitrilo
allylamine alilamina
allylaniline alilanilina
allylene alileno
allylic alílico
allylic rearrangement reordenamiento
    alílico
allylin alilina
allylthiourea aliltiourea
allyltrichlorosilane aliltriclorosilano
almandite almandita
Almen test prueba de Almen
almond oil aceite de almendra
alneon alneón
alnico álnico
aloe áloe
aloeresic acid ácido aloerésico
aloin aloína
aloxiprin aloxiprina
alpax alpax
alpha alfa
alpha cellulose celulosa alfa
alpha decay desintegración alfa
alpha disintegration desintegración alfa
alpha emission emisión alfa
alpha emitter emisor alfa
alpha helix hélice alfa
alpha particle partícula alfa
alpha position posición alfa
alpha radiator radiador alfa

alpha rays rayos alfa
alphahydroxy acid alfahidroxiácido
alphatopic alfatópico
alphyl alfilo
alstonidine alstonidina
alstonine alstonina
alternating polymer polímero alternante
alternation alternación
alternation law ley de alternación
altheine alteína
altrose altrosa
aludel aludel
alum alumbre
alumina alúmina
alumina gel gel de alúmina
alumina trihydrate trihidrato de alúmina
aluminate aluminato
aluminic acid ácido alumínico
aluminiferous aluminífero
aluminite aluminita
aluminium aluminio
aluminized aluminizado
aluminoferric aluminoférrico
aluminosilicate aluminosilicato
aluminum aluminio
aluminum acetate acetato alumínico,
    acetato de aluminio
aluminum acetotartrate acetotartrato
    alumínico
aluminum alkoxide alcóxido alumínico
aluminum alloy aleación alumínica
aluminum ammonium sulfate sulfato
    amónico alumínico
aluminum antimonide antimoniuro
    alumínico
aluminum benzoate benzoato alumínico
aluminum borate borato alumínico
aluminum boride boruro alumínico
aluminum boroformate boroformiato
    alumínico
aluminum borohydride borohidruro
    alumínico
aluminum brass latón de aluminio
aluminum bromate bromato alumínico
aluminum bromide bromuro alumínico
aluminum bronze bronce de aluminio
aluminum butoxide butóxido alumínico
aluminum calcium hydride hidruro
    cálcico alumínico

aluminum carbide   carburo alumínico

aluminum carbonate   carbonato alumínico

aluminum chlorate   clorato alumínico

aluminum chloride   cloruro alumínico

aluminum chlorohydrate   clorohidrato alumínico

aluminum diacetate   diacetato alumínico

aluminum dichromate   dicromato alumínico

aluminum diformate   diformiato alumínico

aluminum distearate   diestearato alumínico

aluminum ethanoate   etanoato alumínico

aluminum ethoxide   etóxido alumínico

aluminum ethylate   etilato alumínico

aluminum fluoride   fluoruro alumínico

aluminum foil   papel de aluminio

aluminum formate   formiato alumínico

aluminum gluconate   gluconato alumínico

aluminum hydrate   hidrato alumínico

aluminum hydride   hidruro alumínico

aluminum hydroxide   hidróxido alumínico

aluminum iodide   yoduro alumínico

aluminum isopropoxide   isopropóxido alumínico

aluminum isopropylate   isopropilato alumínico

aluminum lactate   lactato alumínico

aluminum lithium hydride   hidruro de litio alumínico

aluminum metaphosphate   metafosfato alumínico

aluminum monostearate   monoestearato alumínico

aluminum nitrate   nitrato alumínico

aluminum nitride   nitruro alumínico

aluminum oleate   oleato alumínico

aluminum orthophosphate   ortofosfato alumínico

aluminum oxalate   oxalato alumínico

aluminum oxide   óxido alumínico

aluminum palmitate   palmitato alumínico

aluminum phenolate   fenolato alumínico

aluminum phosphate   fosfato alumínico

aluminum phosphide   fosfuro alumínico

aluminum phosphinate   fosfinato alumínico

aluminum potassium chloride   cloruro potásico alumínico

aluminum potassium sulfate   sulfato potásico alumínico

aluminum potassium tartrate   tartrato potásico alumínico

aluminum propoxide   propóxido alumínico

aluminum resinate   resinato alumínico

aluminum rubidium sulfate   sulfato de rubidio alumínico

aluminum salicylate   salicilato alumínico

aluminum silicate   silicato alumínico

aluminum sodium chloride   cloruro sódico alumínico

aluminum sodium fluoride   fluoruro sódico alumínico

aluminum sodium sulfate   sulfato sódico alumínico

aluminum stearate   estearato alumínico

aluminum sulfate   sulfato alumínico

aluminum sulfide   sulfuro alumínico

aluminum tartrate   tartrato alumínico

aluminum thallium sulfate   sulfato de talio alumínico

aluminum thiocyanate   tiocianato alumínico

aluminum triacetate   triacetato alumínico

aluminum zinc sulfate   sulfato de cinc alumínico

alumite   alumita

alunite   alunita

alunogen   alonógeno

alunogenite   alunogenita

alvite   alvita

Amadori rearrangement   reordenamiento de Amadori

amalgam   amalgama

amalgamation   amalgamación

amalgamation process   proceso de amalgamación

amalgamator   amalgamador

amalinic acid   ácido amalínico

amanitine   amanitina

amaranth   amaranto

amarine   amarina

amaroid   amaroide

amatol amatol
amazonite amazonita
amber ámbar
ambident ambidente
ambident anion anión ambidente
ambient air aire ambiente
ambient pressure presión ambiente
ambient temperature temperatura
   ambiente
amblygonite ambligonita
ambrite ambrita
ameboid ameboide
americium americio
amesite amesita
amethocaine ametocaína
amethyst amatista
amicron amicrón
amidase amidasa
amidation amidación
amide amida
amide chloride cloruro de amida
amide group grupo amida
amidine amidina
amidino amidino
amido amido
amidol amidol
amidone amidona
amidrazone amidrazona
amination aminación
amine amina
aminic acid ácido amínico
aminimide aminimida
amino amino
amino acid aminoácido
amino acid oxidase aminoácido oxidasa
amino alcohol aminoalcohol
amino group grupo amino
amino oxidase aminooxidasa
amino resin aminorresina
aminoacetanilide aminoacetanilida
aminoacetic acid ácido aminoacético
aminoacetone aminoacetona
aminoacetophenone aminoacetofenona
aminoanthraquinone aminoantraquinona
aminoantipyrine aminoantipirina
aminoazo aminoazo
aminoazobenzene aminoazobenceno
aminoazobenzene hydrochloride
   clorhidrato de aminoazobenceno

aminoazonaphthalene
   aminoazonaftaleno
aminoazotoluene aminoazotolueno
aminobenzene aminobenceno
aminobenzenesulfonic acid ácido
   aminobencenosulfónico
aminobenzoic acid ácido aminobenzoico
aminobenzothiazole aminobenzotiazol
aminobiphenyl aminobifenilo
aminobutane aminobutano
aminobutanoic acid ácido
   aminobutanoico
aminobutyric acid ácido aminobutírico
aminocaproic acid ácido aminocaproico
aminocyclohexane aminociclohexano
aminodiethylaniline aminodietilanilina
aminodimethylaniline
   aminodimetilanilina
aminodimethylbenzene
   aminodimetilbenceno
aminodiphenyl aminodifenilo
aminodiphenylamine aminodifenilamina
aminodithioformic acid ácido
   aminoditiofórmico
aminoethane aminoetano
aminoethanol aminoetanol
aminoethyl alcohol alcohol aminoetílico
aminoethylsulfuric acid ácido
   aminoetilsulfúrico
aminoform aminoformo
aminoglutaric acid ácido aminoglutárico
aminohexane aminohexano
aminohexanoic acid ácido
   aminohexanoico
aminohippuric acid ácido aminohipúrico
aminoid aminoide
aminoisobutyric acid ácido
   aminoisobutírico
aminoketone aminocetona
aminomethane aminometano
aminomethylation aminometilación
aminonaphthalene aminonaftaleno
aminonaphthol aminonaftol
aminonaphtholsulfonic acid ácido
   aminonaftolsulfónico
aminonicotinic acid ácido
   aminonicotínico
aminophenol aminofenol
aminophenylacetic acid ácido

aminofenilacético

aminophosphoric acid ácido aminofosfórico

aminophylline aminofilina

aminopicoline aminopicolina

aminopropane aminopropano

aminopropanoic acid ácido aminopropanoico

aminopropanol aminopropanol

aminopropene aminopropeno

aminopropyl alcohol alcohol aminopropílico

aminopropylmorpholine aminopropilmorfolina

aminopterin aminopterina

aminopurine aminopurina

aminopyridine aminopiridina

aminopyrine aminopirina

aminoquinoline aminoquinolina

aminosalicylic acid ácido aminosalicílico

aminothiazole aminotiazol

aminothiopene aminotiopeno

aminothiourea aminotiourea

aminotoluene aminotolueno

aminotriazole aminotriazol

aminourea aminourea

aminoxylene aminoxileno

aminyl aminilo

amithiozone amitiozona

amitrole amitrol

ammeline amelina

ammine amina

ammonation amonación

ammonia amoníaco

ammonia nitrogen nitrógeno amoniacal

ammonia water agua amoniacal

ammoniacal amoniacal

ammoniated iron hierro amoniacal

ammoniated mercury mercurio amoniacal

ammoniated mercury chloride cloruro mercúrico amoniacal

ammoniated superphosphate superfosfato amoniacal

ammonification amonificación

ammonium amonio

ammonium acetate acetato amónico, acetato de amonio

ammonium acid carbonate carbonato

ácido amónico

ammonium acid fluoride fluoruro ácido amónico

ammonium acid phosphate fosfato ácido amónico

ammonium acid tartrate tartrato ácido amónico

ammonium aldehyde aldehído amónico

ammonium alginate alginato amónico

ammonium alum alumbre amónico

ammonium aluminum sulfate sulfato de aluminio amónico

ammonium aminosulfonate aminosulfonato amónico

ammonium antimoniate antimoniato amónico

ammonium arsenate arseniato amónico

ammonium arsenite arsenito amónico

ammonium benzenesulfonate bencenosulfonato amónico

ammonium benzoate benzoato amónico

ammonium bicarbonate bicarbonato amónico

ammonium bichromate bicromato amónico

ammonium bifluoride bifluoruro amónico

ammonium bioxalate bioxalato amónico

ammonium biphosphate bifosfato amónico

ammonium bisulfate bisulfato amónico

ammonium bisulfite bisulfito amónico

ammonium bitartrate bitartrato amónico

ammonium borate borato amónico

ammonium bromide bromuro amónico

ammonium caprylate caprilato amónico

ammonium carbamate carbamato amónico

ammonium carbazotate carbazotato amónico

ammonium carbonate carbonato amónico

ammonium chlorate clorato amónico

ammonium chloride cloruro amónico

ammonium chromate cromato amónico

ammonium chromium sulfate sulfato de cromo amónico

ammonium citrate citrato amónico

ammonium cuprate cuprato amónico

ammonium cyanide cianuro amónico
ammonium dichromate dicromato amónico
ammonium dithiocarbamate ditiocarbamato amónico
ammonium dithionate ditionato amónico
ammonium ferrocyanide ferrocianuro amónico
ammonium fluoride fluoruro amónico
ammonium fluoroborate fluoroborato amónico
ammonium fluosilicate fluosilicato amónico
ammonium formate formiato amónico
ammonium gallate galato amónico
ammonium gluconate gluconato amónico
ammonium glutamate glutamato amónico
ammonium hexabromoplatinate hexabromoplatinato amónico
ammonium hexachloroplatinate hexacloroplatinato amónico
ammonium hexachlorostannate hexacloroestannato amónico
ammonium hexacyanoferrate hexacianoferrato amónico
ammonium hexafluoroaluminate hexafluoroaluminato amónico
ammonium hexafluorophosphate hexafluorofosfato amónico
ammonium hexafluorosilicate hexafluorosilicato amónico
ammonium hydrate hidrato amónico
ammonium hydrogen carbonate carbonato de hidrógeno amónico
ammonium hydrogen difluoride difluoruro de hidrógeno amónico
ammonium hydrogen fluoride fluoruro de hidrógeno amónico
ammonium hydrogen sulfate sulfato de hidrógeno amónico
ammonium hydrogen sulfide sulfuro de hidrógeno amónico
ammonium hydrogen tartrate tartrato de hidrógeno amónico
ammonium hydrosulfide hidrosulfuro amónico
ammonium hydroxide hidróxido amónico
ammonium hypophosphate hipofosfato amónico
ammonium iodate yodato amónico
ammonium iodide yoduro amónico
ammonium ion ion amónico
ammonium iron bromide bromuro de hierro amónico
ammonium iron chloride cloruro de hierro amónico
ammonium iron chromate cromato de hierro amónico
ammonium iron citrate citrato de hierro amónico
ammonium iron oxalate oxalato de hierro amónico
ammonium iron sulfate sulfato de hierro amónico
ammonium iron tartrate tartrato de hierro amónico
ammonium lactate lactato amónico
ammonium laurate laurato amónico
ammonium linoleate linoleato amónico
ammonium magnesium arsenate arseniato magnésico amónico
ammonium magnesium chloride cloruro magnésico amónico
ammonium magnesium phosphate fosfato magnésico amónico
ammonium magnesium sulfate sulfato magnésico amónico
ammonium manganese phosphate fosfato de manganeso amónico
ammonium manganese sulfate sulfato de manganeso amónico
ammonium mercuric chloride cloruro mercúrico amónico
ammonium metaphosphate metafosfato amónico
ammonium metavanadate metavanadato amónico
ammonium molybdate molibdato amónico
ammonium nickel chloride cloruro de níquel amónico
ammonium nickel sulfate sulfato de níquel amónico
ammonium nitrate nitrato amónico
ammonium nitrite nitrito amónico

ammonium oleate oleato amónico
ammonium oxalate oxalato amónico
ammonium oxamate oxamato amónico
ammonium palmitate palmitato amónico
ammonium pentaborate pentaborato
    amónico
ammonium perchlorate perclorato
    amónico
ammonium perchromate percromato
    amónico
ammonium permanganate
    permanganato amónico
ammonium peroxoborate peroxoborato
    amónico
ammonium peroxodisulfate
    peroxodisulfato amónico
ammonium persulfate persulfato
    amónico
ammonium phenolsulfonate
    fenolsulfonato amónico
ammonium phosphate fosfato amónico
ammonium phosphite fosfito amónico
ammonium phosphomolybdate
    fosfomolibdato amónico
ammonium phosphotungstate
    fosfotungstato amónico
ammonium picrate picrato amónico
ammonium polyphosphate polifosfato
    amónico
ammonium polysulfide polisulfuro
    amónico
ammonium ricinoleate ricinoleato
    amónico
ammonium salicylate salicilato amónico
ammonium salt sal amónica
ammonium selenate seleniato amónico
ammonium selenite selenito amónico
ammonium sodium phosphate fosfato
    sódico amónico
ammonium sodium sulfate sulfato
    sódico amónico
ammonium stearate estearato amónico
ammonium succinate succinato amónico
ammonium sulfamate sulfamato
    amónico
ammonium sulfate sulfato amónico
ammonium sulfide sulfuro amónico
ammonium sulfite sulfito amónico
ammonium tartrate tartrato amónico

ammonium tellurate telurato amónico
ammonium thioacetate tioacetato
    amónico
ammonium thiocyanate tiocianato
    amónico
ammonium thiostanate tioestannato
    amónico
ammonium thiosulfate tiosulfato
    amónico
ammonium trinitrophenolate
    trinitrofenolato amónico
ammonium tungstate tungstato amónico
ammonium valerate valerato amónico
ammonium vanadate vanadato amónico
ammonium zirconyl carbonate
    carbonato de zirconilo amónico
ammonolysis amonólisis
amobarbital amobarbital
amorphism amorfismo
amorphous amorfo
amorphous polymer polímero amorfo
amorphous state estado amorfo
amount of substance cantidad de
    substancia
ampelite ampelita
amperometric titration titulación
    amperométrica, valoración
    amperométrica
amperometry amperometría
amphetamine anfetamina
amphibole anfibol
amphibolite anfibolita
amphichroic anficroico
amphichromatic anficromático
amphiphatic anfifático
amphiphile anfífilo
amphiphilic anfifílico
amphiprotic anfiprótico
ampholyte anfolito
ampholytoid anfolitoide
amphoteric anfótero
amphoteric electrolyte electrólito
    anfótero
amphoteric hydroxide hidróxido
    anfótero
amphoteric oxide óxido anfótero
amphoterism anfoterismo
ampicillin ampicilina
ampul ampolla

ampule ampolla
amsonic acid ácido amsónico
amyctic amíctico
amygdalic acid ácido amigdálico
amygdalin amigdalina
amyl amilo
amyl acetate acetato de amilo
amyl acid phosphate fosfato ácido de amilo
amyl alcohol alcohol amílico
amyl aldehyde aldehído amílico
amyl benzoate benzoato de amilo
amyl butyrate butirato de amilo
amyl chloride cloruro de amilo
amyl ethanoate etanoato de amilo
amyl ether éter amílico
amyl formate formiato de amilo
amyl hydrate hidrato de amilo
amyl hydride hidruro de amilo
amyl nitrate nitrato de amilo
amyl nitrite nitrito de amilo
amyl propionate propionato de amilo
amyl salicylate salicilato de amilo
amyl sulfide sulfuro de amilo
amyl xanthate xantato de amilo
amylacetic ester éster amilacético
amylamine amilamina
amylase amilasa
amylate amilato
amylbenzene amilbenceno
amylcarbinol amilcarbinol
amylcinnamic alcohol alcohol amilcinámico
amylcinnamic aldehyde aldehído amilcinámico
amylene amileno
amylene dichloride dicloruro de amileno
amylene hydrate hidrato de amileno
amyloid amiloide
amylolysis amilólisis
amylolytic amilolítico
amylopectin amilopectina
amylose amilosa
amylphenol amilfenol
amyltrichlorosilane amiltriclorosilano
amyrin amirina
anabasine anabasina
anabolic anabólico
anabolism anabolismo

anacardic acid ácido anacárdico
anaerobe anaerobio
anaerobic anaerobio
analog display visualizador analógico
analysis análisis
analytical analítico
analytical balance balanza analítica
analytical chemistry química analítica
analytical distillation destilación analítica
analytical extraction extracción analítica
analytical reaction reacción analítica
analyzer analizador
anamorphism anamorfismo
anaphoresis anaforesis
anaphylaxis anafilaxis
anarcotine anarcotina
anatase anatasa
anchoic acid ácido ancoico
andalusite andalucita
andesine andesina
androgen andrógeno
androstane androstano
androsterone androsterona
anemometer anemómetro
anemonic acid ácido anemónico
anemosite anemosita
anesthesia anestesia
anesthetic anestésico
anethole anetol
angelic acid ácido angélico
angelica lactone lactona de angélica
angle ángulo
anglesite anglesita
angosturine angosturina
angstrom angstrom
angstrom unit unidad angstrom
angular angular
anharmonic anarmónico
anharmonicity anarmonicidad
anhydride anhídrido
anhydrite anhidrita
anhydro anhidro
anhydrone anhidrona
anhydrosynthesis anhidrosíntesis
anhydrous anhidro
anhydrous alcohol alcohol anhidro
anhydrous ammonia amoníaco anhidro
anhydrous hydrogen chloride cloruro de

hidrógeno anhidro
anhydrous phosphoric acid ácido
    fosfórico anhidro
anhydrous sodium sulfate sulfato sódico
    anhidro
anibine anibina
anilide anilida
anilinate anilinato
aniline anilina
aniline acetate acetato de anilina
aniline black negro de anilina
aniline blue azul de anilina
aniline chloride cloruro de anilina
aniline dye colorante de anilina
aniline fluoride fluoruro de anilina
aniline hydrochloride clorhidrato de
    anilina
aniline oxalate oxalato de anilina
aniline point punto de anilina
aniline salt sal de anilina
aniline sulfate sulfato de anilina
aniline yellow amarillo de anilina
anilinesulfonic acid ácido
    anilinosulfónico
anilinium anilinio
anilino anilino
anion anión
anionic aniónico
anionic polymerization polimerización
    aniónica
anionic resin resina aniónica
anionotropy anionotropía
anisal anisal
anisaldehyde anisaldehído
anise alcohol alcohol anísico
anise oil aceite de anís
anisic anísico
anisic acid ácido anísico
anisic alcohol alcohol anísico
anisic aldehyde aldehído anísico
anisidine anisidina
anisole anisol
anisomeric anisomérico
anisonitrile anisonitrilo
anisotonic anisotónico
anisotonic solution solución anisotónica
anisotropic anisótropo, anisotrópico
anisoyl anisoilo
anisoyl chloride cloruro de anisoilo

anisyl anisilo
anisyl acetate acetato de anisilo
anisyl alcohol alcohol anisílico
anisylacetone anisilacetona
annealing recocido
annular anular
annulene anuleno
anode ánodo
anode coating revestimiento anódico
anode effect efecto anódico
anode film película anódica
anode layer capa anódica
anode mud residuos anódicos, lodo
    anódico
anode oxidation oxidación anódica
anode polarization polarización anódica
anode slime residuos anódicos, lodo
    anódico
anode solution solución anódica
anodic anódico
anodic coating revestimiento anódico
anodic effect efecto anódico
anodic film película anódica
anodic layer capa anódica
anodic mud residuos anódicos, lodo
    anódico
anodic oxidation oxidación anódica
anodic polarization polarización anódica
anodic slime residuos anódicos, lodo
    anódico
anodic solution solución anódica
anodize anodizar
anodizing anodización
anolyte anolito
anomalous anómalo
anomalous dispersion dispersión
    anómala
anomaly anomalía
anomer anómero
anomeric anomérico
anorganic anorgánico
anorthic anórtico
anorthite anortita
antacid anitácido
antagonism antagonismo
antagonist antagonista
antalkaline antalcalino
anthemol antemol
anthion antión

anthocyanidin antocianidina
anthocyanin antocianina
anthophyllite antofilita
anthracene antraceno
anthracene blue azul de antraceno
anthracene oil aceite de antraceno
anthracene violet violeta de antraceno
anthracite antracita
anthradiol antradiol
anthragallol antragalol
anthrahydroquinone antrahidroquinona
anthraldehyde antraldehído
anthralin antralina
anthranil antranil
anthranilic acid ácido antranílico
anthranol antranol
anthranone antranona
anthraparazene antraparaceno
anthrapurpurin antrapurpurina
anthraquinone antraquinona
anthraquinone dye colorante de
    antraquinona
anthrarufin antrarrufina
anthratetrol antratetrol
anthratriol antratriol
anthroic acid ácido antroico
anthrol antrol
anthrone antrona
anthryl antrilo
anthrylene antrileno
antiaromatic antiaromático
antibacterial antibacterial, antibacteriano
antibiotic antibiótico
antiblock agent agente anibloqueo
antibody anticuerpo
antibonding antienlazado
antibonding orbital orbital de
    antienlazado, orbital antienlazante
anticatalyzer anticatalizador
anticathode anticátodo
antichlor anticloro
anticholingeric anticolinérgico
anticoagulant anticoagulante
anticomplement anticomplemento
anticorrosive anticorrosivo
antidote antídoto
antienzyme antienzima
antiferromagnetic antiferromagnético
antiferromagnetism

antiferromagnetismo
antifluorite structure estructura
    antifluorita
antifoam agent agente antiespumante
antifoaming agent agente antiespumante
antiformin antiformina
antifreeze anticongelante
antigen antígeno
antiglobulin antiglobulina
antigorite antigorita
antihistamine antihistamina
antiisomer antiisómero
antiisomorphism antiisomorfismo
antiknock antigolpeteo, antidetonante
antiknock additive aditivo antigolpeteo
antiknock agent agente antigolpeteo
antimatter antimateria
antimer antímero
antimetabolite antimetabolito
antimonate antimoniato
antimonic antimónico
antimonic acid ácido antimónico
antimonic anhydride anhídrido
    antimónico
antimonic chloride cloruro antimónico
antimonic compound compuesto
    antimónico
antimonic fluoride fluoruro antimónico
antimonic sulfide sulfuro antimónico
antimonite antimonita
antimonous antimonioso
antimonous bromide bromuro
    antimonioso
antimonous chloride cloruro
    antimonioso
antimonous compound compuesto
    antimonioso
antimonous fluoride fluoruro
    antimonioso
antimonous hydride hidruro antimonioso
antimonous iodide yoduro antimonioso
antimonous oxalate oxalato antimonioso
antimonous oxide óxido antimonioso
antimonous oxychloride oxicloruro
    antimonioso
antimonous oxysulfide oxisulfuro
    antimonioso
antimonous sulfate sulfato antimonioso
antimonous sulfide sulfuro antimonioso

antimony antimonio

antimony anhydride anhídrido de antimonio

antimony arsenide arseniuro de antimonio

antimony black negro de antimonio

antimony bromide bromuro de antimonio

antimony chloride cloruro de antimonio

antimony dichlorotrifluoride diclorotrifluoruro de antimonio

antimony fluoride fluoruro de antimonio

antimony hydride hidruro de antimonio

antimony iodide yoduro de antimonio

antimony lactate lactato de antimonio

antimony orange anaranjado de antimonio

antimony oxide óxido de antimonio

antimony oxychloride oxicloruro de antimonio

antimony pentachloride pentacloruro de antimonio

antimony pentafluoride pentafluoruro de antimonio

antimony pentasulfide pentasulfuro de antimonio

antimony pentoxide pentóxido de antimonio

antimony persulfide persulfuro de antimonio

antimony potassium oxalate oxalato potásico de antimonio

antimony potassium tartrate tartrato potásico de antimonio

antimony red rojo de antimonio

antimony salt sal de antimonio

antimony sodium tartrate tartrato sódico de antimonio

antimony sulfate sulfato de antimonio

antimony sulfide sulfuro de antimonio

antimony tribromide tribromuro de antimonio

antimony trichloride tricloruro de antimonio

antimony trifluoride trifluoruro de antimonio

antimony triiodide triyoduro de antimonio

antimony trioxide trióxido de antimonio

antimony trisulfate trisulfato de antimonio

antimony trisulfide trisulfuro de antimonio

antimony yellow amarillo de antimonio

antimonyl antimonilo

antimycin antimicina

antineutron antineutrón

antioxidant antioxidante

antioxygen antioxígeno

antiozonant antiozonizante

antiparallel spin espín antiparalelo

antiparasitic antiparasítico

antiparticle antipartícula

antiperspirant antiperspirante

antiproton antiprotón

antipsychotic antipsicótico

antipyrene antipireno

antipyrene benzoate benzoato de antipireno

antipyrene monobromide monobromuro de antipireno

antipyretic antipirético

antipyrine antipirina

antipyryl antipirilo

antiquinol antiquinol

antiradiation antirradiación

antiseptic antiséptico

antistatic antiestático

antistatic agent agente antiestático

antitoxic antitóxico

antitoxin antitoxina

antitussive antitusivo

antiviral antiviral

antizymotic antizimótico

antozone antozono

anytol anitol

apatite apatita

aphrodine afrodina

apiolic acid ácido apiólico

apionol apionol

apoatropine apoatropina

apocarotenal apocarotenal

apocholic acid ácido apocólico

apocodeine apocodeína

apolar adsorption adsorción apolar

apomorphine apomorfina

apophyllite apofilita

apparatus aparato

apparent density densidad aparente
apple acid ácido de manzana
apple oil aceite de manzana
applied chemistry química aplicada
applied research investigación aplicada
approved name nombre aprobado
approximate aproximado
aprotic aprótico
aprotic solvent solvente aprótico
aprotic substance substancia aprótica
aqua regia agua regia
aquation acuación
aqueous acuoso
aqueous solution solución acuosa
aqueous solvent solvente acuoso
aqueous tension tensión acuosa
arabic gum goma arábiga
arabin arabina
arabinose arabinosa
arabinoside arabinosida
arabite arabita
arabitol arabitol
arabonic acid ácido arabónico
arachidic acid ácido araquídico
arachidonic acid ácido araquidónico
aragonite aragonita
aralkyl aralquilo
aramid aramida
arborescent arborescente
arbutin arbutina
arc spectrum espectro de arco
areametric analysis análisis areamétrico
arecaidine arecaidina
arecaine arecaína
arecoline arecolina
arene areno
areometer areómetro
argenate argenato
argentamine argentamina
argentan argentán
argentic argéntico
argentic compound compuesto argéntico
argentic oxide óxido argéntico
argentite argentita
argentocyanide argentocianuro
argentometry argentometría
argentopyrite argentopirita
arginase arginasa
arginine arginina

argol argol
argon argón
argyrodite argirodita
aribine aribina
armored thermometer termómetro armado
Armstrong acid ácido de Armstrong
Arndt alloy aleación de Arndt
Arndt-Eistert synthesis síntesis de Arndt-Eistert
Arndt tube tubo de Arndt
arnicin arnicina
aromatic aromático
aromatic acid ácido aromático
aromatic alcohol alcohol aromático
aromatic aldehyde aldehído aromático
aromatic amine amina aromática
aromatic compound compuesto aromático
aromatic hydrocarbon hidrocarburo aromático
aromatic ketone cetona aromática
aromatic nucleus núcleo aromático
aromatic series serie aromática
aromaticity aromaticidad
aromatization aromatización
aroyl aroilo
arrhenic acid ácido arrénico
Arrhenius equation ecuación de Arrhenius
arrowroot arrurruz
arsacetin arsacetina
arsanilate arsanilato
arsanilic acid ácido arsanílico
arsenate arseniato
arsenenic acid ácido arsenénico
arsenic arsénico
arsenic acid ácido arsénico
arsenic anhydride anhídrido arsénico
arsenic bromide bromuro de arsénico
arsenic chloride cloruro de arsénico
arsenic diiodide diyoduro de arsénico
arsenic disulfide disulfuro de arsénico
arsenic fluoride fluoruro de arsénico
arsenic hydride hidruro de arsénico
arsenic iodide yoduro de arsénico
arsenic oxide óxido de arsénico
arsenic pentafluoride pentafluoruro de arsénico

arsenic pentasulfide pentasulfuro de arsénico
arsenic pentoxide pentóxido de arsénico
arsenic selenide seleniuro de arsénico
arsenic sulfide sulfuro de arsénico
arsenic tetrasulfide tetrasulfuro de arsénico
arsenic thioarsenate tioarseniato de arsénico
arsenic thiocyanate tiocianato de arsénico
arsenic tribromide tribromuro de arsénico
arsenic trichloride tricloruro de arsénico
arsenic trifluoride trifluoruro de arsénico
arsenic trioxide trióxido de arsénico
arsenic trisulfide trisulfuro de arsénico
arsenical arsenical
arsenide arseniuro
arsenidine arsenidina
arsenious arsenioso
arsenite arsenito
arseno arseno
arsenobenzene arsenobenceno
arsenolite arsenolita
arsenopyrite arsenopirita
arsenous arsenioso
arsenous acid ácido arsenioso
arsenous bromide bromuro arsenioso
arsenous chloride cloruro arsenioso
arsenous fluoride fluoruro arsenioso
arsenous hydride hidruro arsenioso
arsenous iodide yoduro arsenioso
arsenous oxide óxido arsenioso
arsenous oxychloride oxicloruro arsenioso
arsenous phosphide fosfuro arsenioso
arsenous selenide seleniuro arsenioso
arsenous sulfide sulfuro arsenioso
arsenyl arsenilo
arsine arsina
arsinic acid ácido arsínico
arsinoso arsinoso
arsonate arsonato
arsonic acid ácido arsónico
arsonium arsonio
arsono arsono
arsonoso arsonoso

arsonoyl arsonoilo
arthranitin artranitín
artificial artificial
artificial element elemento artificial
artificial radioactive element elemento radiactivo artificial
artificial radioactivity radiactividad artificial
artificial rubber caucho artificial
artinite artinita
aryl arilo
arylalkyl arilalquilo
arylamine arilamina
arylarsonate arilarsonato
arylene arileno
arylide ariluro
aryne arino
asarone asarona
asaronic acid ácido asarónico
asaryl asarilo
asbestine asbestina
asbestos asbesto
asbolane asbolana
asbolite asbolita
ascaridic acid ácido ascarídico
ascaridole ascaridol
ascaridolic acid ácido ascaridólico
ascorbic acid ácido ascórbico
ascorbic acid oxidase oxidasa de ácido ascórbico
aseptol aseptol
aseptoline aseptolina
ash ceniza
ashing incineración
asiatic acid ácido asiático
asiminine asiminina
askarel askarel
asparagic acid ácido asparágico
asparaginase asparaginasa
asparagine asparagina
asparaginic acid ácido asparagínico
asparamide asparamida
aspariginyl aspariginilo
aspartame aspartama
aspartamic acid ácido aspartámico
aspartamide aspartamida
aspartase aspartasa
aspartate aspartato
aspartic acid ácido aspártico

aspartoyl aspartoilo
aspartyl aspartilo
aspergillic acid ácido aspergílico
asphalt asfalto
asphaltene asfalteno
asphaltic bitumen bitumen asfáltico
asphaltite asfaltita
asphyxiant gas gas asfixiante
aspidinol aspidinol
aspidosamine aspidosamina
aspidospermine aspidospermina
aspirator aspirador
aspirin aspirina
assay ensayo
assimilation asimilación
associated liquids líquidos asociados
association asociación
astacene astaceno
astatic astátco
astatine astatino, ástato
asterisk asterisco
Aston rule regla de Aston
Aston spectrum espectro de Aston
astorism astorismo
astrochemistry astroquímica
astrophyllite astrofilita
asymmetric asimétrico
asymmetric atom átomo asimétrico
asymmetric carbon carbono asimétrico
asymmetric carbon atom átomo de
    carbono asimétrico
asymmetric compound compuesto
    asimétrico
asymmetric induction inducción
    asimétrica
asymmetric structure estructura
    asimétrica
asymmetric synthesis síntesis asimétrica
asymmetrical asimétrico
asymmetry asimetría
atacamite atacamita
atactic atáctico
atactic polymer polímero atáctico
ate ato
atenolol atenolol
atisine atisina
atmolysis atmólisis
atmometer atmómetro
atmosphere atmósfera

atmospheric atmosférico
atmospheric gas gas atmosférico
atmospheric humidity humedad
    atmosférica
atmospheric pollution contaminación
    atmosférica
atmospheric pressure presión
    atmosférica
atom átomo
atomic atómico
atomic absorption absorción atómica
atomic absorption coefficient coeficiente
    de absorción atómica
atomic absorption spectroscopy
    espectroscopia de absorción atómica
atomic attraction atracción atómica
atomic charge carga atómica
atomic constant constante atómica
atomic diamagnetism diamagnetismo
    atómico
atomic diameter diámetro atómico
atomic disintegration desintegración
    atómica
atomic distance distancia atómica
atomic electricity electricidad atómica
atomic emission spectroscopy
    espectroscopia de emisión atómica
atomic energy energía atómica
atomic energy level nivel de energía
    atómica
atomic excitation excitación atómica
atomic field campo atómico
atomic fission fisión atómica
atomic force microscope microscopio de
    fuerza atómica
atomic fragment fragmento atómico
atomic frequency frecuencia atómica
atomic fusion fusión atómica
atomic group grupo atómico
atomic heat calor atómico
atomic hydrogen hidrógeno atómico
atomic hydrogen welding soldadura con
    hidrógeno atómico
atomic level nivel atómico
atomic magnetic moment momento
    magnético atómico
atomic mass masa atómica
atomic mass number número de masa
    atómica

atomic mass unit   unidad de masa atómica

atomic migration   migración atómica

atomic model   modelo atómico

atomic nucleus   núcleo atómico

atomic number   número atómico

atomic orbit   órbita atómica

atomic orbital   orbital atómico

atomic oscillation   oscilación atómica

atomic particle   partícula atómica

atomic plane   plano atómico

atomic polarization   polarización atómica

atomic potential   potencial atómico

atomic properties   propiedades atómicas

atomic radius   radio atómico

atomic refraction   refracción atómica

atomic resonance   resonancia atómica

atomic scale   escala atómica

atomic scattering factor   factor de dispersión atómica

atomic species   especie atómica

atomic spectroscopy   espectroscopia atómica

atomic spectrum   espectro atómico

atomic structure   estructura atómica

atomic susceptibility   susceptibilidad atómica

atomic symbol   símbolo atómico

atomic theory   teoría atómica

atomic transformation   transformación atómica

atomic transmutation   transmutación atómica

atomic unit   unidad atómica

atomic vibration   vibración atómica

atomic volume   volumen atómico

atomic weight   peso atómico

atomic weight unit   unidad de peso atómico

atomically   atómicamente

atomicity   atomicidad

atomization   atomización

atomizer   atomizador

atomology   atomología

atomsite   atomsita

atopite   atopita

atrazine   atrazina

atrolactic acid   ácido atroláctico

atronene   atroneno

atronic acid   ácido atrónico

atropic acid   ácido atrópico

atropine   atropina

atropine nitrate   nitrato de atropina

atropine salicylate   salicilato de atropina

atropine stearate   estearato de atropina

atropine sulfate   sulfato de atropina

atropine valerate   valerato de atropina

atropoyl   atropoilo

attapulgite   attapulgita

attar   aceite esencial

attenuation   atenuación

attenuation coefficient   coeficiente de atenuación

attenuation factor   factor de atenuación

attraction   atracción

Atwater calorimeter   calorímetro de Atwater

aubepine   aubepina

Auger effect   afecto de Auger

Auger electron   electrón de Auger

augite   augita

auramine   auramina

aurate   aurato

aurelite   aurelita

aureolin   aureolina

auribromhydric acid   ácido auribromhídrico

auribromide   auribromuro

auric   áurico

auric acid   ácido áurico

auric bromide   bromuro áurico

auric chloride   cloruro áurico

auric compound   compuesto áurico

auric cyanide   cianuro áurico

auric hydroxide   hidróxido áurico

auric iodide   yoduro áurico

auric oxide   óxido áurico

auric sulfate   sulfato áurico

auric sulfide   sulfuro áurico

aurichalcite   auricalcita

aurichloride   auricloruro

auriferous   aurífero

aurin   aurina

auroauric   auroáurico

aurobromide   aurobromuro

aurochloride   aurocloruro

auromine   auromina

aurone   aurona

aurous auro
aurous bromide aurobromuro
aurous chloride aurocloruro
aurous cyanide aurocianuro
aurous iodide auroyoduro
aurous sulfide aurosulfuro
austenite austenita
autocatalysis autocatálisis
autocatalytic autocatalítico
autoclave autoclave
autodecomposition autodescomposición
autoexcitation autoexcitación
autoignition point punto de autoignición
autoionization autoionización
autoluminescence autoluminiscencia
autolysis autólisis
automatic automático
automatic buret bureta automática
automatic control control automático
automatic pipet pipeta automática
automatic titration titulación
    automática, valoración automática
automation automatización
autooxidation autooxidación
autoprotolysis autoprotólisis
autunite autunita
auxiliary auxiliar
auxin auxina
auxochrome auxocromo
availability disponibilidad
available disponible
available electron electrón disponible
avalite avalita
avenine avenina
average molecular weight peso
    molecular medio
avidin avidina
avocado oil aceite de aguacate
Avogadro constant constante de
    Avogadro
Avogadro hypothesis hipótesis de
    Avogadro
Avogadro law ley de Avogadro
Avogadro number número de Avogadro
Avogadro theory teoría de Avogadro
axial axial
axis eje
axis of symmetry eje de simetría
azaguanine azaguanina

azauridine azauridina
azelaic acid ácido azelaico
azelate azelato
azeotrope azeótropo
azeotropic azeotrópico
azeotropic distillation destilación
    azeotrópica
azeotropic mixture mezcla azeotrópica
azetidine azetidina
azide azida
azido azido
azimide azimida
azimido azimido
azimidobenzene azimidobenceno
azimino azimino
aziminobenzene aziminobenceno
azimuthal quantum number número
    cuántico acimutal
azine azina
azine dye colorante de azina
azino azino
aziridine aziridina
azlon azlón
azo azo
azo dye colorante azoico, azocolorante
azobenzene azobenceno
azobenzoic acid ácido azobenzoico
azobisisobutyronitrile
    azobisisobutironitrilo
azocyanide azocianuro
azodicarbonamide azodicabornamida
azodine azodina
azoic azoico
azoimide azoimida
azole azol
azomethine azometina
azonaphthalene azonaftaleno
azonium azonio
azophenol azofenol
azophenyl azofenilo
azophenylene azofenileno
azorite azorita
azosulfamide azosulfamida
azotoluene azotolueno
azotometer azotómetro
azoxy azoxi
azoxybenzene azoxibenceno
azoxynaphthalene azoxinaftaleno
azoxytoluidine azoxitoluidina

azulene azuleno
azulin azulina

azurite azurita

B

babassu oil aceite de babasú
Babbitt metal metal de Babbitt
Babcock test prueba de Babcock
Babo absorption tube tubo de absorción de Babo
Babo law ley de Babo
baccarin bacarina
bacillus bacilo
bacitracin bacitracina
back donation retrodonación
back-extraction retroextracción
Back-Goudsmit effect efecto de Back-Goudsmit
back titration retrotitulación, retrovaloración
background radiation radiación de fondo
backscatter retrodispersión
bacteria bacterias
bacterial bacterial, bacteriano
bactericidal batericida
bactericide bactericida
bacterin bacterina
bacteriocidal batericida
bacteriological bacteriológico
bacteriolysis bacteriólisis
bacteriophage bacteriófago
bacteriostatic bacteriostático
bacterium bacteria
baddeckite badequita
baddeleyite badeleyita
Badger rule regla de Badger
baeumlerite beumlerita
Baeyer strain theory teoría de deformación de Baeyer
Baeyer-Villiger reactions reacciones de Baeyer-Villiger
baffle bafle
bagasse bagazo
bagrationite bagrationita
baicalein baicaleína
baicalin baicalina

baikalite baikalita
bainite bainita
Baker-Nathan effect efecto de Baker-Nathan
Baker-Venkataraman effect efecto de Baker-Venkataraman
bakerite baquerita
baking finish acabado con horneado
baking powder polvo de hornear
baking soda bicarbonato sódico, bicarbonato de sosa
balance balance, equilibrio, balanza
balanced balanceado
balanced reaction reacción balanceada
balata balata
Baldwin phosphorus fósforo de Baldwin
Baldwin rules reglas de Baldwin
ball mill molino de bolas
balloelectric baloeléctrico
Balmer formula fórmula de Balmer
Balmer lines líneas de Balmer
Balmer series serie de Balmer
balometer balómetro
balsam bálsamo
balsam of Peru bálsamo de Perú
balsam of Tolu bálsamo de Tolú
Bamberger formula fórmula de Bamberger
banalsite banalsita
banana oil aceite de plátano
Banbury mixer mezclador de Banbury
band banda
band spectrum espectro de bandas
Bandrowski base base de Bandrowski
banisterine banisterina
baptisin baptisina
baptitoxine baptitoxina
bar barra
barbaloin barbaloína
barban barbana
barberite barberita
Barbier-Wieland degradation degradación de Barbier-Wieland

barbital barbital
barbitone barbitona
barbiturate barbiturato
barbituric acid ácido barbitúrico
barcenite barcenita
Barff process proceso de Barff
Barfoed reagent reactivo de Barfoed
Barfoed test prueba de Barfoed
barite barita
barium bario
barium acetate acetato bárico, acetato de
    bario
barium aluminate aluminato bárico
barium arsenate arseniato bárico
barium azide azida bárica
barium benzoate benzoato bárico
barium bicarbonate bicarbonato bárico
barium bichromate bicromato bárico
barium binoxide binóxido bárico
barium bioxalate bioxalato bárico
barium bisulfate bisulfato bárico
barium borate borato bárico
barium boride boruro bárico
barium borotungstate borotungstato
    bárico
barium bromate bromato bárico
barium bromide bromuro bárico
barium butyrate butirato bárico
barium carbide carburo bárico
barium carbonate carbonato bárico
barium chlorate clorato bárico
barium chloride cloruro bárico
barium chloroplatinate cloroplatinato
    bárico
barium chromate cromato bárico
barium citrate citrato bárico
barium cyanate cianato bárico
barium cyanide cianuro bárico
barium cyanoplatinite cianoplatinita
    bárica
barium dichromate dicromato bárico
barium dioxide dióxido bárico
barium diphenylamine sulfonate
    difenilaminosulfonato bárico
barium dithionate ditionato bárico
barium diurinate diuriniato bárico
barium ethylsulfate etilsulfato bárico
barium ferrate ferrato bárico
barium ferrite ferrita bárica

barium ferrocyanide ferrocianuro bárico
barium fluoride fluoruro bárico
barium fluosilicate fluosilicato bárico
barium formate formiato bárico
barium hexachloroplatinate
    hexacloroplatinato bárico
barium hexacyanoferrate
    hexacianoferrato bárico
barium hexafluorogermantane
    hexafluorogermantano bárico
barium hexanitride hexanitruro bárico
barium hydrate hidrato bárico
barium hydride hidruro bárico
barium hydrosulfide hidrosulfuro bárico
barium hydroxide hidróxido bárico
barium hypophosphate hipofosfato
    bárico
barium hypophosphite hipofosfito
    bárico
barium hyposulfate hiposulfato bárico
barium hyposulfite hiposulfito bárico
barium iodate yodato bárico
barium iodide yoduro bárico
barium lactate lactato bárico
barium malate malato bárico
barium malonate malonato bárico
barium manganate manganato bárico
barium mercuric iodide yoduro
    mercúrico bárico
barium mercury bromide bromuro
    mercúrico bárico
barium mercury iodide yoduro
    mercúrico bárico
barium metaphosphate metafosfato
    bárico
barium metasilicate metasilicato bárico
barium methylsulfate metilsulfato bárico
barium molybdate molibdato bárico
barium monohydrate monohidrato
    bárico
barium monophosphate monofosfato
    bárico
barium monosulfide monosulfuro bárico
barium monoxide monóxido bárico
barium nitrate nitrato bárico
barium nitrite nitrito bárico
barium octahydrate octahidrato bárico
barium oleate oleato bárico
barium orthoperiodate ortoperyodato

bárico
barium orthophosphate ortofosfato
bárico
barium oxalate oxalato bárico
barium oxide óxido bárico
barium pentahydrate pentahidrato
bárico
barium pentylsulfate pentilsulfato bárico
barium perchlorate perclorato bárico
barium periodate peryodato bárico
barium permanganate permanganato
bárico
barium peroxide peróxido bárico
barium peroxodisulfate peroxodisulfato
bárico
barium phosphate fosfato bárico
barium phosphide fosfuro bárico
barium phosphinate fosfinato bárico
barium phosphosilicate fosfosilicato
bárico
barium potassium chlorate clorato
potásico bárico
barium potassium chromate cromato
potásico bárico
barium propionate propionato bárico
barium protoxide protóxido bárico
barium pyrophosphate pirofosfato
bárico
barium salicylate salicilato bárico
barium selenate seleniato bárico
barium selenide seleniuro bárico
barium silicate silicato bárico
barium silicide siliciuro bárico
barium silicofluoride silicofluoruro
bárico
barium stannate estannato bárico
barium stearate estearato bárico
barium succinate succinato bárico
barium sulfate sulfato bárico
barium sulfide sulfuro bárico
barium sulfite sulfito bárico
barium sulfocyanide sulfocianuro bárico
barium superoxide superóxido bárico
barium tartrate tartrato bárico
barium tetrasulfide tetrasulfuro bárico
barium thiocyanate tiocianato bárico
barium thiosulfate tiosulfato bárico
barium titanate titanato bárico
barium triphosphate trifosfato bárico

barium trisulfide trisulfuro bárico
barium trithionate tritionato bárico
barium tungstate tunsgstato bárico
barium value valor de bario
barium zirconate zirconato bárico
barkevite barkevita
Barlow rule regla de Barlow
barn barn
Barnett acetylation method método de
acetilación de Barnett
baroluminescence baroluminiscencia
barometer barómetro
barometric pressure presión barométrica
barophoresis baroforesis
baroscope baroscopio
barosimin barosimina
barostat barostato
barrandite barrandita
barrel barril
barrier barrera
barrier layer capa de barrera
Bart reaction reacción de Bart
barthrin bartrina
Barton reaction reacción de Barton
barye baria
baryon barión
barysilite barisilita
baryte barita
basanite basanita
base base
base dissociation constant constante de
dissociación básica
base metal metal de base
base number número de base
base unit unidad básica
basic básico
basic anhydride anhídrido básico
basic capacity capacidad básica
basic group grupo básico
basic lining revestimiento básico
basic oxide óxido básico
basic salt sal básica
basic solution solución básica
basic solvent solvente básico
basicity basicidad
basicity constant constante de basicidad
basis metal metal base
bassanite basanita
bassisterol basisterol

bassorin basorina
bastnasite bastnasita
bath baño
bathochromatic batocromático
bathochromatic shift deslizamiento
    batocromático
bathochromic batocrómico
bathocuproine batocuproína
bathophenanthroline batofenantrolina
batrachotoxinin batracotoxinina
battery batería
battery acid ácido de batería
battery electrolyte electrólito de batería
batu batu
batyl alcohol alcohol batílico
Baudisch reaction reacción de Baudisch
bauxite bauxita
Bayer acid ácido de Bayer
Bayer process proceso de Bayer
bayerite bayerita
bead test prueba de perla
beaker vaso de laboratorio
beam attenuator atenuador de haz
beam balance balanza de astil
Beattie-Bridgman equation ecuación de
    Beattie-Bridgman
beccarite becarita
Béchamp reaction reacción de Béchamp
Béchamp reduction reducción de
    Béchamp
Bechhold filter filtro de Bechhold
Becke line línea de Becke
beckelite bequelita
Beckmann reaction reacción de
    Beckmann
Beckmann rearrangement
    reordenamiento de Beckmann
Beckmann thermometer termómetro de
    Beckmann
becquerel becquerel
Becquerel effect efecto de Becquerel
Becquerel rays rayos de Becquerel
becquerelite becquerelita
beehive colmena
beer cerveza
Beer-Lambert law ley de Beer-Lambert
Beer law ley de Beer
beeswax cera de abejas
beet sugar azúcar de remolacha

behenic acid ácido behénico
behenolic acid ácido behenólico
behenolyl behenolilo
behenone behenona
behenyl alcohol alcohol behenílico
Beilby layer capa de Beilby
Beilstein test prueba de Beilstein
bel bel
beldongrite beldongrita
belite belita
bell glass campana
bell jar campana
bell metal bronce de campana
belladonna belladona
belladonnine belladonina
Bellier index índice de Bellier
Bellier number número de Bellier
belonesite belonesita
Belousov-Zhabotinskii reaction reacción
    de Belousov-Zhabotinskii
bemberg bemberg
Bénard cell célula de Bénard
Benary reaction reacción de Benary
benazolin benazolina
bench test prueba de banco
bendiocarb bendiocarb
Benedict solution solución de Benedict
Benedict test prueba de Benedict
beneficiation beneficio
benefin benefina
Benfield process proceso de Benfield
benomyl benomilo
bensulide bensulida
bentazone bentazona
bentonite bentonita
benzacetin benzacetina
benzaconine benzaconina
benzacridine benzacridina
benzal benzal
benzal chloride benzalcloruro
benzalacetone benzalacetona
benzalazine benzalazina
benzalcohol benzalcohol
benzaldehyde benzaldehído
benzaldehyde azide azida de
    benzaldehído
benzaldehyde cyanohydrin cianhidrina
    de benzaldehído
benzaldehyde green verde de

benzaldehído

benzaldehyde phenylhydrazone
fenilhidrazona de benzaldehído

benzaldoxime benzaldoxima

benzaldoxime acetate acetato de
benzaldoxima

benzalkonium benzalconio

benzalkonium chloride cloruro de
benzalconio

benzalphthalide benzalftalida

benzamide benzamida

benzamide oxime oxima de benzamida

benzamidine benzamidina

benzamido benzamido

benzamidoacetic acid ácido
benzamidoacético

benzaminoacetic acid ácido
benzaminoacético

benzanilide benzanilida

benzanthracene benzantracina

benzanthrone benzantrona

benzathine benzatina

benzazide benzazida

benzazimide benzazimida

benzazine benzazina

benzene benceno

benzene azimide bencenoazimida

benzene dibromide dibromuro de
benceno

benzene dichloride dicloruro benceno

benzene hexabromide hexabromuro de
benceno

benzene hexachloride hexacloruro de
benceno

benzene ring anillo bencénico

benzene series serie de benceno

benzene structure estructura de benceno

benzeneazoanilide bencenoazoanilida

benzeneazoaniline bencenoazoanilina

benzeneazobenzene bencenoazobenceno

benzenecarbaldehyde
bencenocarbaldehído

benzenecarbinol bencenocarbinol

benzenecarbonitrile bencenocarbonitrilo

benzenecarbonyl chloride cloruro de
bencenocarbonilo

benzenecarboxylate bencenocarboxilato

benzenecarboxylic acid ácido
bencenocarboxílico

benzenediazonium chloride cloruro de
bencenodiazonio

benzenediazonium hydroxide hidróxido
de bencenodiazonio

benzenedicarbinol bencenodicarbinol

benzenedicarbonal bencenodicarbonal

benzenedicarbonitrile
bencenodicarbonitrilo

benzenedicarboxylic acid ácido
bencenodicarboxílico

benzenedisulfonic acid ácido
bencenodisulfónico

benzenedisulfoxide bencenodisulfóxido

benzenedithiol bencenoditiol

benzenemonosulfonic acid ácido
bencenomonosulfónico

benzenephosphinic acid ácido
bencenofosfínico

benzenephosphonic acid ácido
bencenofosfónico

benzenesulfamide bencenosulfamida

benzenesulfide bencenosulfuro

benzenesulfinic acid ácido
bencenosulfínico

benzenesulfonamide
bencenosulfonamida

benzenesulfonic acid ácido
bencenosulfónico

benzenesulfonyl bencenosulfonilo

benzenesulfoxide bencenosulfóxido

benzenetetracarboxylic acid ácido
bencenotetracarboxílico

benzenetriol bencenotriol

benzenetrisulfonic acid ácido
bencenotrisulfónico

benzenide bencenida

benzenium bencenio

benzenoid bencenoide

benzenone bencenona

benzenyl bencenilo

benzenyl trichloride tricloruro de
bencenilo

benzethonium chloride cloruro de
bencetonio

benzfuran benzfurano

benzhydrazoin benzhidrazoína

benzhydrol benzhidrol

benzhydryl benzhidrilo

benzhydryl bromide bromuro de

benzhidrilo

benzhydryl chloride cloruro de
benzhidrilo

benzidine bencidina

benzidine conversion conversión de
bencidina

benzidine dye colorante de bencidina

benzidine sulfate sulfato de bencidina

benzidine test prueba de bencidina

benzidinedicarboxylic acid ácido
bencidinodicarboxílico

benzidino bencidino

benzil bencilo

benzilic acid ácido bencílico

benzilidene bencilideno

benzimidazole bencimidazol

benzimidazolyl bencimidazolilo

benzimidoyl bencimidoilo

benzin bencina

benzindole bencindol

benzine bencina

benzisoxasole bencisoxasol

benzoate benzoato

benzocaine benzocaína

benzocarbolic acid ácido benzocarbólico

benzodianthrone benzodiantrona

benzodiazine benzodiazina

benzodiazole benzodiazol

benzodihydropyrone benzodihidropirona

benzodioxazine benzodioxazina

benzofluorene benzofluoreno

benzofuran benzofurano

benzofuranone benzofuranona

benzofuranyl benzofuranilo

benzoglycolic acid ácido benzoglicólico

benzoguanamine benzoguanamina

benzohydrol benzohidrol

benzohydroximic acid ácido
benzohidroxímico

benzohydryl benzohidrilo

benzoic acid ácido benzoico

benzoic acid anhydride anhídrido de
ácido benzoico

benzoic alcohol alcohol benzoico

benzoic aldehyde aldehído benzoico

benzoic anhydride anhídrido benzoico

benzoic sulfimide sulfimido benzoico

benzoic trichloride tricloruro benzoico

benzoilacetone benzoilacetona

benzoin benzoína

benzoin condensation condensación de
benzoína

benzoin oxime oxima de benzoína

benzol benzol

benzol bichloride bicloruro de benzol

benzol bromide bromuro de benzol

benzol chloride cloruro de benzol

benzomorpholine benzomorfolina

benzonaphthene benzonafteno

benzonaphthol benzonaftol

benzonitrile benzonitrilo

benzopentazole benzopentazol

benzoperoxide benzoperóxido

benzophenol benzofenol

benzophenone benzofenona

benzophenone oxide óxido de
benzofenona

benzophenone sulfide sulfuro de
benzofenona

benzopinacol benzopinacol

benzopurpurine benzopurpurina

benzopyran benzopirano

benzopyranyl benzopiranilo

benzopyrene benzopireno

benzopyridine benzopiridina

benzopyrone benzopirona

benzopyrrole benzopirrol

benzoquinol benzoquinol

benzoquinone benzoquinona

benzoquinone monoxime monoxima de
benzoquinona

benzoselenofuran benzoselenofurano

benzosulfimide benzosulfimida

benzotetrazine benzotetrazina

benzotetrazole benzotetrazol

benzothiazine benzotiazina

benzothiazole benzotiazol

benzothiazyl benzotiacilo

benzothiazyl disulfide disulfuro de
benzotiacilo

benzothiofuran benzotiofurano

benzothiopyran benzotiopirano

benzothiopyrone benzotiopirona

benzotoluide benzotoluida

benzotriazepine benzotriazepina

benzotriazine benzotriazina

benzotriazole benzotriazol

benzotrichloride benzotricloruro

benzotrifluoride benzotrifluoruro
benzotrifuran benzotrifurano
benzoxanthene benzoxanteno
benzoxazine benzoxazina
benzoxazole benzoxazol
benzoxazolyl benzoxazolilo
benzoxozolone benzoxazolona
benzoxtetrazine benzoxtetrazina
benzoxtriazine benzoxtriazina
benzoyl benzoilo
benzoyl acetate acetato de benzoilo
benzoyl benzoate benzoato de benzoilo
benzoyl bromide bromuro de benzoilo
benzoyl chloride cloruro de benzoilo
benzoyl cyanide cianuro de benzoilo
benzoyl disulfide disulfuro de benzoilo
benzoyl fluoride fluoruro de benzoilo
benzoyl iodide yoduro de benzoilo
benzoyl oxide óxido de benzoilo
benzoyl peroxide peróxido de benzoilo
benzoyl sulfide sulfuro de benzoilo
benzoylacetaldehyde benzoilacetaldehído
benzoylacetic acid ácido benzoilacético
benzoylacrylic acid ácido benzoilacrílico
benzoylamide benzoilamida
benzoylamino benzoilamino
benzoylaminoacetic acid ácido
    benzoilaminoacético
benzoylaniline benzoilanilina
benzoylation benzoilación
benzoylazide benzoilazida
benzoylbenzoic acid ácido
    benzoilbenzoico
benzoylcyclopropane
    benzoilciclopropano
benzoylferrocene benzoilferroceno
benzoylglycin benzoilglicina
benzoylhydrazine benzoilhidrazina
benzoyloxy benzoiloxi
benzoylpropionic acid ácido
    benzoilpropiónico
benzoylpyridine benzoilpiridina
benzozone benzozono
benzpyrazole benzpirazol
benzthiophene benztiofeno
benztropine mesylate mesilato de
    benztropina
benzyl bencilo
benzyl abietate abietato de bencilo

benzyl acetate acetato de bencilo
benzyl alcohol alcohol bencílico
benzyl azide azida de bencilo
benzyl benzoate benzoato de bencilo
benzyl bichloride bicloruro de bencilo
benzyl bromide bromuro de bencilo
benzyl butyrate butirato de bencilo
benzyl carbinol bencilcarbinol
benzyl chloride cloruro de bencilo
benzyl chlorocarbonate clorocarbonato
    de bencilo
benzyl chloroformate cloroformiato de
    bencilo
benzyl cinnamate cinamato de bencilo
benzyl cyanide cianuro de bencilo
benzyl dichloride dicloruro de bencilo
benzyl disulfide disulfuro de bencilo
benzyl ether éter bencílico
benzyl ethyl ether éter etílico bencílico
benzyl ethyl ketone cetona etílica
    bencílica
benzyl ethylsalicylate etilsalicilato de
    bencilo
benzyl fluoride fluoruro de bencilo
benzyl formate formiato de bencilo
benzyl fumarate fumarato de bencilo
benzyl iodide yoduro de bencilo
benzyl isoeugenol bencilisoeugenol
benzyl isothiocyanate bencilisotiocianato
benzyl mercaptan bencilmercaptano
benzyl nitrile nitrilo de bencilo
benzyl pelargonate pelargonato de
    bencilo
benzyl phenylacetate fenilacetato de
    bencilo
benzyl propionate propionato de bencilo
benzyl salicylate salicilato de bencilo
benzyl succinate succinato de bencilo
benzyl sulfide sulfuro de bencilo
benzyl thiocyanate tiocianato de bencilo
benzylacetamide bencilacetamida
benzylamine bencilamina
benzylaniline bencilanilina
benzylation bencilación
benzylbenzene bencilbenceno
benzylcyanamide bencilcianamida
benzyldianil bencildianilo
benzyldimethylamine bencildimetilamina
benzylene bencileno

benzylethanolamine benciletanolamina
benzylformamide bencilformamida
benzylhydrazine bencilhidrazina
benzylhydroquinone bencilhidroquinona
benzylidene bencilideno
benzylidene dibromide dibromuro de
    bencilideno
benzylidene dichloride dicloruro de
    bencilideno
benzylideneacetone bencilidenacetona
benzylideneacetophenone
    bencilidenacetofenona
benzylidiethanolamine
    bencilidietanolamina
benzylidine bencilidina
benzylisopropylamine
    bencilisopropilamina
benzylmethylamine bencilmetilamina
benzylnaphthalene bencilnaftaleno
benzyloxyphenol benciloxifenol
benzylphenanthrene bencilfenantreno
benzylphenol bencilfenol
benzylphenylamine bencilfenilamina
benzylpyridine bencilpiridina
benzylsulfonic acid ácido bencilsulfónico
benzylthiol benciltiol
benzyltrimethylammonium chloride
    cloruro de benciltrimetilamonio
benzyltrimethylammonium methoxide
    metóxido de benciltrimetilamonio
benzyne bencino
berberine berberina
berberonic acid ácido berberónico
beresovite beresovita
bergamot oil aceite de bergamota
Bergius process proceso de Bergius
Bergmann degradation degradación de
    Bergmann
Berkefeld filter filtro de Berkefeld
berkelium berkelio
Berthelot equation ecuación de Berthelot
Berthelot reaction reacción de Berthelot
berthierine bertierina
bertholite bertolita
berthollide bertolida
bertrandite bertrandita
beryl berilo
beryllate berilato
beryllia berilia

beryllide beriliuro
beryllium berilio
beryllium acetate acetato de berilio
beryllium acetylacetonate
    acetilacetonato de berilio
beryllium bromide bromuro de berilio
beryllium carbide carburo de berilio
beryllium carbonate carbonato de berilio
beryllium chloride cloruro de berilio
beryllium-copper berilio-cobre
beryllium fluoride fluoruro de berilio
beryllium hydrate hidrato de berilio
beryllium hydroxide hidróxido de berilio
beryllium iodide yoduro de berilio
beryllium metaphosphate metafosfato de
    berilio
beryllium nitrate nitrato de berilio
beryllium nitride nitruro de berilio
beryllium oxalate oxalato de berilio
beryllium oxide óxido de berilio
beryllium potassium fluoride fluoruro
    potásico berílico
beryllium potassium sulfate sulfato
    potásico berílico
beryllium silicate silicato de berilio
beryllium sodium fluoride fluoruro
    sódico berílico
beryllium sulfate sulfato de berilio
beryllonite berilonita
Bessemer process proceso de Bessemer
Bessonoff reagent reactivo de Bessonoff
beta beta
beta acid ácido beta
beta activity actividad beta
beta decay desintegración beta
beta disintegration desintegración beta
beta emitter emisor beta
beta iron hierro beta
beta lactone lactona beta
beta particle partícula beta
beta position posición beta
beta radiation radiación beta
beta ray rayo beta
beta-ray spectrometer espectrómetro de
    rayos beta
beta sheet lámina beta
betaine betaína
betaine hydrochloride clorhidrato de
    betaína

betaine phosphate fosfato de betaína
betamethasone betametasona
betatopic betatópico
betatron betatrón
Bettendorf reagent reactivo de
    Bettendorf
Bettendorf test prueba de Bettendorf
Betterton-Kroll process proceso de
    Betterton-Kroll
Betti reaction reacción de Betti
Betts process proceso de Betts
betula oil aceite de betula
betulin betulina
betulinic acid ácido betulínico
betulinol betulinol
beyrichite beiriquita
bi bi
biacene biaceno
biacetylene biacetileno
biallyl bialilo
bianiline bianilina
bianisidine bianisidina
bianthryl biantrilo
biaxial biaxial
bibenzyl bibencilo
biborate biborato
bicalcium phosphate fosfato bicálcico
bicarbonate bicarbonato
bicarbonate of potash bicarbonato
    potásico
bicarbonate of soda bicarbonato sódico,
    bicarbonato de sosa
bichloride bicloruro
bicyclic bicíclico
bicyclic compound compuesto bicíclico
bicyclo numbering numeración bicíclica
bicyclodecane biciclodecano
bicyclohexyl biciclohexilo
bieberite bieberita
biethylene bietileno
biflavonyl biflavonilo
biformin biformina
bifurcation bifurcación
Biginelli reaction reacción de Biginelli
biguanide biguanida
bihexyl bihexilo
biimino biimino
bilateral bilateral
bile acid ácido biliar

bile salt sal biliar
bilifuscin bilifuscina
bilirubin bilirrubina
bilirubinic acid ácido bilirrubínico
bimetal bimetal
bimolecular bimolecular
bimolecular reaction reacción
    bimolecular
binapacryl binapacrilo
binaphthalene binaftaleno
binaphthalenyl binaftalenilo
binaphthyl binaftilo
binary binario
binary acid ácido binario
binary alloy aleación binaria
binary compound compuesto binario
binary mixture mezcla binaria
binary salt sal binaria
binary system sistema binario
bind fijar, aglutinar
binder aglutinante, ligador
binding aglutinación, ligadura
binding energy energía de separación
binding site lugar de unión
binitro binitro
binnite binita
binoxide binóxido
binuclear binuclear
bioaccumulation bioacumulación
bioactivation bioactivación
bioanalysis bioanálisis
bioassay bioensayo
biocatalyst biocatalizador
biochemical bioquímico
biochemistry bioquímica
biochrome biocromo
biocide biocida
biocolloid biocoloide
bioconversion bioconversión
biocytin biocitina
biodegradable biodegradable
bioelectrochemistry bioelectroquímica
bioelement bioelemento
bioengineering bioingeniería
bioethics bioética
bioflavonoid bioflavonoide
biogas biogas
biogenic biogénico
biogeochemistry biogeoquímica

bioinorganic chemistry química
  bioinorgánica
biological biológico
biological availability disponibilidad
  biológica
biology biología
bioluminescence bioluminiscencia
biomass biomasa
biomaterial biomaterial
biomimetic chemistry química
  biomimética
biomineral biomineral
biomolecule biomolécula
biophyl biofila
biopolymer biopolímero
bioreactor biorreactor
bioresmethrin biorresmetrina
biorization biorización
biose biosa
biosphere biosfera
biostat biostático
biosterol biosterol
biosynthesis biosíntesis
biotechnology biotecnología
biotic biótico
biotin biotina
biotite biotita
bioxyl bioxilo
biperiden biperiden
biphenyl bifenilo
biphenylamine bifenilamina
biphenyldiamine bifenildiamina
biphenylene bifenileno
biphenylene oxide óxido de bifenileno
biphenylyl bifenililo
biphenylylamine bifenililamina
bipropenyl bipropenilo
bipyridyl bipiridilo
biquinoline biquinolina
biquinolyl biquinolilo
biradical birradical
birectification birrectificación
birefringence birrefrigencia
birefringent birrefringente
Birge-Sponer extrapolation
  extrapolación de Birge-Sponer
Birkeland-Eyde process proceso de
  Birkeland-Eyde
birotation birrotación

bis bis
bisabolene bisaboleno
bisamide bisamida
bisazo bizaso
bisdiazo bisdiazo
bishydroxymethylurea
  bishidroximetilurea
bismanol bismanol
bismite bismita
bismuth bismuto
bismuth acetate acetato de bismuto
bismuth antimonide antimoniuro de
  bismuto
bismuth borate borato de bismuto
bismuth bromide bromuro de bismuto
bismuth carbonate carbonato de bismuto
bismuth chloride cloruro de bismuto
bismuth chromate cromato de bismuto
bismuth citrate citrato de bismuto
bismuth dichloride dicloruro de bismuto
bismuth ditannate ditanato de bismuto
bismuth gallate galato de bismuto
bismuth hydrate hidrato de bismuto
bismuth hydroxide hidróxido de bismuto
bismuth iodide yoduro de bismuto
bismuth nitrate nitrato de bismuto
bismuth oleate oleato de bismuto
bismuth oxalate oxalato de bismuto
bismuth oxide óxido de bismuto
bismuth oxobromide oxobromuro de
  bismuto
bismuth oxochloride oxocloruro de
  bismuto
bismuth oxofluoride oxofluoruro de
  bismuto
bismuth oxoiodide oxoyoduro de
  bismuto
bismuth oxybromide oxibromuro de
  bismuto
bismuth oxycarbonate oxicarbonato de
  bismuto
bismuth oxychloride oxicloruro de
  bismuto
bismuth oxyhydrate oxihidrato de
  bismuto
bismuth oxynitrate oxinitrato de bismuto
bismuth pentafluoride pentafluoruro de
  bismuto
bismuth pentoxide pentóxido de bismuto

bismuth permanganate permanganato de bismuto

bismuth peroxide peróxido de bismuto

bismuth phosphate fosfato de bismuto

bismuth potassium iodide yoduro potásico de bismuto

bismuth potassium tartrate tartrato potásico de bismuto

bismuth propionate propionato de bismuto

bismuth selenide seleniuro de bismuto

bismuth stannate estannato de bismuto

bismuth subcarbonate subcarbonato de bismuto

bismuth subchloride subcloruro de bismuto

bismuth subgallate subgalato de bismuto

bismuth subnitrate subnitrato de bismuto

bismuth subsalicylate subsalicilato de bismuto

bismuth sulfate sulfato de bismuto

bismuth sulfide sulfuro de bismuto

bismuth sulfite sulfito de bismuto

bismuth tannate tanato de bismuto

bismuth telluride telururo de bismuto

bismuth tetraoxide tetróxido de bismuto

bismuth tribromide tribromuro de bismuto

bismuth trichloride tricloruro de bismuto

bismuth trihydrate trihidrato de bismuto

bismuth trihydroxide trihidróxido de bismuto

bismuth triiodide triyoduro de bismuto

bismuth trinitrate trinitrato de bismuto

bismuth trioxide trióxido de bismuto

bismuth trisulfide trisulfuro de bismuto

bismuth tritelluride tritelururo de bismuto

bismuth tungstate tungstato de bismuto

bismuth yellow amarillo de bismuto

bismuthide bismutida

bismuthine bismutina

bismuthinite bismutinita

bismuthite bismutita

bismuthous bismutoso

bismuthyl bismutilo

bismuthyl bromide bromuro de bismutilo

bismuthyl carbonate carbonato de bismutilo

bismuthyl chloride cloruro de bismutilo

bismuthyl fluoride fluoruro de bismutilo

bismuthyl hydroxide hidróxido de bismutilo

bismuthyl nitrate nitrato de bismutilo

bisphenol bisfenol

bistability biestabilidad

bistetrazole bistetrazol

bistriazole bistriazol

bistridecyl phthalate ftalato de bistridecilo

bisulfate bisulfato

bisulfite bisulfito

bitartrate bitartrato

bithionol bitionol

bitrex bitrex

bitter amargo

bitter principle principio amargo

bittern bittern

bitumen bitumen

bituminous bituminoso

bituminous coal carbón bituminoso

bituminous plastic plástico bituminoso

biurea biurea

biuret biuret

biuret reaction reacción de biuret

biuret test prueba de biuret

bivalence bivalencia

bivalent bivalente

bivinyl bivinilo

black alum alumbre negro

black cyanide cianuro negro

black lead plomo negro

black powder polvo negro

Blagden law ley de Blagden

Blaise reaction reacción de Blaise

blanc fixe blanco fijo

Blanc reaction reacción de Blanc

Blanc rule regla de Blanc

blast furnace alto horno

blasting gelatin gelatina explosiva

bleach blanquear

bleaching blanqueo

bleaching material material de blanqueo

bleaching powder polvo de blanqueo

bleed sangrar, drenar

blend mezcla, combinación
blende blenda
bleomycin bleomicina
blister packaging empaquetado ampollado
Bloch theorem teorema de Bloch
block bloque, bloqueo
block copolymer copolímero de bloques
bloom capa fina superficial
blooming capa fina superficial
blown oil aceite soplado
blowpipe soplete
blue copper cobre azul
blue copperas caparrosa azul
blue lead plomo azul
blue powder polvo azul
blue salt sal azul
blue verdigris cardenillo azul
blue vitriol vitriolo azul
blueing azulamiento
blush capa fina superficial
boat artesa
boat conformation conformación en bote
Bodroux reaction reacción de Bodroux
body cuerpo
body-centered cubic cúbico centrado en el cuerpo
body-centered cubic structure estructura cúbica centrada en el cuerpo
body-centered lattice red centrada en el cuerpo
body-centered structure estructura centrada en el cuerpo
boehmite boehmita
Boeseken method método de Boeseken
Boettger test prueba de Boettger
Bogert-Cook synthesis síntesis de Bogert-Cook
boghead coal boghead
Bohn-Schmidt reaction reacción de Bohn-Schmidt
Bohr atom átomo de Bohr
Bohr magneton magnetón de Bohr
Bohr orbit órbita de Bohr
Bohr radius radio de Bohr
Bohr-Sommerfeld theory teoría de Bohr-Sommerfeld
Bohr theory teoría de Bohr

boil hervir
boiled oil aceite cocido
boiler incrustation incrustación de caldera
boiler scale incrustaciones de caldera
boiler stone incrustación de caldera
boiling ebullición
boiling chips gravilla de ebullición
boiling point punto de ebullición
boiling-point apparatus aparato de punto de ebullición
boiling-point diagram diagrama de puntos de ebullición
boiling-point elevation elevación de punto de ebullición
boiling-point range intervalo de puntos de ebullición
boiling-point reduction reducción de punto de ebullición
boiling stones gravilla de ebullición
boleite boleíta
bolometer bolómetro
Bolton reagent reactivo de Bolton
Boltzmann constant constante de Boltzmann
Boltzmann equation ecuación de Boltzmann
Boltzmann law ley de Boltzmann
bomb bomba
bomb calorimeter bomba calorimétrica
bombardment bombardeo
bond enlace
bond angle ángulo de enlace
bond distance distancia de enlace
bond energy energía de enlace
bond hybridization hibridización de enlace
bonding enlazado
bonding distance distancia de enlazado
bonding electron electrón de enlazado
bonding orbital orbital de enlazado, orbital enlazante
bondur bondur
bone ash ceniza de huesos
bone black negro de huesos
bone oil aceite de huesos
bone phosphate fosfato de huesos
boracic acid ácido borácico
boracite boracita

borane borano
boranediyl boranodiilo
boranetriyl broranotriilo
borate borato
borax bórax
borax-bead test prueba de perla con
    bórax
borax glass vidrio de bórax
borazine borazina
borazole borazol
Borcher metal metal de Borcher
Borcher process proceso de Borcher
boric acid ácido bórico
boric acid ester éster de ácido bórico
boric anhydride anhídrido bórico
boric oxide óxido bórico
boride boruro
borine borina
Born-Haber cycle ciclo de Born-Haber
Born-Oppenheimer approximation
    aproximación de Born-Oppenheimer
Born-Oppenheimer method método de
    Born-Oppenheimer
bornane bornano
bornene borneno
borneol borneol
bornite bornita
bornyl bornilo
bornyl acetate acetato de bornilo
bornyl alcohol alcohol bornílico
bornyl formate formiato de bornilo
bornyl isovalerate isovalerato de bornilo
bornyl salicylate salicilato de bornilo
boroethane boroetano
borohydride borohidruro
borol borol
boron boro
boron alloy aleación de boro
boron bromide bromuro de boro
boron carbide carburo de boro
boron chloride cloruro de boro
boron fluoride fluoruro de boro
boron fuel combustible de boro
boron hydride hidruro de boro
boron hydroxide hidróxido de boro
boron nitride nitruro de boro
boron oxide óxido de boro
boron phosphate fosfato de boro
boron phosphide fosfuro de boro

boron steel acero al boro
boron sulfide sulfuro de boro
boron tribromide tribromuro de boro
boron trichloride tricloruro de boro
boron trifluoride trifluoruro de boro
boron trisulfide trisulfuro de boro
borophosphoric acid ácido borofosfórico
borosalicylate borosalicilato
borosilicate borosilicato
borosilicate glass vidrio de borosilicato
borotungstic acid ácido borotúngstico
boroxine boroxina
bort bort
Bosch process proceso de Bosch
Bose-Einstein statistics estadística de
    Bose-Einstein
bottle botella
Bouchardat solution solución de
    Bouchardat
Bouguer-Lambert law ley de Bouguer-
    Lambert
Bouin fluid fluido de Bouin
bound electron electrón ligado
bound state estado ligado
bound water agua retenida
Bouveault aldehyde synthesis síntesis de
    aldehídos de Bouveault
Bouveault-Blanc method método de
    Bouveault-Blanc
Bouveault-Blanc reduction reduccion de
    Bouveault-Blanc
Bowen tube tubo de Bowen
Boyce burner mechero de Boyce
Boyle law ley de Boyle
Brackett series serie de Brackett
brackish water agua salobre
Bradsher reaction reacción de Bradsher
Brady reagent reactivo de Brady
Bragg angle ángulo de Bragg
Bragg crystallogram cristalograma de
    Bragg
Bragg curve curva de Bragg
Bragg equation ecuación de Bragg
Bragg ionization curve curva de
    ionización de Bragg
Bragg law ley de Bragg
Bragg method método de Bragg
Bragg peak pico de Bragg
Bragg rule regla de Bragg

Bragg spectrometer espectrómetro de Bragg
bran oil aceite de salvado
branched ramificado
branched chain cadena ramificada
branched-chain reaction reacción de cadena ramificada
branching ramificación
brasilin brasilina
brass latón
brassidic acid ácido brasídico
brassylic acid ácido brasílico
braunite braunita
Brazil wax cera de Brasil
brazing soldadura fuerte
Bredt rule regla de Bredt
breeder reactor engendrador
breeze cisco
bretonite bretonita
breunnerite breunerita
brewer's yeast levadura de cerveza
brewing fabricación de cerveza
Brewster angle ángulo de Brewster
Brewster process proceso de Brewster
bridge puente
bridgehead molecule molécula con puente
Briet-Wigner formula fómula de Briet-Wigner
bright-line spectrum espectro de líneas brillantes
brightener abrillantador
brightening abrillantamiento
brightening agent agente de abrillantamiento
brilliant brillante
brilliant crocein croceína brillante
brilliant green verde brillante
brilliant yellow amarillo brillante
brimstone piedra de azufre
Brin process proceso de Brin
brine salmuera
Brinell hardness dureza de Brinell
Brinell hardness test prueba de dureza de Brinell
britholite britolita
brittle quebradizo
brittle point punto de destrozo
Brix degree grado de Brix

brochanite brocanita
Brodie solution solución de Brodie
Broenner acid ácido de Broenner
bromacetate bromacetato
bromacil bromacil
bromal bromal
bromalin bromalina
bromate bromato
bromated brominado
bromelain bromelaína
bromelin bromelina
bromeosin bromeosina
bromethyl brometilo
bromic acid ácido brómico
brominated brominado
brominated camphor alcanfor brominado
bromination brominación
bromine bromo
bromine azide azida de bromo
bromine chloride cloruro de bromo
bromine cyanide cianuro de bromo
bromine fluoride fluoruro de bromo
bromine hydride hidruro de bromo
bromine iodide yoduro de bromo
bromine pentafluoride pentafluoruro de bromo
bromine trifluoride trifluoruro de bromo
bromine water agua de bromo
bromoacetamide bromoacetamida
bromoacetate bromoacetato
bromoacetic acid ácido bromoacético
bromoacetone bromoacetona
bromoacetone cyanohydrin cianhidrina de bromoacetona
bromoallylene bromoalileno
bromoaniline bromoanilina
bromoanthrallic acid ácido bromoantrálico
bromoauric acid ácido bromoaúrico
bromobenzaldehyde bromobenzaldehído
bromobenzene bromobenceno
bromobenzenesulfonic acid ácido bromobencenosulfónico
bromobenzoic acid ácido bromobenzoico
bromobenzyl cyanide cianuro de bromobencilo
bromobutane bromobutano
bromobutyric acid ácido bromobutírico

bromochloroethane bromocloroetano
bromochloromethane bromoclorometano
bromochloropropane
    bromocloropropano
bromocresol green verde de bromocresol
bromocresol purple púrpura de
    bromocresol
bromocyclopentane bromociclopentano
bromododecane bromododecano
bromoethane bromoetano
bromoethyl alcohol alcohol bromoetílico
bromofluorobenzene
    bromofluorobenceno
bromoform bromoformo
bromohexane bromohexano
bromol bromol
bromomethane bromometano
bromometry bromometría
bromonaphthalene bromonaftaleno
bromopentane bromopentano
bromophenol bromofenol
bromophenol blue azul de bromofenol
bromophenol red rojo de bromofenol
bromophenylacetonitrile
    bromofenilacetonitrilo
bromophosgene bromofosgeno
bromopicrin bromopicrina
bromopropene bromopropeno
bromopropionic acid ácido
    bromopropiónico
bromopyridine bromopiridina
bromopyrine bromopirina
bromostyrene bromoestireno
bromosuccinic acid ácido
    bromosuccínico
bromosuccinimide bromosuccinimida
bromothiophenol bromotiofenol
bromothymol blue azul de bromotimol
bromotoluene bromotolueno
bromotrichloromethane
    bromotriclorometano
bromotrifluoroethylene
    bromotrifluoroetileno
bromotrifluoromethane
    bromotrifluorometano
Brönner acid ácido de Brönner
Brönsted acid ácido de Brönsted
Brönsted-Lowry theory teoría de
    Brönsted-Lowry

Brönsted theory teoría de Brönsted
bronze bronce
bronze blue azul de bronce
bronze orange anaranjado de bronce
brosylate ester éster de brosilato
Brown-Boverti test prueba de Brown-
    Boverti
brown-ring test prueba de anillo pardo
Brownian motion movimiento
    browniano
Brownian movement movimiento
    browniano
browning reaction reacción de dorado
brucine brucina
brucine nitrate nitrato de brucina
brucine sulfate sulfato de brucina
brucite brucita
Bruehl receiver receptor de Bruehl
bubble burbuja
bubble counter contador de burbujas
Bucherer-Bergs reaction reacción de
    Bucherer-Bergs
Bucherer reaction reacción de Bucherer
Buchner funnel embudo de Buchner
buckyballs fulereno
buclizine hydrochloride clorhidrato de
    buclizina
buffer tampón, amortiguador
buffer action acción amortiguadora
buffer capacity capacidad amortiguadora
buffer index índice amortiguador
buffer range intervalo amortiguador
buffer salt sal amortiguadora
buffer solution solución amortiguadora
buffer value valor de amortiguador
buffering amortiguamiento
bufotanine bufotanina
bufotenine bufotenina
bulb matraz, cubeta
bumping ebullición despareja
bunamiodyl bunamiodilo
Bunsen burner mechero de Bunsen
Bunsen cell pila de Bunsen
Bunsen clamp abrazadera de Bunsen
Bunsen eudiometer eudiómetro de
    Bunsen
Bunsen flame llama de Bunsen
Bunsen funnel embudo de Bunsen
Bunsen-Kirchhoff law ley de Bunsen-

Kirchhoff
bunsenite bunsenita
buoyancy flotabilidad
buphanine bufanina
buret bureta
buret clamp abrazadera de bureta
burette bureta
burner mechero
burner attachment aditamento de mechero
burnt quemado, cocido
burnt lime cal viva
bursting disk disco de estallido
butadiene butadieno
butaldehyde butaldehído
butanal butanal
butane butano
butanedial butanodial
butanediamide butanodiamida
butanediamine butanodiamina
butanediol butanodiol
butanediolamine butanodiolamina
butanedione butanodiona
butanenitrile butanonitrilo
butanethiol butanotiol
butanetriol butanotriol
butanoic acid ácido butanoico
butanol butanol
butanol acetate acetato de butanol
butanone butanona
butanoyl chloride cloruro de butanoilo
butenal butenal
butene buteno
butenoic acid ácido butenoico
butenyl butenilo
butenylidene butenilideno
butesin butesina
butethal butetal
butethamine hydrochloride clorhidrato de butetamina
Butler-Volmer equation ecuación de Butler-Volmer
butonate butonato
butopyronoxyl butopironoxilo
butoxy butoxi
butoxyethanol butoxietanol
butoxyethyl laurate laurato de butoxietilo
butoxyethyl oleate oleato de butoxietilo

butoxyethyl stearate estearato de butoxietilo
butoxyphenol butoxifenol
butoxypropanol butoxipropanol
butoxytriglycol butoxitriglicol
butter mantequilla
butter yellow amarillo de mantequilla
butyl butilo
butyl acetate acetato de butilo
butyl acetoacetate acetoacetato de butilo
butyl acetoxystearate acetoxiestearato de butilo
butyl acetylene acetileno de butilo
butyl acrylate acrilato de butilo
butyl alcohol alcohol butílico
butyl aldehyde aldehído butílico
butyl anthranilate antralinato de butilo
butyl benzoate benzoato de butilo
butyl borate borato de butilo
butyl bromide bromuro de butilo
butyl butanoate butanoato de butilo
butyl butyrate butirato de butilo
butyl carbamate carbamato de butilo
butyl carbinol butilcarbinol
butyl chloral hydrate hidrdato de butilcloral
butyl chloride cloruro de butilo
butyl chromate cromato de butilo
butyl citrate citrato de butilo
butyl crotonate crotonato de butilo
butyl cyanide cianuro de butilo
butyl diethanolamine dietanolamina de butilo
butyl diglyme diglime de butilo
butyl dodecanoate dodecanoato de butilo
butyl epoxystearate epoxiestearato de butilo
butyl ether éter butílico
butyl formate formiato de butilo
butyl furoate furoato de butilo
butyl iodide yoduro de butilo
butyl lactate lactato de butilo
butyl myristate miristato de butilo
butyl nitrate nitrato de butilo
butyl nitrite nitrito de butilo
butyl nonanoate nonanoato de butilo
butyl octadecanoate octadecanoato de butilo
butyl oleate oleato de butilo

butyl pelargonate pelargonato de butilo

butyl peracetate peracetato de butilo

butyl perbenzoate perbenzoato de butilo

butyl peroxide peróxido de butilo

butyl peroxyacetate peroxiacetato de butilo

butyl peroxybenzoate peroxibenzoato de butilo

butyl peroxyisobutyrate peroxiisobutirato de butilo

butyl phthalate ftalato de butilo

butyl propionate propionato de butilo

butyl ricinoleate ricinoleato de butilo

butyl rubber caucho butilo

butyl sebacate sebacato de butilo

butyl sorbate sorbato de butilo

butyl stearamide estearamida de butilo

butyl stearate estearato de butilo

butyl sulfide sulfuro de butilo

butyl thiocyanate tiocianiato de butilo

butyl titanate titanato de butilo

butyl valerate valerato de butilo

butylacrylamide butilacrilamida

butylamine butilamina

butylaminoethanol butilaminoetanol

butylaniline butilanilina

butylanthraquinone butilantraquinona

butylated hydroxyanisole hidroxianisol butilado

butylated hydroxytoluene hidroxitolueno butilado

butylbenzene butilbenceno

butylbenzenesulfonamide butilbencenosulfonamida

butylbenzyl phthalate ftalato de butilbencilo

butyldiamylamine butildiamilamina

butylene butileno

butylene dimethacrylate dimetracrilato de butileno

butylene glycol butilenglycol

butylene oxide óxido de butileno

butylethanolamine butiletanolamina

butylethylacetaldehyde butiletilacetaldehído

butylformamide butilformamida

butylhydroquinone butilhidroquinona

butylidene butilideno

butylin trichloride tricloruro de butilina

butylisocyanate butilisocianato

butyllithium butil-litio

butylmagnesium chloride cloruro de butilmagnesio

butylmercaptan butilmercaptano

butylmethacrylate butilmetacrilato

butylparaben butilparabeno

butylperoxypivalate butilperoxipivalato

butylphenol butilfenol

butylphenyl ether éter butilfenílico

butylstannic acid ácido butilestánnico

butylthiophenol butiltiofenol

butyltoluene butiltolueno

butyltrichlorosilane butiltriclorosilano

butylurea butilurea

butyne butino

butynediol butinodiol

butyraldehyde butiraldehído

butyric butírico

butyric acid ácido butírico

butyric alcohol alcohol butírico

butyric aldehyde aldehído butírico

butyric anhydride anhídrido butírico

butyrin butirina

butyrolactam butirolactama

butyrolactone butirolactona

butyrone butirona

butyronitrile butironitrilo

butyroyl chloride cloruro de butiroilo

butyryl butirilo

butyryl chloride cloruro de butirilo

by-product producto secundario

C

Cabannes factor factor de Cabannes
cabrerite cabrerita
cacao butter manteca de cacao
cacao syrup jarabe de cacao
cacodyl cacodilo
cacodyl chloride cloruro de cacodilo
cacodyl oxide óxido de cacodilo
cacodylate cacodilato
cacodylic acid ácido cacodílico
cacotheline cacotelina
cactine cactina
cadalene cadaleno
cadaverine cadaverina
cadinene cadineno
cadion cadión
cadmium cadmio
cadmium acetate acetato de cadmio
cadmium amalgam amalgama de cadmio
cadmium ammonium bromide bromuro
    amónico de cadmio
cadmium antimonide antimoniuro de
    cadmio
cadmium borotungstate borotungstato
    de cadmio
cadmium bromate bromato de cadmio
cadmium bromide bromuro de cadmio
cadmium carbonate carbonato de
    cadmio
cadmium cell pila de cadmio
cadmium chlorate clorato de cadmio
cadmium chloride cloruro de cadmio
cadmium chloroacetate cloroacetato de
    cadmio
cadmium cinnamate cinamato de cadmio
cadmium cyanide cianuro de cadmio
cadmium diethyldithiocarbamate
    dietilditiocarbamato de cadmio
cadmium electrode electrodo de cadmio
cadmium fluoride fluoruro de cadmio
cadmium formate formiato de cadmio
cadmium fumarate fumarato de cadmio
cadmium halide haluro de cadmio
cadmium hexafluorosilicate

hexafluorurosilicato de cadmio
cadmium hydroxide hidróxido de
    cadmio
cadmium iodate yodato de cadmio
cadmium iodide yoduro de cadmio
cadmium lactate lactato de cadmio
cadmium molybdate molybdato de
    cadmio
cadmium nitrate nitrato de cadmio
cadmium oxalate oxalato de cadmio
cadmium oxide óxido de cadmio
cadmium permanganate permanganato
    de cadmio
cadmium phosphate fosfato de cadmio
cadmium pigment pigmento de cadmio
cadmium potassium cyanide cianuro
    potásico de cadmio
cadmium potassium iodide yoduro
    potásico de cadmio
cadmium propionate propionato de
    cadmio
cadmium red rojo de cadmio
cadmium ricinoleate ricinoleato de
    cadmio
cadmium selenate seleniato de cadmio
cadmium selenide seleniuro de cadmio
cadmium stearate estearato de cadmio
cadmium succinate succinato de cadmio
cadmium sulfate sulfato de cadmio
cadmium sulfide sulfuro de cadmio
cadmium tartrate tartrato de cadmio
cadmium telluride telururo de cadmio
cadmium tungstate tungstato de cadmio
cadmium valerate valerato de cadmio
cadmium yellow amarillo de cadmio
cadmous cadmoso
caesium cesio
caffeic acid ácido cafeico
caffeine cafeína
caffeine benzoate benzoato de cafeína
caffeine citrate citrato de cafeína
caffeine hydrochloride clorhidrato de
    cafeína

caffeine salicylate salicilato de cafeína
caffetannic acid ácido cafetánico
cage compound compuesto en jaula
cage zeolite zeolita en jaula
calabarine calabarina
calamine calamina
calamus oil aceite de cálamo
calaverite calaverita
calcein calceína
calciferol calciferol
calcimeter calcímetro
calcimine calcimina
calcination calcinación
calcine calcina
calcined calcinado
calcined bauxite bauxita calcinada
calcinite calcinita
calcinol calcinol
calcite calcita
calcium calcio
calcium abietate abietato cálcico
calcium acetate acetato cálcico, acetato
  de calcio
calcium acetylide acetiluro cálcico
calcium acetylsalicylate acetilsalicilato
  cálcico
calcium acrylate acrilato cálcico
calcium alginate alginato cálcico
calcium aluminate aluminato cálcico
calcium aluminum hydride hidruro
  alumínico cálcico
calcium aluminum nitrate nitrato
  alumínico cálcico
calcium arsenate arseniato cálcico
calcium arsenite arsenito cálcico
calcium ascorbate ascorbato cálcico
calcium benzoate benzoato cálcico
calcium biophosphate biofosfato cálcico
calcium bisulfide bisulfuro cálcico
calcium bisulfite bisulfito cálcico
calcium borate borato cálcico
calcium bromate bromato cálcico
calcium bromide bromuro cálcico
calcium butanoate butanoato cálcico
calcium carbide carburo cálcico
calcium carbonate carbonato cálcico
calcium caseinate caseinato cálcico
calcium chlorate clorato cálcico
calcium chloride cloruro cálcico

calcium chlorite clorito cálcico
calcium chromate cromato cálcico
calcium cinnamate cinamato cálcico
calcium citrate citrato cálcico
calcium cyanamide cianamida cálcica
calcium cyanide cianuro cálcico
calcium cyclamate ciclamato cálcico
calcium dehydroacetate deshidroacetato
  cálcico
calcium diacetate diacetato cálcico
calcium dicarbide dicarburo cálcico
calcium dichromate dicromato cálcico
calcium dioxide dióxido cálcico
calcium disodium edetate edetato
  disódico cálcico
calcium ethanoate etanoato cálcico
calcium ethylhexoate etilhexoato cálcico
calcium ferrocyanide ferrocianuro
  cálcico
calcium fluoride fluoruro cálcico
calcium fluorophosphate fluorofosfato
  cálcico
calcium fluorosilicate fluorosilicato
  cálcico
calcium formate formiato cálcico
calcium fumarate fumarato cálcico
calcium gluconate gluconato cálcico
calcium glutamate glutamato cálcico
calcium glycerophosphate glicerofosfato
  cálcico
calcium glycolate glicolato cálcico
calcium hexafluorosilicate
  hexafluorosilicato cálcico
calcium hydrate hidrato cálcico
calcium hydride hidruro cálcico
calcium hydrogen sulfite sulfito de
  hidrógeno cálcico
calcium hydrosulfide hidrosulfuro
  cálcico
calcium hydroxide hidróxido cálcico
calcium hypochlorite hipoclorito cálcico
calcium hypophosphite hipofosfito
  cálcico
calcium iodate yodato cálcico
calcium iodide yoduro cálcico
calcium iodobehenate yodobehenato
  cálcico
calcium isobutyrate isobutirato cálcico
calcium lactate lactato cálcico

calcium malate malato cálcico
calcium metaphosphate metafosfato
    cálcico
calcium metasilicate metasilicato cálcico
calcium molybdate molibdato cálcico
calcium naphthenate naftenato cálcico
calcium nitrate nitrato cálcico
calcium nitride nitruro cálcico
calcium nitrite nitrito cálcico
calcium octoate octoato cálcico
calcium oleate oleato cálcico
calcium orthophosphate ortofosfato
    cálcico
calcium orthotungstate ortotungstato
    cálcico
calcium oxalate oxalato cálcico
calcium oxide óxido cálcico
calcium oxychloride oxicloruro cálcico
calcium palmitate palmitato cálcico
calcium pantothenate pantotenato
    cálcico
calcium perborate perborato cálcico
calcium perchlorate perclorato cálcico
calcium permanganate permanganato
    cálcico
calcium peroxide peróxido cálcico
calcium phenolate fenolato cálcico
calcium phenylate fenilato cálcico
calcium phosphate fosfato cálcico
calcium phosphide fosfuro cálcico
calcium phosphinate fosfinato cálcico
calcium phosphite fosfito cálcico
calcium phytate fitato cálcico
calcium plumbate plumbato cálcico
calcium polysilicate polisilicato cálcico
calcium propionate propionato cálcico
calcium pyrophosphate pirofosfato
    cálcico
calcium resinate resinato cálcico
calcium ricinoleate ricinoleato cálcico
calcium selenate seleniato cálcico
calcium silicate silicato cálcico
calcium silicide siliciuro cálcico
calcium silicofluoride silicofluoruro
    cálcico
calcium sorbate sorbato cálcico
calcium stannate estannato cálcico
calcium stearate estearato cálcico
calcium succinate succinato cálcico

calcium sulfamate sulfamato cálcico
calcium sulfate sulfato cálcico
calcium sulfhydrate sulfhidrato cálcico
calcium sulfide sulfuro cálcico
calcium sulfite sulfito cálcico
calcium sulfocyanate sulfocianato
    cálcico
calcium superoxide superóxido cálcico
calcium superphosphate superfosfato
    cálcico
calcium tannate tanato cálcico
calcium tartrate tartrato cálcico
calcium thiocyanate tiocianato cálcico
calcium titanate titanato cálcico
calcium tungstate tungstato cálcico
calcium undecylenate undecilenato
    cálcico
calcium valerate valerato cálcico
calcium zirconate zirconato cálcico
caledonite caledonita
calender calandria
calibrate calibrar
calibrated calibrado
calibration calibración
calibration error error de calibración
caliche caliche
californite californita
californium californio
calixarene calixareno
calomel calomel
calomel electrode electrodo de calomel
calorie caloría
calorific value valor calorífico
calorimeter calorímetro
calorimetric calorimétrico
calorimetry calorimetría
calorizing calorización
calotropin calotropina
calutron calutrón
calycanthine calicantina
cambopinic acid ácido cambopínico
Campbell synthesis síntesis de Campbell
camphane canfano
camphanol canfanol
camphanone canfanona
camphene canfeno
camphocarboxylic acid ácido
    canfocarboxílico
campholic acid ácido canfólico

camphor alcanfor
camphor bromate bromato de alcanfor
camphor imide imida de alcanfor
camphor oxime oxima de alcanfor
camphoric acid ácido canfórico
camphoroquinone canforoquinona
camphoroyl canforoilo
Camps quinoline synthesis síntesis de
    quinolinas de Camps
canadine canadina
canal rays rayos canales
cananga oil aceite de cananga
canavanine canavanina
cancrinite cancrinita
candelilla wax cera de candelilla
cane sugar azúcar de caña
cannabidiol cannabidiol
cannabinol cannabinol
cannabis cannabis
cannel coal carbón de bujía
Cannizzaro reaction reacción de
    Cannizzaro
cannonite canonita
canola aceite de colza
canonical form forma canónica
canphane canfano
cantharidin cantaridina
canthaxanthin cantaxantina
capacity capacidad
capillarity capilaridad
capillary condensation condensación
    capilar
capillary pipet pipeta capilar
capraldehyde capraldehído
caprate caprato
capreomycin capreomicina
capric acid ácido cáprico
capric aldehyde aldehído cáprico
caproic acid ácido caproico
caproic aldehyde aldehído caproico
caprolactam caprolactama
caprolactone caprolactona
capronyl capronilo
caproyl caproilo
capryl caprilo
caprylamide caprilamida
caprylate caprilato
caprylic acid ácido caprílico
caprylic anhydride anhídrido caprílico

caprylyl caprililo
capsule cápsula
captafol captafol
captan captano
capture captura
caracolite caracolita
caramel caramelo
caramelan caramelano
caramelization caramelización
carane carano
carat quilate
caraway oil aceite de alcaravea
carbaborane carbaborano
carbachol carbacol
carbaldehyde carbaldehído
carbamate carbamato
carbamic carbámico
carbamic acid ácido carbámico
carbamide carbamida
carbamide chloride cloruro de
    carbamida
carbamide peroxide peróxido de
    carbamida
carbamidine carbamidina
carbamido carbamido
carbamite carbamita
carbamoyl carbamoilo
carbamyl carbamilo
carbamylurea carbamilurea
carbanil carbanilo
carbanilic acid ácido carbanílico
carbanilide carbanilida
carbanion carbanión
carbaryl carbarilo
carbate carbato
carbazic acid ácido carbázico
carbazide carbazida
carbazole carbazol
carbazolyl carbazolilo
carbazone carbazona
carbazotic acid ácido carbazótico
carbazoyl carbazoilo
carbene carbeno
carbenium carbenio
carbenium ion ion carbenio
carbenoid carbenoide
carbethoxycyclohexanone
    carbetoxiciclohexanona
carbethoxycyclopentanone

carbetoxiciclopentanona
carbethoxypiperazine
  carbetoxipiperazina
carbide carburo
carbinol carbinol
carbitol carbitol
carbobenzoyl carbobenzoilo
carbocation carbocatión
carbocyclic carbocíclico
carbodiazone carbodiazona
carbodihydrazide carbodihidrazida
carbodiimide carbodiimida
carbodithioic carboditioico
carbofuran carbofurano
carbohydrase carbohidrasa
carbohydrate carbohidrato, hidrato de
  carbono
carbohydrazide carbohidrazida
carbohydride carbohidruro
carboid carboide
carbolfuchsin carbolfucsina
carbolic acid ácido carbólico
carbolic oil aceite carbólico
carbomer carbómero
carbometer carbonómetro
carbomethene carbometeno
carbomethoxy carbometoxi
carbomycin carbomicina
carbon carbono, carbón
carbon assimilation asimilación de
  carbono
carbon atom átomo de carbono
carbon bisulfide bisulfuro de carbono
carbon black negro de carbón
carbon bond enlace de carbono
carbon chain cadena de carbonos
carbon compound compuesto de carbono
carbon cycle ciclo de carbono
carbon dating datación por carbono
carbon dichloride dicloruro de carbono
carbon dioxide dióxido de carbono
carbon dioxide snow nieve de dióxido de
  carbono
carbon disulfide disulfuro de carbono
carbon electrode electrodo de carbono
carbon fiber fibra de carbono
carbon fluoride fluoruro de carbono
carbon hexachloride hexacloruro de
  carbono

carbon monosulfide monosulfuro de
  carbono
carbon monoxide monóxido de carbono
carbon nitride nitruro de carbono
carbon-nitrogen cycle ciclo de carbono-
  nitrógeno
carbon oxide óxido de carbono
carbon oxybromide oxibromuro de
  carbono
carbon oxychloride oxicloruro de
  carbono
carbon oxycyanide oxicianuro de
  carbono
carbon oxyfluoride oxifluoruro de
  carbono
carbon oxysulfide oxisulfuro de carbono
carbon ratio razón de carbono
carbon steel acero al carbono
carbon suboxide subóxido de carbono
carbon tetrabromide tetrabromuro de
  carbono
carbon tetrachloride tetracloruro de
  carbono
carbon tetrafluoride tetrafluoruro de
  carbono
carbon tetraiodide tetrayoduro de
  carbono
carbon trichloride tricloruro de carbono
carbonado carbonado
carbonatation carbonatación
carbonate carbonato
carbonate mineral mineral de carbonato,
  carbonatomineral
carbonate rock roca de carbonato, roca
  carbonatada
carbonation carbonación
carbonic carbónico
carbonic acid ácido carbónico
carbonic anhydrase anhidrasa carbónica
carbonic ester éster carbónico
carbonic ether éter carbónico
carbonimidoyl carbonimidoilo
carbonite carbonita
carbonitrile carbonitrilo
carbonium carbonio
carbonium ion ion carbonio
carbonization carbonización
carbonize carbonizar
carbonized carbonizado

carbonohydrazide carbonohidrazida
carbonohydrazido carbonohidrazido
carbonyl carbonilo
carbonyl bromide bromuro de carbonilo
carbonyl chloride cloruro de carbonilo
carbonyl cyanide cianuro de carbonilo
carbonyl fluoride fluoruro de carbonilo
carbonyl group grupo carbonilo
carbonyl halide haluro de carbonilo
carbonyl hydride hidruro de carbonilo
carbonyl sulfide sulfuro de carbonilo
carbonylate carbonilato
carbonylation carbonilación
carbonyldiimidazole carbonildiimidazol
carbonyldioxy carbonldioxi
carbophenothion carbofenotión
carborane carborano
carbostyril carboestirilo
carbothioic carbotioico
carbothionic carbotiónico
carboxamide carboxamida
carboxamidine carboxamidina
carboxy carboxi
carboxybenzene carboxibenceno
carboxyl carboxilo
carboxyl group grupo carboxilo
carboxylase carboxilasa
carboxylation carboxilación
carboxylic carboxílico
carboxylic acid ácido carboxílico
carboxymethylcellulose
      carboximetilcelulosa
carboxypeptidase carboxipeptidasa
carburite carburita
carburization carburización
carbylamine carbilamina
carbylic acid ácido carbílico
carbyloxime carbiloxima
carbyne carbino
carcinogen carcinógeno
carcinomic acid ácido carcinómico
cardamom oil aceite de cardamomo
cardenolide cardenolida
carene careno
Carius method método de Carius
Carius tube tubo de Carius
Carl solution solución de Carl
carlic acid ácido cárlico
carmine carmín

carminic acid ácido carmínico
carminite carminita
carnallite carnalita
carnauba wax cera de carnauba
carnaubic acid ácido carnáubico
carnegine carnegina
carnine carnina
carnitine carnitina
carnosine carnosina
Carnot cycle ciclo de Carnot
Carnot reagent reactivo de Carnot
carnotite carnotita
Carnoy fluid fluido de Carnoy
Caro acid ácido de Caro
Caro reagent reactivo de Caro
carobic acid ácido caróbico
carobine carobina
carolic acid ácido carólico
carone carona
carotene caroteno
carotenoid carotenoide
carpain carpaína
carrageenan carraginina
carreen carreno
carrier portador
carrier gas gas portador
Carroll reaction reacción de Carroll
carthamic acid ácido cartámico
carthamin cartamina
carvacrol carvacrol
carveol carveol
carvestrene carvestreno
carvone carvona
caryophyllene cariofileno
caryophyllic acid ácido cariofílico
cascade cascada
cascade process proceso en cascada
cascarilla oil aceite de cascarilla
casein caseína
caseinase caseinasa
cashew oil aceite de anacardo
Cassella acid ácido de Cassella
cassia oil aceite de casia
cassiterite casiterita
cast iron hierro fundido
Castner-Kellner cell pila de Castner-
      Kellner
Castner process proceso de Castner
castor oil aceite de castor

castor oil acid ácido de aceite de castor
catabiotic catabiótico
catabolic catabólico
catabolism catabolismo
catalase catalasa
catalysis catálisis
catalyst catalizador
catalyst carrier portador de catalizador
catalyst promotor promotor de catalizador
catalyst selectivity selectividad de catalizador
catalytic catalítico
catalytic action acción catalítica
catalytic converter convertidor catalítico
catalytic cracking craqueo catalítico
catalytic force fuerza catalítica
catalytic poison veneno catalítico
catalytic polymerization polimerización catalítica
catalytic reforming reformación catalítica
catalyzer catalizador
cataphoresis cataforesis
catechin catequina
catechol catecol
catecholborane catecolborano
catena catena
catenane catenano
catenary catenario
catenation catenación
catenyl catenilo
catharometer catarómetro
cathartin catartina
cathetometer catetómetro
cathine catina
cathode cátodo
cathode coating revestimiento catódico
cathode evaporation evaporación catódica
cathode layer capa catódica
cathode polarization polarización catódica
cathode protection protección catódica
cathode rays rayos catódicos
cathode sputtering pulverización catódica
cathodic catódico
cathodic coating revestimiento catódico

cathodic evaporation evaporación catódica
cathodic layer capa catódica
cathodic polarization polarización catódica
cathodic protection protección catódica
cathodic rays rayos catódicos
cathodic sputtering pulverización catódica
catholyte católito
cation catión
cation exchange intercambio de cationes
cationic catiónico
cationic polymerization polimerización catiónica
cationic reagent reactivo catiónico
cationotropy cationotropía
cationtrophy cationtrofía
catlinite catlinita
caustic cáustico
caustic alcohol alcohol cáustico
caustic antimony antimonio cáustico
caustic baryta barita cáustica
caustic cracking fragilización cáustica
caustic embrittlement fragilización cáustica
caustic lime cal cáustica
caustic potash potasa cáustica
caustic soda sosa cáustica
caustic wash lavado cáustico
causticity causticidad
causticized ash ceniza caustificada
cavitation cavitación
cedrin cedrina
cedrol cedrol
cedryl acetate acetato de cedrilo
celestine celestina
celestite celestita
cell célula, pila
cell constant constante de célula
cellase celasa
cellobiose celobiosa
cellodextrin celodextrina
celloidin celoidina
cellophane celofán
cellosteric effect efecto celostérico
cellular plastic plástico celular
cellulase celulasa
cellulose celulosa

cellulose acetate   acetato de celulosa
cellulose acetate butyrate   acetatobutirato de celulosa
cellulose acetatobutyrate   acetatobutirato de celulosa
cellulose diacetate   diacetato de celulosa
cellulose ether   éter celulósico
cellulose fiber   fibra de celulosa
cellulose gum   goma de celulosa
cellulose monoacetate   monoacetato de celulosa
cellulose nitrate   nitrato de celulosa
cellulose propionate   propionato de celulosa
cellulose triacetate   triacetato de celulosa
cellulose trinitrate   trinitrato de celulosa
cellulose xanthate   xantato de celulosa
cellulosic   celulósico
cellulosic plastic   plástico celulósico
cellulosic resin   resina celulósica
cellulosic thiocarbonate   tiocarbonato celulósico
Celsius scale   escala Celsius
cement   cemento
cementation   cementación
cementite   cementita
center of symmetry   centro de simetría
centi   centi
centigrade   centígrado
centigrade scale   escala centígrada
centinormal   centinormal
centinormal solution   solución centinormal
centrifugal   centrífugo
centrifugal force   fuerza centrífuga
centrifugal pump   bomba centrífuga
centrifugal separation   separación centrífuga
centrifugation   centrifugación
centrifuge   centrífuga
centripetal   centrípeto
centroid diagram   diagrama centroide
centron   centrón
cephalin   cefalina
cephalosporin   cefalosporina
cephamycin   cefamicina
ceramic   cerámica
ceramic alloy   aleación cerámica
ceramide   ceramida

cerate   cerato
cerebronic acid   ácido cerebrónico
cerebrose   cerebrosa
cerebrosides   cerebrósidos
Cerenkov effect   efecto de Cerenkov
ceresin   ceresina
ceresin wax   cera de ceresina
ceria   ceria
ceric   cérico
ceric ammonium nitrate   nitrato amónico cérico
ceric fluoride   fluoruro cérico
ceric hydroxide   hidróxido cérico
ceric nitrate   nitrato cérico
ceric oxide   óxido cérico
ceric sulfate   sulfato cérico
ceric sulfide   sulfuro cérico
cerin   cerina
ceritamic acid   ácido ceritámico
cerite   cerita
cerium   cerio
cerium carbonate   carbonato de cerio
cerium chloride   cloruro de cerio
cerium dioxide   dióxido de cerio
cerium hydrate   hidrato de cerio
cerium nitrate   nitrato de cerio
cerium oxalate   oxalato de cerio
cerium oxide   óxido de cerio
cerium stearate   estearato de cerio
cerium sulfate   sulfato de cerio
cermet   cermet
ceroplastic acid   ácido ceroplástico
cerosin   cerosina
cerosinyl   cerosinilo
cerotene   ceroteno
cerotic acid   ácido cerótico
cerotinic acid   ácido cerotínico
cerous   ceroso
cerous acetate   acetato ceroso
cerous benzoate   benzoato ceroso
cerous bromate   bromato ceroso
cerous bromide   bromuro ceroso
cerous carbonate   carbonato ceroso
cerous chloride   cloruro ceroso
cerous citrate   citrato ceroso
cerous fluoride   fluoruro ceroso
cerous hydroxide   hidróxido ceroso
cerous iodide   yoduro ceroso
cerous lactate   lactato ceroso

cerous nitrate nitrato ceroso
cerous oxalate oxalato ceroso
cerous phosphate fosfato ceroso
cerous sulfate sulfato ceroso
cerous valerate valerato ceroso
certified color color certificado
cerulean blue azul cerúleo
cerussite cerusita
cervantite cervantita
ceryl cerilo
ceryl alcohol alcohol cerílico
cerylic cerílico
cesium cesio
cesium acetate acetato de cesio
cesium alum alumbre de cesio
cesium aluminum sulfate sulfato
    alumínico de cesio
cesium antimonide antimoniuro de cesio
cesium arsenide arseniuro de cesio
cesium benzoate benzoato de cesio
cesium bicarbonate bicarbonato de cesio
cesium bromate bromato de cesio
cesium bromide bromuro de cesio
cesium carbonate carbonato de cesio
cesium chloride cloruro de cesio
cesium chromate cromato de cesio
cesium cyanide cianuro de cesio
cesium dichromate dicromato de cesio
cesium dioxide dióxido de cesio
cesium fluoride fluoruro de cesio
cesium hexachloroplatinate
    hexacloroplatinato de cesio
cesium hexafluorosilicate
    hexafluorosilicato de cesio
cesium hydrate hidrato de cesio
cesium hydrogensulfate
    hidrogenosulfato de cesio
cesium hydroxide hidróxido de cesio
cesium iodate yodato de cesio
cesium iodide yoduro de cesio
cesium nitrate nitrato de cesio
cesium oxalate oxalato de cesio
cesium perchlorate perclorato de cesio
cesium periodate peryodato de cesio
cesium permanganate permanganato de
    cesio
cesium peroxide peróxido de cesio
cesium phosphide fosfuro de cesio
cesium salicylate salicilato dc cesio

cesium sulfate sulfato de cesio
cesium sulfite sulfito de cesio
cesium tetraoxide tetróxido de cesio
cesium trioxide trióxido de cesio
cesium trisulfide trisulfuro de cesio
cespitine cespitina
cetane cetano
cetane number número de cetano
cetene ceteno
cetin cetina
cetol cetol
cetolic acid ácido cetólico
cetrimide cetrimida
cetyl cetilo
cetyl alcohol alcohol cetílico
cetyl bromide bromuro de cetilo
cetyl ether éter cetílico
cetyl mercaptan cetilmercaptano
cetyl palmitate palmitato de cetilo
cetylate cetilato
cetylbetamine cetilbetamina
cetylene cetileno
cetylic acid ácido cetílico
cetylic alcohol alcohol cetílico
cetylpyridinium bromide bromuro de
    cetilpiridinio
cetylpyridinium chloride cloruro de
    cetilpiridinio
cetyltrimethylammonium bromide
    bromuro de cetiltrimetilamonio
cetyltrimethylammonium chloride
    cloruro de cetiltrimetilamonio
chabazite chabasita
Chadwick-Goldhaber effect efecto de
    Chadwick-Goldhaber
chain cadena
chain conformation conformación en
    cadena
chain decay desintegración en cadena
chain disintegration desintegración en
    cadena
chain initiation inicio de cadena
chain isomerization isomerización de
    cadena
chain mechanism mecanismo en cadena
chain polymer polímero de cadena
chain reaction reacción en cadena
chain termination terminación de cadena
chain transfer transferencia de cadena

chair conformation conformación en silla
chalcanthite calcantita
chalcedony calcedonia
chalcocite calcosina
chalcogen calcógeno
chalcone calcona
chalcophanite calcofanita
chalcopyrite calcopirita
chalcosiderite calcosiderita
chalk creta
chamber cámara
chamber acid ácido de cámara
chamomile oil aceite de manzanilla
Chance-Claus process proceso de Chance-Claus
change of phase cambio de fase
change of state cambio de estado
channeling canalización
Channing solution solución de Channing
chaos caos
chaotic reaction reacción caótica
Chapman rearrangement reordenamiento de Chapman
charcoal carbón
charge carga
charge density densidad de carga
charge independence independencia de carga
charge-mass ratio razón de carga a masa
charge of electron carga de electrón
charge transfer transferencia de carga
charged cargado
charged particle partícula cargada
Charles law ley de Charles
chaulmoogric acid ácido chaulmúgrico
chavicol chavicol
chebulinic acid ácido quebulínico
cheiramidine queiramidina
chelate quelato
chelate effect efecto de quelato
chelating agent agente quelante
chelation quelación
chelerythrine queleritrina
chelidonic acid ácido quelidónico
chemical substancia química, producto químico
chemical action acción química
chemical activity actividad química

chemical affinity afinidad química
chemical agent agente químico
chemical analysis análisis químico
chemical bond enlace químico
chemical burn quemadura química
chemical change cambio químico
chemical combination combinación química
chemical compound compuesto químico
chemical constant constante química
chemical constitution constitución química
chemical corrosion corrosión química
chemical data datos químicos
chemical dating datación química
chemical deposition deposición química
chemical detector detector químico
chemical effect efecto químico
chemical efficiency eficiencia química
chemical element elemento químico
chemical energy energía química
chemical engineering ingeniería química
chemical entity entidad química
chemical equation ecuación química
chemical equilibrium equilibrio químico
chemical equivalent equivalente químico
chemical erosion erosión química
chemical exchange intercambio químico
chemical flooding inundación química
chemical formula fórmula química
chemical garden jardín químico
chemical incompatibility incompatibilidad química
chemical indicator indicador químico
chemical inhibitor inhibidor químico
chemical kinetics cinética química
chemical laws leyes químicas
chemical light luz química
chemical load carga química
chemical microscopy microscopia química
chemical milling molienda química
chemical modification modificación química
chemical nomenclature nomenclatura química
chemical oxygen demand demanda química de oxígeno
chemical passivation pasivación química

chemical passivity pasividad química
chemical polarity polaridad química
chemical pollution contaminación
    química
chemical potential potencial químico
chemical process proceso químico
chemical processing procesamiento
    químico
chemical product producto químico
chemical property propiedad química
chemical protector protector químico
chemical purity pureza química
chemical reaction reacción química
chemical reactivity reactividad química
chemical reduction reducción química
chemical research investigación química
chemical sediment sedimento químico
chemical series serie química
chemical shift deslizamiento químico
chemical smoke humo químico
chemical solution solución química
chemical solvent solvente químico
chemical specialty especialidad química
chemical spectrum espectro químico
chemical standard patrón químico
chemical substance substancia química
chemical symbol símbolo químico
chemical synthesis síntesis química
chemical technology tecnología química
chemical testing pruebas químicas
chemical thermodynamics
    termodinámica química
chemical warfare guerra química
chemical waste desperdicios químicos
chemically induced químicamente
    inducido
chemically pure químicamente puro
chemiluminescence quimioluminiscencia
chemisorption quimisorción
chemist químico
chemistry química
chemodynamics quimiodinámica
chemolysis quimiólisis
chemometrics quimiometría
chemonuclear quimionuclear
chemonuclear production producción
    quimionuclear
chemosmosis quimiósmosis
chemosmotic quimiosmótico

chemosorption quimiosorción
chemostat quimióstato
chemosynthesis quimiosíntesis
chemotaxis quimiotaxis
chemotaxonomy quimiotaxonomía
chemotherapy quimioterapia
chemotropism quimiotropismo
chenopodium oil aceite de quenopodio
chevkinite quevkinita
chi chi
chi acid ácido chi
Chicago acid ácido de Chicago
Chichibabin pyridine synthesis síntesis
    de piridinas de Chichibabin
Chichibabin reaction reacción de
    Chichibabin
chicle chicle
Chinese blue azul de China
Chinese wax cera de China
Chinese white blanco de China
chinic acid ácido quínico
chinoidine quinoidina
chinone quinona
chiral quiral
chiral center centrol quiral
chirality quiralidad
chirooptical quiroóptico
chirooptical spectroscopy espectroscopia
    quiroóptica
chitin quitina
chitinase quitinasa
chitosan quitosana
chloral cloral
chloral amide amida de cloral
chloral hydrate hidrato de cloral
chloralkali clorálcali
chloralose cloralosa
chloramine cloramina
chloramphenicol cloranfenicol
chloranil cloranil
chloranilate cloranilato
chlorapatite clorapatita
chlorate clorato
chloraurite cloraurita
chlorazine clorazina
chlorbenside clorbensida
chlorbutanol clorbutanol
chlordane clordano
chlordiazepoxide hydrochloride

clorhidrato de clordiazepóxido
chlordimeform clordimeformo
chlorendic anhydride anhídrido
   cloréndico
chlorfenvinphos clorfenvinfos
chlorhexidine clorhexidina
chloric acid ácido clórico
chloride cloruro
chloridion cloridión
chlorimide clorimida
chlorinated clorado
chlorinated acetone acetona clorada
chlorinated camphene canfeno clorado
chlorinated diphenyl difenilo clorado
chlorinated hydrocarbon hidrocarburo
   clorado
chlorinated isocyanuric acid ácido
   isocianúrico clorado
chlorinated lime cal clorada
chlorinated naphthalene naftaleno
   clorado
chlorinated paraffin parafina clorada
chlorinated polyether poliéter clorado
chlorinated polyolefin poliolefina
   clorada
chlorinated rubber caucho clorado
chlorinated solvent solvente clorado
chlorinated trisodium phosphate fosfato
   trisódico clorado
chlorination clorinación
chlorine cloro
chlorine bromide bromuro de cloro
chlorine cyanide cianuro de cloro
chlorine dioxide dióxido de cloro
chlorine halide haluro de cloro
chlorine heptoxide heptóxido de cloro
chlorine hydrate hidrato de cloro
chlorine monofluoride monofluoruro de
   cloro
chlorine monoxide monóxido de cloro
chlorine oxide óxido de cloro
chlorine trifluoride trifluoruro de cloro
chlorine water agua de cloro
chlorinity clorinidad
chlorinolysis clorinólisis
chlorite clorito
chlorition cloritión
chlormethazanone clormetazanona
chloroacetaldehyde cloroacetaldehído

chloroacetamide cloroacetamida
chloroacetate cloroacetato
chloroacetic acid ácido cloroacético
chloroacetic anhydride anhídrido
   cloroacético
chloroacetoacetanilide
   cloroacetoacetanilida
chloroacetone cloroacetona
chloroacetonitrile cloroacetonitrilo
chloroacetophenone cloroacetofenona
chloroacetyl cloroacetilo
chloroacetyl chloride cloruro de
   cloroacetilo
chloroacetylurethane cloroacetiluretano
chloroacrolein cloroacroleína
chloroacrylate cloroacrilato
chloroacrylonitrile cloroacrilonitrilo
chloroaluminum diisopropoxide
   diisopropóxido de cloroaluminio
chloroaniline cloroanilina
chloroanthraquinone cloroantraquinona
chloroargentate cloroargentato
chloroauric acid ácido cloroáurico
chloroazotic acid ácido cloroazótico
chlorobenzal clorobenzal
chlorobenzaldehyde clorobenzaldehído
chlorobenzanthrone clorobenzantrona
chlorobenzene clorobenceno
chlorobenzenesulfonamide
   clorobencenosulfonamida
chlorobenzenethiol clorobencenotiol
chlorobenzhydrol clorobenzhidrol
chlorobenzoate clorobenzoato
chlorobenzohydrol clorobenzohidrol
chlorobenzoic acid ácido clorobenzoico
chlorobenzophenone clorobenzofenona
chlorobenzotrichloride
   clorobenzotricloruro
chlorobenzotrifluoride
   clorobenzotrifluoruro
chlorobenzoyl clorobenzoilo
chlorobenzoyl chloride cloruro de
   clorobenzoilo
chlorobenzoyl peroxide peróxido de
   clorobenzoilo
chlorobenzyl chloride cloruro de
   clorobecilo
chlorobenzyl cyanate cianato de
   clorobecilo

chlorobenzylpyridine clorobecilpiridina
chlorobromo clorobromo
chlorobutadiene clorobutadieno
chlorobutane clorobutano
chlorobutanol clorobutanol
chlorobutyryl clorobutirilo
chlorocarbon clorocarburo
chlorocarbonate clorocarbonato
chlorocarbonic clorocarbónico
chlorocarbonyl ferrocene ferroceno de clorocarbonilo
chlorochromate clorocromato
chlorochromic anhydride anhídrido clorocrómico
chlorocosane clorocosano
chlorocoumarin clorocumarina
chlorocyanide clorocianuro
chlorocyanogen clorocianógeno
chlorodecone clorodecona
chlorodifluoroacetic acid ácido clorodifluoroacético
chlorodifluoroethane clorodifluoroetano
chlorodifluoromethane clorodifluorometano
chlorodinitrobenzene clorodinitrobenceno
chlorodiphenyl clorodifenilo
chloroethanamide cloroetanamida
chloroethane cloroetano
chloroethanoic acid ácido cloroetanoico
chloroethanoic anhydride anhídrido cloroetanoico
chloroethanol cloroetanol
chloroethene cloroeteno
chloroethyl alcohol alcohol cloroetílico
chloroethylchloroformate cloroetilcloroformiato
chloroethylene cloroetileno
chloroethylphosphonic acid ácido cloroetilfosfónico
chlorofenethol clorofenetol
chlorofluoride clorofluoruro
chlorofluorocarbon clorofluorocarburo
chloroform cloroformo
chloroformate cloroformiato
chloroformic acid ácido clorofórmico
chloroformoxime cloroformoxima
chloroformyl chloride cloruro de cloroformilo

chlorogenic acid ácido clorogénico
chlorohydrin clorohidrina
chlorohydrin rubber caucho de clorohidrina
chlorohydroquinone clorohidroquinona
chlorohydroxybenzene clorohidroxibenceno
chlorohydroxytoluene clorohidroxitolueno
chloroiodide cloroyoduro
chloroisopropyl alcohol alcohol cloroisopropílico
chloromadinone acetate acetato de cloromadinona
chloromaleic anhydride anhídrido cloromaleico
chloromalonic acid ácido cloromalónico
chloromercuriferrocene clomercuriferroceno
chloromethane clorometano
chloromethyl clorometilo
chloromethyl cyanide cianuro de clorometilo
chloromethylaniline clorometilanilina
chloromethylated diphenyl oxide óxido de difenilo clorometilado
chloromethylation clorometilación
chloromethylbenzene clorometilbenceno
chloromethylchloroformate clorometilcloroformiato
chloromethylchlorosulfonate clorometilclorosulfonato
chloromethylnaphthalene clorometilnaftaleno
chloromethylphosphonic acid ácido clorometilfosfónico
chloromethylphosphonic dichloride dicloruro clorometilfosfónico
chloronaphthalene cloronaftaleno
chloronaphthalene oil aceite de cloronaftaleno
chloronitric acid ácido cloronítrico
chloronitroaniline cloronitroanilina
chloronitrobenzene cloronitrobenceno
chloronitrobenzoic acid ácido cloronitrobenzoico
chloronitrotoluene cloronitrotolueno
chloronitrous acid ácido cloronitroso
chloronium ion ion cloronio

chloropentafluoroacetone cloropentafluoroacetona

chloropentane cloropentano

chloroperbenzoic acid ácido cloroperbenzoico

chlorophenol clorofenol

chlorophenol red rojo de clorofenol

chlorophenyl isocyanate isocianato de clorofenilo

chlorophenyltrichlorosilane clorofeniltriclorosilano

chlorophoenicite clorofenicita

chlorophyll clorofila

chlorophyllin clorofilina

chloropicrin cloropicrina

chloroplatinate cloroplatinato

chloroplatinic acid ácido cloroplatínico

chloroprene clopreno

chloropropane cloropropano

chloropropene cloropropeno

chloropropionic acid ácido cloropropiónico

chloropropionitrile cloropropionitrilo

chloropropylene cloropropileno

chloropyridine cloropiridina

chloroquinaldine cloroquinaldina

chloroquine cloroquina

chlorosalicylanilide clorosalicilanilida

chlorosalicylic acid ácido clorosalicílico

chlorostyrene cloroestireno

chlorosuccinimide clorosuccinimida

chlorosulfonic acid ácido clorosulfónico

chlorosulfonylbenzoic acid ácido clorosulfonilbenzoico

chlorosulfuric acid ácido clorosulfúrico

chlorosyl clorosilo

chlorothiazide clorotiazida

chlorothiophenol clorotiofenol

chlorothymol clorotimol

chlorotoluene clorotolueno

chlorotrifluoromethane clorotrifluorometano

chlorotrifluoroethylene clorotrifluoroetileno

chlorotrifluoroethylene polymer polímero de clorotrifluoroetileno

chlorous cloroso

chlorous acid ácido cloroso

chlorovaleric acid ácido clorovalérico

chlorovinyldichloroarsine clorovinildicloroarsina

chloroxine cloroxina

chloroxylene cloroxileno

chlorpromazine clorpromazina

chlorquinaldol clorquinaldol

chlortetracycline clortetraciclina

chloryl clorilo

cholaic acid ácido colaico

cholane colano

cholanthrene colantreno

cholecalciferol colecalciferol

choleic acid ácido coleico

cholesteric colestérico

cholesterol colesterol

cholestyramine colestiramina

cholic acid ácido cólico

choline colina

choline base base de colina

choline bicarbonate bicarbonato de colina

choline bitartrate bitartrato de colina

choline chloride cloruro de colina

cholinesterase colinesterasa

cholinesterase inhibitor inhibidor de colinesterasa

chorionic gonadotropin gonadotropina coriónica

chorismic acid ácido corísmico

choritoid coritoide

chroma croma

chroman cromano

chromate cromato

chromatic cromático

chromatic aberration aberración cromática

chromatic scale escala cromática

chromaticity cromaticidad

chromatin cromatina

chromatogram cromatograma

chromatographic adsorption adsorción cromatográfica

chromatography cromatografía

chrome cromo

chrome alum alumbre de cromo

chrome dye colorante de cromo

chrome green verde de cromo

chrome orange anaranjado de cromo

chrome pigment pigmento de cromo

chrome red rojo de cromo
chrome steel acero al cromo
chrome yellow amarillo de cromo
chromic crómico
chromic acetate acetato crómico
chromic acid ácido crómico
chromic anhydride anhídrido crómico
chromic bromide bromuro crómico
chromic carbide carburo crómico
chromic chloride cloruro crómico
chromic fluoride fluoruro crómico
chromic hydroxide hidróxido crómico
chromic iodide yoduro crómico
chromic nitrate nitrato crómico
chromic oxide óxido crómico
chromic oxychloride oxicloruro crómico
chromic phosphate fosfato crómico
chromic silicide siliciuro crómico
chromic sulfate sulfato crómico
chromic tartrate tartrato crómico
chrominance crominancia
chromite cromita
chromitite cromitita
chromium cromo
chromium acetate acetato de cromo
chromium acetylacetonate
    acetilacetonato de cromo
chromium ammonium sulfate sulfato
    amónico de cromo
chromium arsenide arseniuro de cromo
chromium boride boruro de cromo
chromium bromide bromuro de cromo
chromium carbide carburo de cromo
chromium carbonyl carbonilo de cromo
chromium chloride cloruro de cromo
chromium-copper cromo-cobre
chromium dioxide dióxido de cromo
chromium fluoride fluoruro de cromo
chromium hexacarbonyl hexacarbonilo
    de cromo
chromium hydrate hidrato de cromo
chromium hydroxide hidróxido de
    cromo
chromium manganese antimonide
    antimoniuro de manganeso crómico
chromium naphthenate naftenato de
    cromo
chromium nitrate nitrato de cromo
chromium oxide óxido de cromo

chromium oxychloride oxicloruro de
    cromo
chromium oxyfluoride oxifluoruro de
    cromo
chromium phosphate fosfato de cromo
chromium phosphide fosfuro de cromo
chromium potassium oxalate oxalato
    potásico crómico
chromium potassium sulfate sulfato
    potásico crómico
chromium sesquichloride sesquicloruro
    de cromo
chromium steel acero al cromo
chromium sulfate sulfato de cromo
chromium tetrasulfide tetrasulfuro de
    cromo
chromium trichloride tricloruro de
    cromo
chromium trifluoride trifluoruro de
    cromo
chromium trioxide trióxido de cromo
chromo cromo
chromogen cromógeno
chromogenic cromogénico
chromoisomer cromoisómero
chromone cromona
chromophilic cromofílico
chromophore cromóforo
chromophoric cromofórico
chromosan cromosana
chromosome cromosoma
chromotropic acid ácido cromotrópico
chromotropy cromotropía
chromous cromoso
chromous acetate acetato cromoso
chromous acid ácido cromoso
chromous bromide bromuro cromoso
chromous carbonate carbonato cromoso
chromous chloride cloruro cromoso
chromous fluoride fluoruro cromoso
chromous formate formiato cromoso
chromous hydroxide hidróxido cromoso
chromous oxalate oxalato cromoso
chromous oxide óxido cromoso
chromous sulfate sulfato cromoso
chromyl cromilo
chromyl chloride cloruro de cromilo
chromyl fluoride fluoruro de cromilo
chronopotentiometry

cronopotenciometría
chrysalicic acid ácido crisalícico
chrysaniline crisanilina
chrysanisic acid ácido crisanísico
chrysanthemummonocarboxylic acid
    ácido crisantemomonocarboxílico
chrysene criseno
chrysolite crisólito
chrysophanic acid ácido crisofánico
chrysoquinone crisoquinona
chrysotile crisotilo
Chugaev reaction reacción de Chugaev
chymosin quimosina
chymotrypsin quimotripsina
chymotrypsinogen quimotripsinógeno
Ciamician-Dennstedt rearrangement
    reordenamiento de Ciamician-
    Dennstedt
cicutoxine cicutoxina
ciguatoxin ciguatoxina
cimetidine cimetidina
cinchocain cincocaína
cinchomeronic acid ácido cincomerónico
cinchona bark quina
cinchonamine cinconamina
cinchonidin cinconidina
cinchonine cinconina
cinchonine nitrate nitrato de cinconina
cinchoninic acid ácido cinconínico
cinene cineno
cineol cineol
cinerin cinerina
cinerolone cinerolona
cinnabar cinabrio
cinnamaldehyde cinamaldehído
cinnamamide cinamamida
cinnamate cinamato
cinnamein cinameína
cinnamene cinameno
cinnamenyl cinamenilo
cinnamic acid ácido cinámico
cinnamic alcohol alcohol cinámico
cinnamic aldehyde aldehído cinámico
cinnamic anhydride anhídrido cinámico
cinnamide cinamida
cinnamilidene cinamilideno
cinnamon oil aceite de canela
cinnamoyl cinamoilo
cinnamoyl chloride cloruro de cinamoilo

cinnamyl acetate acetato de cinamilo
cinnamyl alcohol alcohol cinamílico
cinnamyl aldehyde aldehído cinamílico
cinnamyl cinnamate cinamato de
    cinamilo
cinnamyl hydride hidruro de cinamilo
cinnamylidene cinamilideno
cinnoline cinolina
circular birefringence birrefrigencia
    circular
circular chromatography cromatografía
    circular
circular dichroism dicroismo circular
circular polarization polarización
    circular
cis cis
cis-isomer isómero cis
cis-trans isomerism isomerismo cis-trans
cisoid cisoide
citiolone citiolona
citraconic acid ácido citacrónico
citraconic anhydride anhídrido
    citacrónico
citraconoyl citraconoilo
citral citral
citramalic acid ácido citramálico
citramide citramida
citrate citrato
citrene citreno
citresia citresia
citric acid ácido cítrico
citric acid cycle ciclo de ácido cítrico
citridic acid ácido citrídico
citrinin citrinina
citronella oil aceite de citronela
citronellal citronelal
citronellic acid ácido citronélico
citronellol citronelol
citronyl citronilo
citrulline citrulina
civet civeto
civettal civetal
cladding chapado, revestimiento
Claisen condensation condensación de
    Claisen
Claisen flask matraz de Claisen
Claisen reaction reacción de Claisen
Claisen rearrangement reordenamiento
    de Claisen

Claisen-Schmidt condensation
  condensación de Claisen-Schmidt
clamp abrazadera
Clapeyron-Clausius equation ecuación
  de Clapeyron-Clausius
Clapeyron equation ecuación de
  Clapeyron
clarification clarificación
clarifier clarificador
Clark cell pila de Clark
Clark electrode electrodo de Clark
Clark-Lubs indicators indicadores de
  Clark-Lubs
Clark process proceso de Clark
classical electron radius radio de
  electrón clásico
classification clasificación
clathrate clatrato
Claude process proceso de Claude
Claude system sistema de Claude
Clausius law ley de Clausius
Clausius-Mosotti equation ecuación de
  Clausius-Mosotti
Clausius-Mosotti law ley de Clausius-
  Mosotti
clay arcilla
clean room sala limpia
cleavage división, hendidura
clebrium clebrio
Cleland reagent reactivo de Cleland
Clemmensen reduction reducción de
  Clemmensen
Cleve acid ácido de Cleve
clinical chemistry química clínica
clinohedrite clinohedrita
clinoptilolite clinoptilolita
clioquinol clioquinol
clomiphene clomifeno
clonazepam clonazepam
clopidol clopidol
close-packed structure estructura
  compacta
closed chain cadena cerrada
closo closo
cloudy turbio
clove oil aceite de clavero
Clusius column columna de Clusius
cluster compound compuesto de grupos
coacervation coacervación

coagulant coagulante
coagulation coagulación
Coahran process proceso de Coahran
coal hulla, carbón
coal gas gas de hulla
coal oil aceite de hulla
coal tar alquitrán de hulla
coal-tar distillate destilado de alquitrán
  de hulla
coal-tar dye colorante de alquitrán de
  hulla
coal-tar oil aceite de alquitrán de hulla
coal-tar pitch brea de alquitrán de hulla
coalescence coalescencia
coalescent coalescente
coated cubierto, revestido
coating revestimiento
cobalamin cobalamina
cobalt cobalto
cobalt acetate acetato de cobalto
cobalt alloy aleación de cobalto
cobalt aluminate aluminato de cobalto
cobalt ammonium sulfate sulfato
  amónico de cobalto
cobalt arsenate aresniato de cobalto
cobalt black negro de cobalto
cobalt blue azul de cobalto
cobalt bromide bromuro de cobalto
cobalt carbonate carbonato de cobalto
cobalt carbonyl carbonilo de cobalto
cobalt chloride cloruro de cobalto
cobalt chromate cromato de cobalto
cobalt cyanide cianuro de cobalto
cobalt difluoride difluoruro de cobalto
cobalt ethylhexoate etilhexoato de
  cobalto
cobalt fluorosilicate fluorosilicato de
  cobalto
cobalt green verde de cobalto
cobalt halide haluro de cobalto
cobalt hydrate hidrato de cobalto
cobalt hydrocarbonyl hidrocarbonilo de
  cobalto
cobalt hydroxide hidróxido de cobalto
cobalt iodide yoduro de cobalto
cobalt linoleate linoleato de cobalto
cobalt molybdate molibdato de cobalto
cobalt monoxide monóxido de cobalto
cobalt naphthenate naftenato de cobalto

cobalt nitrate nitrato de cobalto
cobalt nitride nitruro de cobalto
cobalt oleate oleato de cobalto
cobalt oxide óxido de cobalto
cobalt phosphate fosfato de cobalto
cobalt phosphide fosfuro de cobalto
cobalt potassium nitrate nitrato potásico de cobalto
cobalt potassium sulfate sulfato potásico de cobalto
cobalt resinate resinato de cobalto
cobalt selenite selenito de cobalto
cobalt silicide siliciuro de cobalto
cobalt sulfate sulfato de cobalto
cobalt tetracarbonyl tetracarbonilo de cobalto
cobalt trifluoride trifluoruro de cobalto
cobalt tungstate tungstato de cobalto
cobalt violet violeta de cobalto
cobalt yellow amarillo de cobalto
cobalt zincate cincato cobalto
cobaltammine cobaltamina
cobaltic cobáltico
cobaltic boride boruro cobáltico
cobaltic chloride cloruro cobáltico
cobaltic fluoride fluoruro cobáltico
cobaltic hydroxide hidróxido cobáltico
cobaltic oxide óxido cobáltico
cobaltic potassium nitrite nitrito potásico cobáltico
cobaltic sulfate sulfato cobáltico
cobaltic sulfide sulfuro cobáltico
cobaltite cobaltita
cobaltocene cobaltoceno
cobaltous acetate acetato cobaltoso
cobaltous aluminate aluminato cobaltoso
cobaltous ammonium phosphate fosfato amónico cobaltoso
cobaltous ammonium sulfate sulfato amónico cobaltoso
cobaltous arsenate arseniato cobaltoso
cobaltous benzoate benzoato cobaltoso
cobaltous bromate bromato cobaltoso
cobaltous bromide bromuro cobaltoso
cobaltous butyrate butirato cobaltoso
cobaltous carbonate carbonato cobaltoso
cobaltous chlorate clorato cobaltoso
cobaltous chloride cloruro cobaltoso
cobaltous chromate cromato cobaltoso

cobaltous citrate citrato cobaltoso
cobaltous cyanide cianuro cobaltoso
cobaltous ferrite ferrita cobaltosa
cobaltous fluoride fluoruro cobaltoso
cobaltous formate formiato cobaltoso
cobaltous hexafluorosilicate hexafluorosilicato cobaltoso
cobaltous hydroxide hidróxido cobaltoso
cobaltous iodide yoduro cobaltoso
cobaltous linoleate linoleato cobaltoso
cobaltous naphthenate naftenato cobaltoso
cobaltous nitrate nitrato cobaltoso
cobaltous oleate oleato cobaltoso
cobaltous oxalate oxalato cobaltoso
cobaltous oxide óxido cobaltoso
cobaltous perchlorate perclorato cobaltoso
cobaltous phosphate fosfato cobaltoso
cobaltous propionate propionato cobaltoso
cobaltous resinate resinato cobaltoso
cobaltous silicate silicato cobaltoso
cobaltous stannate estannato cobaltoso
cobaltous succinate succinato cobaltoso
cobaltous sulfate sulfato cobaltoso
cobaltous sulfide sulfuro cobaltoso
cobaltous tartrate tartrato cobaltoso
cobaltous tungstate tungstato cobaltoso
cobaltouscobaltic oxide óxido colbatosocobáltico
coca coca
cocaine cocaína
cocaine benzoate benzoato de cocaína
cocaine borate borato de cocaína
cocaine chloride cloruro de cocaína
cocaine hydrochloride clorhidrato de cocaína
cocaine nitrate nitrato de cocaína
cocaine sulfate sulfato de cocaína
cocarboxylase cocarboxilasa
cocculin coculina
coccus coco
cochineal cochinilla
cochinilin cochinilina
cochrome cocromo
cocoa cacao
cocoa butter manteca de cacao
cocoa oil aceite de cacao

coconut acid ácido de nuez de coco
coconut oil aceite de nuez de coco
cod-liver oil aceite de hígado de bacalao
codeine codeína
codeine citrate citrato de codeína
codeine hydrochloride clorhidrato de
    codeína
codeine phosphate fosfato de codeína
codeine salicylate salicilato de codeína
codeine sulfate sulfato de codeína
coefficient coeficiente
Coehn law ley de Coehn
coenzyme coenzima
coercivity coercitividad
cofactor cofactor
coherence coherencia
coherent coherente
coherent light luz coherente
coherent precipitate precipitado
    coherente
cohesion cohesión
cohobation cohobación
coil condenser condensador en espiral
coiled condenser condensador en espiral
coke coque
coke gas gas de coque
coke oven horno de coque
coking coquización
cola cola
colamine colamina
colchicine colquicina
cold flow flujo en frío
cold light luz fría
cold rubber caucho frío
colemanite colemanita
colestipol colestipol
colistin colistina
collagen colágeno
collagenase colagenasa
collargol colargol
collateral series serie colateral
collector colector
collidine colidina
colligate coligar
colligative coligativo
colligative property propiedad coligativa
collision colisión
collision density densidad de colisiones
collision excitation excitación por

colisión
collision ionization ionización por
    colisión
Collman reagent reactivo de Collman
collodion colodión
colloid coloide
colloid equivalent equivalente coloidal
colloid mill molino coloidal
colloidal coloidal
colloidal dispersion dispersión coloidal
colloidal electrolyte electrólito coloidal
colloidal metal metal coloidal
colloidal solution solución coloidal
colloidal state estado coloidal
colloidal suspension suspensión coloidal
colloidal system sistema coloidal
colopholic acid ácido colofólico
colophonic acid ácido colofónico
colophonite colofonita
colophony colofonia
color comparator comparador de colores
color indicator indicador de color
color standard patrón de color
colorant colorante
coloration coloración
colorimeter colorímetro
colorimetric colorimétrico
colorimetric analysis análisis
    colorimétrico
colorimetric purity pureza colorimétrica
colorimetry colorimetría
colorization colorización
colorless dye colorante incoloro
columbate columbato
columbite columbita
columbium columbio
column columna
colza colza
Combes quinoline synthesis síntesis de
    quinolinas de Combes
combination combinación
combined combinado
combined carbon carbono combinado
combining number número de
    combinación
combining volume volumen de
    combinación
combining weight peso de combinación
combustibility combustibilidad

combustible combustible
combustible material material combustible
combustion combustión
combustion boat artesa de combustión
combustion capsule cápsula de combustión
combustion furnace horno de combustión
combustion rate velocidad de combustión
combustion tube tubo de combustión
comenic acid ácido coménico
comicellization comicelización
comirin comirina
comminution pulverización
common-ion effect efecto de ion común
common salt sal común
comparator comparador
compatibility compatibilidad
compatible compatible
complex complejo
complex acid ácido complejo
complex compound compuesto complejo
complex ion ion complejo
complex reaction reacción compleja
complex salt sal compleja
complexation complejación
complexing complejación
complexing agent agente complejante
complexometric analysis análisis complejométrico
complexometric titration titulación complejométrica, valoración complejométrica
complexone complexona
component componente
composite compuesto
composition composición
compost abono orgánico
compound compuesto
compound molecule molécula compuesta
compound nucleus núcleo compuesto
compressed comprimido
compressed gas gas comprimido
Compton effect efecto de Compton
Compton rule regla de Compton
computational chemistry química computacional

concave cóncavo
concentrate concentrar
concentrated concentrado
concentrated solution solución concentrada
concentration concentración
concentration gradient gradiente de concentración
concentration limit límite de concentración
concentrator concentrador
conchiolin conquiolina
conchoidal concoidal
concrete concreto
condensation condensación
condensation polymer polímero de condensación
condensation reaction reacción de condensación
condensation temperature temperatura de condensación
condensed condensado
condenser condensador
condistillation condestilación
conductance conductancia
conduction conducción
conduction band banda de conducción
conduction electron electrón de conducción
conductivity conductividad
conductivity apparatus aparato de conductividad
conductometric conductométrico
conductometric analysis análisis conductométrico
conductometric method método conductométrico
conductometric titration titulación conductiométrica, valoración conductométrica
conductor conductor
Condy fluid fluido de Condy
cone cono
conessine conesina
configuration configuración
configuration space espacio de configuración
conformation conformación
conformational analysis análisis

conformacional
congealing congelación
congelation congelación
conglomerate conglomerado
congo blue azul congo
congo red rojo congo
congo yellow amarillo congo
conicein coniceína
coniferin coniferina
coniferol coniferol
coniferyl coniferilo
coniferyl alcohol alcohol coniferilo
conine conina
conjuctive nomenclature nomenclatura
    conjuntiva
conjugate acid ácido conjugado
conjugate base base conjugada
conjugated conjugado
conjugated double bonds enlaces dobles
    conjugados
conjugated fatty acid ácido graso
    conjugado
conjugated system sistema conjugado
conjugation conjugación
conproportionation conproporción
Conrad-Limpach reaction reacción de
    Conrad-Limpach
conservation law ley de conservación
conservation of energy conservación de
    energía
conservation of matter conservación de
    materia
consistence consistencia
consistency consistencia
consistometer consistómetro
consolute consoluto
constant constante
constant-boiling mixture mezcla de
    ebullición constante
constant composition composición
    constante
constant current corriente constante
constant pressure presión constante
constant proportion proporción
    constante
constant temperature temperatura
    constante
constantan constantán
constituent constituyente

constitutional constitucional
constitutional formula fórmula
    constitucional
contact contacto
contact acid ácido de contacto
contact potential potencial de contacto
contact process proceso de contacto
contact resin resina de contacto
contact substance substancia de contacto
container contenedor, receptáculo
contaminant contaminante
contamination contaminación
continuity of state continuidad de estado
continuous continuo
continuous distillation destilación
    continua
continuous phase fase continua
continuous spectrum espectro continuo
contraction contracción
contrast contraste
control control
controlled atmosphere atmósfera
    controlada
controlled substance substancia
    controlada
convection convección
convergence limit límite de convergencia
conversion conversión
conversion coefficient coeficiente de
    conversión
conversion electron electrón de
    conversión
conversion factor factor de conversión
conversion process proceso de
    conversión
conversion ratio razón de conversión
converter convertidor
convex convexo
Cook formula fórmula de Cook
cooking cocido
coolant refrigerante
Coolidge tube tubo de Coolidge
cooling enfriamiento
coordinate coordinado
coordinate bond enlace coordinado
coordinate covalence covalencia
    coordinada
coordinated complex complejo
    coordinado

coordination coordinación
coordination compound compuesto de coordinación
coordination isomerism isomerismo por coordinación
coordination number número de coordinación
coordination polymer polímero de coordinación
copaiba oil aceite de copaiba
copaiba resin resina de copaiba
copal copal
Cope elimination reaction reacción de eliminación de Cope
Cope rearrangement reordenamiento de Cope
copigment copigmento
copolymer copolímero
copolymerization copolimerización
copper cobre
copper abietate abietato de cobre
copper acetate acetato de cobre
copper acetoarsenite acetoarsenito de cobre
copper alloy aleación de cobre
copper amalgam amalgama de cobre
copper arsenate arseniato de cobre
copper arsenite arsenito de cobre
copper benzoate benzoato de cobre
copper blue azul de cobre
copper borate borato de cobre
copper bromide bromuro de cobre
copper carbonate carbonato de cobre
copper chloride cloruro de cobre
copper chromate cromato de cobre
copper cyanide cianuro de cobre
copper ethylacetoacetate etilacetoacetato de cobre
copper ferrocyanide ferrocianuro de cobre
copper fluoride fluoruro de cobre
copper gluconate gluconato de cobre
copper glycinate glicinato de cobre
copper green verde de cobre
copper hexacyanoferrate hexacianoferrato de cobre
copper hydroxide hidróxido de cobre
copper iodide yoduro de cobre
copper lactate lactato de cobre

copper metaborate metaborato de cobre
copper molybdate molibdato de cobre
copper monoxide monóxido de cobre
copper naphthenate naftenato de cobre
copper nitrate nitrato de cobre
copper nitrite nitrito de cobre
copper number número de cobre
copper oleate oleato de cobre
copper oxalate oxalato de cobre
copper oxide black óxido negro de cobre
copper oxychloride oxicloruro de cobre
copper phenolsulfonate fenolsulfonato de cobre
copper phosphate fosfato de cobre
copper phosphide fosfuro de cobre
copper resinate resinato de cobre
copper silicate silicato de cobre
copper stearate estearato de cobre
copper sulfate sulfato de cobre
copper sulfide sulfuro de cobre
copper trifluoroacetylacetonate trifluoroacetilacetonato de cobre
copper tungstate tungstato de cobre
copperas caparrosa
Coppet law ley de Coppet
coprastanol coprastanol
coprecipitation coprecipitación
coprosterol coprosterol
copyrine copirina
coral coral
cordite cordita
coriandrol coriandrol
cork corcho
corn oil aceite de maíz
corn sugar azúcar de almidón de maíz
corn syrup jarabe de maíz
cornstarch almidón de maíz
corona corona
coronene coroneno
correlation correlación
correlation diagram diagrama de correlación
correlation function función de correlación
correlation spectroscopy espectroscopia de correlación
correspondence principle principio de correspondencia
corresponding state estado

correspondiente
corresponding temperature temperatura
  correspondiente
corrosion corrosión
corrosive corrosivo
corrosive material material corrosivo
corrosive substance substancia corrosiva
corsite corsita
corticoid corticoide
corticoid hormone hormona corticoide
corticosterone corticosterona
corticotropin corticotropina
cortisol cortisol
cortisone cortisona
corundum corindón
cosanoic acid ácido cosanoico
cosanol cosanol
cosmetic cosmético
cosyl cosilo
cotton algodón
Cotton effect efecto de Cotton
cottonseed semilla de algodón
cottonseed meal harina de semilla de
  algodón
cottonseed oil aceite de semilla de
  algodón
coulomb coulombio
Coulomb barrier barrera de Coulomb
Coulomb excitation excitación de
  Coulomb
Coulomb explosion explosión de
  Coulomb
coumalic acid ácido cumálico
coumaric acid ácido cumárico
coumarin cumarina
coumarone cumarona
count cuenta
countercurrent contracorriente
counterion contraion
counterstain contracolorante
coupling acoplamiento
coupling constant constante de
  acoplamiento
covalence covalencia
covalent covalente
covalent bond enlace covalente
covalent crystal cristal covalente
covalent radius radio covalente
coveline covelina

covellite covelita
Cox chart diagrama de Cox
crack craquear, fraccionar
cracking craqueo, fraccionamiento
Craig method método de Craig
Cram rule regla de Cram
cream of tartar crémor tártaro
creatinase creatinasa
creatine creatina
creatininase creatininasa
creatinine creatinina
creeping precipitación ascendente
Creighton process proceso de Creighton
creosol creosol
creosote creosota
cresol cresol
cresol purple púrpura de cresol
cresol red rojo de cresol
cresolphthalein cresolftaleína
cresorcin cresorcina
cresotic acid ácido cresótico
cresyl cresilo
cresyl acetate acetato de cresilo
cresyl blue azul de cresilo
cresyl phosphate fosfato de cresilo
cresylate cresilato
cresylic acid ácido cresílico
Criegee reaction reacción de Criegee
cristobalite cristobalita
critical crítico
critical assembly conjunto crítico
critical coefficient coeficiente crítico
critical conditions condiciones críticas
critical constant constante crítica
critical density densidad crítica
critical humidity humedad crítica
critical mass masa crítica
critical point punto crítico
critical potential potencial crítico
critical pressure presión crítica
critical solution temperature
  temperatura de solución crítica
critical state estado crítico
critical temperature temperatura crítica
critical volume volumen crítico
crocetin crocetina
crocidolite crocidolita
croconic acid ácido crocónico
Crookes tube tubo de Crookes

cross-linkage encadenamiento cruzado
cross-linking encadenamiento cruzado
cross-section sección transversal
crossed-beam reaction reacción de haces
    cruzadas
croton oil aceite de crotón
crotonal crotonal
crotonaldehyde crotonaldehído
crotonic acid ácido crotónico
crotonic aldehyde aldehído crotónico
crotonoid crotonoide
crotonoyl crotonoilo
crotonyl crotonilo
crotonylene crotonileno
crotoxin crotoxina
crotyl alcohol alcohol crotílico
crown ether éter en corona
crucible crisol
crucible holder sujetador de crisol
crucible tongs tenazas de crisol
crucible triangle triángulo de crisol
crude crudo, bruto
crude oil petróleo bruto
crude petroleum petróleo bruto
cryochemistry crioquímica
cryogenic criogénico
cryogenic pump bomba criogénica
cryogenics criogénica, criogenia
cryogenin criogenina
cryohydrate criohidrato
cryolite criolita
cryometer criómetro
cryoscopic constant constante
    crioscópica
cryoscopy crioscopia
cryostat crióstato
cryptand criptando
cryptate criptato
cryptidine criptidina
cryptocyanine criptocianina
cryptohalite criptohalita
cryptoxanthin criptoxantina
crystal cristal
crystal-field theory teoría de campo
    cristalino
crystal form forma cristalina
crystal growth crecimiento de cristal
crystal lattice red cristalina
crystal structure estructura cristalina

crystal symmetry simetría de cristal
crystal system sistema cristalográfico
crystalline cristalino
crystalline state estado cristalino
crystallite cristalito
crystallization cristalización
crystallize cristalizar
crystallogram cristalograma
crystallographic system sistema
    cristalográfico
crystallography cristalografía
crystalloid cristaloide
crystalloluminescence
    cristaloluminiscencia
cube cubo
cubic-centered centrado en el cubo
cubic close packing estructura compacta
    cúbica
cubic crystal cristal cúbico
cubic system sistema cúbico
cubical cúbico
cumal cumal
cumaldehyde cumaldehído
cumene cumeno
cumene hydroperoxide hidroperóxido de
    cumeno
cumene process proceso de cumeno
cumenyl cumenilo
cumic acid ácido cúmico
cumic alcohol alcohol cúmico
cumic aldehyde aldehído cúmico
cumidine cumidina
cuminic acid ácido cumínico
cuminic alcohol alcohol cumínico
cuminic aldehyde aldehído cumínico
cuminol cuminol
cumol cumol
cumulalive double bonds enlaces dobles
    acumulativos
cumulene cumuleno
cumyl cumilo
cupellation copelación
cupellation process proceso de
    copelación
cupferron cupferrón
cuprate cuprato
cupreine cupreína
cuprene cupreno
cupric cúprico

cupric acetate acetato cúprico
cupric arsenate arseniato cúprico
cupric benzoate benzoato cúprico
cupric bromide bromuro cúprico
cupric carbonate carbonato cúprico
cupric chloride cloruro cúprico
cupric chromate cromato cúprico
cupric citrate citrato cúprico
cupric cyanide cianuro cúprico
cupric fluoride fluoruro cúprico
cupric hexafluorosilicate hexafluorosilicato cúprico
cupric hydroxide hidróxido cúprico
cupric iodide yoduro cúprico
cupric nitrate nitrato cúprico
cupric nitrite nitrito cúprico
cupric oxalate oxalato cúprico
cupric oxide óxido cúprico
cupric phosphate fosfato cúprico
cupric selenate seleniato cúprico
cupric subcarbonate subcarbonato cúprico
cupric sulfate sulfato cúprico
cupric sulfide sulfuro cúprico
cupric sulfite sulfito cúprico
cupric tartrate tartrato cúprico
cuprite cuprita
cuprocupric cuprocúprico
cuprocyanide cuprocianuro
cuproine cuproína
cupronickel cuproníquel
cuprotungsten cuprotungsteno
cuprous cuproso
cuprous acetylide acetiluro cuproso
cuprous bromide bromuro cuproso
cuprous carbonate carbonato cuproso
cuprous chloride cloruro cuproso
cuprous cyanide cianuro cuproso
cuprous hexafluorosilicate hexafluorosilicato cuproso
cuprous hydroxide hidróxido cuproso
cuprous iodide yoduro cuproso
cuprous oxide óxido cuproso
cuprous phosphide fosfuro cuproso
cuprous potassium cyanide cianuro potásico cuproso
cuprous selenide seleniuro cuproso
cuprous sulfate sulfato cuproso
cuprous sulfide sulfuro cuproso

cuprous sulfite sulfito cuproso
cuprous thiocyanate tiocianato cuproso
curare curare
curative curativo
curcumin curcumina
curie curie
Curie law ley de Curie
Curie point punto de Curie
Curie temperature temperatura de Curie
Curie-Weiss law ley de Curie-Weiss
curing curado
curing agent agente de curado
curium curio
current density densidad de corriente
Curtius reaction reacción de Curtius
Curtius rearrangement reordenamiento de Curtius
Curtius transformation transformación de Curtius
cusparine cusparina
cyanamide cianamida
cyanamide process proceso de cianamida
cyanate cianato
cyanato cianato
cyanaurite cianaurita
cyanazine cianazina
cyanoacetamide cianoacetamida
cyanic acid ácido ciánico
cyanide cianuro
cyanide process proceso de cianuro
cyanidine cianidina
cyanine cianina
cyanine dye colorante de cianina
cyanite cianita
cyano ciano
cyanoacetic acid ácido cianoacético
cyanoacetic ester éster cianoacético
cyanoacrylate cianoacrilato
cyanoaldehyde cianoaldehído
cyanoaniline cianoanilina
cyanobenzamide cianobenzamida
cyanobenzene cianobenceno
cyanobenzyl cianobencilo
cyanocarbon cianocarburo
cyanocarbonic acid ácido cianocarbónico
cyanocobalamin cianocobalamina
cyanoethanoic acid ácido cianoetanoico
cyanoethyl acryate acrilato de cianoetilo
cyanoethylation cianoetilación

cyanoferrate cianoferrato
cyanoformate cianoformiato
cyanoformic chloride cloruro
　　cianofórmico
cyanogen cianógeno
cyanogen azide azida de cianógeno
cyanogen bromide bromuro de
　　cianógeno
cyanogen chloride cloruro de cianógeno
cyanogen fluoride fluoruro de cianógeno
cyanogen iodide yoduro de cianógeno
cyanogen sulfide sulfuro de cianógeno
cyanohydrin cianhidrina
cyanomethyl acetate acetato de
　　cianometilo
cyanoplatinate cianoplatinato
cyanopyridine cianopiridina
cyanuric cianúrico
cyanuric acid ácido cianúrico
cyanuric chloride cloruro cianúrico
cyanuric ester éster cianúrico
cybotactic cibotáctico
cyclamate ciclamato
cyclamic acid ácido ciclámico
cycle ciclo
cyclene cicleno
cyclic cíclico
cyclic action acción cíclica
cyclic adenosine monophosphate
　　monofosfato de adenosina cíclico
cyclic amide amida cíclica
cyclic anhydride anhídrido cíclico
cyclic compound compuesto cíclico
cyclic hydrocarbon hidrocarburo cíclico
cyclic process proceso cíclico
cyclitol ciclitol
cyclization ciclización
cyclizine hydrochloride clorhidrato de
　　ciclizina
cyclo ciclo
cycloaddition cicloadición
cycloaliphatic cicloalifático
cycloalkane cicloalcano
cycloalkene cicloalqueno
cycloalkyl cicloalquilo
cyclobarbital ciclobarbital
cyclobarbitone ciclobarbitona
cyclobutadiene ciclobutadieno
cyclobutane ciclobutano

cyclobutanol ciclobutanol
cyclobutene ciclobuteno
cyclobutyl ciclobutilo
cycloheptadecene cicloheptadeceno
cycloheptane cicloheptano
cycloheptanol cicloheptanol
cycloheptanone cicloheptanona
cycloheptene ciclohepteno
cycloheptyl cicloheptilo
cyclohexane ciclohexano
cyclohexanecarboxylic acid ácido
　　ciclohexanocarboxílico
cyclohexanediol ciclohexanodiol
cyclohexanol ciclohexanol
cyclohexanol acetate acetato de
　　ciclohexanol
cyclohexanone ciclohexanona
cyclohexanone peroxide peróxido de
　　ciclohexanona
cyclohexene ciclohexeno
cyclohexene oxide óxido de ciclohexeno
cyclohexenol ciclohexenol
cyclohexenyltrichlorosilane
　　ciclohexeniltriclorosilano
cycloheximide cicloheximida
cyclohexyl bromide bromuro de
　　ciclohexilo
cyclohexyl chloride cloruro de
　　ciclohexilo
cyclohexyl isocyanate isocianato de
　　ciclohexilo
cyclohexyl stearate estearato de
　　ciclohexilo
cyclohexyl trichlorosilane triclorosilano
　　de ciclohexilo
cyclohexylamine ciclohexilamina
cyclohexylbenzene ciclohexilbenceno
cyclohexylidene ciclohexilideno
cyclohexylphenol ciclohexilfenol
cyclohexylpiperdine ciclohexilpiperdina
cyclone ciclón
cyclonite ciclonita
cyclononane ciclononano
cyclooctadiene ciclooctadieno
cyclooctane ciclooctano
cyclooctatetraene ciclooctatetraeno
cyclooctene ciclooocteno
cycloolefin cicloolefina
cycloparaffin cicloparafina

cyclopentadecanone ciclopentadecanona
cyclopentadiene ciclopentadieno
cyclopentane ciclopentano
cyclopentanol ciclopentanol
cyclopentanone ciclopentanona
cyclopentene ciclopenteno
cyclopentenyl ciclopentenilo
cyclopentenylacetone
      ciclopentenilacetona
cyclopentyl ciclopentilo
cyclopentylacetone ciclopentilacetona
cyclopentylene ciclopentileno
cyclopentylpropionic acid ácido
      ciclopentilpropiónico
cyclophane ciclofano
cyclophosphamide ciclofosfamida
cyclopropane ciclopropano
cyclopropyl ciclopropilo
cyclosilane ciclosilano

cyclotetraene ciclotetraeno
cyclotron ciclotrón
cylinder cilindro
cylindrite cilindrita
cymene cimeno
cymyl cimilo
cyprite ciprita
cystamine cistamina
cysteine cisteína
cystine cistina
cytidine citidina
cytidylic acid ácido citidílico
cytochemistry citoquímica
cytochrome citocromo
cytoplasm citoplasma
cytosine citosina
cytotoxic citotóxico
cytotoxic agent agente citotóxico
cytotoxin citotoxina

D

dacite dacita
dactinomycin dactinomicina
dactylin dactilina
Dahl acid ácido de Dahl
dahlin dalina
dahllite dahllita
dahmenite damenita
Dakin reaction reacción de Dakin
Dakin solution solución de Dakin
dalapon dalapón
dalton dalton
Dalton atomic theory teoría atómica de Dalton
Dalton law ley de Dalton
damascinine damascinina
dambonite dambonita
damourite damourita
damping period periodo de amortiguamiento
danaite danaíta
danalite danalita
Daniell cell pila de Daniell
dansyl dansilo
daphnetin dafnetina
daphnite dafnita
dapsone dapsona
darapskite darapskita
dark reaction reacción obscura
Darzens condensation condensación de Darzens
Darzens procedure procedimiento de Darzens
Darzens reaction reacción de Darzens
Darzens synthesis síntesis de Darzens
dating datación
dating technique técnica de datación
dative bond enlace dativo
dative covalent bond enlace covalente dativo
datolite datolita
daturic acid ácido datúrico
daubreelite daubreelita
daunomycin daunomicina

daviesite daviesita
dawsonite dawsonita
de Broglie equation ecuación de de Broglie
de Broglie wavelength longitud de onda de de Broglie
De Mayo reaction reacción de De Mayo
deacetylation desacetilación
Deacon process proceso de Deacon
deactivating collision colisión desactivante
deactivation desactivación
deactivator desactivador
deamidation desamidación
deamination desaminación
deanol deanol
debye debye
Debye characteristic temperature temperatura característica de Debye
Debye-Hückel theory teoría de Debye-Hückel
Debye-Scherrer method método de Debye-Scherrer
decaborane decaborano
decahydrate decahidrato
decahydronaphthalene decahidronaftaleno
decalcification descalcificación
decalin decalina
decamethrin decametrina
decamethyltetrasiloxane decametiltetrasiloxano
decanal decanal
decanamide decanamida
decane decano
decanecarboxylic acid ácido decanocarboxílico
decanedioic acid ácido decanodioico
decanediol decanodiol
decanenitrile decanonitrilo
decanoate decanoato
decanoic acid ácido decanoico
decanoic anhydride anhídrido decanoico

decanol decanol
decanoyl decanoilo
decanoyl chloride cloruro de decanoilo
decanoyl peroxide peróxido de decanoilo
decant decantar
decantation decantación
decarbonize descarbonizar
decarboxylase descarboxilasa
decarboxylate descarboxilar
decarboxylation descarboxilación
decarboxyrissic acid ácido
    descarboxirrísico
decarburization descarburización
decarburize descarburizar
decatoic acid ácido decatoico
decatyl decatilo
decatyl alcohol alcohol decatílico
decay desintegración, decaimiento
decay chain cadena de desintegración
decay constant constante de
    desintegración
decay family familia de desintegración
decay mode modo de desintegración
decay particle partícula de
    desintegración
decay product producto de
    desintegración
decay rate velocidad de desintegración
decay series serie de desintegración
decene deceno
dechlorination desclorinación
decicaine decicaína
decimolar decimolar
decimolar solution solución decimolar
decinormal decinormal
decinormal solution solución decinormal
decoction decocción
decoic acid ácido decoico
decolorant decolorante
decoloration decoloración
decolorization decolorización
decolorizing agent agente decolorante
decompose descomponer
decomposition descomposición
decomposition point punto de
    descomposición
decontamination descontaminación
decouple desacoplar
decrepitate decrepitar

decrepitation decrepitación
decyclic acid ácido decíclico
decyl decilo
decyl acetate acetato de decilo
decyl alcohol alcohol decílico
decyl aldehyde aldehído decílico
decyl hydride hidruro de decilo
decyl mercaptan decilmercaptano
decylamide decilamida
decylamine decilamina
decylene decileno
decylethylene deciletileno
decylic acid ácido decílico
decyltrichlorosilane deciltriclorosilano
decyne decino
defect defecto
defect structure estructura con defectos
defervescence defervescencia
definite composition law ley de
    composición definida
definite proportions proporciones
    definidas
definite proportions law ley de
    proporciones definidas
deflagration deflagración
deflocculant desfloculante
deflocculation desfloculación
defoamer desespumante
defoaming agent agente desespumante
deformability deformabilidad
deformation energy energía de
    deformación
degasification desgasificación
degassing desgaseamiento
Degener indicator indicador de Degener
degenerate degenerado
degenerate orbital orbital degenerado
degenerate system sistema degenerado
degeneration degeneración
degradation degradación
degrade degradar
degree grado
degree of dilution grado de dilución
degree of dissociation grado de
    disociación
degree of hardness grado de dureza
degree of hydrolysis grado de hidrólisis
degree of polymerization grado de
    polimerización

degree of substitution grado de substitución
degree of temperature grado de temperatura
degree of vacuum grado de vacío
degrees of freedom grados de libertad
dehumidification deshumidificación
dehydrant deshidratante
dehydrated deshidratado
dehydration deshidratación
dehydrator deshidratador
dehydro deshidro
dehydroabietic acid ácido deshidroabiético
dehydroacetic acid ácido deshidroacético
dehydroascorbic acid ácido deshidroascórbico
dehydrocholesterol deshidrocolesterol
dehydrocholic acid ácido deshidrocólico
dehydrocyclization deshidrociclización
dehydrogenase deshidrogenasa
dehydrogenation deshidrogenación
dehydroisoandrosterone deshidroisoandrosterona
dehydrolysis deshidrólisis
deionization desionización
deionized desionizado
deionizing desionización
delanium delanio
delayed neutron neutrón retardado
delayed proton protón retardado
Delbrück scattering dispersión de Delbrück
delcosine delcosina
Delepine reaction reacción de Delepine
delessite delessita
deliquescence delicuescencia
deliquescent delicuescente
deliquescent substance substancia delicuescente
delocalization deslocalización
delphinic acid ácido delfínico
delphinin delfinina
delta delta
delta acid ácido delta
delta bond enlace delta
delta bonding enlazado delta
delta iron hierro delta
delta orbital orbital delta

delta ray rayo delta
delta value valor delta
delvauxite delvauxita
demagnetization desmagnetización
demal demal
demeclocycline demeclociclina
demethylation desmetilación
demethylene demetileno
demeton demetona
demineralization desmineralización
Demjanov rearrangement reordenamiento de Demjanov
demolization demolización
demulsibility demulsibilidad
demulsification desemulsificación
demuriation desmuriación
denatonium benzoate benzoato de denatonio
denaturant desnaturalizante
denaturation desnaturalización
denature desnaturalizar
denatured desnaturalizado
denatured alcohol alcohol desnaturalizado
dendrite dendrita
dendrobine dendrobina
Denigè reagent reactivo de Denigè
denitration desnitración
denitrification desnitrificación
denitrogenation desnitrogenación
densitometer densitómetro
density densidad
density comparator comparador de densidad
density-gradient column columna de gradiente de densidad
dental amalgam amalgama dental
deodorant desodorante
deoxidant desoxidante
deoxidation desoxidación
deoxidize desoxidar
deoxidizer desoxidante
deoxy desoxi, deoxi
deoxyanisoin desoxianisoína
deoxybenzoin desoxibenzoína
deoxycholic acid ácido desoxicólico
deoxygenation desoxigenación
deoxyribonuclease desoxirribonucleasa
deoxyribonucleic acid ácido

desoxirribonucleico
deoxyribose desoxirribosa
dephlegmation deflegmación
depleted uranium uranio agotado
depolarization despolarización
depolarizer despolarizador
depolymerization despolmerización
deposit depósito, precipitado
deposition deposición
deposition potential potencial de
deposición
depression depresión
depression of freezing point depresión
de punto de congelación
depressor depresor
depside depsida
depurator depurador
derivation derivación
derivative derivado
derived unit unidad derivada
derric acid ácido dérrico
derritol derritol
des des
desalination desalinación
descending chromatography
cromatografía descendente
desclizite desclizita
desiccant desecante
desiccated desecado
desiccating desecante
desiccator desecador
deslanoside deslanosida
desmin desmina
desmotrope desmótropo
desmotropism desmotropismo
desorption desorción
desoxy desoxi
desoxycholic acid ácido desoxicólico
Despretz law ley de Despretz
destructive distillation destilación
destructiva
desulfuration desulfuración
desyl desilo
detergent detergente
determinate error error determinado
determination determinación
detonating gas gas detonante
detonation detonación
detonator detonador

detoxification destoxificación
deuterate deuterar
deuterated deuterado
deuteration deuteración
deuteric acid ácido deutérico
deuteride deuteruro
deuterio deuterio
deuterium deuterio
deuterium oxide óxido de deuterio
deutero deutero
deuteron deuterón
deuteroxyl deuteroxilo
Devarda alloy aleación de Devarda
Devarda metal metal de Devarda
developed dye colorante desarrollado
developer revelador, desarrollador
development desarrollo
devitrification desvitrificación
dew point punto de rocío
Dewar flask recipiente de Dewar
Dewar structure estructura de Dewar
deweylite deweylita
dexamphetamine dexanfetamina
dextran dextrano
dextranase dextranasa
dextrin dextrina
dextrinose dextrinosa
dextro dextro
dextrogyric dextrógiro
dextrorotary dextrógiro
dextrorotatory dextrógiro
dextrose dextrosa
di di
diabase diabasa
diacetamide diacetamida
diacetanilide diacetanilida
diacetate diacetato
diacetic acid ácido diacético
diacetic ether éter diacético
diacetin diacetina
diacetonamine diacetonamina
diacetone diacetona
diacetone acrylamide
diacetonaacrilamida
diacetone alcohol diacetonaalcohol
diacetyl diacetilo
diacetylamide diacetilamida
diacetylanilide diacetilanilida
diacetylene diacetileno

diacetylethane diacetiletano
diacetylmorphine diacetilmorfina
diacetylperoxide diacetilperóxido
diacetylurea diacetilurea
diacid diácido
diacolation diacolación
diadochite diadoquita
diagenesis diagénesis
diagonal relationship relación diagonal
dialdehyde dialdehído
dialkene dialqueno
dialkyl dialquilo
dialkylene dialquileno
diallyl dialilo
diallyl adipate adipato de dialilo
diallyl chlorendate clorendato de dialilo
diallyl cyanamide cianamida de dialilo
diallyl isophthalate isoftalato de dialilo
diallyl maleate maleato de dialilo
diallyl phosphite fosfito de dialilo
diallyl phthalate ftalato de dialilo
diallyl sulfide sulfuro de dialilo
diallylamine dialilamina
diallylmelamine dialilmelanina
diallylurea dialilurea
dialozite dialozita
dialuric acid ácido dialúrico
dialysis diálisis
dialyzed dializado
dialyzer dializador
diamagnetism diamagnetismo
diamide diamida
diamide hydrate hidrato de diamida
diamidine diamidina
diamido diamido
diamine diamina
diamino diamino
diaminoacridine diaminoacridina
diaminoazobenzene diaminoazobenceno
diaminoazoxytoluene
    diaminoazoxitolueno
diaminobenzene diaminobenceno
diaminobenzidine diaminobencidina
diaminobenzophenone
    diaminobenzofenona
diaminobutane diaminobutano
diaminocaproic acid ácido
    diaminocaproico
diaminodiphenic acid ácido

    diaminodifénico
diaminodiphenyl diaminodifenilo
diaminodiphenylamine
    diaminodifenilamina
diaminodiphenylmethane
    diaminodifenilmetano
diaminoethane diaminoetano
diaminohexane diaminohexano
diaminopentane diaminopentano
diaminophenazine diaminofenazina
diaminophenol diaminofenol
diaminophenol hydrochloride
    clorhidrato de diaminofenol
diaminopropane diaminopropano
diaminotoluene diaminotolueno
diammonium diamonio
diammonium phosphate fostato
    diamónico
diamond diamante
diamorphine diamorfina
diamyl diamilo
diamyl phenol diamilfenol
diamyl sulfide sulfuro de diamilo
diamylamine diamilamina
diamylaniline diamilanilina
diamylene diamileno
dianiline dianilina
dianilinoethane dianilinoetano
dianisidine dianisidina
diaphorite diaforita
diaphragm diafragma
diarabinose diarabinosa
diarsane diarsano
diarsenite diarsenito
diarsyl diarsilo
diasolysis diasólisis
diaspore diásporo
diastase diastasa
diastereoisomer diaestereoisómero
diastereomer diaestereoisómero
diastereotopic diaestereotópico
diathermy diatermia
diatomaceous earth tierra de diatomeas
diatomic diatómico
diatomic molecule molécula diatómica
diatomite diatomita
diazacyclo diazaciclo
diazepam diazepam
diazine diazina

diazinon diazinón
diazo diazo
diazo dye colorante diazoico
diazoacetate diazoacetato
diazoacetic ester éster diazoacético
diazoamine diazoamina
diazoamino diazoamino
diazoaminobenzene diazoaminobenceno
diazoaminonaphthalene
    diazoaminonaftaleno
diazoate diazoato
diazobenzene diazobenceno
diazobenzene chloride cloruro de
    diazobenceno
diazobenzene hydroxide hidróxido de
    diazobenceno
diazobenzene nitrate nitrato de
    diazobenceno
diazobenzenesulfonic acid ácido
    diazobencenosulfónico
diazoethane diazoetano
diazohydrate diazohidrato
diazohydroxide diazohidróxido
diazoic acid ácido diazoico
diazoimide diazoimida
diazoimido diazoimido
diazole diazol
diazomethane diazometano
diazonitrophenol diazonitrofenol
diazonium diazonio
diazonium salt sal de diazonio
diazophenol diazofenol
diazosulfonate diazosulfonato
diazosulfonic acid ácido diazosulfónico
diazotization diazotización
diazoxide diazóxido
diazoxy diazoxi
dibasic dibásico
dibasic acid ácido dibásico
dibenalacetone dibenalacetona
dibenzamide dibenzamida
dibenzanthracene dibenzantraceno
dibenzanthrone dibenzantrona
dibenzanthronyl dibenzantronilo
dibenzene dibenceno
dibenzenechromium dibencenocromo
dibenzenyl dibencenilo
dibenzo dibenzo
dibenzoanthracene dibenzoantraceno

dibenzocyclohepatadienone
    dibenzociclohepatadienona
dibenzopyrone dibenzopirona
dibenzopyrrole dibenzopirrol
dibenzothiazine dibenzotiacina
dibenzoyl dibenzoilo
dibenzoyl peroxide peróxido de
    dibenzoilo
dibenzoylethane dibenzoiletano
dibenzoylethylene dibenzoiletileno
dibenzyl dibencilo
dibenzyl disulfide disulfuro de dibencilo
dibenzylamine dibencilamina
dibenzylether dinenciléter
dibenzylidene dibencilideno
dibenzylidyne dibencilidino
dibenzylmethylamine dibencilmetilamina
diborane diborano
dibromide dibromuro
dibromo dibromo
dibromoacetylene dibromoacetileno
dibromoanthracene dibromoantraceno
dibromobenzene dibromobenceno
dibromochloromethane
    dibromoclorometano
dibromochloropropane
    dibromocloropropano
dibromodiethyl sulfide sulfuro de
    dibromodietilo
dibromodifluoromethane
    dibromodifluorometano
dibromoethane dibromoetano
dibromofluorobenzene
    dibromofluorobenceno
dibromomalonic acid ácido
    dibromomalónico
dibromomethane dibromometano
dibromopentane dibromopentano
dibromopropane dibromopropano
dibromopropanol dibromopropanol
dibucaine dibucaína
dibutoxyaniline dibutoxianilina
dibutyl dibutilo
dibutyl chlorophosphate clorofosfato de
    dibutilo
dibutyl fumarate fumarato de dibutilo
dibutyl maleate maleato de dibutilo
dibutyl oxalate oxalato de dibutilo
dibutyl phosphate fosfato de dibutilo

dibutyl phosphite fosfito de dibutilo
dibutyl phthalate ftalato de dibutilo
dibutyl tartrate tartrato de dibutilo
dibutylamine dibultilamina
dibutylbutyl phosphonate fosfonato de dibutilbutilo
dibutylthiourea dibutiltiourea
dibutyltin dibutilestaño
dibutyltin diacetate diacetato de dibutilestaño
dibutyltin dichloride dicloruro de dibutilestaño
dibutyltin dilaurate dilaurato de dibutilestaño
dibutyltin oxide óxido de dibutilestaño
dibutyltin sulfide sulfuro de dibutilestaño
dicalcium orthophosphate ortofosfato dicálcico
dicalcium phosphate fosfato dicálcico
dicapryl dicaprilo
dicarbide dicarburo
dicarbocyanine dicarbocianina
dicarbonate dicarbonato
dicarboxyl dicarboxilo
dicarboxylic acid ácido dicarboxílico
dicarbylamine dicarbilamina
dicetyl dicetilo
dichlone diclona
dichloramine dicloramina
dichloride dicloruro
dichlorine dicloro
dichlorine oxide óxido de dicloro
dichloro dicloro
dichloroacetal dicloroacetal
dichloroacetic acid ácido dicloroacético
dichloroacetyl chloride cloruro de dicloroacetilo
dichloroacetylene dicloroacetileno
dichloroaniline dicloroanilina
dichloroanthracene dicloroantraceno
dichlorobenzaldehyde diclorobenzaldehído
dichlorobenzene diclorobenceno
dichlorobenzidine diclorobencidina
dichlorobenzoic acid ácido diclorobenzoico
dichlorobenzoyl chloride cloruro de diclorobenzoilo

dichlorobenzyl chloride cloruro de diclorobencilo
dichlorobutane diclorobutano
dichlorobutene diclorobuteno
dichlorocarbene diclorocarbeno
dichlorodiethyl sulfide sulfuro de diclorodietilo
dichlorodifluoromethane diclorodifluorurometano
dichloroethane dicloroetano
dichloroethanoic acid ácido dicloroetanoico
dichloroethene dicloroeteno
dichloroethyl acetate acetato de dicloroetilo
dichloroethyl ether éter dicloroetílico
dichloroethyl oxide óxido de dicloroetilo
dichloroethylene dicloroetileno
dichlorohydrin diclorohidrina
dichloroisopropyl alcohol alcohol dicloroisopropílico
dichloromethane diclorometano
dichloromethylsilane diclorometilsilano
dichloronaphthalene dicloronaftaleno
dichloronitrobenzene dicloronitrobenceno
dichloropentane dicloropentano
dichlorophene diclorofeno
dichlorophenol diclorofenol
dichlorophenoxyacetic acid ácido diclorofenoxiacético
dichloropropane dicloropropano
dichloropropene dicloropropeno
dichloropropionic acid ácido dicloropropiónico
dichloropropyl alcohol alcohol dicloropropílico
dichlorothiophene diclororiofeno
dichlorotoluene diclorotolueno
dichroic dicroico
dichroism dicroismo
dichromate dicromato
dichromatic dicromático
dichromic dicrómico
dichromic acid ácido dicrómico
dicryl dicrilo
dicyandiamide diciandiamida
dicyanide dicianuro
dicyano diciano

dicyanogen dicianógeno
dicyclic dicíclico
dicyclohexyl diciclohexilo
dicyclohexylamine diciclohexilamina
dicyclohexylcarbodiimide
    diciclohexilcarbodiimida
dicyclopentadiene diciclopentadieno
dicyclopentadienyl diciclopentadienilo
didecyl adipate adipato de didecilo
didecyl ether éter didecílico
didecyl sulfide sulfuro de didecilo
didymium didimio
didymium chloride cloruro de didimio
didymium nitrate nitrato de didimio
didymium oxide óxido de didimio
didymium sulfate sulfato de didimio
didymolite didimolita
Dieckmann reaction reacción de
    Dieckmann
diectyl diectilo
diectyl ether éter diectílico
diectyl sulfide sulfuro de diectilo
dieldrin dieldrina
dielectric dieléctrico
dielectric cohesion cohesión dieléctrica
dielectric constant constante dieléctrica
dielectric strength fuerza dieléctrica
Diels-Alder reaction reacción de Diels-
    Alder
diene dieno
dienestrol dienestrol
dienone-phenol rearrangement
    reordenamiento de dienona-fenol
dienophile dienófilo
diester diéster
Dieterici equation ecuación de Dieterici
Dieterici rule regla de Dieterici
diethanolamine dietanolamina
diether diéter
diethoxyaniline dietoxianilina
diethoxyethane dietoxietano
diethyl dietilo
diethyl carbonate carbonato de dietilo
diethyl chlorophosphate clorosfosfato de
    dietilo
diethyl diethylmalonate dietilmalonato
    de dietilo
diethyl disulfide dislfuro de dietilo
diethyl ether éter dietílico

diethyl ethylmalonate etilmalonato de
    dietilo
diethyl ketone dietilcetona
diethyl malate malato de dietilo
diethyl maleate maleato de dietilo
diethyl oxalate oxalato de dietilo
diethyl oxide oxido de dietilo
diethyl peroxide peroxido de dietilo
diethyl phosphine fosfina de dietilo
diethyl phosphite fosfito de dietilo
diethyl phthalate ftalato de dietilo
diethyl selenide seleniuro de dietilo
diethyl succinate succinato de dietilo
diethyl sulfate sulfato de dietilo
diethyl sulfide sulfuro de dietilo
diethyl sulfite sulfito de dietilo
diethyl tartrate tartrato de dietilo
diethylacetal dietilacetal
diethylacetic acid ácido dietilacético
diethylaluminum chloride cloruro
    dietilalumínico
diethylamine dietilamina
diethylamino dietilamino
diethylaminobenzaldehyde
    dietilaminobenzaldehído
diethylaminoethanol dietilaminoetanol
diethylaminopropylamine
    dietilaminopropilamina
diethylaniline dietilanilina
diethylbarbituric acid ácido
    dietilbarbitúrico
diethylbenzene dietilbenceno
diethylcadmium dietilcadmio
diethylcarbamazine dietilcarbamazina
diethylcarbinol dietilcarbinol
diethylcyclohexylamine
    dietilciclohexilamina
diethylene dietileno
diethylene dioxide dióxido de dietileno
diethylene glycol dietilenglicol
diethylene glycol dinitrate dinitrato de
    dietilenglicol
diethylene glycol ethyl ether éter etílico
    de dietilenglicol
diethylene glycol stearate estearato de
    dietilenglicol
diethylenediamine dietilendiamina
diethylenetriamine dietilenotriamina
diethylethylene dictilctilcno

diethylidene dietilideno
diethylmethylmethane dietilmetilmetano
diethyloxamide dietiloxamida
diethylstilbestrol dietilestilbestrol
diethylurea dietilurea
diethylzinc dietilcinc
dietzeite dietzeita
differential adsorption adsorción
     diferencial
differential ebulliometer ebullómetro
     diferencial
differential reduction reducción
     diferencial
differential scanning calorimetry
     calorimetría de exploración diferencial
differential spectrophotometry
     espectrofotometría diferencial
differential thermal analysis análisis
     térmico diferencial
differential titration titulación
     diferencial, valoración diferencial
diffraction difracción
diffraction grating rejilla de difracción
diffraction pattern patrón de difracción
diffraction spectrum espectro de
     difracción
diffuse difuso
diffusion difusión
diffusion analysis análisis por difusión
diffusion constant constante de difusión
diffusion current corriente de difusión
diffusion law ley de difusión
diffusion layer capa de difusión
diffusion limited aggregation agregación
     limitada por difusión
diffusion pump bomba de difusión
difluoride difluoruro
difluoro difluoro
difluoroamino difluoroamino
difluorobenzene difluorobenceno
difluorodiazine difluorodiazina
difluoroethane difluoroetano
difluoromethane difluorometano
digallic acid ácido digálico
digestion digestión
digitalin digitalina
digitalis digitalis
digitalose digitalosa
digitonin digitonina

digitoxigenin digitoxigenina
digitoxin digitoxina
diglyceride diglicérido
diglycerol diglicerol
diglycol diglicol
diglycol carbamate carbamato de
     diglicol
diglycol laurate laurato de diglicol
diglycol oleate oleato de diglicol
diglycol stearate estearato de diglicol
diglycolic acid ácido diglicólico
diglycolide diglicólido
diglyme diglime
digol digol
digoxin digoxina
dihedral dihedral
dihedral angle ángulo dihedral
dihedron dihedrón
dihexyl dihexilo
dihexylamine dihexilamina
dihydrate dihidrato
dihydrazine sulfate sulfato de
     dihidrazina
dihydrazone dihidrazona
dihydric alcohol alcohol dihídrico
dihydride dihidruro
dihydro dihidro
dihydroazirine dihidroazirina
dihydrobenzene dihidrobenceno
dihydrobromide dibromhidrato
dihydrochalcone dihidrocalcona
dihydrochloride diclorhidrato
dihydrocholesterol dihidrocolesterol
dihydronaphthalene dihidronaftaleno
dihydropropene dihidropropeno
dihydropyran dihidropirano
dihydroxy dihidroxi
dihydroxyacetone dihidroxiacetona
dihydroxyamine dihidroxiamina
dihydroxyanthraquinone
     dihidroxiantraquinona
dihydroxybenzene dihidroxibenceno
dihydroxybenzoic acid ácido
     dihidroxibenzoico
dihydroxybutanedioic acid ácido
     dihidroxibutanodioico
dihydroxycinnamic acid ácido
     dihidroxicinámico
dihydroxydiphenylsulfone

dihidroxidifenilsulfona
dihydroxyethane dihidroxietano
dihydroxymalonic acid ácido
  dihidroximalónico
dihydroxynaphthalene
  dihidroxinaftaleno
dihydroxypropane dihidroxipropano
dihydroxystearic acid ácido
  dihidroxiesteárico
dihydroxysuccinic acid ácido
  dihidroxisuccínico
diimide diimida
diimine diimina
diimino diimino
diiodide diyoduro
diiodo diyodo
diiodoacetic acid ácido diyodoacético
diiodoacetylene diyodoacetileno
diiodomethane diyodometano
diisoamyl diisoamilo
diisobutyl diisobutilo
diisobutyl ketone diisobutilcetona
diisobutylamine diisobutilamina
diisobutylene diisobutileno
diisobutylphenol diisobutilofenol
diisocyanate diisocianato
diisooctyl adipate adipato de diisooctilo
diisopentyl diisopentilo
diisopropyl diisopropilo
diisopropyl ether éter diisopropílico
diisopropylbenzene diisopropilbenceno
diketene dicetena
diketone dicetona
diketopiperazine dicetopiperazina
diketopyrrolidine dicetopirrolidina
dilactone dilactona
dilatation dilatación
dilation dilatación
dilatometer dilatómetro
dilauryl dilaurilo
dilauryl peroxide dilaurilperóxido
dilauryl phosphite dilaurilfosfito
dilauryl sulfide dilaurilsulfuro
dilaurylamine dilaurilamina
dilead diplomo
diluent diluyente
dilute diluido
dilute acid ácido diluido
dilute base base diluida

dilute solution solución diluida
dilution dilución
dilution law ley de dilución
dilution ratio razón de dilución
dilution rule regla de dilución
dimedone dimedona
dimer dímero
dimeric dimérico
dimerization dimerización
dimethano dimetano
dimethoate dimetoato
dimethoxy dimetoxi
dimethoxyaniline dimetoxianilina
dimethoxybenzene dimetoxibenceno
dimethoxyethane dimetoxietano
dimethoxymethane dimetoximetano
dimethoxytetraglycol dimetoxitetraglicol
dimethyl dimetilo
dimethyl cadmium cadmio de dimetilo
dimethyl cyanamide cianamida de
  dimetilo
dimethyl disulfide disulfuro de dimetilo
dimethyl ether éter dimetílico
dimethyl malonate malonato de dimetilo
dimethyl phosphate fosfato de dimetilo
dimethyl phthalate ftalato de dimetilo
dimethyl sebacate sebacato de dimetilo
dimethyl selenide seleniuro de dimetilo
dimethyl sulfate sulfato de dimetilo
dimethyl sulfide sulfuro de dimetilo
dimethyl sulfoxide sulfóxido de dimetilo
dimethyl telluride telururo de dimetilo
dimethyl terephthalate tereftalato de
  dimetilo
dimethylacetal dimetilacetal
dimethylamine dimetilamina
dimethylamino dimetilamino
dimethylaminobenzoic acid ácido
  dimetilaminobenzoico
dimethylaminoethanol
  dimetilaminoetanol
dimethylaminopropanol
  dimetilaminopropanol
dimethylaniline dimetilanilina
dimethylanthracene dimetilantraceno
dimethylarsine dimetilarsina
dimethylarsinic acid ácido
  dimetilarsínico
dimethylbenzene dimetilbenceno

dimethylbenzyl chloride cloruro de
dimetilbencilo
dimethylberyllium dimetilberilio
dimethylbutadiene dimetilbutadieno
dimethylbutane dimetilbutano
dimethylbutene dimetilbuteno
dimethylcarbinol dimetilcarbinol
dimethylchloroacetal dimetilcloroacetal
dimethylcyclohexane dimetilciclohexano
dimethylcyclopentane
dimetilciclopentano
dimethyldichlorosilane
dimetildiclorosilano
dimethylene dimetileno
dimethylethylene dimetiletileno
dimethylformamide dimetilformamida
dimethylglyoxime dimetilglioxima
dimethylheptene dimetilhepteno
dimethylhexadiene dimetilhexadieno
dimethylhexanediol dimetilhexanodiol
dimethylhexynediol dimetilhexinodiol
dimethylhydantoin dimetilhidantoína
dimethylhydrazine dimetilhidrazina
dimethylhydroxybenzene
dimetilhidroxibenceno
dimethylisopropanolamine
dimetilisopropanolamina
dimethylmercury dimetilmercurio
dimethylmethane dimetilmetano
dimethylnitrosamine dimetilnitrosamina
dimethyloctadiene dimetiloctadieno
dimethylpentaldehyde
dimetilpentaldehído
dimethylpentane dimetilpentano
dimethylphenol dimetilfenol
dimethylphosphine dimetilfosfina
dimethylpropane dimetilpropano
dimethyltin dichloride dicloruro de
dimetilestaño
dimethylurea dimetilurea
dimethylxanthine dimetilxantina
dimetilan dimetilán
dimetric dimétrico
dimorphic dimórfico
dimorphism dimorfismo
dimyristyl sulfide sulfuro de dimiristilo
dinaphtho dinafto
dinaphthol dinaftol
dinaphthyl dinaftilo

dinaphthylene dinaftileno
dineric dinérico
dineutron dineutrón
dinitrate dinitrato
dinitrite dinitrito
dinitro dinitro
dinitroaniline dinitroanilina
dinitroanthraquinone
dinitroantraquinona
dinitrobenzene dinitrobenceno
dinitrobenzoyl chloride cloruro de
dinitrobenzoilo
dinitrocresol dinitrocresol
dinitrofluorobenzene
dinitrofluorobenceno
dinitrogen dinitrógeno
dinitrogen oxide óxido de dinitrógeno
dinitrogen tetraoxide tetróxido de
dinitrógeno
dinitronaphthalene dinitronaftaleno
dinitronaphthol dinitronaftol
dinitrophenol dinitrofenol
dinitrophenylhydrazine
dinitrofenilhidrazina
dinitrosalicylic acid ácido
dinitrosalicílico
dinitroso dinitroso
dinitrotoluene dinitrotolueno
dinonyl phthalate ftalato de dinonilo
dinucleotide dinucleótido
dioctyl dioctilo
dioctyl chlorophosphate clorofosfato de
dioctilo
dioctyl ether éter dioctílico
dioctyl phosphite fosfito de dioctilo
dioctyl phthalate ftalato de dioctilo
dioctyl sebacate sebacato de dioctilo
dioctylamine dioctilamina
dioctylmethane dioctilmetano
diol diol
diolefin diolefina
dione diona
diopside diópsido
dioptase dioptasa
diorite diorita
dioxadiene dioxadieno
dioxalane dioxalano
dioxan dioxana
dioxane dioxano

dioxazole dioxazol
dioxdiazole dioxdiazol
dioxide dióxido
dioxime dioxima
dioxin dioxina
dioxo dioxo
dioxolane dioxolano
dioxonitric acid ácido dioxonítrico
dioxy dioxi
dioxygenyl dioxigenilo
dipalmitate dipalmitato
dipalmitylamine dipalmitilamina
dipentaerythritol dipentaeritritol
dipentene dipenteno
dipentene dioxide dióxido de dipenteno
dipentene glycol dipentenoglicol
dipentene hydrochloride clorhidrato de
      dipenteno
dipentene monoxide monóxido de
      dipenteno
dipentyl dipentilo
dipentyl sulfide sulfuro de dipentilo
dipeptide dipéptido
diphenic acid ácido difénico
diphenol difenol
diphenolate difenolato
diphenyl difenilo
diphenyl carbonate carbonato de difenilo
diphenyl ether éter difenílico
diphenyl ketone difenilcetona
diphenyl oxide óxido de difenilo
diphenyl phosphite fosfito de difenilo
diphenyl phthalate ftalato de difenilo
diphenylamine difenilamina
diphenylamino difenilamino
diphenylbenzene difenilbenceno
diphenylcarbazide difenilcarbazida
diphenylchloroarsine difenilcloroarsina
diphenyldichlorosilane
      difenildiclorosilano
diphenyldiimide difenildiimida
diphenylene difenileno
diphenylethane difeniletano
diphenylethylene difeniletileno
diphenylglyoxal difenilglioxal
diphenylguanidine difenilguanidina
diphenylmethane difenilmetano
diphenylmethyl difenilmetilo
diphenylmethyl bromide bromuro de

      difenilmetilo
diphenylsulfone difenilsulfona
diphenyltin difenilestaño
diphenylurea difenilurea
diphosgene difosgeno
diphosphane difosfano
diphosphate difosfato
diphosphine difosfina
diphosphoglyceric acid ácido
      difosfoglicérico
diphosphoric acid ácido difosfórico
dipolar dipolar
dipolar ion ion dipolar
dipole dipolo
dipole moment momento dipolar
dipole radiation radiación dipolar
dipropyl dipropilo
dipropyl ether éter dipropílico
dipropyl sulfide sulfuro de dipropilo
dipropylberyllium dipropilberilio
dipropylene glycol dipropilenglicol
diproton diprotón
dipyridine dipiridina
dipyridyl dipiridilo
diradical dirradical
direct dye colorante directo
direct nuclear reaction reacción nuclear
      directa
diresorcinol dirresorcina
disaccharide disacárido
disagglomeration desaglomeración
disalicylic acid ácido disalicílico
disassociation disociación
discaryl disacrilo
discharge descarga
Dische reaction reacción de Dische
discontinuous phase fase discontinua
discrete spectrum espectro discreto
diselane diselano
diselenide deseleniuro
disilane disilano
disilanyl disinalilo
disilicate disilicato
disilicide disiliciuro
disiloxane disiloxano
disilyl disililo
disinfectant desinfectante
disinfection desinfección
disintegration desintegración

disintegration chain cadena de desintegración

disintegration constant constante de desintegración

disintegration family familia de desintegración

disintegration mode modo de desintegración

disintegration particle partícula de desintegración

disintegration product producto de desintegración

disintegration rate velocidad de desintegración

disintegration series serie de desintegración

dislocation dislocación

dismutation dismutación

disodium citrate citrato disódico

disodium diphosphate difosfato disódico

disodium hydrogenphosphate hidrógenofosfato disódico

disodium phosphate fosfato disódico

dispersator dispersador

disperse dispersar

dispersed disperso

dispersed phase fase dispersa

dispersed system sistema disperso

dispersing dispersante

dispersing agent agente dispersante

dispersion dispersión

dispersion forces fuerzas de dispersión

dispersion medium medio de dispersión

dispersoid dispersoide

displacement desplazamiento

displacement chromatography cromatografía por desplazamiento

displacement law ley de desplazamiento

displacement reaction reacción de desplazamiento

displacement series serie de desplazamiento

display indicador visual, indicación visual

disproportionation desproporción

dissipation disipación

dissipative system sistema disipativo

dissociate disociar

dissociated disociado

dissociated state estado disociado

dissociation disociación

dissociation constant constante de disociación

dissociation energy energía de disociación

dissociation pressure presión de disociación

dissociation reaction reacción de disociación

dissolution disolución

dissolve disolver

dissolved disuelto

dissolved oxygen oxígeno disuelto

dissolvent disolvente

dissolving disolvente

dissolving agent agente disolvente

dissymmetrical disimétrico

dissymmetry disimetría

distance distancia

distearyl sulfide sulfuro de diestearilo

distillate destilado

distillation destilación

distillation apparatus aparato de destilación

distillation column columna de destilación

distillation flask matraz de destilación

distillation residue residuo de destilación

distillation tower torre de destilación

distilled destilado

distilled water agua destilada

distiller destilador

distribution coefficient coeficiente de distribución

distribution law ley de distribución

distribution principle principio de distribución

distyrene diestireno

disulfate disulfato

disulfide disulfuro

disulfiram disulfiram

disulfite disulfito

disulfo disulfo

disulfonate disulfonato

disulfonic acid ácido disulfónico

disulfuric acid ácido disulfúrico

diterpene diterpeno

diterpenoid diterpenoide

dithiane ditiano
dithiazine ditiazina
dithiene ditieno
dithienyl ditienilo
dithio ditio
dithiobenzoic acid ácido ditiobenzoico
dithiocarbamate ditiocarbamato
dithiocarbamic acid ácido
     ditiocarbámico
dithiocarbonate ditiocarbonato
dithiocarboxy ditiocarboxi
dithiol ditiol
dithiolene ditioleno
dithionate ditionato
dithionic acid ácido ditiónico
dithionite ditionita
dithiooxamide ditiooxamida
dithizone ditizona
ditolyl ditolilo
ditungsten carbide carburo de
     ditungsteno
divalence divalencia
divalent divalente
divalent carbon carbono divalente
divinyl divinilo
divinyl ether éter divinílico
divinyl oxide óxido de divinilo
divinyl sulfide sulfuro de divinilo
divinylacetylene divinilacetileno
divinylbenzene divinilbenceno
dixylyl dixililo
dixylylethane dixililetano
Dobbin reagent reactivo de Dobbin
docosane docosano
docosanoic acid ácido docosanoico
docosanol docosanol
docosenoic acid ácido docosenoico
dodecahedral dodecaedral
dodecahedro dodecaedro
dodecahedron dodecaedro
dodecahydrate dodecahidrato
dodecanal dodecanal
dodecane dodecano
dodecanoic acid ácido dodecanoico
dodecanol dodecanol
dodecanoyl dodecanoilo
dodecenal dodecenal
dodecene dodeceno
dodecenylsuccinic acid ácido

dodecenilsuccínico
dodecyl dodecilo
dodecyl alcohol alcohol dodecílico
dodecylamine dodecilamina
dodecylbenzene dodecilbenceno
dodecylene dodecileno
dodecylphenol dodecilfenol
dodecyltrichlorosilane
     dodeciltriclorosilano
Doebner-Miller reaction reacción de
     Doebner-Miller
Doebner reaction reacción de Doebner
dolerite dolerita
dolichol dolicol
dolomite dolomita
dolomol dolomol
dome cúpula
domesticine domesticina
domingite dominguita
donarite donarita
Donnan equilibrium equilibrio de
     Donnan
Donnan hydrolysis hidrólisis de Donnan
donor donador
donor-acceptor pair par donador-aceptor
donor atom átomo donador
dope dopante
doped dopado
doping dopado
dotriacontane dotriacontano
double-beam spectrophotometer
     espectrofotómetro de doble haz
double bond enlace doble
double-bond isomerism isomerismo de
     enlace doble
double decomposition descomposición
     doble
double displacement desplazamiento
     doble
double layer capa doble
double refraction refracción doble
double replacement reemplazo doble
double salt sal doble
double weighing pesaje doble
doublet doblete
Downs process proceso de Downs
draconyl draconilo
Dragendorff reaction reacción de
     Dragendorff

Dragendorff reagent reactivo de Dragendorff
Draper effect efecto de Draper
Draper law ley de Draper
Drew number número de Drew
drimin drimina
dropping bottle botella para gotear
drug droga
dry seco
dry acid ácido seco
dry cell pila seca
dry chemical producto químico seco
dry deposition deposición seca
dry distillation destilación seca
dry ice hielo seco
dry point punto seco
dryer secador
drying secado
drying agent agente secante
drying oil aceite secante
drying tube tubo secante
ductile dúctil
ductility ductilidad
Duff reaction reacción de Duff
dufrenite dufrenita
dufrenoysite dufrenoisita
Duhem equation ecuación de Duhem
Dühring rule regla de Dühring
dulcitol dulcitol
Dumas method método de Dumas
dumortierite dumortierita

dunnite dunita
duplet duplete
duralumin duraluminio
durene dureno
durol durol
duryl durilo
durylene durileno
durylic acid ácido durílico
dust polvo
Dutt-Wormall reaction reacción de Dutt-Wormall
dye colorante, tinte
dye-retarding agent agente retardante de teñido
dyeing assistant asistente de teñido
dyestuff materia colorante
dynad dinada
dynamic equilibrium equilibrio dinámico
dynamic isomerism isomerismo dinámico
dynamic system sistema dinámico
dynamite dinamita
dypnone dipnona
dyscrasite discrasita
Dyson notation notación de Dyson
dysprosium disprosio
dysprosium nitrate nitrato de disprosio
dysprosium oxide óxido de disprosio
dysprosium sulfate sulfato de disprosio
dystectic mixture mezcla distéctica

E

earth tierra
earth alkali metal metal alcalinotérreo
earth wax cera de tierra, ozocerita
easin easina
ebonite ebonita
ebullator ebullidor
ebulliometer ebulliómetro
ebulliometry ebullometría
ebullioscope ebulloscopio
ebullioscopic constant constante
    ebulloscópica
ebullioscopy ebulloscopia
ebullition ebullición
eccentric excéntrico
ecdysone ecdisona
ecgonine ecgonina
echelon grating rejilla escalonada
echinomycin equinomicina
echinopsine equinopsina
eclipsed eclipsado
eclipsed conformation conformación
    eclipsada
eclogite eclogita
ecology ecología
economizer economizador
ecosystem ecosistema
edeate edeato
Edeleanu process proceso de Edeleanu
edenite edenita
Eder solution solución de Eder
edestin edestina
edetate edetato
edetate disodium edetato disódico
edge filtration filtración por bordes
edible oil aceite comestible
edingtonite edingtonita
Edison cell pila de Edison
Edman degradation degradación de
    Edman
effect efecto
effective atomic number número
    atómico efectivo
effective charge carga efectiva

effective molecular diameter diámetro
    molecular efectivo
effective permeability permeabilidad
    efectiva
effective temperature temperatura
    efectiva
effervescence efervescencia
effervescent efervescente
effervescent mixture mezcla
    efervescente
efflorescence eflorescencia
effluent efluente
effusiometer efusiómetro
effusion efusión
egg oil aceite de huevo
Ehrenfest classification clasificación de
    Ehrenfest
Ehrlich-Sachs reaction reacción de
    Ehrlich-Sachs
eicosane eicosano
eicosanoic acid ácido eicosanoico
eicosanoid eicosanoide
eicosanol eicosanol
eigenfunction autofunción
eigenvalue autovalor
Einchluss thermometer termómetro de
    Einchluss
Einhorn-Brunner reaction reacción de
    Einhorn-Brunner
einstein einstein
Einstein coefficient coeficiente de
    Einstein
Einstein law ley de Einstein
einsteinium einstenio
eka eka
ekahafnium ekahafnio
elaboration elaboración
elaidic acid ácido elaídico
elaidinization elaidinización
elastase elastasa
elastic elástico
elastic collision colisión elástica
elastic constant constante elástica

elasticity elasticidad
elastin elastina
elastomer elastómero
elateric acid ácido elatérico
elaterite elaterita
Elbs reaction reacción de Elbs
electret electreto
electric eléctrico
electric conductivity conductividad
    eléctrica
electric current corriente eléctrica
electric double layer capa doble eléctrica
electric field effect efecto de campo
    eléctrico
electric furnace horno eléctrico
electric steel acero de horno eléctrico
electride electruro
electrification electrificación
electroanalysis electroanálisis
electrochemical electroquímico
electrochemical cell célula
    electroquímica
electrochemical corrosion corrosión
    electroquímica
electrochemical deposition deposición
    electroquímica
electrochemical deterioration deterioro
    electroquímico
electrochemical diffusion difusión
    electroquímica
electrochemical effect efecto
    electroquímico
electrochemical equivalent equivalente
    electroquímico
electrochemical hardening
    endurecimiento electroquímico
electrochemical measurement medida
    electroquímica
electrochemical migration migración
    electroquímica
electrochemical oxidation oxidación
    electroquímica
electrochemical passivation pasivación
    electroquímica
electrochemical polarization
    polarización electroquímica
electrochemical potential potencial
    electroquímico
electrochemical process proceso

electroquímico
electrochemical reduction reducción
    electroquímica
electrochemical series serie
    electroquímica
electrochemical spectrum espectro
    electroquímico
electrochemistry electroquímica
electrochromatography
    electrocromatografía
electrocoating electrorrevestimiento
electrocratic electrocrático
electrocyclic reaction reacción
    electrocíclica
electrode electrodo
electrode characteristic característica de
    electrodo
electrode conductance conductancia de
    electrodo
electrode current corriente de electrodo
electrode dissipation disipación de
    electrodo
electrode potential potencial de
    electrodo
electrode voltage tensión de electrodo
electrodeposit electrodepósito
electrodeposition electrodeposición
electrodeposition analysis análisis por
    electrodeposición
electrodialysis electrodiálisis
electrodialyzer electodializador
electrodispersion electrodispersión
electrodissolution electrodisolución
electroflotation electroflotación
electrofocusing electroenfoque
electroforming electroformación
electrographic analysis análisis
    electrográfico
electrographite electrografito
electrogravimetry electrogravimetría
electrokinetic potential potencial
    electrocinético
electrokinetics electrocinética
electroluminescence electroluminiscencia
electrolysis electrólisis
electrolyte electrólito
electrolyte acid ácido electrolítico
electrolytic electrolítico
electrolytic action acción electrolítica

electrolytic analysis análisis electrolítico
electrolytic cathode cátodo electrolítico
electrolytic cell celda electrolítica
electrolytic cleaning limpieza electrolítica
electrolytic conductance conductancia electrolítica
electrolytic conduction conducción electrolítica
electrolytic conductivity conductividad electrolítica
electrolytic corrosion corrosión electrolítica
electrolytic decomposition descomposición electrolítica
electrolytic deposition deposición electrolítica
electrolytic dissociation disociación electrolítica
electrolytic gas gas electrolítico
electrolytic ionization ionización electrolítica
electrolytic iron hierro electrolítico
electrolytic metallization metalización electrolítica
electrolytic oxidation oxidación electrolítica
electrolytic polarization polarización electrolítica
electrolytic potential potencial electrolítico
electrolytic process proceso electrolítico
electrolytic reduction reducción electrolítica
electrolytic separation separación electrolítica
electrolytic solution solución electrolítica
electrolyze electrolizar
electrolyzer electrolizador
electromagnetic electromagnético
electromagnetic radiation radiación electromagnética
electromagnetic separation separación electromagnética
electromagnetic spectrum espectro electromagnético
electromerism electromerismo
electrometallurgy electrometalurgia
electrometer electrómetro

electrometric titration titulación electrométrica, valoración electrométrica
electromigration electromigración
electromotive force fuerza electromotriz
electromotive series serie electromotriz
electron electrón
electron accelerator acelerador electrónico, acelerador de electrones
electron acceptor aceptor electrónico
electron affinity afinidad electrónica
electron avalanche avalancha electrónica
electron band banda electrónica
electron beam haz electrónico
electron bombardment bombardeo electrónico
electron bunching agrupamiento electrónico
electron capture captura electrónica
electron charge carga electrónica
electron-charge density densidad de carga electrónica
electron cloud nube electrónica
electron collection recolección electrónica
electron collision colisión electrónica
electron concentration concentración electrónica
electron conduction conducción electrónica
electron conductor conductor electrónico
electron configuration configuración electrónica
electron crystallography cristalografía electrónica
electron current corriente electrónica
electron deficiency deficiencia electrónica
electron density densidad electrónica
electron diffraction difracción electrónica
electron dipole moment momento dipolar electrónico
electron distribution distribución electrónica
electron donor donador electrónico
electron drift flujo electrónico
electron emission emisión electrónica
electron energy energía electrónica

electron energy level nivel de energía electrónica

electron exchange intercambio electrónico

electron flow flujo electrónico

electron flux flujo electrónico

electron formula fórmula electrónica

electron gas gas electrónico

electron image imagen electrónica

electron impact impacto electrónico

electron linear accelerator acelerador lineal electrónico

electron magnetic moment momento magnético electrónico

electron magnetic resonance resonancia magnética electrónica

electron mass masa de electrón

electron microanalyzer microanalizador electrónico

electron microprobe microsonda electrónica

electron microscope microscopio electrónico

electron microscopy microscopia electrónica

electron motion moción electrónica

electron multiplication multiplicación electrónica

electron multiplicity multiplicidad electrónica

electron multiplier multiplicador electrónico

electron octet octeto electrónico

electron orbit órbita electrónica

electron orbital orbital electrónico

electron oscillator oscilador electrónico

electron pair par de electrones

electron-pair bond enlace de par de electrones

electron paramagnetic resonance resonancia paramagnética electrónica

electron physics física electrónica

electron probe sonda electrónica

electron probe analysis análisis con sonda electrónica

electron resonance resonancia electrónica

electron rest mass masa en reposo de electrón

electron shell capa electrónica

electron source fuente de electrones

electron specific charge carga específica de electrón

electron spectroscopy espectroscopia electrónica

electron spectrum espectro electrónico

electron spin espín de electrón

electron-spin resonance resonancia de espín electrónico

electron stream corriente electrónica

electron trajectory trayectoria electrónica

electron transfer transferencia electrónica

electron transport chain cadena de transporte electrónico

electron velocity velocidad de electrón

electron-volt electrón-volt

electronegative electronegativo

electronegative element elemento electronegativo

electronegative potential potencial electronegativo

electronegative radical radical electronegativo

electronegativity electronegatividad

electronic configuration configuración electrónica

electronic energy curve curva de energía electrónica

electronic transition transición electrónica

electroorganic reaction reacción electroorgánica

electroosmosis electroósmosis

electrophile electrófilo

electrophilic electrofílico

electrophilic addition adición electrofílica

electrophilic reagent reactivo electrofílico

electrophilic substitution substitución electrofílica

electrophoresis electroforesis

electrophoretic electroforético

electrophoretic effect efecto electroforético

electrophoretogram electroforetograma

electrophorogram electroforograma
electroplating galvanoplastia,
    electrorrecubrimiento
electropolishing electropulido
electropositive electropositivo
electropositive element elemento
    electropositivo
electropositive potential potencial
    electropositivo
electropotential electropotencial
electroscope electroscopio
electroscopy electroscopia
electrosol electrosol
electrostatic bond enlace electrostático
electrostatic precipitator precipitador
    electrostático
electrostriction electroestricción
electrosynthesis electrosíntesis
electrovalence electrovalencia
electrovalent electrovalente
electrovalent bond enlace electrovalente
electrovalent compound compuesto
    electrovalente
electroviscosity electroviscocidad
electrowinning extracción electrolítica
electrum electrum
element elemento
element of symmetry elemento de
    simetría
elementary analysis análisis elemental
elementary charge carga elemental
elementary molecule molécula elemental
elementary particle partícula elemental
elementary process proceso elemental
elementary space espacio elemental
elemonic acid ácido elemónico
eleonorite eleonorita
eleostearic acid ácido eleoesteárico
elevation of boiling point elevación de
    punto de ebullición
elimination reaction reacción de
    eliminación
elixir elíxir
ellagic acid ácido elágico
Ellingham diagram diagrama de
    Ellingham
elpidite elpidita
Eltekoff reaction reacción de Eltekoff
eluant eluyente

eluate eluido
elution elución
elutriation elutriación
emanation emanación
embelic acid ácido embélico
embolite embolita
embonic acid ácido embónico
embrittlement fragilización
Emde degradation degradación de Emde
emerald esmeralda
emerald green verde esmeralda
emery esmeril
emetamine emetamina
emetine emetina
emission emisión
emission lines líneas de emisión
emission spectroscopy espectroscopia de
    emisión
emission spectrum espectro de emisión
Emmert reaction reacción de Emmert
emodin emodina
empirical empírico
empirical formula fórmula empírica
emplectite emplectita
emulgator emulagador
emulsification emulsionamiento
emulsifier emulsionador
emulsify emulsionar
emulsifying agent agente emulsionante
emulsion emulsión
emulsion breaker rompedor de emulsión
emulsion breaking rompimiento de
    emulsión
emulsion polymerization polimerización
    en emulsión
emulsion stabilizer estabilizador de
    emulsión
enamel esmalte
enamine enamina
enanthic acid ácido enántico
enanthol enantol
enanthyl enantilo
enantiomer enantiómero
enantiomeric enantiomérico
enantiomorph enantiomorfo
enantiomorphic enantiomórfico
enantiomorphism enantiomorfismo
enantiotopic enantiotópico
enantiotropic enantiótropo,

enantiotrópico

enantiotropy enantiotropía

enargite enargita

encapsulation encapsulación

enclosure compound compuesto de encerramiento

end point punto final

endo endo

endogenous endógeno

endomycin endomicina

endopeptidase endopeptidasa

endorphin endorfina

endosmosis endósmosis

endothermic endotérmico

endothermic reaction reacción endotérmica

endrin endrina

ene eno

enediol enodiol

energy energía

energy balance equilibrio energético

energy band banda de energía

energy change cambio de energía

energy conservation conservación de energía

energy conversion conversión de energía

energy converter convertidor de energía

energy density densidad de energía

energy level nivel de energía

energy-level diagram diagrama de niveles de energía

energy of activation energía de activación

energy quantum cuanto de energía

energy source fuente de energía

enflurane enflurano

Engel salt sal de Engel

enhanced line línea realzada

enkephalin encefalina

enol enol

enolase enolasa

enriched uranium uranio enriquecido

enrichment enriquecimiento

enstatite enstatita

enthalpy entalpía

enthalpy of evaporation entalpía de evaporación

enthalpy of fusion entalpía de fusión

enthalpy of reaction entalpía de reacción

enthalpy titration titulación entálpica, valoración entálpica

entropy entropía

entropy of activation entropía de activación

entropy of mixing entropía de mezcla

entropy of transition entropía de transición

environmental chemistry química ambiental

environmental conditions condiciones ambientales

environmental contamination contaminación ambiental

environmental factors factores ambientales

environmental protection protección ambiental

environmental requirements requisitos ambientales

environmental stability estabilidad ambiental

enyl enilo

enzyme enzima

enzyme kinetics cinética de enzima

enzyme-substrate complex complejo enzima-substrato

eosin eosina

ephedrine efedrina

ephedrine hydrochloride clorhidrato de efedrina

ephedrine sulfate sulfato de efedrina

epi epi

epicamphor epialcanfor

epichlorite epiclorita

epichlorohydrin epiclorohidrina

epicyanohydrin epicianhidrina

epididymite epididimita

epidioxide epidióxido

epidioxy epidioxi

epidithio epiditio

epigenite epigenita

epihydrin epihidrina

epimer epímero

epimerase epimerasa

epimeric epimérico

epimeride epimerida

epimerism epimerismo

epimerization epimerización

epinephrine epinefrina
epistilbite epiestilbita
epitaxy epitaxia
epithio epitio
epithioximino epitioximino
epoxidation epoxidación
epoxide epóxido
epoxy epoxi
epoxy polymer polímero epóxido
epoxy resin resina epóxica
epoxybutane epoxibutano
epoxyethane epoxietano
epoxyimino epoxiimino
epoxynitrilo epoxinitrilo
epoxypropane epoxipropano
epoxypropanol epoxipropanol
epoxythio epoxitio
epsilon epsilón
epsilon acid ácido epsilón
Epsom salts sales de Epsom
epsomite epsomita
equation ecuación
equation of state ecuación de estado
equatorial ecuatorial
equilenin equilenina
equilibrium equilibrio
equilibrium constant constante de equilibrio
equilibrium diagram diagrama de equilibrio
equilibrium dialysis diálisis de equilibrio
equilibrium law ley de equilibrio
equilibrium point punto de equilibrio
equilibrium potential potencial de equilibrio
equilibrium ratio razón de equilibrio
equilibrium solubility solubilidad de equilibrio
equipartition equipartición
equipartition of energy equipartición de energía
equipotential energy energía equipotencial
equivalence equivalencia
equivalence point punto de equivalencia
equivalent equivalente
equivalent charge carga equivalente
equivalent conductance conductancia equivalente

equivalent conductivity conductividad equivalente
equivalent electrons electrones equivalentes
equivalent nuclei núcleos equivalentes
equivalent proportions proporciones equivalentes
equivalent weight peso equivalente
erbia erbia
erbium erbio
erbium acetate acetato de erbio
erbium halide haluro de erbio
erbium nitrate nitrato de erbio
erbium oxalate oxalato de erbio
erbium oxide óxido de erbio
erbium sulfate sulfato de erbio
erbon erbón
Erdmann reagent reactivo de Erdmann
Erdmann salt sal de Erdmann
ergine ergina
ergocalciferol ergocalciferol
ergodic ergódico
ergometrine ergometrina
ergonovine ergonovina
ergosinine ergosinina
ergosterol ergosterol
ergot cornezuelo
ergotamine ergotamina
ergotinine ergotinina
ergotoxine ergotoxina
erilite erilita
erinite erinita
eriodictyol eriodictiol
Erlenmeyer flask matraz de Erlenmeyer
Erlenmeyer rule regla de Erlenmeyer
Erlenmeyer synthesis síntesis de Erlenmeyer
erosion erosión
erubescite erubescita
erucic acid ácido erúcico
erucyl alcohol alcohol erucílico
erythorbic acid ácido eritórbico
erythrin eritrina
erythrite eritrita
erythritol eritritol
erythritol anhydride anhídrido de eritritol
erythrityl eritritilo
erythro eritro

erythroidine eritroidina
erythrol eritrol
erythromycin eritromicina
erythrose eritrosa
erythrosiderite eritrosiderita
erythrosin eritrosina
Eschka mixture mezcla de Eschka
Eschweiler-Clarke modification
    modificación de Eschweiler-Clarke
eserine eserina
essential esencial
essential oil aceite esencial
essential salt sal esencial
ester éster
ester gum goma de éster
ester interchange cambio de éster
ester number número de éster
esterase esterasa
esterification esterificación
esterification law ley de esterificación
estersil estersil
estradiol estradiol
estragole estragol
estrane estrano
estriol estriol
estrogen estrógeno
estrone estrona
eta eta
Étard reaction reacción de Étard
Étard salt sal de Étard
ethacetic acid ácido etacético
ethal etal
ethaldehyde etaldehído
ethamine etamina
ethanal etanal
ethanaldehyde etanaldehído
ethanamide etanamida
ethanamidine etanamidina
ethane etano
ethane hydrate hidrato de etano
ethanedial etanodial
ethanediamide etanodiamida
ethanediamine etanodiamina
ethanedinitrile etanodinitrilo
ethanedioic acid ácido etanodioico
ethanedithiol etanoditiol
ethanenitrile etanonitrilo
ethanesulfonic acid ácido etanosulfónico
ethanethial etanotial

ethanethiol etanotiol
ethanethiolic acid ácido etanotiólico
ethanite etanita
ethanoate etanoato
ethanoic acid ácido etanoico
ethanoic anhydride anhídrido etanoico
ethanol etanol
ethanol formamide formamida de etanol
ethanol hydrazine hidrazina de etanol
ethanolamine etanolamina
ethanolate etanolato
ethanolurea etanolurea
ethanoyl etanoilo
ethanoyl chloride cloruro de etanoilo
ethanoylation etanoilación
ethanyl etanilo
ethanyl alcohol alcohol etanílico
ethene eteno
ethene polymer polímero de eteno
ethenium etenio
ethenol etenol
ethenone etenona
ethenyl etenilo
ether éter
ethereal etéreo
etheric acid ácido etérico
etherification eterificación
ethide etida
ethidine etidina
ethine etina
ethinoic acid ácido etinoico
ethinyl etinilo
ethinylation etinilación
ethion etión
ethionamide etionamida
ethisterone etisterona
ethohexadiol etohexadiol
etholide etolida
ethoxalyl etoxalilo
ethoxide etóxido
ethoxy etoxi
ethoxyacetanilide etoxiacetanilida
ethoxyacetone etoxiacetona
ethoxyaniline etoxianilina
ethoxycarbonyl etoxicarbonilo
ethoxycarbonyl isothiocyanate
    isotiocianato de etoxicarbonilo
ethoxyethanol etoxietanol
ethoxyl etoxilo

ethoxymethylenemalononitrile
  etoximetilenomalononitrilo

ethoxyphenol etoxifenol

ethoxytriglycol etoxitriglicol

ethyl etilo

ethyl abietate abietato de etilo

ethyl acetate acetato de etilo

ethyl acetoacetate acetoacetato de etilo

ethyl acrylate acrilato de etilo

ethyl alcohol alcohol etílico

ethyl amyl ketone etilamilcetona

ethyl benzoate benzoato de etilo

ethyl benzoylacetate benzoilacetato de
  etilo

ethyl bromide bromuro de etilo

ethyl bromoacetate bromoacetato de
  etilo

ethyl butanoate butanoato de etilo

ethyl butyrate butirato de etilo

ethyl caprate caprato de etilo

ethyl caproate caproato de etilo

ethyl caprylate caprilato de etilo

ethyl carbamate carbamato de etilo

ethyl carbazate carbazato de etilo

ethyl carbonate carbonato de etilo

ethyl chloride cloruro de etilo

ethyl chloroacetal cloroacetal de etilo

ethyl chloroacetate cloroacetato de etilo

ethyl chloroacetoacetate
  cloroacetoacetato de etilo

ethyl chlorocarbonate clorocarbonato de
  etilo

ethyl chloroformate cloroformiato de
  etilo

ethyl chloropropionate cloropropionato
  de etilo

ethyl chlorosulfonate clorosulfonato de
  etilo

ethyl cinnamate cinamato de etilo

ethyl citrate citrato de etilo

ethyl crotonate crotonato de etilo

ethyl cyanate cianato de etilo

ethyl cyanide cianuro de etilo

ethyl cyanoacetate cianoacetato de etilo

ethyl cyanoethanoate cianoetanoato de
  etilo

ethyl decanoate decanoato de etilo

ethyl diazoacetate diazoacetato de etilo

ethyl diazoethanoate diazoetanoato de
  etilo

ethyl dibromoacetate dibromoacetato de
  etilo

ethyl dichloroacetate dicloroacetato de
  etilo

ethyl ethanoate etanoato de etilo

ethyl ether éter etílico

ethyl ferrocyanate ferrocianato de etilo

ethyl formamide formamida de etilo

ethyl formate formiato de etilo

ethyl furoate furoato de etilo

ethyl glycerate glicerato de etilo

ethyl glycolate glicolato de etilo

ethyl heptanoate heptanoato de etilo

ethyl hexanoate hexanoato de etilo

ethyl hydrosulfide hidrosulfuro de etilo

ethyl hydroxyethylcellulose
  hidroxietilcelulosa de etilo

ethyl iodide yoduro de etilo

ethyl iodoacetate yodoacetato de etilo

ethyl isobutyrate isobutirato de etilo

ethyl isocyanate isocianato de etilo

ethyl isocyanide isocianuro de etilo

ethyl isophthalate isoftalato de etilo

ethyl isovalerate isovalerato de etilo

ethyl lactate lactato de etilo

ethyl malate malato de etilo

ethyl malonate malonato de etilo

ethyl mercaptan etilmercaptano

ethyl methacrylate metacrilato de etilo

ethyl methanoate metanoato de etilo

ethyl methyl ether éter etilmetílico

ethyl methyl ketone etilmetilcetona

ethyl monotartrate monotartrato de etilo

ethyl myristate miristato de etilo

ethyl nitrate nitrato de etilo

ethyl nitrite nitrito de etilo

ethyl nonanoate nonanoato de etilo

ethyl octanoate octanoato de etilo

ethyl octoate octoato de etilo

ethyl oleate oleato de etilo

ethyl orthophosphate ortofosfato de etilo

ethyl orthopropionate ortopropionato de
  etilo

ethyl orthosilicate ortosilicato de etilo

ethyl oxalate oxalato de etilo

ethyl oxide óxido de etilo

ethyl palmitate palmitato de etilo

ethyl perchlorate perclorato de etilo

ethyl pelargonate pelargonato de etilo
ethyl peroxide peróxido de etilo
ethyl phenyl ketone etilfenilcetona
ethyl phenylacetate fenilacetato de etilo
ethyl phosphate fosfato de etilo
ethyl phosphine fosfina de etilo
ethyl phthalate ftalato de etilo
ethyl propiolate propiolato de etilo
ethyl propionate propionato de etilo
ethyl pyrophosphate pirofosfato de etilo
ethyl pyruvate piruvato de etilo
ethyl salicylate salicilato de etilo
ethyl silicate silicato de etilo
ethyl succinate succinato de etilo
ethyl sulfate sulfato de etilo
ethyl sulfhydrate sulfhidrato de etilo
ethyl sulfide sulfuro de etilo
ethyl thiocyanate tiocianato de etilo
ethyl thioethane tioetano de etilo
ethyl thioethanol tioetanol de etilo
ethyl triacetylgallate triacetilgalato de
    etilo
ethyl valerate valerato de etilo
ethyl vanillate vanilato de etilo
ethyl vanillin vanilina de etilo
ethyl vinyl ether etilviniléter
ethylacetamide etilacetamida
ethylacetanilide etilacetanilida
ethylacetic acid ácido etilacético
ethylacetone etilacetona
ethylacetylene etilacetileno
ethylaluminum dichloride dicloruro de
    etilaluminio
ethylaluminum sesquichloride
    sesquicloruro de etilaluminio
ethylamine etilamina
ethylamine hydrobromide bromhidrato
    de etilamina
ethylamino etilamino
ethylaminoacetate etilaminoacetato
ethylaniline etilanilina
ethylanthraquinone etilantraquinona
ethylarsenious oxide óxido etilarsenioso
ethylate etilato
ethylation etilación
ethylaziridine etilaziridina
ethylbenzene etilbenceno
ethylbenzenesulfonic acid ácido
    etilbencenosulfónico

ethylbenzyl alcohol alcohol etilbencílico
ethylbenzyl chloride cloruro de
    etilbencilo
ethylbenzylaniline etilbencilanilina
ethylbutanol etilbutanol
ethylbutene etilbuteno
ethylbutyl etilbutilo
ethylbutyl acetate acetato de etilbutilo
ethylbutyl alcohol alcohol etilbutílico
ethylbutyl carbonate carbonato de
    etilbutilo
ethylbutyl ketone etilbutilcetona
ethylbutyl silicate silicato de etilbutilo
ethylbutylamine etilbutilamina
ethylbutyraldehyde etilbutiraldehído
ethylbutyric acid ácido etilbutírico
ethylcarbazole etilcarbazol
ethylcarbinol etilcarbinol
ethylcellulose etilcelulosa
ethylcyclohexane etilciclohexano
ethylcyclohexylamine etilciclohexilamina
ethylcyclopentane etilciclopentano
ethylcyclopentanone etilciclopentanona
ethyldichlorosilane etildiclorosilano
ethyldiethanolamine etildietanolamina
ethylene etileno
ethylene benzoate benzoato de etileno
ethylene bromide bromuro de etileno
ethylene bromohydrin bromohidrina de
    etileno
ethylene carbonate carbonato de etileno
ethylene chloride cloruro de etileno
ethylene chlorobromide clorobromuro
    de etileno
ethylene chlorohydrin clorohidrina de
    etileno
ethylene cyanide cianuro de etileno
ethylene cyanohydrin cianhidrina de
    etileno
ethylene diacetate diacetato de etileno
ethylene dibromide dibromuro de etileno
ethylene dichloride dicloruro de etileno
ethylene dicyanide dicianuro de etileno
ethylene diiodide diyoduro de etileno
ethylene dinitrate dinitrato de etileno
ethylene glycol etilenglicol
ethylene glycol diacetate diacetato de
    etilenglicol
ethylene glycol dibutyrate dibutirato de

etilenglicol

ethylene glycol diformate diformiato de
   etilenglicol

ethylene glycol dinitrate dinitrato de
   etilenglicol

ethylene glycol monoacetate
   monoacetato de etilenglicol

ethylene glycol monobutyl ether éter
   monobutílico de etilenglicol

ethylene glycol monoethyl ether éter
   monoetílico de etilenglicol

ethylene glycol monomethyl ether éter
   monometílico de etilenglicol

ethylene hydrate hidrato de etileno

ethylene hydride hidruro de etileno

ethylene nitrate nitrato de etileno

ethylene oxide óxido de etileno

ethylene-propylene etileno-propileno

ethylene resin resina etilénica

ethylene tetrabromide tetrabromuro de
   etileno

ethylene tetrachloride tetracloruro de
   etileno

ethylene thiourea tiourea de etileno

ethylene trichloride tricloruro de etileno

ethylene urea urea de etileno

ethyleneamine etilenamina

ethylenediamine etilendiamina

ethylenediaminetetraacetic acid ácido
   etilendiaminotetraacético

ethylenediaminetetraacetonitrile
   etilendiaminotetraacetonitrilo

ethylenedithiol etilenditiol

ethyleneimine etilenimina

ethylenenaphthalene etilenonaftaleno

ethylenesulfonic acid ácido
   etilensulfónico

ethylethanolamine etiletanolamina

ethylethylene etiletileno

ethylethyleneimine etiletilenimina

ethylfluorosulfonate etilfluorosulfonato

ethylfuran etilfurano

ethylheptane etilheptano

ethylhexaldehyde etilhexaldehído

ethylhexanal etilhexanal

ethylhexanediol etilhexanodiol

ethylhexanol etilhexanol

ethylhexenal etilhexenal

ethylhexoic acid ácido etilhexoico

ethylhexyl etilhexilo

ethylhexyl acetate acetato de etilhexilo

ethylhexyl acrylate acrilato de etilhexilo

ethylhexyl alcohol alcohol etilhexílico

ethylhexyl bromide bromuro de
   etilhexilo

ethylhexyl chloride cloruro de etilhexilo

ethylhexyl cyanoacetate cianoacetato de
   etilhexilo

ethylhexylamine etilhexilamina

ethylhexylaniline etilhexilanilina

ethylhydrazine etilhidrazina

ethylidene etilideno

ethylidene chloride cloruro de etilideno

ethylidene diacetate diacetato de
   etilideno

ethylidene dibromide dibromuro de
   etilideno

ethylidene diiodide diyoduro de etilideno

ethylidyne etilidino

ethylin etilina

ethylisobutylmethane etilisobutilmetano

ethylisohexanol etilisohexanol

ethyllithium etil-litio

ethylmagnesium bromide bromuro de
   etilmagnesio

ethylmagnesium chloride cloruro de
   etilmagnesio

ethylmethylacetylene etilmetilacetileno

ethylmorphine hydrochloride
   clorhidrato de etilmorfina

ethylmorpholine etilmorfolina

ethylnaphthylamine etilnaftilamina

ethyloic etiloico

ethylol etilol

ethylparaben etilparabeno

ethylpentane etilpentano

ethylphenol etilfenol

ethylphosphoric acid ácido etilfosfórico

ethylsulfuric acid ácido etisulfúrico

ethyltrichlorosilane etiltriclorosilano

ethylurethane etiluretano

ethyne etino

ethynyl etinilo

ethynylation etinilación

ethynylestradiol etinilestradiol

ettringite ettringita

eucairite eucairita

eucalyptol eucaliptol

eucarvone eucarvona
euchroite eucroíta
eucryptite eucriptita
eudalene eudaleno
eudialite eudialita
eudiometer eudiómetro
eugenic acid ácido eugénico
eugenol eugenol
eulytin eulitina
eupittonic acid ácido eupitónico
eupyrine eupirina
europium europio
europium chloride cloruro de europio
europium fluoride fluoruro de europio
europium nitrate nitrato de europio
europium oxalate oxalato de europio
europium oxide óxido de europio
europium sulfate sulfato de europio
eutectic eutéctico
eutectic alloy aleación eutéctica
eutectic mixture mezcla eutéctica
eutectic point punto eutéctico
eutectic system sistema eutéctico
eutectic temperature temperatura
   eutéctica
eutectoid eutectoide
eutrophication eutroficación
eutropic series serie eutrópica
eutropy eutropía
euxenite euxenita
evansite evansita
evaporate evaporar
evaporating dish plato de evaporación
evaporation evaporación
evaporation dish plato de evaporación
evaporator evaporador
evaporimeter evaporímetro
evaporite evaporita
even-even nucleus núcleo par-par
even-odd nucleus núcleo par-impar
evolution evolución
evolved gas analysis análisis de gases
   despedidos
exchange intercambio
exchange collision colisión de
   intercambio
exchange reaction reacción de
   intercambio
excimer excímero

exciplex complejo excitado
excitation excitación
excitation curve curva de excitación
excitation function función de excitación
excitation index índice de excitación
excitation spectrum espectro de
   excitación
excite excitar
excited excitado
excited atom átomo excitado
excited ion ion excitado
excited nucleus núcleo excitado
excited state estado excitado
exciter excitador
exciton excitón
exclusion exclusión
exclusion principle principio de
   exclusión
exo exo
exocarbon exocarburo
exocondensation exocondensación
exogenous exógeno
exopeptidase exopeptidasa
exosmosis exósmosis
exothermic exotérmico
exothermic compound compuesto
   exotérmico
exothermic reaction reacción exotérmica
exotic atom átomo exótico
expand expandirse
expander expansor
expansion expansión
experiment experimento
experimental experimental
experimentation experimentación
explode explotar
explosion explosión
explosion spectrum espectro de
   explosión
explosive explosivo
explosive reaction reacción explosiva
explosive substance substancia explosiva
exponential decay desintegración
   exponencial
exponential disintegration
   desintegración exponencial
exposure exposición
expression expresión
extender extensor, diluente

extinction extinción

extinction coefficient coeficiente de extinción

extract extracto

extractant extractante

extraction extracción

extraction apparatus aparato de extracción

extraction coefficient coeficiente de extracción

extractive distillation destilación extractiva

extractor extractor

extranuclear extranuclear

extrapolation extrapolación

extrusion extrusión

exude exudar

eye-wash lavador de ojos

Eyring equation ecuación de Eyring

F

fabianol fabianol
fabulite fabulita
fac fac
face-centered cubic cúbico centrado en las caras
face-centered cubic structure estructura cúbica centrada en las caras
face-centered lattice red centrada en las caras
face-centered structure estructura centrada en las caras
factor factor
factor quantity cantidad de factor
fagarine fagarina
Fahrenheit scale escala Fahrenheit
Fajans rules reglas de Fajans
Fajans-Soddy law ley de Fajans-Soddy
Fament process proceso de Fament
farad farad
faraday faraday
Faraday constant constante de Faraday
Faraday effect efecto de Faraday
Faraday law ley de Faraday
Faraday tube tubo de Faraday
faradiol faradiol
farinose farinosa
farnesene farneseno
farnesol farnesol
Farrar process proceso de Farrar
fast fijo, rápido
fast-atom bombardment bombardeo con átomos rápidos
fast chemical reaction reacción química rápida
fast neutron neutrón rápido
fast-neutron fission fisión por neutrones rápidos
fast-neutron spectrometry espectrometría de neutrones rápidos
fast reaction reacción rápida
fat grasa
fat hardening endurecimiento de grasa
fat soluble soluble en grasa

fat splitting división de grasas
fatty acid ácido graso
fatty alcohol alcohol graso
fatty amine amina grasa
fatty ester éster graso
fatty nitrile nitrilo graso
faujasite faujasita
Favorskii-Babayan synthesis síntesis de Favorskii-Babayan
Favorskii rearrangement reordenamiento de Favorskii
fayalite fayalita
feedback retroalimentación
feeder alimentador
feedstock materia prima para proceso
Fehling reagent reactivo de Fehling
Fehling solution solución de Fehling
Fehling test prueba de Fehling
Feiser solution solución de Feiser
Feist-Benary synthesis síntesis de Feist-Benary
feldspar feldespato
feldspathoid feldespatoide
felite felita
felspar feldespato
felvation felvación
femic fémico
femtochemistry femtoquímica
fen fen
fenchane fencano
fenchanol fencanol
fenchanone fencanona
fenchene fenqueno
fenchol fencol
fencholic acid ácido fencólico
fenchone fencona
fenchoxime fencoxima
fenchyl fenquilo
fenchyl alcohol alcohol fenquílico
fenitrothion fenitrotión
fennel oil aceite de hinojo
fensulfothion fensulfotiona
fentanyl citrate citrato de fentanilo

fenthion fentión
Fenton reaction reacción de Fenton
Fenton reagent reactivo de Fenton
ferberite ferberita
fergusonite fergusonita
ferment fermentar
fermentable fermentable
fermentation fermentación
fermentation tube tubo de fermentación
Fermi constant constante de Fermi
Fermi-Dirac statistics estadística de
    Fermi-Dirac
Fermi level nivel de Fermi
Fermi resonance resonancia de Fermi
Fermi surface superficie de Fermi
Fermi transition transición de Fermi
fermium fermio
Ferrario reaction reacción de Ferrario
ferrate ferrato
ferredoxin ferredoxina
ferri ferri
ferric férrico
ferric acetate acetato férrico
ferric acetylacetonate acetilacetonato
    férrico
ferric ammonium amonio férrico
ferric ammonium alum alumbre
    amónico férrico
ferric ammonium citrate citrato
    amónico férrico
ferric ammonium oxalate oxalato
    amónico férrico
ferric ammonium sulfate sulfato
    amónico férrico
ferric arsenate arseniato férrico
ferric arsenite arsenito férrico
ferric benzoate benzoato férrico
ferric bromide bromuro férrico
ferric chloride cloruro férrico
ferric chromate cromato férrico
ferric citrate citrato férrico
ferric dichromate dicromato férrico
ferric ferricyanide ferricianuro férrico
ferric ferrocyanide ferrocianuro férrico
ferric fluoride fluoruro férrico
ferric fluoroborate fluoroborato férrico
ferric fluorosilicate fluorosilicato férrico
ferric formate formiato férrico
ferric glycerophosphate glicerofosfato

férrico
ferric hydrate hidrato férrico
ferric hydroxide hidróxido férrico
ferric hypophosphite hipofosfito férrico
ferric iodide yoduro férrico
ferric nitrate nitrato férrico
ferric octoate octoato férrico
ferric oleate oleato férrico
ferric oxalate oxalato férrico
ferric oxide óxido férrico
ferric oxide red rojo de óxido férrico
ferric perchlorate perclorato férrico
ferric perchloride percloruro férrico
ferric phenolate fenolato férrico
ferric phosphate fosfato férrico
ferric pyrophosphate pirofosfato férrico
ferric resinate resinato férrico
ferric sesquibromide sesquibromuro
    férrico
ferric sodium oxalate oxalato sódico
    férrico
ferric steareate estearato férrico
ferric subcarbonate subcarbonato férrico
ferric subsulfate subsulfato férrico
ferric sulfate sulfato férrico
ferric sulfide sulfuro férrico
ferric tallate talato férrico
ferric tannate tanato férrico
ferric thiocyanate tiocianato férrico
ferric tribromide tribromuro férrico
ferric trichloride tricloruro férrico
ferric trioxide trióxido férrico
ferric trisulfate trisulfato férrico
ferric vanadate vanadato férrico
ferrichrome ferricromo
ferricyanic acid ácido ferriciánico
ferricyanide ferricianuro
ferriferrous ferriferroso
ferrimagnetism ferrimagnetismo
ferrimanganese ferrimanganeso
ferrimanganic ferrimangánico
ferripotassium ferripotasio
ferripotassium cyanide cianuro
    ferripotásico
ferripyrine ferripirina
ferrisodium ferrisodio
ferrite ferrita
ferritic ferrítico
ferro ferro

ferroalloy ferroaleación
ferroaluminum ferroaluminio
ferroammonium ferroamonio
ferroboron ferroboro
ferrocene ferroceno
ferrocenoyl ferrocenoilo
ferrocenoyl chloride cloruro de
    ferrocenoilo
ferrocenoyl dichloride dicloruro de
    ferrocenoilo
ferrocenylborane ferrocenilborano
ferrocerium ferrocerio
ferrochrome ferrocromo
ferrochromium ferrocromo
ferrocyanic acid ácido ferrociánico
ferrocyanide ferrocianuro
ferrodolomite ferrodolomita
ferrodynamic ferrodinámico
ferroelectric ferroeléctrico
ferroelectricity ferroelectricidad
ferroferric ferroférrico
ferroferricyanide ferroferricianuro
ferroferrocyanide ferroferrocianuro
ferroin ferroína
ferromagnesite ferromagnesita
ferromagnesium ferromagnesio
ferromagnetic ferromagnético
ferromagnetic oxide óxido
    ferromagnético
ferromagnetism ferromagnetismo
ferromanganese ferromanganeso
ferromanganic ferromangánico
ferromanganous ferromanganoso
ferromolybdenum ferromolibdeno
ferron ferrón
ferronickel ferroníquel
ferroniobium ferroniobio
ferrophosphorus ferrofósforo
ferropotassium ferropotasio
ferropyrine ferropirina
ferrosilicon ferrosilicio
ferrosilite ferrosilita
ferrosoferric ferrosoférrico
ferrosoferric oxide óxido ferrosoférrico
ferrotitanium ferrotitanio
ferrotungsten ferrotungsteno
ferrous ferroso
ferrous acetate acetato ferroso
ferrous ammonium amonio ferroso

ferrous ammonium sulfate sulfato
    amónico ferroso
ferrous arsenate arseniato ferroso
ferrous bromide bromuro ferroso
ferrous carbonate carbonato ferroso
ferrous chloride cloruro ferroso
ferrous ferricyanide ferricianuro ferroso
ferrous ferrite ferrita ferrosa
ferrous ferrocyanide ferrocianuro
    ferroso
ferrous fluoride fluoruro ferroso
ferrous fluorosilicate fluorosilicato
    ferroso
ferrous fumarate fumarato ferroso
ferrous gluconate gluconato ferroso
ferrous hydroxide hidróxido ferroso
ferrous hypophosphite hipofosfito
    ferroso
ferrous hyposulfate hiposulfato ferroso
ferrous iodide yoduro ferroso
ferrous lactate lactato ferroso
ferrous naphthenate naftenato ferroso
ferrous nitrate nitrato ferroso
ferrous octoate octoato ferroso
ferrous oxalate oxalato ferroso
ferrous oxide óxido ferroso
ferrous perchlorate perclorato ferroso
ferrous phosphate fosfato ferroso
ferrous phosphide fosfuro ferroso
ferrous selenide seleniuro ferroso
ferrous sulfate sulfato ferroso
ferrous sulfide sulfuro ferroso
ferrous tantalate tantalato ferroso
ferrous tartrate tartrato ferroso
ferrous thiocyanate tiocianato ferroso
ferrous titanate titanato ferroso
ferrous tungstate tungstato ferroso
ferrovanadium ferrovanadio
ferroverdin ferroverdina
ferroxyl ferroxilo
ferrozirconium ferrozirconio
ferruginous ferruginoso
ferrum ferrum
fertile material material fértil
fertilizer fertilizante
ferulaldehyde ferulaldehído
ferulene feruleno
ferulic acid ácido ferúlico
fervanite fervanita

Feulgen reaction reacción de Feulgen
fiber fibra
fiber-optic fibroóptico
fiberglass fibra de vidrio
fibrid fibrido
fibril fibrilla
fibrin fibrina
fibrinogen fibrinógeno
fibroin fibroína
fibrolite fibrolita
fichtelite fictelita
ficin ficina
Fick diffusion law ley de difusión de Fick
fictile fíctil
field campo
field-emission microscope microscopio de emisión de campo
field intensity intensidad de campo
field-ion emission emisión iónica de campo
field-ion microscope microscopio iónico de campo
field ionization ionización de campo
field strength fuerza de campo
filament filamento
filicic acid ácido filícico
filixic acid ácido filíxico
filler relleno
film película
film badge dosímetro de película
filter filtro
filter aid facilitador de filtración
filter alum alumbre de filtro
filter bag bolsa de filtración
filter cone cono de filtración
filter crucible crisol de filtración
filter cylinder cilindro de filtración
filter discrimination discriminación de filtro
filter effectiveness efectividad de filtro
filter flask matraz de filtración
filter funnel embudo de filtración
filter medium medio para filtración
filter paper papel de filtro
filter photometry fotometría con filtro
filter pump bomba para filtración
filter sand arena de filtración
filter selectivity selectividad de filtro

filter spectrophotometer espectrofotómetro de filtro
filter tube tubo de filtro
filtered filtrado
filtering filtración
filtering flask matraz de filtración
filtrate filtrado
filtration filtración
filtration under pressure filtración bajo presión
fine chemical producto químico fino
fine structure estructura fina
fineness pureza, ley
finishing compound compuesto de acabado
finishing material material de acabado
Finkelstein reaction reacción de Finkelstein
fire alarm alarma de fuego
fire detector detector de fuego
fire extinguisher extintor de fuego
fire-extinguisher system sistema extintor de fuego
fire point punto de llama
fire protection protección contra fuego
fire resistance resistencia al fuego
fire resistant resistente al fuego
fireclay arcilla refractaria
fireproof a prueba de fuego
fireproof material material a prueba de fuego
first law of thermodynamics primera ley de termodinámica
first-order reaction reacción de primer orden
first-order spectrum espectro de primer orden
Fischer-Hepp rearrangement reordenamiento de Fischer-Hepp
Fischer indole synthesis síntesis de indoles de Fischer
Fischer oxazole synthesis síntesis de oxazoles de Fischer
Fischer phenylhydrazine synthesis síntesis de fenilhidrazinas de Fischer
Fischer polypeptide synthesis síntesis de polipéptidos de Fischer
Fischer projection proyección de Fischer
Fischer reagent reactivo de Fischer

Fischer salt sal de Fischer
Fischer solution solución de Fischer
Fischer-Tropsch process proceso de
    Fischer-Tropsch
Fischer-Tropsch reaction reacción de
    Fischer-Tropsch
Fischer-Tropsch synthesis síntesis de
    Fischer-Tropsch
fisetin fisetina
fish oil aceite de pescado
fissile físil
fissiochemistry fisioquímica
fission fisión
fission chamber cámara de fisión
fission cross-section sección transversal
    de fisión
fission energy energía de fisión
fission neutron neutrón de fisión
fission product producto de fisión
fission rate velocidad de fisión
fission spectrum espectro de fisión
fission threshold umbral de fisión
fission yield rendimiento de fisión
fissionable fisionable
Fittig reaction reacción de Fittig
Fittig synthesis síntesis de Fittig
fix fijar
fixation fijación
fixative fijativo
fixed carbon carbono fijo
fixed oil aceite fijo
fixed proportion proporción fija
fixing fijación
fixing agent agente fijador
flagstaffite flagstafita
flame llama
flame alarm alarma de llama
flame control control de llama
flame cutting oxicorte
flame detector detector de llama
flame-emission spectroscopy
    espectroscopia de emisión de llama
flame hardening endurecimiento por
    llama
flame-ionization detector detector de
    ionización por llama
flame photometer fotómetro de llama
flame photometry fotometría de llama
flame reaction reacción con llama

flame-resistant ignífugo
flame-retarding retardante de llama
flame spectrometer espectrómetro de
    llama
flame spectrometry espectrometría de
    llama
flame spectrophotometer
    espectrofotómetro de llama
flame spectrophotometry
    espectrofotometría de llama
flame spectrum espectro de llama
flame test prueba con llama
flameproof antideflagrante,
    incombustible
flameproof material material
    antideflagrante, material
    incombustible
flammability inflamabilidad
flammability limit límite de
    inflamabilidad
flammable inflamable
flammable material material inflamable
flammable substance substancia
    inflamable
flash destello, llama súbita luminosa,
    rebaba
flash distillation destilación instantánea
flash drying secado instantáneo
flash photolysis fotólisis instantánea
flash point punto de inflamabilidad
flash spectroscopy espectroscopia de
    destello
flask matraz
flat-bottom flask matraz de base plana
flavaniline flavanilina
flavanol flavanol
flavanone flavanona
flavazine flavazina
flavianic acid ácido flaviánico
flavin flavina
flavine flavina
flavoglaucin flavoglaucina
flavol flavol
flavone flavona
flavonoid flavonoide
flavonol flavonol
flavopannin flavopanina
flavophenine flavofenina
flavoprotein flavoproteína

flavopurpurin flavopurpurina
flavor sabor
flavoxanthin flavoxantina
flavylium flavilio
flaxseed oil aceite de linaza
Fleming tube tubo de Fleming
Fletcher furnace horno de Fletcher
flint pedernal
floatation separación por chorro de aire
floc masa flocosa
flocculant floculante
flocculate floculado
flocculating agent agente floculante
flocculation floculación
flocculent floculento
Flood equation ecuación de Flood
Flood reaction reacción de Flood
Flory temperature temperatura de Flory
flotation flotación
flotation activator activador de flotación
flotation agent agente de flotación
flow flujo
flow diagram diagrama de flujo
flower flor
flox flox
flucloxacillin flucloxacilina
fluctuation fluctuación
flue humero
fluellite fluellita
fluid fluido
fluidity fluidez
fluidization fluidización
fluo fluo
fluoboric acid ácido fluobórico
fluocerite fluocerita
fluoflavine fluoflavina
fluohydric acid ácido fluohídrico
fluor fluor
fluoracetamide fluoracetamida
fluoran fluorán
fluorandiol fluorandiol
fluoranthene fluoranteno
fluoranthraquinone fluorantraquinona
fluorapatite fluorapatita
fluoration fluoración
fluorene fluoreno
fluorenic acid ácido fluorénico
fluorenol fluorenol
fluorenone fluorenona

fluorenyl fluorenilo
fluorenylamine fluorenilamina
fluorenylidene fluorenilideno
fluorescein fluoresceína
fluorescence fluorescencia
fluorescence analysis análisis por fluorescencia
fluorescence efficiency eficiencia de fluorescencia
fluorescence indicator indicador de fluorescencia
fluorescence microscopy microscopia de fluorescencia
fluorescence spectroscopy espectroscopia de fluorescencia
fluorescence yield rendimiento de fluorescencia
fluorescent fluorescente
fluorescent dye colorante fluorescente
fluorescent pigment pigmento fluorescente
fluorescent radiation radiación fluorescente
fluorescin fluorescina
fluorgermanate fluorgermanato
fluorgermanic acid ácido fluorgermánico
fluoridation fluorización
fluoride fluoruro
fluorimeter fluorímetro
fluorimetry fluorimetría
fluorination fluorización
fluorine flúor
fluorine cyanide cianuro de flúor
fluorine halide haluro de flúor
fluorine hydride hidruro de flúor
fluorine nitrate nitrato de flúor
fluorine oxide óxido de flúor
fluorine polymer polímero con flúor
fluorinion fluorinión
fluoriodine fluoryodo
fluorion fluorión
fluorite fluorita
fluorization fluorización
fluoro fluoro
fluoroacetamide fluoroacetamida
fluoroacetanilide fluoroacetanilida
fluoroacetate fluoroacetato
fluoroacetic acid ácido fluoroacético
fluoroacetophenone fluoroacetofenona

fluoroalkane fluoroalcano
fluoroaniline fluoroanilina
fluoroapatite fluoroapatita
fluorobenzene fluorobenceno
fluoroborate fluoroborato
fluoroboric acid ácido fluorobórico
fluorocarbon fluorocarburo
fluorocarbon fiber fibra de
    fluorocarburo
fluorocarbon polymer polímero de
    fluorocarburo
fluorocarbon resin resina de
    fluorocarburo
fluorochemical fluoroquímico
fluorochromate fluorocromato
fluorodichloromethane
    fluorodiclorometano
fluoroelastomer fluoroelastómero
fluoroethanoic acid ácido fluoroetanoico
fluoroform fluoroformo
fluorogen fluorógeno
fluorohydrocarbon fluorohidrocarburo
fluoroiodate fluoroyodato
fluorometer fluorómetro
fluoromethane fluorometano
fluorometholone fluorometolona
fluorometric analysis análisis
    fluorométrico
fluorometry fluorometría
fluorone fluorona
fluoronium fluoronio
fluorophenol fluorofenol
fluorophore fluoróforo
fluorophosphonate fluorofosfonato
fluorophosphoric acid ácido
    fluorofosfórico
fluorophotometer fluorofotómetro
fluoroplumbic acid ácido fluoroplúmbico
fluoropolymer fluoropolímero
fluoroscope fluoroscopio
fluorosilicate fluorosilicato
fluorosilicic acid ácido fluorosilícico
fluorosulfonic acid ácido fluorosulfónico
fluorosulfuric acid ácido fluorosulfúrico
fluorothene fluoroteno
fluorotrichloromethane
    fluorotriclorometano
fluorspar espato flúor
fluoryl fluorilo

fluorylidine fluorilidina
fluosilicate fluosilicato
fluosilicic acid ácido fluosilícico
fluosulfonic acid ácido fluosulfónico
fluoxymesterone fluoximesterona
fluted filter paper papel de filtro
    plegado
flux flujo, fundente
fly ash ceniza de humero
foam espuma
fog niebla
foil hojuela, hoja
Fokker-Planck equation ecuación de
    Fokker-Planck
folacin folacina
folic acid ácido fólico
Folin apparatus aparato de Folin
Folin solution solución de Folin
folinic acid ácido folínico
fomometer fomómetro
fongisterol fongisterol
food additive aditivo de alimento
food color color de alimento
foots sedimentos de aceite
forbidden band banda prohibida
forbidden energy band banda de energía
    prohibida
forbidden line línea prohibida
forbidden transition transición prohibida
force constant constante de fuerza
forceps pinzas
forensic analysis análisis forense
forensic chemistry química forense
formal formal
formal charge carga formal
formaldehyde formaldehído
formaldehyde aniline anilina de
    formaldehído
formaldoxime formaldoxima
formalin formalina
formality formalidad
formamide formamida
formamidine formamidina
formamido formamido
formamine formamina
formamyl formamilo
Formanek indicator indicador de
    Formanek
formanilide formanilida

formaniline formanilina
formate formiato, formato
formation formación
formation constant constante de
    formación
formazan formazán
formazyl formacilo
formazyl hydride hidruro de formacilo
formiate formiato
formic acid ácido fórmico
formic aldehyde aldehído fórmico
formic ether éter fórmico
formic nitrile nitrilo fórmico
formimidoyl formimidoilo
formine formina
formohydrazide formohidrazida
formol formol
formolite formolita
formonitrile formonitrilo
formonitrolic acid ácido formonitrólico
formothion formotión
formoxime formoxima
formoxy formoxi
formoxyl formoxilo
formoxyl hydride hidruro de formoxilo
formoyl formoilo
formula fórmula
formula weight peso de fórmula, peso
    molecular
formulation formulación
formyl formilo
formyl bromide bromuro de formilo
formyl chloride cloruro de formilo
formyl fluoride fluoruro de formilo
formyl trichloride tricloruro de formilo
formyl triiodide triyoduro de formilo
formylacetic acid ácido formilacético
formylamide formilamida
formylamine formilamina
formylamino formilamino
formylation formilación
formylene formileno
formylhydrazine formilhidrazina
formylic acid ácido formílico
formyloxy formiloxi
formylpiperidine formilpiperidina
Forster reaction reacción de Forster
forsterite forsterita
fosfomycin fosfomicina

foshagite foshagita
fossil fósil
fossil fuel combustible fósil
fostion fostión
Foulger test prueba de Foulger
Fourier transform transformación de
    Fourier
Fourier transform infrared infrarrojo
    con transformación de Fourier
fowlerite fowlerita
foyaite foyaíta
fraction fracción
fractional fraccionario
fractional combustion combustión
    fraccionaria
fractional condensation condensación
    fraccionaria
fractional condensation tube tubo de
    condensación fraccionaria
fractional crystallization cristalización
    fraccionaria
fractional distillation destilación
    fraccionaria
fractional distillation tube tubo de
    destilación fraccionaria
fractional expression expresión
    fraccionaria
fractional filtration filtración
    fraccionaria
fractional precipitation precipitación
    fraccionaria
fractionating fraccionante
fractionating column columna
    fraccionante
fractionation fraccionación
fracture fractura
fragility fragilidad
fragrance fragancia
Franchimont reaction reacción de
    Franchimont
francium francio
Franck-Condon principle principio de
    Franck-Condon
franckeite franqueíta
francolite francolita
frangulic acid ácido frangúlico
frangulin frangulina
frangulinic acid ácido frangulínico
Frankland-Duppa reaction reacción de

Frankland-Duppa

Frankland method método de Frankland

Frankland notation notación de Frankland

Frankland reaction reacción de Frankland

Frankland synthesis síntesis de Frankland

franklinite franklinita

Frary metal metal de Frary

Frasch process proceso de Frasch

Fraude reagent reactivo de Fraude

fraunhofer fraunhofer

Fraunhofer lines líneas de Fraunhofer

Fraunhofer spectrum espectro de Fraunhofer

fraxetin fraxetina

fraxin fraxina

fraxinol fraxinol

free atom átomo libre

free charge carga libre

free electron electrón libre

free energy energía libre

free ion ion libre

free molecule molécula libre

free path recorrido libre

free radical radical libre

free sulfur azufre libre

freeze congelar

freeze drying liofilización

freezing congelación

freezing mixture mezcla para congelación

freezing point punto de congelación

freezing-point apparatus aparato de punto de congelación

freezing-point depression depresión de punto de congelación

freezing-point diagram diagrama de puntos de congelación

freezing-point elevation elevación de punto de congelación

freezing-point range intervalo de puntos de congelación

freezing-point reduction reducción de punto de congelación

freibergite freibergita

freierslebenite freierslebenita

Fremy salt sal de Fremy

Frenkel defect defecto de Frenkel

Frenkel-Kontorowa model modelo de Frenkel-Kontorowa

frequency frecuencia

Fresenius desiccator desecador de Fresenius

Freund acid ácido de Freund

Freund method método de Freund

Freund synthesis síntesis de Freund

Freundlich isotherm isoterma de Freundlich

freyalite freyalita

friction fricción

Friedel-Crafts condensation condensación de Friedel-Crafts

Friedel-Crafts reaction reacción de Friedel-Crafts

Friedlander synthesis síntesis de Friedlander

Friedrichs condenser condensador de Friedrichs

Fries rearrangement reordenamiento de Fries

Fries rule regla de Fries

frigorific frigorífico

frigorific mixture mezcla frigorífica

frit frita

fritting fritado

Froehde reagent reactivo de Froehde

frontier orbital orbital fronterizo

froth espuma

fructofuranosidase fructofuranosidasa

fructose fructosa

fructose diphosphate difosfato de fructosa

fructoside fructosida

fuberidazole fuberidazol

fuchsin fucsina

fuchsite fuchsita

fucitol fucitol

fucose fucosa

fucosite fucosita

fucusine fucusina

fucusol fucusol

fuel combustible

fuel cell pila de combustible

fuel element elemento combustible

fuel oil aceite combustible

fugacity fugacidad

Fujimoto-Belleau reaction reacción de
    Fujimoto-Belleau
Fujiwara reaction reacción de Fujiwara
fulgenic acid ácido fulgénico
fulguration fulguración
fulgurator fulgurador
fulgurite fulgurita
fuller's earth tierra de batán
fullerene fulereno
fullerite fulerita
fulminate fulminato
fulminic acid ácido fulmínico
fulminuric acid ácido fulminúrico
fulvalene fulvaleno
fulvene fulveno
fumagillin fumagilina
fumaramic acid ácido fumarámico
fumaramide fumaramida
fumarate fumarato
fumarhydrazide fumarhidrazida
fumaric acid ácido fumárico
fumaroyl fumaroilo
fumaroyl chloride cloruro de fumaroilo
fumaryl fumarilo
fumaryl chloride cloruro de fumarilo
fume humo
fume cupboard alacena protectora contra
    humo, campana de humos
fume hood alacena protectora contra
    humo, campana de humos
fumigant fumigante
fuming humeante, fumante
fuming nitric acid ácido nítrico fumante
fuming sulfuric acid ácido sulfúrico
    fumante
functional group grupo funcional
fundamental fundamental
fundamental constant constante
    fundamental
fundamental particle partícula
    fundamental
fundamental research investigación
    fundamental
fundamental series serie fundamental
fundamental unit unidad fundamental
fungicide fungicida
fungus hongo
funnel embudo
funnel tube tubo con embudo

furacrylic acid ácido furacrílico
fural fural
furaldehyde furaldehído
furamide furamida
furan furano
furancarbinol furancarbinol
furancarboxylic acid ácido
    furancarboxílico
furandione furandiona
furanmethanethiol furanmetanotiol
furanmethylamine furanmetilamina
furano furano
furanose furanosa
furanoside furanosida
furanylmethyl furanilmetilo
furazan furazán
furazanyl furazanilo
furazolidone furazolidona
furfural furfural
furfuraldehyde furfuraldehído
furfuramide furfuramida
furfuran furfurano
furfurol furfurol
furfuryl furfurilo
furfuryl acetate acetato de furfurilo
furfuryl alcohol alcohol furfurílico
furfurylamide furfurilamida
furfurylamine furfurilamina
furfurylidene furfurilideno
furil furil
furildioxime furildioxima
furnace horno, crisol
furnace black negro de horno
furnace oil aceite de horno
furoamide furoamida
furoate furoato
furoic acid ácido furoico
furoin furoína
furol furol
furonic acid ácido furónico
furoyl furoilo
furoyl chloride cloruro de furoilo
furyl furilo
furylcarbinol furilcarbinol
furylidine furilidina
furylmethyl furilmetilo
fuse fundir
fused fundido
fused electrolyte electrólito fundido

fused quartz cuarzo fundido
fused ring anillo fundido
fused salt sal fundida
fused silica sílice fundida
fusel oil aceite de fusel
fusibility fusibilidad
fusible fusible
fusible alloy aleación fusible

fusidic acid ácido fusídico
fusion fusión
fusion energy energía de fusión
fusion heat calor de fusión
fusion point punto de fusion
fusion product producto de fusión
fusion tube tubo de fusión
fusionable fusionable

G

Gabbet solution solución de Gabbet
Gabriel-Colman rearrangement
    reordenamiento de Gabriel-Colman
Gabriel reaction reacción de Gabriel
Gabriel synthesis síntesis de Gabriel
gadinine gadinina
gadoleic acid ácido gadoleico
gadolinite gadolinita
gadolinium gadolinio
gadolinium bromide bromuro de
    gadolinio
gadolinium chloride cloruro de gadolinio
gadolinium fluoride fluoruro de
    gadolinio
gadolinium hydroxide hidróxido de
    gadolinio
gadolinium nitrate nitrato de gadolinio
gadolinium oxalate oxalato de gadolinio
gadolinium oxide óxido de gadolinio
gadolinium sulfate sulfato de gadolinio
gage medidor
gahnite gahnita
gaidic acid ácido gaídico
galactan galactano
galactaric acid ácido galactárico
galactosamine hydrochloride clorhidrato
    de galactosamina
galactose galactosa
galactosidase galactosidasa
galacturonic acid ácido galacturónico
galena galena
galenite galenita
gallacetophenone galacetofenona
gallamine galamina
gallate galato
gallein galeína
gallic gálico
gallic acid ácido gálico
gallic bromide bromuro gálico
gallic chloride cloruro gálico
gallic hydroxide hidróxido gálico
gallic iodide yoduro gálico

gallic oxide óxido gálico
gallin galina
gallium galio
gallium acetate acetato de galio
gallium antimonide antimoniuro de galio
gallium arsenide arseniuro de galio
gallium bromide bromuro de galio
gallium dibromide dibromuro de galio
gallium dichloride dicloruro de galio
gallium diiodide diyoduro de galio
gallium halide haluro de galio
gallium hydride hidruro de galio
gallium hydroxide hidróxido de galio
gallium iodide yoduro de galio
gallium monoxide monóxido de galio
gallium nitrate nitrato de galio
gallium oxide óxido de galio
gallium phosphate fosfato de galio
gallium phosphide fosfuro de galio
gallium sesquioxide sesquióxido de galio
gallium sulfate sulfato de galio
gallium sulfide sulfuro de galio
gallium trichloride tricloruro de galio
gallocyanine galocianina
gallogen galógeno
gallotannic acid ácido galotánico
gallotannin galotanino
gallous galoso
gallous bromide bromuro galoso
gallous oxide óxido galoso
galloyl galoilo
galvanic galvánico
galvanic action acción galvánica
galvanic anode ánodo galvánico
galvanic bath baño galvánico
galvanic cell pila galvánica
galvanic corrosion corrosión galvánica
galvanic current corriente galvánica
galvanic electricity electricidad
    galvánica
galvanic metallization metalización
    galvánica

galvanic series serie galvánica
galvanism galvanismo
galvanization galvanización
galvanize galvanizar
galvanized galvanizado
galvanized iron hierro galvanizado
galvanizing galvanización
galvanomagnetic galvanomagnético
galvanometer galvanómetro
galvanometric galvanométrico
galvanometry galvanometría
galvanoplasty galvanoplastia
gamma gamma
gamma acid ácido gamma
gamma-aminobutyric acid ácido gammaaminobutírico
gamma cross-section sección transversal gamma
gamma decay desintegración gamma
gamma disintegration desintegración gamma
gamma emission emisión gamma
gamma emitter emisor gamma
gamma ferric oxide óxido férrico gamma
gamma-globulin gammaglobulina
gamma iron hierro gamma
gamma radiation radiación gamma
gamma-ray source fuente de rayos gamma
gamma-ray spectrometer espectrómetro de rayos gamma
gamma-ray spectroscopy espectroscopia de rayos gamma
gamma-ray spectrum espectro de rayos gamma
gamma rays rayos gamma
gamma spectroscopy espectroscopia gamma
gamma transition transición gamma
gammexane gamexano
gammil gamilo
Gamow barrier barrera de Gamow
Gamow-Teller interaction interacción de Gamow-Teller
gangue ganga
ganomalite ganomalita
ganomatite ganomatita
ganophyllite ganofilita

garlic oil aceite de ajo
garnet granate
garnierite garnierita
gas gas
gas absorption absorción de gas
gas-admittance valve válvula de admisión de gas
gas admixture ratio razón de mezcla de gases
gas adsorption adsorción de gas
gas analysis análisis de gases
gas-analysis apparatus aparato de análisis de gases
gas black negro de gas
gas cell pila de gas
gas chromatography cromatografía de gases
gas chromatography infrared infrarrojo de cromatografía de gases
gas cleanup limpieza de gas
gas constant constante de gas
gas current corriente de gas
gas cylinder cilíndro de gas
gas density densidad de gas
gas detector detector de gases
gas discharge descarga gaseosa
gas electrode electrodo con gas
gas equation ecuación de gases
gas escape escape de gas
gas filter filtro de gas
gas flow flujo de gas
gas-flow control control de flujo de gas
gas generator generador de gas
gas hydrate hidrato de gas
gas inlet entrada de gas
gas-inlet valve válvula de entrada de gas
gas ionization ionización de gas
gas laws leyes de gases
gas leak fuga de gas
gas leakage fuga de gas
gas-liquid chromatography cromatografía gas-líquido
gas oil gasóleo
gas outlet salida de gas, toma de gas
gas pressure presión de gas
gas regulator regulador de gas
gas sniffer detector de gases
gas solubility solubilidad de gas
gas thermometer termómetro de gas

gaseous gaseoso
gaseous diffusion difusión gaseosa
gaseous discharge descarga gaseosa
gaseous mixture mezcla gaseosa
gaseous state estado gaseoso
gasification gasificación
gasohol gasohol
gasol gasol
gasoline gasolina
gasometric gasométrico
gasometric method método gasométrico
gasproof a prueba de gas
gassing desprendimiento de gas,
    absorción de gas, gaseamiento
Gastaldi synthesis síntesis de Gastaldi
gastric juice jugo gástrico
gastrin gastrina
Gattermann aldehyde synthesis síntesis
    de aldehídos de Gattermann
Gattermann-Koch reaction reacción de
    Gattermann-Koch
Gattermann-Koch synthesis síntesis de
    Gattermann-Koch
Gattermann reaction reacción de
    Gattermann
Gattermann synthesis síntesis de
    Gattermann
gauge medidor
gaultherin gaulterina
Gay-Lussac law ley de Gay-Lussac
Gay-Lussac tower torre de Gay-Lussac
gaylussite gailusita
gehlenite gehlenita
Geiger counter contador de Geiger
Geiger-Müller counter contador de
    Geiger-Müller
Geiger-Müller region región de Geiger-
    Müller
Geiger-Nutall rule regla de Geiger-
    Nutall
geissine geisina
gel gel
gel chromatography cromatografía de
    gel
gel electrophoresis electroforesis de gel
gel filtration filtración de gel
gel-permeation chromatography
    cromatografía de permeación en gel
gelatin gelatina

gelatin filter filtro de gelatina
gelatinization gelatinización
gelatinize gelatinizar
gelatinous gelatinoso
gelignite gelignita
gelling gelificación
gem gema, piedra preciosa
geminal geminal
geminal coupling acoplamiento geminal
gemstone piedra preciosa
genalkaloid genalcaloide
gene gen
general formula fórmula general
generate generar
generation generación
generation time tiempo de generación
generic genérico
generic formula fórmula genérica
generic name nombre genérico
genetic genético
Geneva system sistema de Ginebra
genistin genistina
gentamicin gentamicina
gentiobiose gentiobiosa
gentisic acid ácido gentísico
geochemistry geoquímica
geocronite geocronita
geometric conversion conversión
    geométrica
geometric isomer isómero geométrico
geometrical isomerism isomerismo
    geométrico
geranial geranial
geranialdehyde geranialdehído
geranic acid ácido geránico
geraniol geraniol
geranyl geranilo
geranyl acetate acetato de geranilo
geranyl butyrate butirato de geranilo
geranyl formate formiato de geranilo
geranyl propionate propionato de
    geranilo
Gerard reagent reactivo de Gerard
germanate germanato
germane germano
germanic germánico
germanic chloride cloruro germánico
germanide germaniuro
germanite germanita

germanium germanio
germanium dibromide dibromuro de germanio
germanium dichloride dicloruro de germanio
germanium diiodide diyoduro de germanio
germanium dioxide dióxido de germanio
germanium disulfide disulfuro de germanio
germanium halide haluro de germanio
germanium hydride hidruro de germanio
germanium hydroxide hidróxido de germanio
germanium monoxide monóxido de germanio
germanium oxide óxido de germanio
germanium potassium fluoride fluoruro potásico de germanio
germanium sulfide sulfuro de germanio
germanium telluride telururo de germanio
germanium tetrabromide tetrabromuro de germanio
germanium tetrachloride tetracloruro de germanio
germanium tetrafluoride tetrafluoruro de germanio
germanium tetrahydride tetrahidruro de germanio
germanium tetraiodide tetrayoduro de germanio
germanoformic acid ácido germanofórmico
germanous germanoso
germicidal germicida
germicide germicida
germination germinación
getter rarefactor, depurador de gas
gibberellic acid ácido giberélico
gibberellin giberelina
Gibbs adsorption equation ecuación de adsorción de Gibbs
Gibbs adsorption isotherm isoterma de adsorción de Gibbs
Gibbs-Donnan equilibrium equilibrio de Gibbs-Donnan
Gibbs-Duhem equation ecuación de Gibbs-Duhem

Gibbs free energy energía libre de Gibbs
Gibbs function función de Gibbs
Gibbs-Helmholtz equation ecuación de Gibbs-Helmholtz
Gibbs phase rule regla de fases de Gibbs
Gibbs-Poynting equation ecuación de Gibbs-Poynting
Gibbs rule regla de Gibbs
gibbsite gibbsita
gibrel gibrel
gigantolite gigatolita
Giles flask matraz de Giles
gilpinite gilpinita
gilsonite gilsonita
ginger oil aceite de jengibre
Girard reagent reactivo de Girard
gismondite gismondita
gitonin gitonina
gitoxin gitoxina
glacial glacial
glacial acetic acid ácido acético glacial
glacial ethanoic acid ácido etanoico glacial
gladiolic acid ácido gladiólico
Glaser coupling acoplamiento de Glaser
glaserite glaserita
glass vidrio
glass beads perlas de vidrio
glass-ceramic vidrio cerámico
glass electrode electrodo de vidrio
glass enamel esmalte de vidrio
glass fiber fibra de vidrio
glass stirrer agitador de vidrio
glass stopper tapón de vidrio
glass-transition temperature temperatura de transición de vidrio
glass tube tubo de vidrio
glass tubing tubería de vidrio
glass wool lana de vidrio
Glauber salt sal de Glauber
glauberite glauberita
glaucochroite glaucocroíta
glaucodot glaucodot
glauconite glauconita
glaucophane glaucofana
glaucopicrine glaucopicrina
glaze barniz
gliadin gliadina

globulin globulina
glove box caja con guantes
glow luminiscencia
glucagon glucagón
glucan glucano
glucase glucasa
glucinium glucinio
glucogen glucógeno
gluconate gluconato
gluconic acid ácido glucónico
glucopyranose glucopiranosa
glucosamine glucosamina
glucosazone glucosazona
glucose glucosa
glucose isomerase isomerasa de glucosa
glucose oxidase oxidasa de glucosa
glucosidase glucosidasa
glucoside glucósido
glucosin glucosina
glucuronic acid ácido glucurónico
glucuronidase glucuronidasa
glucuronide glucuronida
glucuronolactone glucuronolactona
glue cola
gluside glusida
glutaconic acid ácido glutacónico
glutamic acid ácido glutámico
glutamine glutamina
glutamoyl glutamoilo
glutamyl glutamilo
glutamylcysteinylglycine
    glutamilcisteinilglicina
glutaraldehyde glutaraldehído
glutaric glutárico
glutaric acid ácido glutárico
glutaric anhydride anhídrido glutárico
glutaronitrile glutaronitrilo
glutathione glutationa
glutelin glutelina
gluten gluten
glutenin glutenina
glutethimide glutetimida
glutinic acid ácido glutínico
glutose glutosa
glycal glical
glyceral glyceral
glyceraldehyde gliceraldehído
glycerate glicerato
glyceric glicérico

glyceric acid ácido glicérico
glyceric aldehyde aldehído glicérico
glyceride glicérido
glycerin glicerina
glycerin carbonate carbonato de
    glicerina
glycerinate glicerinato
glycerine glicerina
glycero glicero
glycerol glicerol, glicerina
glycerol diacetate diacetato de glicerol,
    diacetato de glicerina
glycerol dichlorohydrin diclorohidrina
    de glicerol
glycerol distearate diesterato de glicerol
glycerol monoacetate monoacetato de
    glicerol
glycerol monochlorohydrin
    monoclorohidrina de glicerol
glycerol monolaurate monolaurato de
    glicerol
glycerol monooleate monooleato de
    glicerol
glycerol monostearate monoestearato de
    glicerol
glycerol trinitrate trinitrato de glicerol
glycerol tristearate triestearato de
    glicerol
glycerophosphate glicerofosfato
glycerophosphoric acid ácido
    glicerofosfórico
glycerose glicerosa
glyceroyl gliceroilo
glyceryl glicerilo
glyceryl abietate abietato de glicerilo
glyceryl chloride cloruro de glicerilo
glyceryl diacetate diacetato de glicerilo
glyceryl monoacetate monoacetato de
    glicerilo
glyceryl monolaurate monolaurato de
    glicerilo
glyceryl phthalate ftalato de glicerilo
glyceryl triacetate triacetato de glicerilo
glyceryl trinitrate trinitrato de glicerilo
glyceryl tripalmitate tripalmitato de
    glicerilo
glyceryl tristearate triestearato de
    glicerilo
glycide glicida

glycidic acid ácido glicídico
glycidol glicidol
glycidyl glicidilo
glycidyl ether éter glicidílico
glycine glicina
glycinium glicinio
glyco glico
glycocholic acid ácido glicocólico
glycogen glicógeno, glucógeno
glycogenase glicogenasa
glycogenesis glicogénesis
glycogenic acid ácido glicogénico
glycol glicol
glycol acetate acetato de glicol
glycol bromohydrin bromohidrina de glicol
glycol carbonate carbonato de glicol
glycol dichloride dicloruro de glicol
glycol diformate diformiato de glicol
glycol dimercaptoacetate dimercaptoacetato de glicol
glycol dimethyl ether éter dimetílico de glicol
glycol dinitrate dinitrato de glicol
glycol ester éster glicólico
glycol ether éter glicólico
glycol monoacetate monoacetato de glicol
glycol propionate propionato de glicol
glycol stearate estearato de glicol
glycol sulfhydrate sulfhidrato de glicol
glycolal glicolal
glycolaldehyde glicolaldehído
glycolic acid ácido glicólico
glycolic aldehyde aldehído glicólico
glycolide glicólido
glycolipid glicolípido
glycolonitrile glicolonitrilo
glycoloyl glicoloilo
glycolyl glicolilo
glycolysis glicólisis
glyconic acid ácido glicónico
glyconitrile gliconitrilo
glycoprotein glicoproteína
glycosaminoglycan glicosaminoglicano
glycosidase glicosidasa
glycoside glicósido, glucósido
glycosidic glicosídico
glycosidic bond enlace glicosídico

glycosylation glicosilación
glycyl glicilo
glycyl alcohol alcohol glicílico
glycylglycine glicilglicina
glyme glime
glyoxal glioxal
glyoxalase glioxalasa
glyoxalic acid ácido glioxálico
glyoxaline glioxalina
glyoxime glioxima
glyoxylase glioxilasa
glyoxylate cycle ciclo de glioxilato
glyoxylic acid ácido glioxílico
glyphtal gliftal
gmelinite gmelinita
gneiss gneis
gnoscopine gnoscopina
goethite goetita
goggles gafas de protección
Gogte synthesis síntesis de Gogte
gold oro
gold alloy aleación de oro
gold bromide bromuro de oro
gold bronze bronce de oro
gold chloride cloruro de oro
gold coating revestimiento de oro
gold cyanide cianuro de oro
gold dichloride dicloruro de oro
gold doping dopado con oro
gold foil hoja de oro
gold halide haluro de oro
gold hydroxide hidróxido de oro
gold iodide yoduro de oro
gold leaf hoja de oro, pan de oro
gold monobromide monobromuro de oro
gold monochloride monocloruro de oro
gold monoiodide monoyoduro de oro
gold number número de oro
gold oxide óxido de oro
gold pentafluoride pentafluoruro de oro
gold-plated dorado
gold-plating dorado
gold salt sal de oro
gold sulfide sulfuro de oro
gold tribromide tribromuro de oro
gold trichloride tricloruro de oro
gold tricyanide tricianuro de oro
gold triiodide triyoduro de oro
gold trioxide trióxido de oro

gold trisulfide trisulfuro de oro
Goldschmidt law ley de Goldschmidt
Goldschmidt process proceso de
 Goldschmidt
Goldschmidt reaction reacción de
 Goldschmidt
Gomberg-Bachmann reaction reacción
 de Gomberg-Bachmann
Gomberg reaction reacción de Gomberg
Gooch crucible crisol de Gooch
goslarite goslarita
gossypol gosipol
Göulard extract extracto de Göulard
Gould-Jacobs reaction reacción de
 Gould-Jacobs
Gouy-Chapman model modelo de Gouy-
 Chapman
Gouy layer capa de Gouy
grade grado
gradient gradiente
grading gradación
gradual gradual
graduate cilindro graduado
graduated graduado
graduated buret bureta graduada
graduated cylinder cilindro graduado
graduated pipet pipeta graduada
graduation graduación
Graebe-Ullman reaction reacción de
 Graebe-Ullman
Graebe-Ullman synthesis síntesis de
 Graebe-Ullman
graft copolymer copolímero con injerto
graft polymer polímero con injerto
grafting injerto
Graham law ley de Graham
Graham salt sal de Graham
grain alcohol alcohol de grano
gram gramo
gram-atom átomo-gramo
gram-atomic weight peso atómico-gramo
gram-equivalent equivalente-gramo
gram-equivalent weight peso
 equivalente-gramo
gram-ion ion-gramo
gram-molecular solution solución
 molecular-gramo
gram-molecular volume volumen
 molecular-gramo

gram-molecular weight peso molecular-
 gramo
gram-molecule molécula-gramo
gram-negative gramnegativo
gram-positive grampositivo
gramicidin gramicidina
gramine gramina
granite granito
granular granular
granularity granulosidad
granulation granulación
granule gránulo
granulose granulosa
grape sugar azúcar de uva
graphic formula fórmula gráfica
graphite grafito
graphite fiber fibra de grafito
graphitic acid ácido grafítico
graphitic oxide óxido grafítico
graphitization grafitización
grating rejilla, retículo
gravimetric gravimétrico
gravimetric analysis análisis
 gravimétrico
gravimetry gravimetría
gray gray
grease grasa
green salt sal verde
green vitriol vitriolo verde
greenhouse effect efecto de invernadero
greenockite greenockita
grid rejilla
grid spectrometer espectrómetro de
 rejilla
Griess reaction reacción de Griess
Griess reagent reactivo de Griess
Grignard degradation degradación de
 Grignard
Grignard reaction reacción de Grignard
Grignard reagent reactivo de Grignard
Grignard synthesis síntesis de Grignard
grind moler
grinding molido
griphite grifita
Grob fragmentation reaction reacción
 de fragmentación de Grob
grossularite grosularita
Grotrian diagram diagrama de Grotrian
Grottius-Draper law ley de Grottius-

Draper
ground molido, esmerilado, fundamental
ground glass vidrio esmerilado
ground-glass neck cuello de vidrio
    esmerilado
ground state estado fundamental
group grupo
group of elements grupo de elementos
group properties propiedades de grupo
Grove synthesis síntesis de Grove
grown crystal cristal cultivado
Grundmann synthesis síntesis de
    Grundmann
grunerite grunerita
gryolite griolita
guaiacol guayacol
guaiacol acetate acetato de guayacol
guaiol guaiol
guanamine guanamina
guanase guanasa
guanazyl guanacilo
guanidine guanidina
guanidine carbonate carbonato de
    guanidina
guanidine hydrochloride clorhidrato de
    guanidina
guanidine nitrate nitrato de guanidina
guanidine thiocyanate tiocianato de

guanidina
guanidinium guanidinio
guanidino guanidino
guanido guanido
guanine guanina
guano guano
guanosine guanosina
guanosine monophosphate monofosfato
    de guanosina
guanylic acid ácido guanílico
guanylurea sulfate sulfato de guanilurea
guar gum goma guar
Guareschi-Thorpe condensation
    condensación de Guareschi-Thorpe
Guerbet reaction reacción de Guerbet
Guldberg rule regla de Guldberg
gum acacia goma de acacia
gum arabic goma arábiga
gun metal bronce de cañón
guncotton algodón polvora, pólvora de
    algodón
gunpowder pólvora
Günzberg reagent reactivo de Günzberg
gutta-percha gutapercha
gynocardine ginocardina
gypsum yeso
gyrolite girolita
gyromagnetic ratio razón giromagnética

H

Haber process proceso de Haber
habit hábito
hafnium hafnio
hafnium boride boruro de hafnio
hafnium carbide carburo de hafnio
hafnium disulfide disulfuro de hafnio
hafnium hydroxide hidróxido de hafnio
hafnium nitride nitruro de hafnio
hafnium oxide óxido de hafnio
hafnium sulfate sulfato de hafnio
Haggenmacher equation ecuación de Haggenmacher
halazone halazona
half-cell semicelda, semipila
half-life media vida
half-reaction media reacción
halide haluro
halite halita
Hall effect efecto de Hall
Hall process proceso de Hall
Haller-Bauer reaction reacción de Haller-Bauer
halloysite halloysita
hallucinogen alucinógeno
halo halo
haloalkane haloalcano
haloamine haloamina
halocarbon halocarburo
halocarbon plastic plástico de halocarburo
halocarbon resin resina de halocarburo
halochromism halocromismo
haloform haloformo
halogen halógeno
halogen counter contador con halógeno
halogen lamp lámpara de halógeno
halogenate halogenar
halogenated halogenado
halogenating agent agente halogenante
halogenation halogenación
halogenide halogeniuro
halogenous halógeno
halohydrin halohidrina

haloid haloide
halonium halonio
haloperidol haloperidol
halothane halotano
Halphen reagent rectivo de Halphen
hamartite hamartita
Hammett equation ecuación de Hammett
Hammick-Illingworth rules reglas de Hammick-Illingworth
Hammick reaction reacción de Hammick
handedness asimetría, quiralidad
hanksite hanksita
Hantach pyrrole synthesis síntesis de pirroles de Hantach
Hantzsch pyridine synthesis síntesis de piridinas de Hantzsch
Hanus reagent reactivo de Hanus
Hanus solution solución de Hanus
hapto hapto
hard duro
hard acid ácido duro
hard base base dura
hard brass latón duro
hard bronze bronce duro
hard water agua dura
hardener endurecedor
hardening endurecimiento
hardness dureza
hardness of water dureza de agua
Hargreaves process proceso de Hargreaves
harmful radiation radiación dañina
harmful substance substancia dañina
harmine harmina
harmonic oscillator oscilador armónico
harmotome harmotoma
Harries ozonide reaction reacción de ozónidos de Harries
hartite hartita
hashish hachís
Hass chlorination rules reglas de clorinación de Hass**

hausmannite hausmannita
hauyne hauyna
Haworth methylation metilación de
    Haworth
Haworth projection proyección de
    Haworth
Haworth synthesis síntesis de Haworth
Hayashi rearrangement reordenamiento
    de Hayasi
hazard warning advertencia de peligro
hazardous peligroso, perjudicial
hazardous area área peligrosa
hazardous chemical producto químico
    peligroso
hazardous environment ambiente
    peligroso
hazardous material material peligroso
hazardous substance substancia
    peligrosa
hazardous waste desperdicios peligrosos
heat calor
heat absorption absorción térmica,
    absorción de calor
heat balance equilibrio térmico
heat capacity capacidad térmica
heat conductivity conductividad térmica
heat content contenido térmico
heat cycle ciclo térmico
heat detector detector térmico
heat dissipation disipación térmica
heat effect efecto térmico
heat energy energía térmica
heat equivalent equivalente térmico
heat exchange intercambio térmico
heat exchanger intercambiador térmico
heat flow flujo térmico
heat index índice térmico
heat insulation termoaislamiento
heat of absorption calor de absorción
heat of activation calor de activación
heat of adhesion calor de adhesión
heat of adsorption calor de adsorción
heat of aggregation calor de agregación
heat of association calor de asociación
heat of atomization calor de atomización
heat of combination calor de
    combinación
heat of combustion calor de combustión
heat of compression calor de compresión

heat of condensation calor de
    condensación
heat of cooling calor de enfriamiento
heat of crystallization calor de
    cristalización
heat of decomposition calor de
    descomposición
heat of dilution calor de dilución
heat of dissociation calor de disociación
heat of dissolution calor de disolución
heat of emission calor de emisión
heat of evaporation calor de evaporación
heat of explosion calor de explosión
heat of formation calor de formación
heat of fuel calor de combustible
heat of fusion calor de fusión
heat of hydration calor de hidratación
heat of ionization calor de ionización
heat of isomerization calor de
    isomerización
heat of neutralization calor de
    neutralización
heat of oxidation calor de oxidación
heat of racemization calor de
    racemización
heat of reaction calor de reacción
heat of solidification calor de
    solidificación
heat of solution calor de solución
heat of sublimation calor de sublimación
heat of transition calor de transición
heat of vaporization calor de
    vaporización
heat transfer transferecia térmica
heating calentamiento, caldeo
heavy pesado
heavy alloy aleación pesada
heavy atom átomo pesado
heavy chemical producto químico pesado
heavy electron electrón pesado
heavy element elemento pesado
heavy hydrogen hidrógeno pesado
heavy isotope isótopo pesado
heavy metal metal pesado
heavy nucleus núcleo pesado
heavy oil aceite pesado
heavy oxygen oxígeno pesado
heavy spar espato pesado
heavy water agua pesada

hectorite hectorita
hedenbergite hedenbergita
hedonal hedonal
Hehner number número de Hehner
Heisenberg uncertainty principle
    principio de incertidumbre de
    Heisenberg
helcosol helcosol
helenine helenina
helenite helenita
Helferich method método de Helferich
helical helicoidal
helium helio
helium mass spectrometer
    espectrómetro de masa de helio
helium spectrometer espectrómetro de
    helio
helix hélice
Hell-Volhard-Zelinsky reaction reacción
    de Hell-Volhard-Zelinsky
Helmholtz equation ecuación de
    Helmholtz
Helmholtz free energy energía libre de
    Helmholtz
Helmholtz layer capa de Helmholtz
Helmholtz model modelo de Helmholtz
helvite helvita
hematein hemateína
hematin hematina
hematite hematita
hematoxylin hematoxilina
heme hem
hemel hemel
hemellitic acid ácido hemelítico
hemi hemi
hemiacetal hemiacetal
hemicellulose hemicelulosa
hemihedral hemihedral
hemimellitene hemimeliteno
hemin hemina
hemiquinonoid hemiquinonoide
hemlock cicuta
hemoglobin hemoglobina
hemp cáñamo
hempa hempa
hempseed oil aceite de cañamón
hendecane hendecano
Henderson-Hasselbach equation
    ecuación de Henderson-Hasselbach

heneicosane heneicosano
heneicosanoic acid ácido heneicosanoico
Henkel reaction reacción de Henkel
henna alheña
Henry law ley de Henry
Henry reaction reacción de Henry
hentriacontane hentriacontano
heparin heparina
hepta hepta
heptabarbital heptabarbital
heptachlor heptacloro
heptachlorepoxide heptaclorepóxido
heptacosane heptacosano
heptacosanoic acid ácido heptacosanoico
heptadecane heptadecano
heptadecanoic acid ácido heptadecanoico
heptadecanol heptadecanol
heptadecanone heptadecanona
heptadecylglyoxalidine
    heptadecilglioxalidina
heptadiene heptadieno
heptafluorobutyric acid ácido
    heptafluorobutírico
heptahydrate heptahidrato
heptaldehyde heptaldehído
heptalene heptaleno
heptamethyldisilazane
    heptametildisilazano
heptamethylene heptametileno
heptamethylnonane heptametilnonano
heptanal heptanal
heptane heptano
heptanedicarboxylic acid ácido
    heptanodicarboxílico
heptanedioic acid ácido heptanodioico
heptanedioyl heptanodioilo
heptanoate heptanoato
heptanoic acid ácido heptanoico
heptanol heptanol
heptanone heptanona
heptanoyl heptanoilo
heptanoyl chloride cloruro de heptanoilo
heptavalent heptavalente
heptene hepteno
heptenyl heptenilo
heptoic acid ácido heptoico
heptose heptosa
heptoxide heptóxido
heptyl heptilo

heptyl acetate acetato de heptilo
heptyl alcohol alcohol heptílico
heptyl formate formiato de heptilo
heptyl heptoate heptoato de heptilo
heptyl pelargonate pelargonato de heptilo
heptylamine heptilamina
heptylene heptileno
heptylic acid ácido heptílico
heptyne heptino
herapathite herapatita
herbicide herbicida
hercynite hercinita
herderite herderita
Hermann-Mauguin system sistema de Hermann-Mauguin
hermetic hermético
heroin heroína
herrerite herrerita
Herz compound compuesto de Herz
Herz reaction reacción de Herz
hesperidin hesperidina
Hess law ley de Hess
hessite hessita
hessonite hessonita
hetero hetero
heteroaromatic heteroaromático
heteroatomic heteroatómico
heteroatomic ring anillo heteroatómico
heteroauxin heteroauxina
heteroazeotrope heteroazeótropo
heterobaric heterobárico
heterocyclic heterocíclico
heterocyclic atom átomo heterocíclico
heterocyclic compound compuesto heterocíclico
heterogeneous heterogéneo
heterogeneous catalysis catálisis heterogénea
heterogeneous reaction reacción heterogénea
heterogeneous system sistema heterogéneo
heteroleptic heteroléptico
heterolysis heterólisis
heterolytic heterolítico
heterolytic reaction reacción heterolítica
heterometry heterometría
heteromolybdate heteromolibdato

heteronuclear heteronuclear
heteronuclear molecule molécula heteronuclear
heteropolar heteropolar
heteropolar bond enlace heteropolar
heteropoly heteropoli
heteropolymer heteropolímero
heterotope heterótopo
heterotopic heterotópico
heterotype heterotipo
heulandite heulandita
Heumann-Pfleger synthesis síntesis de Heumann-Pfleger
hexa hexa
hexaamine hexaamina
hexaborane hexaborano
hexabromo hexabromo
hexabromoethane hexabromoetano
hexacalcium phytate fitato hexacálcico
hexachloro hexacloro
hexachloroacetone hexacloroacetona
hexachlorobenzene hexaclorobenceno
hexachlorobutadiene hexaclorobutadieno
hexachlorocyclohexane hexaclorociclohexano
hexachlorocyclopentadiene hexaclorociclopentadieno
hexachlorodiphenyl oxide óxido de hexaclorodifenilo
hexachloroethane hexacloroetano
hexachloromethyl ether éter hexaclorometílico
hexachloromethylcarbonate hexaclorometilcarbonato
hexachloronaphthalene hexacloronaftaleno
hexachlorophene hexaclorofeno
hexachloroplatinate hexacloroplatinato
hexachloropropane hexacloropropano
hexachloropropylene hexacloropropileno
hexacontane hexacontano
hexacosane hexacosano
hexacosanoic acid ácido hexacosanoico
hexacosanol hexacosanol
hexacosyl hexacosilo
hexacosyl alcohol alcohol hexacosílico
hexacyano hexaciano
hexacyanogen hexacianógeno
hexadecanal hexadecanal

hexadecane hexadecano
hexadecanoate hexadecanoato
hexadecanoic acid ácido hexadecanoico
hexadecanol hexadecanol
hexadecanoyl hexadecanoilo
hexadecanoyl chloride cloruro de
hexadecanoilo
hexadecene hexadeceno
hexadecenoic acid ácido hexadecenoico
hexadecenolide hexadecenolida
hexadecyl hexadecilo
hexadecyl alcohol alcohol hexadecílico
hexadecyl iodide yoduro de hexadecilo
hexadecyl mercaptan
hexadecilmercaptano
hexadecyltrichlorosilane
hexadeciltriclorosilano
hexadecyne hexadecino
hexadiene hexadieno
hexadienoic acid ácido hexadienoico
hexadiyne hexadiíno
hexaethyl tetraphosphate tetrafosfato de
hexaetilo
hexafluoroacetone hexafluoroacetona
hexafluorobenzene hexafluorobenceno
hexafluorodisilane hexafluorodisilano
hexafluoroethane hexafluoroetano
hexafluorophosphoric acid ácido
hexafluorofosfórico
hexafluoropropene hexafluoropropeno
hexafluoropropylene
hexafluoropropileno
hexafluoropropylene epoxide epóxido de
hexafluoropropileno
hexafluorosilicate hexafluorosilicato
hexafluorosilicic acid ácido
hexafluorosilícico
hexaglycerol hexaglicerol
hexagon hexágono
hexagonal close packing estructura
compacta hexagonal
hexagonal system sistema hexagonal
hexagonite hexagonita
hexahelicine hexahelicina
hexahydrate hexahidrato
hexahydric hexahídrico
hexahydric acid ácido hexahídrico
hexahydric alcohol alcohol hexahídrico
hexahydro hexahidro

hexahydroaniline hexahidroanilina
hexahydroanthracene
hexahidroantraceno
hexahydrobenzene hexahidrobenceno
hexahydrobenzoic acid ácido
hexahidrobenzoico
hexahydrocresol hexahidrocresol
hexahydromethyl hexahidrometilo
hexahydrophenol hexahidrofenol
hexahydrophthalic acid ácido
hexahidroftálico
hexahydropyridine hexahidropiridina
hexahydrotoluene hexahidrotolueno
hexahydroxy hexahidroxi
hexahydroxycyclohexane
hexahidroxiciclohexano
hexahydroxylene hexahidroxileno
hexaiodo hexayodo
hexaldehyde hexaldehído
hexalin hexalina
hexametaphosphate hexametafosfato
hexamethonium chloride cloruro de
hexametonio
hexamethylbenzene hexametilbenceno
hexamethyldisilane hexametildisilano
hexamethyldisilazane
hexametildisilazano
hexamethylene hexametileno
hexamethylene diisocyanate diisocianato
de hexametileno
hexamethylene glycol hexametilenglicol
hexamethylenediamine
hexametilendiamina
hexamethyleneimine hexametilenimina
hexamethylenetetramine
hexametilentetramina
hexamethylmelamine
hexametilmelamina
hexamine hexamina
hexanal hexanal
hexanaphthene hexanafteno
hexane hexano
hexanediamide hexanodiamida
hexanedinitrile hexanodinitrilo
hexanedioate hexanodioato
hexanedioic acid ácido hexanodioico
hexanediol hexanodiol
hexanedione hexanodiona
hexanelactam hexanolactama

hexanetriol hexanotriol
hexanitro hexanitro
hexanoate hexanoato
hexanoic acid ácido hexanoico
hexanol hexanol
hexanone hexanona
hexanoyl hexanoilo
hexanoyl chloride cloruro de hexanoilo
hexaphenyldisilane hexafenildisilano
hexaphenylethane hexafeniletano
hexaprismo hexaprismo
hexatriacontane hexatriacontano
hexenal hexenal
hexene hexeno
hexenic hexénico
hexenoic acid ácido hexenoico
hexenol hexenol
hexenyl hexenilo
hexestrol hexestrol
hexetidine hexetidina
hexobarbital hexobarbital
hexobarbitone hexobarbitona
hexoic hexoico
hexoic acid ácido hexoico
hexokinase hexoquinasa
hexone hexona
hexose hexosa
hexotriose hexotriosa
hexyl hexilo
hexyl acetate acetato de hexilo
hexyl alcohol alcohol hexílico
hexyl bromide bromuro de hexilo
hexyl chloride cloruro de hexilo
hexyl cinnamaldehyde cinamaldehído de
    hexilo
hexyl ether éter hexílico
hexyl formate formiato de hexilo
hexyl mercaptan hexilmercaptano
hexyl methacrylate metacrilato de hexilo
hexylacetic acid ácido hexilacético
hexylamine hexilamina
hexylene hexileno
hexylene glycol hexilenglicol
hexylic acid ácido hexílico
hexylphenol hexilfenol
hexylresorcinol hexilresorcina
hexyltrichlorosilane hexiltriclorosilano
hexyne hexino
hexynol hexinol

hiddenite hiddenita
hielmite hielmita
high definition alta definición
high density alta densidad
high energy alta energía
high-energy bond enlace de alta energía
high-energy neutron neutrón de alta
    energía
high-energy particle partícula de alta
    energía
**high-performance liquid
    chromatography** cromatografía
    líquida de alto rendimiento
high polymer alto polímero
high pressure alta presión
high temperature alta temperatura
high vacuum alto vacío
high-vacuum distillation destilación a
    alto vacío
higher level nivel superior
highest occupied molecular orbital
    orbital molecular de máxima energía
    ocupado
Hilbert-Johnson reaction reacción de
    Hilbert-Johnson
Hilbert space espacio de Hilbert
Hill reaction reacción de Hill
hindrance impedimento
Hinsberg reaction reacción de Hinsberg
Hinsberg synthesis síntesis de Hinsberg
Hinsberg test prueba de Hinsberg
hippurate hipurato
hippuric acid ácido hipúrico
hippuroyl hipuroilo
histaminase histaminasa
histamine histamina
histidine histidina
histochemistry histoquímica
histone histona
Hoch-Campbell synthesis síntesis de
    Hoch-Campbell
Hoesch synthesis síntesis de Hoesch
Hofmann amine separation separación
    de aminas de Hofmann
Hofmann degradation degradación de
    Hofmann
Hofmann isonitrile synthesis síntesis de
    isonitrilos de Hofmann
Hofmann-Martius rearrangement

reordenamiento de Hofmann-Martius
Hofmann reaction reacción de Hofmann
Hofmann rule regla de Hofmann
Hofmann-Sand reaction reacción de
   Hofmann-Sand
Hofmann transformation
   transformación de Hofmann
Hofmeister series serie de Hofmeister
hogbomite hogbomita
hole hueco, abertura
holmium holmio
holmium chloride cloruro de holmio
holmium fluoride fluoruro de holmio
holmium oxide óxido de holmio
holmium oxalate oxalato de holmio
holocellulose holocelulosa
holography holografía
holohedral holohedral
homatropine homatropina
homilite homilita
homo homo
homoatomic ring anillo homoatómico
homocentric homocéntrico
homochromic homocrómico
homocyclic homocíclico
homogeneous homogéneo
homogeneous catalysis catálisis
   homogénea
homogeneous combustion combustión
   homogénea
homogeneous reaction reacción
   homogénea
homogeneous series serie homogénea
homogeneous system sistema homogéneo
homogenization homogenización
homogentisic acid ácido homogentísico
homoisohydric homoisohídrico
homoleptic homoléptico
homologous series serie homóloga
homolysis homólisis
homolytic fission fisión homolítica
homolytic reaction reacción homolítica
homomorph homomorfo
homonuclear homonuclear
homonuclear molecule molécula
   homonuclear
homophthalic acid ácido homoftálico
homopolar homopolar
homopolar adsorption adsorción

homopolar
homopolar bond enlace homopolar
homopolar crystal cristal homopolar
homopolymer homopolímero
homosalate homosalato
homosalicylic acid ácido homosalicílico
homotope homótopo
honey miel
hood alacena protectora, campana
Hooker reaction reacción de Hooker
hopcalite hopcalita
Hopkins-Cole reaction reacción de
   Hopkins-Cole
hops lúpulo
horizontal chromatography
   cromatografía horizontal
hormone hormona
hornblende hornablenda
horsfordite horsfordita
Hortvet cryoscope crioscopio de Hortvet
Hortvet sublimator sublimador de
   Hortvet
Houben-Fischer synthesis síntesis de
   Houben-Fischer
Houben-Hoesch reaction reacción de
   Houben-Hoesch
Houben-Hoesch synthesis síntesis de
   Houben-Hoesch
Houdriflow process proceso de
   Houdriflow
Houdry process proceso de Houdry
howlite howlita
Huber reagent reactivo de Huber
Hübl reagent reactivo de Hübl
Hübl solution solución de Hübl
hübnerite hubnerita
Hückel approximation aproximación de
   Hückel
Hückel rule regla de Hückel
Hudson isorotation rules reglas de
   isorrotación de Hudson
Hudson lactone rule regla de lactonas de
   Hudson
Hume-Rothery rule regla de Hume-
   Rothery
humectant humectante
humic acid ácido húmico
humidification humidificación
humidimeter humidímetro

humidity humedad
humidity detector detector de humedad
humidity indicator indicador de humedad
humidity meter humidímetro
humidity-proof a prueba de humedad
humidity protection protección contra humedad
humidity sensor sensor de humedad
humus humus
Hund rules reglas de Hund
Hunsdiecker reaction reacción de Hunsdiecker
Hunter process proceso de Hunter
hyalophane hialofano
hyalosiderite hialosiderita
hyaluronic acid ácido hialurónico
hyaluronidase hialuronidasa
hybrid orbital orbital híbrido
hybridization hibridización
hydantoic acid ácido hidantoico
hydantoin hidantoína
hydnocarpic acid ácido hidnocárpico
hydracrylic acid ácido hidracrílico
hydralazine hydrochloride clorhidrato de hidralazina
hydramine hidramina
hydrargillite hidrargillita
hydrase hidrasa
hydrastine hidrastina
hydrate hidrato
hydrated hidratado
hydrated alumina alúmina hidratada
hydrated aluminum oxide óxido alumínico hidratado
hydrated lime cal hidratada
hydrated silica sílice hidratada
hydration hidratación
hydratropic acid ácido hidratrópico
hydraulic fluid fluido hidráulico
hydraulic fracturing fracturación hidráulica
hydrazi hidrazi
hydrazide hidrazida
hydrazidine hidrazidina
hydrazido hidrazido
hydrazine hidrazina
hydrazine chloride cloruro de hidrazina
hydrazine dihydrochloride diclorhidrato

de hidrazina
hydrazine formate formiato de hidrazina
hydrazine hydrate hidrato de hidrazina
hydrazine monobromide monobromuro de hidrazina
hydrazine monochloride monocloruro de hidrazina
hydrazine nitrate nitrato de hidrazina
hydrazine perchlorate perclorato de hidrazina
hydrazine sulfate sulfato de hidrazina
hydrazine tartrate tartrato de hidrazina
hydrazinium hidrazinio
hydrazino hidrazino
hydrazinobenzene hidrazinobenceno
hydrazo hidrazo
hydrazobenzene hidrazobenceno
hydrazoic acid ácido hidrazoico
hydrazone hidrazona
hydrazono hidrazono
hydrazotoluene hidrazotolueno
hydrazyl hidracilo
hydride hidruro
hydridic bridge puente hidrídico
hydrido hidrido
hydrin hidrina
hydriodic acid ácido yodhídrico
hydro hidro
hydroabietyl alcohol alcohol hidroabietílico
hydrobenzoin hidrobenzoína
hydrobiotite hidrobiotita
hydroboration hidroboración
hydrobromic acid ácido bromhídrico
hydrobromide bromhidrato
hydrocarbon hidrocarburo
hydrocarbon gas gas de hidrocarburos
hydrocarbon resin resina de hidrocarburos
hydrocarbon series serie de hidrocarburos
hydrocellulose hidrocelulosa
hydrochloric clorhídrico
hydrochloric acid ácido clorhídrico
hydrochloric ether éter clorhídrico
hydrochloride clorhidrato
hydrocinnamic acid ácido hidrocinámico
hydrocinnamic alcohol alcohol hidrocinámico

hydrocinnamic aldehyde aldehído
  hidrocinámico
hydrocinnamyl acetate acetato de
  hidrocinamilo
hydrocolloid hidrocoloide
hydrocortisone hidrocortisona
hydrocracking hidrocraqueo
hydrocyanic acid ácido cianhídrico
hydrocyanide hidrocianuro
hydrodealkylation hidrodesalquilación
hydrodistillation hidrodestilación
hydrodynamic radius radio
  hidrodinámico
hydroextractor hidroextractor
hydrofining hidrorrefinación
hydroflumethiazide hidroflumetiazida
hydrofluoric acid ácido fluorhídrico
hydrofluorosilicic acid ácido
  hidrofluorosilícico
hydrofluosilicic acid ácido
  hidrofluosilícico
hydroforming hidroformación
hydroformylation hidroformilación
hydrofuramide hidrofuramida
hydrogasification hidrogasificación
hydrogel hidrogel
hydrogen hidrógeno
hydrogen acceptor aceptor de hidrógeno
hydrogen atmosphere atmósfera de
  hidrógeno
hydrogen atom átomo de hidrógeno
hydrogen azide azida de hidrógeno
hydrogen bond enlace de hidrógeno
hydrogen bromide bromuro de
  hidrógeno
hydrogen carbonate carbonato de
  hidrógeno
hydrogen carrier portadora de hidrógeno
hydrogen chloride cloruro de hidrógeno
hydrogen cyanide cianuro de hidrógeno
hydrogen dioxide dióxido de hidrógeno
hydrogen disulfide disulfuro de
  hidrógeno
hydrogen electrode electrodo de
  hidrógeno
hydrogen equivalent equivalente de
  hidrógeno
hydrogen fluoride fluoruro de hidrógeno
hydrogen hexafluorosilicate

  hexafluorosilicato de hidrógeno
hydrogen inlet entrada de hidrógeno
hydrogen iodide yoduro de hidrógeno
hydrogen ion ion hidrógeno
hydrogen-ion concentration
  concentración de ion hidrógeno
hydrogen-ion determination
  determinación de ion hidrógeno
hydrogen-ion indicator indicador de ion
  hidrógeno
hydrogen-ion meter medidor de ion
  hidrógeno
hydrogen line línea de hidrógeno
hydrogen oxide óxido de hidrógeno
hydrogen peroxide peróxido de
  hidrógeno
hydrogen persulfide persulfuro de
  hidrógeno
hydrogen phosphide fosfuro de
  hidrógeno
hydrogen selenide seleniuro de
  hidrógeno
hydrogen spectrum espectro de
  hidrógeno
hydrogen sulfate sulfato de hidrógeno
hydrogen sulfide sulfuro de hidrógeno
hydrogen tellurate telurato de hidrógeno
hydrogen telluride teluluro de hidrógeno
hydrogenate hidrogenar
hydrogenated hidrogenado
hydrogenated oil aceite hidrogenado
hydrogenation hidrogenación
hydrogenolysis hidrogenólisis
hydrohematite hidrohematita
hydrol hidrol
hydrolase hidrolasa
hydroliquefaction hidrolicuefacción
hydrolized hidrolizado
hydrolized protein proteína hidrolizada
hydrolysis hidrólisis
hydrolyte hidrólito
hydrolytic hidrolítico
hydrolytic dissociation disociación
  hidrolítica
hydrolyze hidrolizar
hydromagnesite hidromagnesita
hydrometer hidrómetro
hydrone hidrona
hydronitric acid ácido hidronítrico

hydronium ion ion hidronio
hydroperoxide hidroperóxido
hydrophile hidrófilo
hydrophilic hidrofílico
hydrophilic colloid coloide hidrofílico
hydrophilite hidrofilita
hydrophobic hidrofóbico
hydrophobic bonding enlazado
   hidrofóbico
hydrophobic colloid coloide hidrofóbico
hydropolysulfide hidropolisulfuro
hydroponics hidropónica
hydroquinol hidroquinol
hydroquinone hidroquinona
hydroquinone benzyl ether éter
   bencílico de hidroquinona
hydroquinone diethyl ether éter dietílico
   de hidroquinona
hydroquinone dimethyl ether éter
   dimetílico de hidroquinona
hydroquinone monoethyl ether éter
   monoetílico de hidroquinona
hydroseleno hidroseleno
hydrosol hidrosol
hydrosolvation hidrosolvatación
hydrostatic hidrostático
hydrosulfide hidrosulfuro
hydrosulfite hidrosulfito
hydrosulfite-formaldehyde hidrosulfito-
   formaldehído
hydrosulfuric acid ácido hidrosulfúrico
hydrosulfurous acid ácido
   hidrosulfuroso
hydrotetrazone hidrotetrazona
hydrotrope hidrótropo
hydrous hidratado
hydroxamic acid ácido hidroxámico
hydroxide hidróxido
hydroximic acid ácido hidroxímico
hydroxo hidroxo
hydroxonium hidroxonio
hydroxy hidroxi
hydroxyacetal hidroxiacetal
hydroxyacetanilide hidroxiacetanilida
hydroxyacetic acid ácido hidroxiacético
hydroxyacetone hidroxiacetona
hydroxyacetophenone
   hidroxiacetofenona
hydroxyadipaldehyde

hidroxiadipaldehído
hydroxyalanine hidroxialanina
hydroxyamino hidroxiamino
hydroxyaniline hidroxianilina
hydroxyanisole hidroxianisol
hydroxyanthracene hidroxiantraceno
hydroxyapatite hidroxiapatita
hydroxyazobenzenesulfonic acid ácido
   hidroxiazobencenosulfónico
hydroxybenzaldehyde
   hidroxibenzaldehído
hydroxybenzamide hidroxibenzamida
hydroxybenzene hidroxibenceno
hydroxybenzoic acid ácido
   hidroxibenzoico
hydroxybenzophenone
   hidroxibenzofenona
hydroxybenzyl alcohol alcohol
   hidroxibencílico
hydroxybutanal hidroxibutanal
hydroxybutyraldehyde
   hidroxibutiraldehído
hydroxybutyranilide hidroxibutiranilida
hydroxybutyric acid ácido
   hidroxibutírico
hydroxycamphane hidroxicanfano
hydroxycarotene hidroxicaroteno
hydroxycerussite hidroxicerusita
hydroxycholine hidroxicolina
hydroxycholestane hidroxicolestano
hydroxycitronellal hidroxicitronelal
hydroxycobalamin hidroxicobalamina
hydroxycorticosterone
   hidroxicorticosterona
hydroxydimethylbenzene
   hidroxidimetilbenceno
hydroxydiphenyl hidroxidifenilo
hydroxydiphenylamine
   hidroxidifenilamina
hydroxyethanesulfonic acid ácido
   hidroxietanosulfónico
hydroxyethyl carbamate carbamato de
   hidroxietilo
hydroxyethyl methacrylate metacrilato
   de hidroxietilo
hydroxyethyl piperazine piperazina de
   hidroxietilo
hydroxyethylacetamide
   hidroxietilacetamida

hydroxyethylamine hidroxietilamina
hydroxyethylcellulose hidroxietilcelulosa
hydroxyethylethylenediamine
  hidroxietiletilendiamina
hydroxyethylhydrazine
  hidroxietilhidrazina
hydroxyfenchane hidroxifencano
hydroxyhydrazide hidroxihidrazida
hydroxyimino hidroxiimino
hydroxyl hidroxilo
hydroxyl ion ion hidroxilo
hydroxylamine hidroxilamina
hydroxylamine acid sulfate sulfato ácido
  de hidroxilamina
hydroxylamine hydrochloride
  clorhidrato de hidroxilamina
hydroxylamine sulfate sulfato de
  hidroxilamina
hydroxylammonium chloride cloruro de
  hidroxilamonio
hydroxylation hidroxilación
hydroxymercuricresol
  hidroximercuricresol
hydroxymesitylene hidroximesitileno
hydroxymethane hidroximetano
hydroxymethyl hidroximetilo
hydroxymethylene hidroximetileno
hydroxymethylene carbonate carbonato
  de hidroximetileno
hydroxymethylfuraldehyde
  hidroximetilfuraldehído
hydroxynaphthalene hidroxinaftaleno
hydroxynaphthoic acid ácido
  hidroxinaftoico
hydroxynaphthoic anilide anilida
  hidroxinaftoica
hydroxynaphthoquinone
  hidroxinaftoquinona
hydroxyoleic acid ácido hidroxioleico
hydroxyphenol hidroxifenol
hydroxyphenyl hidroxifenilo
hydroxyphenylalanine
  hidroxifenilalanina
hydroxyphenylmercuric chloride
  cloruro hidroxifenilmercúrico
hydroxypiperidine hidroxipiperidina
hydroxyprogesterone
  hidroxiprogesterona
hydroxyproline hidroxiprolina

hydroxypropanone hidroxipropanona
hydroxypropionic acid ácido
  hidroxipropiónico
hydroxypropionitrile
  hidroxipropionitrilo
hydroxypropyl acrylate acrilato de
  hidroxipropilo
hydroxypropyl cellulose
  hidroxipropilcelulosa
hydroxypropyl methylcellulose
  hidroxipropilmetilcelulosa
hydroxypropylamine hidroxipropilamina
hydroxyquinoline hidroxiquinolina
hydroxyquinoline benzoate benzoato de
  hidroxiquinolina
hydroxyquinoline sulfate sulfato de
  hidroxiquinolina
hydroxysalicylic acid ácido
  hidroxisalicílico
hydroxystearic acid ácido
  hidroxiesterárico
hydroxystearyl alcohol alcohol
  hidroxiestearílico
hydroxysuccinic acid ácido
  hidroxisuccinico
hydroxytoluene hidroxitolueno
hydroxytoluic acid ácido hidroxitoluico
hydroxytriacontane hidroxitriacontano
hydroxytryptamine hidroxitriptamina
hydrozincite hidrocincita
hygric acid ácido hígrico
hygrometer higrómetro
hygrometric higrométrico
hygromycin higromicina
hygroscopic higroscópico
hyoscine hioscina
hyoscyamine hiosciamina
hyper hiper
hyperchromic hipercrómico
hyperconjugation hiperconjugación
hyperfine structure estructura hiperfina
hyperon hiperón
hyperoxide hiperóxido
hypersorption hipersorción
hypertonic hipertónico
hypertonic solution solución hipertónica
hypo hipo
hypoallergenic hipoalergénico
hypobromate hipobromato

hypobromite hipobromito
hypobromous acid ácido hipobromoso
hypochlorite hipoclorito
hypochlorous acid ácido hipocloroso
hypochromic hipocrómico
hypofluorite hipofluorita
hypoiodate hipoyodato
hyponitrile hiponitrilo
hyponitrous acid ácido hiponitroso
hypophamine hipofamina
hypophosphate hipofosfato
hypophosphite hipofosfito

hypophosphoric acid ácido hipofosfórico
hypophosphorous acid ácido
   hipofosforoso
hyposulfate hiposulfato
hyposulfite hiposulfito
hyposulfuric acid ácido hiposulfúrico
hyposulfurous acid ácido hiposulfuroso
hypothesis hipótesis
hypotonic hipotónico
hypotonic solution solución hipotónica
hypoxanthine hipoxantina
hysteresis histéresis

I

ibogaine ibogaína
ibuprofen ibuprofeno
ic ico
ice hielo
ice point punto de congelación
Iceland moss liquen islándico
Iceland spar espato de Islandia
ichthammol ictamol
ichthyocolla ictiocola
ichthyolate ictiolato
icosa icosa
icosahedron icosaedro
icosane icosano
icosanoic acid ácido icosanoico
icosanol icosanol
icosyl icosilo
ide uro, iuro, ido
ideal crystal cristal ideal
ideal gas gas ideal
ideal solution solución ideal
idene ideno
identification limit límite de
    identificación
identity period periodo de identidad
idiochromatic idiocromático
idose idosa
idyne idino
ignition ignición
ignition point punto de ignición
ignition temperature temperatura de
    ignición
Ilkovic equation ecuación de Ilkovic
illinium ilinio
ilmenite ilmenita
ilvaite ilvaíta
imazalil imazalil
imbibition imbibición
imidazole imidazol
imidazoletrione imidazoltriona
imidazolidone imidazolidona
imidazolium imidazolio
imidazolyl imidazolilo
imide imida

imidic acid ácido imídico
imidine imidina
imido imido
imidoyl imodoilo
imine imina
iminio iminio
iminium iminio
imino imino
imino acid iminoácido
iminobispropylamine
    iminobispropilamina
iminodiacetic acid ácido iminodiacético
iminodiacetonitrile iminodiacetonitrilo
iminourea iminourea
immersion inmersión
immiscible inmiscible
immunochemistry inmunoquímica
immunoglobulin inmunoglobulina
impact impacto
impalpable impalpable
impeller impulsor
imperial green verde imperial
implosion implosión
impregnate impregnar
impregnation impregnación
impurity impureza
in vitro in vitro
in vivo in vivo
inactive inactivo
incandescence incandescencia
incandescent incandescente
incendiary incendiario
incendiary gel gel incendiario
incidence incidencia
incineration incineración
inclusion complex complejo de inclusión
inclusion compound compuesto de
    inclusión
incompatible incompatible
incomplete equilibrium equilibrio
    incompleto
incomplete reaction reacción incompleta
incompressible incompresible

129

incompressible volume volumen incompresible
incubation incubación
indamine indamina
indan indano
indandione indandiona
indanthrene indantreno
indanthrone indantrona
indanyl indanilo
indazole indazol
indene indeno
indenone indenona
indenyl indenilo
indeterminate indeterminado
index índice
Indian red rojo de la India
indican indicán
indicator indicador
indicator paper papel indicador
indicator range intervalo de indicador
indigo índigo, añil
indigo blue azul de índigo
indigo red rojo de índigo
indigotin indigotina
indirect dye colorante indirecto
indium indio
indium acetylacetonate acetilacetonato de indio
indium antimonide antimoniuro de indio
indium arsenide arseniuro de indio
indium bromide bromuro de indio
indium chloride cloruro de indio
indium dichloride dicloruro de indio
indium hydrogensulfide hidrogenosufuro de indio
indium hydroxide hidróxido de indio
indium nitrate nitrato de indio
indium oxide óxido de indio
indium phosphide fosfuro de indio
indium sulfate sulfato de indio
indium sulfide sulfuro de indio
indium telluride telururo de indio
indium trichloride tricloruro de indio
indogen indógeno
indogenide indogenuro
indole indol
indoleacetic acid ácido indolacético
indolebutyric acid ácido indolbutírico
indolizine indolizina

indolol indolol
indolyl indolilo
indomethacin indometacina
indophenine reaction reacción de indofenina
indophenol indofenol
indoxyl indoxilo
induced emission emisión inducida
induced fission fisión inducida
induced reaction reacción inducida
induction inducción
induction period periodo de inducción
inductive effect efecto inductivo
indurite indurita
industrial industrial
industrial alcohol alcohol industrial
industrial carbon carbón industrial
industrial chemistry química industrial
industrial diamond diamante industrial
industrial waste desperdicios industriales
indyl indilo
ine ina
inelastic inelástico
inelastic collision colisión inelástica
inert inerte
inert atmosphere box caja con atmósfera inerte
inert complex complejo inerte
inert gas gas inerte
inert substance substancia inerte
infinite dilution dilución infinita
inflame inflamar
inflamer inflamador
inflammability inflamabilidad
inflammable inflamable
inflammableness inflamabilidad
infrared infrarrojo
infrared absorption spectrum espectro de absorción de infrarrojo
infrared detector detector de infrarrojo
infrared microscope microscopio de infrarrojo
infrared radiation radiación infrarroja
infrared source fuente de infrarrojo
infrared spectrometer espectrómetro de infrarrojo
infrared spectroscopy espectroscopia de infrarrojo
infrared spectrum espectro de infrarrojo

infusion infusión
ingot iron hierro de lingote
inhibition inhibición
inhibitor inhibidor
initial boiling point punto de ebullición inicial
initiating explosive explosivo iniciador
initiator iniciador
ink tinta
inlet entrada
ino ino
inorganic inorgánico
inorganic acid ácido inorgánico
inorganic chemistry química inorgánica
inorganic polymer polímero inorgánico
inosamine inosamina
inosine inosina
inosinic acid ácido inosínico
inositol inositol
insecticide insecticida
insertion reaction reacción de inserción
insolubility insolubilidad
insoluble insoluble
insolubleness insolubilidad
instrument instrumento
instrumentation instrumentación
insulator aislador
insulin insulina
intensity intensidad
intensive properties propiedades intensivas
interatomic interatómico
interatomic distance distancia interatómica
interatomic space espacio interatómico
intercalation compound compuesto de intercalación
interdiffusion interdifusión
interelectrode interelectródico
interface interfase
interfacial interfacial
interfacial film película interfacial
interfacial surface energy energía superficial interfacial
interference interferencia
interferon interferón
interhalogen interhalógeno
interionic interiónico
interionic attraction atracción

interiónica
intermediate producto intermedio
intermediate bond enlace intermedio
intermetallic intermetálico
intermetallic compound compuesto intermetálico
intermolecular intermolecular
intermolecular force fuerza intermolecular
internal interno
internal compensation compensación interna
internal conversion conversión interna
internal energy energía interna
internal phase fase interna
internal reaction reacción interna
internal symmetry simetría interna
international system sistema internacional
international unit unidad internacional
interstice intersticio
interstitial intersticial
intoxication intoxicación
intraannular intraanular
intraatomic intraatómico
intraatomic space espacio intraatómico
intramolecular intramolecular
intramolecular compensation compensación intramolecular
intramolecular condensation condensación intramolecular
intramolecular reaction reacción intramolecular
intranuclear intranuclear
intrinsic intrínseco
intrinsic viscosity viscosidad intrínseca
introfaction introfacción
intron intrón
inulin inulina
inulinase inulinasa
invariant point punto invariante
invention invención
inversion inversión
inversion point punto de inversión
inversion temperature temperatura de inversión
invert sugar azúcar invertido
invertase invertasa
iodate yodato

iodeosin yodeosina
iodic acid ácido yódico
iodic acid anhydride anhídrido de ácido
    yódico
iodide yoduro
iodimetry yodimetría
iodine yodo
iodine acetate acetato de yodo
iodine bisulfide bisulfuro de yodo
iodine bromide bromuro de yodo
iodine chloride cloruro de yodo
iodine cyanide cianuro de yodo
iodine dioxide dióxido de yodo
iodine disulfide disulfuro de yodo
iodine fluoride fluoruro de yodo
iodine monobromide monobromuro de
    yodo
iodine monochloride monocloruro de
    yodo
iodine number número de yodo
iodine oxide óxido de yodo
iodine pentabromide pentabromuro de
    yodo
iodine pentafluoride pentafluoruro de
    yodo
iodine pentoxide pentóxido de yodo
iodine sulfate sulfato de yodo
iodine test prueba de yodo
iodine tincture tintura de yodo
iodine trichloride tricloruro de yodo
iodine value valor de yodo
iodipamide yodipamida
iodisan yodisán
iodite yodita
iodized yodado
iodized oil aceite yodado
iodoacetic acid ácido yodoacético
iodoalkane yodoalcano
iodoaniline yodoanilina
iodobenzene yodobenceno
iodoethane yodoetano
iodoethylene yodoetileno
iodoform yodoformo
iodogorgoic acid ácido yodogorgoico
iodomethane yodometano
iodometric yodométrico
iodometry yodometría
iodonium yodonio
iodophor yodóforo

iodopropane yodopropano
iodoso yodoso
iodosobenzene yodosobenceno
iodosuccinimide yodosuccinimida
iodosyl yodosilo
iodoxy yodoxi
iodoxyl yodoxilo
iodyl yodilo
iolite iolita
ion ion
ion activity actividad iónica
ion avalanche avalancha iónica
ion charge carga iónica
ion cloud nube iónica
ion detector detector iónico
ion exchange intercambio iónico
ion-exchange chromatography
    cromatografía de intercambio iónico
ion-exchange column columna de
    intercambio iónico
ion-exchange resin resina de intercambio
    iónico
ion exclusion exclusión iónica
ion-exclusion chromatography
    cromatografía de exclusión iónica
ion migration migración iónica
ion pair par iónico
ion pump bomba iónica
ion retardation retardo iónico
ion source fuente iónica
ionic iónico
ionic atmosphere atmósfera iónica
ionic bond enlace iónico
ionic bonding enlazado iónico
ionic charge carga iónica
ionic conductance conductancia iónica
ionic crystal cristal iónico
ionic detergent detergente iónico
ionic equilibrium equilibrio iónico
ionic gel gel iónico
ionic polymerization polimerización
    iónica
ionic product producto iónico
ionic radius radio iónico
ionic solution solución iónica
ionic strength fuerza iónica
ionization ionización
ionization chamber cámara de ionización
ionization coefficient coeficiente de

ionización

ionization constant constante de
  ionización

ionization current corriente de
  ionización

ionization degree grado de ionización

ionization effect efecto de ionización

ionization energy energía de ionización

ionization potential potencial de
  ionización

ionize ionizar

ionized ionizado

ionized gas gas ionizado

ionizing ionizante

ionizing potential potencial de ionización

ionizing radiation radiación de
  ionización

ionizing solvent solvente de ionización

ionogen ionógeno

ionogenic ionogénico

ionography ionografía

ionol ionol

ionomer ionómero

ionomer resin resina de ionómero

ionone ionona

iprodione iprodiona

iridic chloride cloruro irídico

iridium iridio

iridium bromide bromuro de iridio

iridium chloride cloruro de iridio

iridium disulfide disulfuro de iridio

iridium oxide óxido de iridio

iridium potassium chloride cloruro
  potásico de iridio

iridium sesquioxide sesquióxido de
  iridio

iridium tetrabromide tetrabromuro de
  iridio

iridium tetrachloride tetracloruro de
  iridio

iridium tetraiodide tetrayoduro de iridio

iridium tribromide tribromuro de iridio

iridium trichloride tricloruro de iridio

iridium trioxide trióxido de iridio

irido irido

iridosmine iridosmina

Irish moss musgo de Irlanda

iron hierro

iron acetate acetato de hierro

iron ammonium sulfate sulfato amónico
  de hierro

iron arsenate arseniato de hierro

iron benzoate benzoato de hierro

iron black negro de hierro

iron blue azul de hierro

iron bromide bromuro de hierro

iron carbide carburo de hierro

iron carbonate carbonato de hierro

iron carbonyl carbonilo de hierro

iron chloride cloruro de hierro

iron chromate cromato de hierro

iron citrate citrato de hierro

iron dichloride dicloruro de hierro

iron disulfate disulfato de hierro

iron disulfide disulfuro de hierro

iron ethanoate etanoato de hierro

iron ferrocyanide ferrocianuro de hierro

iron fluoride fluoruro de hierro

iron formate formiato de hierro

iron hexacyanoferrate hexacianoferrato
  de hierro

iron hexafluorosilicate hexafluorosilicato
  de hierro

iron hydroxide hidróxido de hierro

iron iodide yoduro de hierro

iron monoxide monóxido de hierro

iron-nickel alloy aleación de hierro-
  níquel

iron nitrate nitrato de hierro

iron octoate octoato de hierro

iron ore mineral de hierro

iron oxalate oxalato de hierro

iron oxide óxido de hierro

iron oxide yellow amarillo de óxido de
  hierro

iron pentacarbonyl pentacarbonilo de
  hierro

iron perchlorate perclorato de hierro

iron phenolate fenolato de hierro

iron phosphate fosfato de hierro

iron phosphide fosfuro de hierro

iron powder polvo de hierro

iron protochloride protocloruro de
  hierro

iron protosulfide protosulfuro de hierro

iron pyrite pirita de hierro

iron pyrolignite pirolignita de hierro

iron pyrophosphate pirofosfato de hierro

iron resinate resinato de hierro
iron sponge esponja de hierro
iron sulfate sulfato de hierro
iron sulfide sulfuro de hierro
iron sulfite sulfito de hierro
iron thiocyanate tiocianato de hierro
iron thiosulfate tiosulfato de hierro
iron trichloride tricloruro de hierro
iron trioxide trióxido de hierro
irone irona
irradiating irradiante
irradiation irradiación
irreversible irreversible
irreversible colloid coloide irreversible
irreversible process proceso irreversible
irreversible reaction reacción
   irreversible
Irving-Williams order órden de Irving-
   Williams
isatin isatina
isatoic anhydride anhídrido isatoico
isazol isazol
isentropic process proceso isentrópico
isethionic acid ácido isetiónico
Ising model modelo de Ising
isinglass colapez, mica
iso iso
isoactivity line línea de isoactividad
isoamyl isoamilo
isoamyl acetate acetato de isoamilo
isoamyl alcohol alcohol isoamílico
isoamyl benzoate benzoato de isoamilo
isoamyl benzyl ether éter bencílico de
   isoamilo
isoamyl butyrate butirato de isoamilo
isoamyl chloride cloruro de isoamilo
isoamyl ether éter isoamílico
isoamyl formate formiato de isoamilo
isoamyl furoate furoato de isoamilo
isoamyl isovalerate isovalerato de
   isoamilo
isoamyl mercaptan isoamilmercaptano
isoamyl nitrite nitrito de isoamilo
isoamyl pelargonate pelargonato de
   isoamilo
isoamyl phthalate ftalato de isoamilo
isoamyl propionate propionato de
   isoamilo
isoamyl salicylate salicilato de isoamilo

isoamyl valerate valerato de isoamilo
isoamyldichloroarsine
   isoamildicloroarsina
isoamylene isoamileno
isoascorbic acid ácido isoascórbico
isobar isóbaro
isobaric spin espín isobárico
isobestic point punto isobéstico
isoborneol isoborneol
isobornyl isobornilo
isobornyl acetate acetato de isobornilo
isobornyl chloride cloruro de isobornilo
isobornyl salicylate salicilato de
   isobornilo
isobornyl thiocyanoacetate
   tiocianoacetato de isobornilo
isobutane isobutano
isobutanol isobutanol
isobutanolamine isobutanolamina
isobutene isobuteno
isobutoxy isobutoxi
isobutyl isobutilo
isobutyl acetate acetato de isobutilo
isobutyl acrylate acrilato de isobutilo
isobutyl alcohol alcohol isobutílico
isobutyl aldehyde aldehído isobutílico
isobutyl benzoate benzoato de isobutilo
isobutyl carbinol isobutilcarbinol
isobutyl chloroformate cloroformiato de
   isobutilo
isobutyl cinnamate cinamato de isobutilo
isobutyl cyanoacrylate cianoacrilato de
   isobutilo
isobutyl ether éter isobutílico
isobutyl furoate furoato de isobutilo
isobutyl isobutyrate isobutirato de
   isobutilo
isobutyl mercaptan isobutilmercaptano
isobutyl methacrylate metacrilato de
   isobutilo
isobutyl phenylacetate fenilacetato de
   isobutilo
isobutyl propionate propionato de
   isobutilo
isobutyl salicylate salicilato de isobutilo
isobutyl stearate estearato de isobutilo
isobutyl valerate valerato de isobutilo
isobutylamine isobutilamina
isobutylaminobenzoate

isobutilaminobenzoato

isobutylbenzene isobutilbenceno

isobutylene isobutileno

isobutyraldehyde isobutiraldehído

isobutyric isobutírico

isobutyric acid ácido isobutírico

isobutyric anhydride anhídrido isobutírico

isobutyronitrile isobutironitrilo

isobutyroyl isobutiroilo

isobutyroyl chloride cloruro de isobutiroilo

isobutyryl isobutirilo

isocetyl laurate laurato de isocetilo

isocetyl myristate miristato de isocetilo

isocetyl oleate oleato de isocetilo

isocetyl stearate estearato de isocetilo

isochore isócora

isocinchomeronic acid ácido isocincomerónico

isocrotonic acid ácido isocrotónico

isocyanate isocianato

isocyanate resin resina de isocianato

isocyanic acid ácido isociánico

isocyanide isocianuro

isocyano isociano

isocyanurate isocianurato

isocyanuric acid ácido isocianúrico

isocyclic isocíclico

isodecaldehyde isodecaldehído

isodecane isodecano

isodecanoic acid ácido isodecanoico

isodecanol isodecanol

isodecyl isodecilo

isodecyl chloride cloruro de isodecilo

isodispersion isodipersión

isodrin isodrina

isodurene isodureno

isodynamic isodinámico

isoelectric isoeléctrico

isoelectric point punto isoeléctrico

isoelectric precipitation precipitación isoeléctrica

isoelectronic isoelectrónico

isoelectronic sequence secuencia isoelectrónica

isoenzyme isoenzima

isoeugenol isoeugenol

isoheptane isoheptano

isohexane isohexano

isolate aislar

isolated aislado

isolated double bond enlace doble aislado

isolation aislamiento

isoleptic isoléptico

isoleucine isoleucina

isolog isólogo

isomer isómero

isomerase isomerasa

isomeric isomérico

isomeric shift deslizamiento isomérico

isomeric transition transición isomérica

isomerism isomerismo

isomerization isomerización

isometric isométrico

isomolecule isomolécula

isomorph isomorfo

isomorphic isomórfico

isomorphism isomorfismo

isonicotinic acid ácido isonicotínico

isonitrile isonitrilo

isonitro isonitro

isonitroso isonitroso

isonitrosoketone isonitrosocetona

isononyl alcohol alcohol isononílico

isooctane isooctano

isooctene isoocteno

isooctyl adipate adipato de isooctilo

isooctyl alcohol alcohol isooctílico

isooctyl palmitate palmitato de isooctilo

isooctyl thioglycolate tioglicolato de isooctilo

isoparaffin isoparafina

isopentaldehyde isopentaldehído

isopentane isopentano

isopentanoic acid ácido isopentanoico

isopentyl alcohol alcohol isopentílico

isophorone isoforona

isophthalic acid ácido isoftálico

isophthaloyl chloride cloruro de isoftaloilo

isopolyacid isopoliácido

isopolyester isopoliéster

isopolymolybdate isopolimolibdato

isopolymorphism isopolimorfismo

isopolytungstate isopolitungstato

isoprene isopreno

isoprene polymer polímero de isopreno
isoprenoid isoprenoide
isopropanol isopropanol
isopropanolamine isopropanolamina
isopropenyl acetate acetato de isopropenilo
isopropenyl chloride cloruro de isopropenilo
isopropoxyethanol isopropoxietanol
isopropyl isopropilo
isopropyl acetate acetato de isopropilo
isopropyl alcohol alcohol isopropílico
isopropyl bromide bromuro de isopropilo
isopropyl butyrate butirato de isopropilo
isopropyl chloride cloruro de isopropilo
isopropyl cyanide cianuro de isopropilo
isopropyl ether éter isopropílico
isopropyl iodide yoduro de isopropilo
isopropyl mercaptan isopropilmercaptano
isopropyl nitrate nitrato de isopropilo
isopropylacetone isopropilacetona
isopropylamine isopropilamina
isopropylaminoethanol isopropilaminoetanol
isopropylaniline isopropilanilina
isopropylbenzene isopropilbenceno
isopropylbenzyl isopropilbencilo
isopropylcarbinol isopropilcarbinol
isopropylcresol isopropilcresol
isopropylethylene isopropiletileno
isopropylidene isopropilideno
isopropylmethylbenzene isopropilmetilbenceno
isopropylnaphthalene isopropilnaftaleno
isopropylphenol isopropilfenol
isopropyltoluene isopropiltolueno
isopulegol isopulegol
isoquinoline isoquinolina
isosafrole isosafrol
isosteric isostérico

isosterism isosterismo
isostructural isoestructural
isotactic isotáctico
isotherm isoterma
isothermal process proceso isotérmico
isothiocyanate isotiocianato
isothiocyanato isotiocianato
isothiocyanic acid ácido isotiociánico
isotone isótono
isotonic isotónico
isotonic solution solución isotónica
isotope isótopo
isotope abundance abundancia isotópica
isotope effect efecto isotópico
isotope separation separación de isótopos
isotope separator separador de isótopos
isotope shift deslizamiento isotópico
isotopic isotópico
isotopic abundance abundancia isotópica
isotopic composition composición isotópica
isotopic indicator indicador isotópico
isotopic mass masa isotópica
isotopic number número isotópico
isotopic ratio razón isotópica
isotopic reaction reacción isotópica
isotopic spin espín isotópico
isotopic tracer trazador isotópico
isotopic weight peso isotópico
isotron isotrón
isotropic isotrópico
isovalent isovalente
isovalent hyperconjugation hiperconjugación isovalente
isovaleraldehyde isovaleraldehído
isovaleric acid ácido isovalérico
isoxazolyl isoxazolilo
itaconic acid ácido itacónico
ite ita, ito
ium io
Ivanov reagent reactivo de Ivanov

J

Jablonski diagram diagrama de Jablonski

jaborine jaborina

Jacobsen rearrangement reordenamiento de Jacobsen

Jacquemart reagent reactivo de Jacquemart

jade jade

jadeite jadeíta

Jahn-Teller effect efecto de Jahn-Teller

Janovsky reaction reacción de Janovsky

Japp-Klingemann reaction reacción de Japp-Klingemann

jasmine oil aceite de jazmín

jasmone jazmona

jasper jaspe

jelly jalea

johimbine johimbina

jojoba oil aceite de jojoba

Jones oxidation oxidación de Jones

Jones reageant reactivo de Jones

Jones reductor reductor de Jones

joseite joseíta

Joule effect efecto de Joule

Joule-Kelvin effect efecto de Joule-Kelvin

Joule law ley de Joule

Joule-Thomas coefficient coeficiente de Joule-Thomas

Joule-Thomson effect efecto de Joule-Thomson

juniper tar aceite de cada

juniperic acid ácido junipérico

K

kainite cainita
kalinite calinita
kanamycin kanamicina
kanamycin sulfate sulfato de kanamicina
kaolin caolin
kaolinite caolinita
kappa kappa
karat quilate
karaya gum goma karaya
Karl Fischer reagent reactivo de Karl
    Fischer
Karplus equation ecuación de Karplus
Keesom relationship relación de Keesom
Kekulé ring anillo de Kekulé
Kekulé structure estructura de Kekulé
Kelvin effect efecto de Kelvin
Kelvin scale escala Kelvin
Kendall-Mattox reaction reacción de
    Kendall-Mattox
keratin queratina
keratinase queratinasa
kernite kernita
kerogen kerógeno
kerosene queroseno
ketal cetal
ketene cetena
ketimide cetimida
ketimine cetimina
ketine cetina
keto ceto
keto acid cetoácido
ketoamine cetoamina
ketobenzotriazine cetobenzotriazina
ketoglutaric acid ácido cetoglutárico
ketohexose cetohexosa
ketol cetol
ketomalonic acid ácido cetomalónico
ketone cetona
ketone body cuerpo cetónico
ketopentose cetopentosa
ketopropionic acid ácido cetopropiónico
ketose cetosa
ketosis cetosis

ketovaleric acid ácido cetovalérico
ketoxime cetoxima
ketyl cetilo
key atom átomo clave
Keyes process proceso de Keyes
Kharasch abnormal addition adición
    anormal de Kharasch
kieselguhr kieselguhr
kieserite kieserita
Kiliani-Fischer synthesis síntesis de
    Kiliani-Fischer
kiln horno
kilogram kilogramo
kimberlite kimberlita
kinetic cinético
kinetic effect efecto cinético
kinetic energy energía cinética
kinetic theory teoría cinética
kinetics cinética
kinic acid ácido quínico
Kipp apparatus aparato de Kipp
Kishner cyclopropane synthesis síntesis
    de ciclopropanos de Kishner
Kjeldahl flask matraz de Kjeldahl
Kjeldahl method método de Kjeldahl
Kjeldahl test prueba de Kjeldahl
Klein reagent reactivo de Klein
kleinite kleinita
knock golpeteo
knocking golpeteo
Knoevenagel reaction reacción de
    Knoevenagel
Knoop hardness dureza de Knoop
Knoop scale escala de Knoop
Knorr pyrazole synthesis síntesis de
    pirazoles de Knorr
Knorr pyrrole synthesis síntesis de
    pirroles de Knorr
Knorr quinoline synthesis síntesis de
    quinolinas de Knorr
Knudsen flow flujo de Knudsen
Koch acid ácido de Koch
Koch-Haaf carboxylation carboxilación

de Koch-Haaf
Kochi reaction reacción de Kochi
Kohlrausch equation ecuación de Kohlrausch
kojic acid ácido cójico
Kolbe method método de Kolbe
Kolbe reaction reacción de Kolbe
Kolbe synthesis síntesis de Kolbe
Komarowsky reaction reacción de Komarowsky
Konowaloff rule regla de Konowaloff
Kopp law ley de Kopp
Kopp rule regla de Kopp
koppite koppita
Körner method método de Körner
kovar kovar
Krafft degradation degradación de Krafft

Kramers theorem teorema de Kramers
Krebs cycle ciclo de Krebs
kremersite kremersita
Kröhnke aldehyde synthesis síntesis de aldehídos de Kröhnke
Kroll process proceso de Kroll
krügite krugita
krypton criptón
Kucherov reaction reacción de Kucherov
Kuhn-Winterstein reaction reacción de Kuhn-Winterstein
Kundt effect efecto de Kundt
Kundt rule regla de Kundt
kunzite kunzita
kyanite quianita
kynurenine quinurenina

L

label marcador, etiqueta
LaBel tube tubo de LaBel
labeled marcado, rotulado
labeled compound compuesto marcado
labeling marcado, rotulación
labile lábil
labile complex complejo lábil
laboratory laboratorio
laboratory apparatus aparato de laboratorio
laboratory conditions condiciones de laboratorio
laboratory device dispositivo de laboratorio
laboratory environment ambiente de laboratorio
laboratory equipment equipo de laboratorio
laboratory experiment experimento de laboratorio
laboratory machinery maquinaria de laboratorio
laboratory standard patrón de laboratorio
labradorite labradorita
laccase lacasa
lachrymator lacrimógeno
lachrymatory lacrimógeno
lacmoid lacmoide
lacquer laca
lactalbumin lactalbúmina
lactam lactama
lactamic acid ácido lactámico
lactamide lactamida
lactase lactasa
lactate lactato
lactic acid ácido láctico
lactic anhydride anhídrido láctico
lactide lactida
lactim lactima
lactobiose lactobiosa
lactogenic lactogénico
lactoglobulin lactoglobulina

lactol lactol
lactolide lactolida
lactone lactona
lactonic acid ácido lactónico
lactonitrile lactonitrilo
lactoprene lactopreno
lactose lactosa
lactoyl lactoilo
lactyl lactilo
ladder polymer polímero en escalera
Ladenburg rearrangement reordenamiento de Ladenburg
lagoriolite lagoriolita
lake laca
lambda lambda
lambda phenomenon fenómeno lambda
lambda point punto lambda
lambert lambert
Lambert-Beer law ley de Lambert-Beer
Lambert law ley de Lambert
laminarin laminarina
laminate laminado
laminated laminado
lampblack negro de humo
lanarkite lanarkita
Landau levels niveles de Landau
Landé interval rule regla de intervalos de Landé
landsbergite landsbergita
langbeinite langbeinita
Langmuir adsorption isotherm isoterma de adsorción de Langmuir
Langmuir-Blodgett film película de Langmuir-Blodgett
Langmuir theory teoría de Langmuir
lanoceric acid ácido lanocérico
lanolin lanolina
lanosterol lanosterol
lansfordite lansfordita
lanthana lantana
lanthanide lantánido
lanthanide contraction contracción lantánida

lanthanide series serie lantánida
lanthanum lantano
lanthanum acetate acetato de lantano
lanthanum ammonium nitrate nitrato amónico de lantano
lanthanum antimonide antimoniuro de lantano
lanthanum arsenide arseniuro de lantano
lanthanum carbonate carbonato de lantano
lanthanum chloride cloruro de lantano
lanthanum chloroanilate cloroanilato de lantano
lanthanum fluoride fluoruro de lantano
lanthanum nitrate nitrato de lantano
lanthanum oxalate oxalato de lantano
lanthanum oxide óxido de lantano
lanthanum phosphide fosfuro de lantano
lanthanum sesquioxide sesquióxido de lantano
lanthanum sulfate sulfato de lantano
lanthanum sulfide sulfuro de lantano
lanthanum trioxide trióxido de lantano
lanthionine lantionina
lapis lazuli lapislázuli
Laporte selection rule regla de selección de Laporte
lard manteca
lard oil aceite de manteca
larvicide larvicida
laser láser
laser spectroscopy espectroscopia de láser
Lassaigne test prueba de Lassaigne
latent energy energía latente
latent heat calor latente
latent solvent solvente latente
lateral chain cadena lateral
laterite laterita
latex látex
lattice red, celosía, retículo
lattice energy energía reticular
laudanidine laudanidina
laudanine laudanina
laudanosine laudanosina
laudanum láudano
Laue pattern patrón de Laue
laughing gas gas hilarante
laumonite laumonita

lauraldehyde lauraldehído
laurel leaves oil aceite de hojas de laurel
Laurent acid ácido de Laurent
lauric acid ácido láurico
lauric aldehyde aldehído láurico
laurite laurita
laurone laurona
lauroyl lauroilo
lauroyl chloride cloruro de lauroilo
lauroyl peroxide peróxido de lauroilo
lauroylsarcosine lauroilsarcosina
lauryl laurilo
lauryl acetate laurilacetato
lauryl alcohol alcohol laurílico
lauryl aldehyde aldehído laurílico
lauryl bromide laurilbromuro
lauryl chloride laurilcloruro
lauryl mercaptan laurilmercaptano
lauryl methacrylate laurilmetacrilato
lauryl sulfate laurilsulfato
laurylpyridinium chloride cloruro de laurilpiridinio
lautal lautal
lavender oil aceite de lavanda
law of alternation ley de alternación
law of conservation ley de conservación
law of definite composition ley de composición definida
law of definite proportions ley de proporciones definidas
law of diffusion ley de difusión
law of dilution ley de dilución
law of displacement ley de desplazamiento
law of distribution ley de distribución
law of equilibrium ley de equilibrio
law of mass action ley de acción de masas
law of mass conservation ley de conservación de masa
law of multiple proportions ley de proporciones múltiples
law of partition ley de partición
law of radioactive displacement ley de desplazamiento radiactivo
law of reciprocal proportions ley de proporciones recíprocas
lawrencium laurencio
laws of thermodynamics leyes de

termodinámica
layer capa
lazulite lazulita
lazurite lazurita
Le Chatelier principle principio de Le
   Chatelier
leaching lixiviación
lead plomo
lead accumulator acumulador de plomo
lead acetate acetato de plomo
lead antimonate antimoniato de plomo
lead arsenate arseniato de plomo
lead arsenite arsenito de plomo
lead azide azida de plomo
lead battery batería de plomo
lead benzoate benzoato de plomo
lead biorthophosphate biortofosfato de
   plomo
lead borate borato de plomo
lead borosilicate borosilicato de plomo
lead bromate bromato de plomo
lead bromide bromuro de plomo
lead butyrate butirato de plomo
lead carbonate carbonato de plomo
lead cell celda de plomo
lead chlorate clorato de plomo
lead chloride cloruro de plomo
lead chromate cromato de plomo
lead citrate citrato de plomo
lead-coated plomado
lead-coating plomado
lead cyanate cianato de plomo
lead cyanide cianuro de plomo
lead dichromate dicromato de plomo
lead dioxide dióxido de plomo
lead dithionate ditionato de plomo
lead ethanoate etanoato de plomo
lead fluoride fluoruro de plomo
lead fluoroborate fluoroborato de plomo
lead fluorosilicate fluorosilicato de
   plomo
lead formate formiato de plomo
lead fumarate fumarato de plomo
lead glass vidrio de plomo
lead halide haluro de plomo
lead hexacyanoferrate hexacianoferrato
   de plomo
lead hexafluorosilicate hexafluorosilicato
   de plomo

lead hydroxide hidróxido de plomo
lead hyposulfite hiposulfito de plomo
lead iodate yodato de plomo
lead iodide yoduro de plomo
lead lactate lactato de plomo
lead laurate laurato de plomo
lead linoleate linoleato de plomo
lead malate malato de plomo
lead maleate maleato de plomo
lead metasilicate metasilicato de plomo
lead molybdate molibdato de plomo
lead mononitroresorcinate
   mononitrorresorcinato de plomo
lead monosulfide monosulfuro de plomo
lead monoxide monóxido de plomo
lead naphthalenesulfonate
   naftalenosulfonato de plomo
lead naphthenate naftenato de plomo
lead nitrate nitrato de plomo
lead nitrite nitrito de plomo
lead octoate octoato de plomo
lead oleate oleato de plomo
lead orthophosphate ortofosfato de
   plomo
lead oxalate oxalato de plomo
lead oxide óxido de plomo
lead palmitate palmitato de plomo
lead pentasulfide pentasulfuro de plomo
lead perchlorate perclorato de plomo
lead peroxide peróxido de plomo
lead phosphate fosfato de plomo
lead phosphinate fosfinato de plomo
lead phosphite fosfito de plomo
lead phthalate ftalato de plomo
lead picrate picrato de plomo
lead propionate propionato de plomo
lead protoxide protóxido de plomo
lead pyrophosphate pirofosfato de
   plomo
lead resinate resinato de plomo
lead salicylate salicilato de plomo
lead selenate seleniato de plomo
lead selenide seleniuro de plomo
lead sesquioxide sesquióxido de plomo
lead silicate silicato de plomo
lead silicofluoride silicofluoruro de
   plomo
lead stannate estannato de plomo
lead stearate estearato de plomo

lead subacetate subacetato de plomo
lead subcarbonate subcarbonato de
    plomo
lead suboxide subóxido de plomo
lead sulfate sulfato de plomo
lead sulfide sulfuro de plomo
lead sulfite sulfito de plomo
lead sulfocyanide sulfocianuro de plomo
lead superoxide superóxido de plomo
lead tallate talato de plomo
lead tartrate tartrato de plomo
lead telluride telururo de plomo
lead tetraacetate tetraacetato de plomo
lead tetrachloride tetracloruro de plomo
lead tetraoxide tetróxido de plomo
lead thiocyanate tiocianato de plomo
lead thiosulfate tiosulfato de plomo
lead titanate titanato de plomo
lead trinitroresorcinate
    trinitrorresorcinato de plomo
lead tungstate tungstato de plomo
lead vanadate vanadato de plomo
lead wolframate wolframato de plomo
leaf hoja
leak fuga
leakage fuga
leakage radiation radiación de fuga
leaky con fugas
leather cuero
Lebedev process proceso de Lebedev
lecithin lecitina
Leclanché cell pila de Leclanché
lectin lectina
Leduc rule regla de Leduc
legal chemistry química legal
Lehmsted-Tanasescu reaction reacción
    de Lehmsted-Tanasescu
lemon oil aceite de limón
lemongrass oil aceite de luisa
lenacil lenacil
lene leno
leonardite leonardita
leonite leonita
lepargylic acid ácido lepargílico
lepidine lepidina
lepidolite lepidolita
lepidone lepidona
lepton leptón
lethal letal

lethal dose dosis letal
Letts synthesis síntesis de Letts
leucine leucina
leucite leucita
Leuckart reaction reacción de Leuckart
leuco leuco
leucoaniline leucoanilina
leucoline leucolina
leuconic acid ácido leucónico
leucovorin leucovorina
leucyl leucilo
leukonite leuconita
levo levo
levorotatory levorrotatorio
levulinic acid ácido levulínico
levulinic aldehyde aldehído levulínico
levulose levulosa
Lewis acid ácido de Lewis
Lewis atom átomo de Lewis
Lewis base base de Lewis
Lewis electron theory teoría electrónica
    de Lewis
Lewis metal metal de Lewis
Lewis structure estructura de Lewis
Lewis theory teoría de Lewis
lewisite lewisita
liberation liberación
licanic acid ácido licánico
lid tapa
lidocaine lidocaína
Lieben reaction reacción de Lieben
Liebermann reaction reacción de
    Liebermann
Liebig condenser condensador de Liebig
ligand ligando
ligand-field theory teoría de campo de
    ligandos
ligase ligasa
light emission emisión de luz
light flash destello de luz
light hydrocarbon hidrocarburo ligero
light hydrogen hidrógeno ligero
light metal metal ligero
light oil aceite ligero
light quantum cuanto de luz
light scattering dispersión de luz
light-sensitive sensible a la luz
light spectrum espectro de luz
light water agua ligera

lignin lignina
lignite lignita
lignocaine lignocaína
lignoceric acid ácido lignocérico
lignosulfonate lignosulfonato
ligroin ligroína
lime cal
lime oil aceite de lima
lime water agua de cal
limestone caliza
limited stability estabilidad limitada
limiting density densidad limitadora
limiting factor factor limitador
limonene limoneno
limonite limonita
linalool linalol
linalyl linalilo
linalyl acetate acetato de linalilo
linalyl formate formiato de linalilo
linalyl isobutyrate isobutirato de linalilo
linalyl propionate propionato de linalilo
linamarin linamarina
linarite linarita
lincomycin lincomicina
lindane lindano
Linde process proceso de Linde
lindgrenite lindgrenita
line spectrum espectro de líneas
linear lineal
linear molecule molécula lineal
linear polymer polímero lineal
linkage enlace, encadenamieto
linnaeite linnaeíta
linoleate linoleato
linoleic acid ácido linoleico
linolenic acid ácido linolénico
linolenyl alcohol alcohol linolenílico
linolic acid ácido linólico
linseed oil aceite de linaza
linseed oil meal harina de aceite de
    linaza
lionite lionita
lipase lipasa
lipid lípido
lipolysis lipólisis
lipophilic lipofílico
lipoprotein lipoproteína
lipotropic agent agente lipotrópico
lipoxidase lipoxidasa

lipoxygenase lipoxigenasa
liquation licuación
liquefaction licuefacción
liquefied licuado
liquefied gas gas licuado
liquefied natural gas gas natural licuado
liquefied petroleum gas gas de petróleo
    licuado
liqueur licor aromático
liquid líquido
liquid air aire líquido
liquid chromatography cromatografía
    líquida
liquid crystal cristal líquido
liquid dioxide dióxido líquido
liquid gas gas líquido
liquid glass vidrio líquido
liquid-liquid extraction extracción
    líquido-líquido
liquid nitrogen nitrógeno líquido
liquid oxygen oxígeno líquido
liquid-solid equilibrium equilibrio
    líquido-sólido
liquid state estado líquido
liquid-vapor equilibrium equilibrio
    líquido-vapor
liquor licor
liroconite liroconita
liter litro
litharge litarge
lithia litina
lithic acid ácido lítico
lithium litio
lithium acetate acetato de litio
lithium aluminate aluminato de litio
lithium aluminum deuteride deuteruro
    alumínico de litio
lithium aluminum hydride hidruro
    alumínico de litio
lithium amide amida de litio
lithium arsenate arseniato de litio
lithium benzoate benzoato de litio
lithium bicarbonate bicarbonato de litio
lithium borate borato de litio
lithium borohydride borohidruro de litio
lithium bromide bromuro de litio
lithium carbide carburo de litio
lithium carbonate carbonato de litio
lithium chlorate clorato de litio

lithium chloride cloruro de litio
lithium chromate cromato de litio
lithium citrate citrato de litio
lithium deuteride deuteruro de litio
lithium dichromate dicromato de litio
lithium fluoride fluoruro de litio
lithium fluorophosphate fluorofosfato de litio
lithium germanate germanato de litio
lithium halide haluro de litio
lithium hexachloroplatinate hexacloroplatinato de litio
lithium hexafluorosilicate hexafluorosilicato de litio
lithium hydride hidruro de litio
lithium hydrogen carbonate carbonato de hidrógeno de litio
lithium hydroxide hidróxido de litio
lithium hypochlorite hipoclorito de litio
lithium iodate yodato de litio
lithium iodide yoduro de litio
lithium lactate lactato de litio
lithium metaborate metaborato de litio
lithium metasilicate metasilicato de litio
lithium methoxide metóxido de litio
lithium methylate metilato de litio
lithium molybdate molibdato de litio
lithium myristate miristato de litio
lithium niobate niobato de litio
lithium nitrate nitrato de litio
lithium nitride nitruro de litio
lithium orthophosphate ortofosfato de litio
lithium oxalate oxalato de litio
lithium oxide óxido de litio
lithium palmitate palmitato de litio
lithium perchlorate perclorato de litio
lithium peroxide peróxido de litio
lithium phosphate fosfato de litio
lithium ricinoleate ricinoleato de litio
lithium salicylate salicilato de litio
lithium silicate silicato de litio
lithium stearate estearato de litio
lithium sulfate sulfato de litio
lithium sulfide sulfuro de litio
lithium sulfite sulfito de litio
lithium tetraborate tetraborato de litio
lithium tetrahydroaluminate tetrahidroaluminato de litio

lithium thiocyanate tiocianato de litio
lithium titanate titanato de litio
lithium tungstate tungstato de litio
lithium vanadate vanadato de litio
lithium zirconate zirconato de litio
lithocholic acid ácido litocólico
litidionite litidionita
litmus tornasol
litmus paper papel de tornasol
litre litro
livingstoneite livingstonita
lixiviation lixiviación
lobeline lobelina
localization localización
locant ubicador
logwood palo de campeche
London formula fórmula de London
lone pair par sin compartir
long-necked flask matraz de cuello largo
lorandite lorandita
lorazepam lorazepam
Loschmidt number número de Loschmidt
Lossen rearrangement reordenamiento de Lossen
low energy baja energía
low-energy electron diffraction difracción electrónica de baja energía
low-energy neutron neutrón de baja energía
low-energy particle partícula de baja energía
lowest occupied molecular orbital orbital molecular de mínima energía ocupado
Lowry-Brønsted theory teoría de Lowry-Brønsted
lubricant lubricante
lubricating oil aceite lubricante
lubrication lubricación
Lucas theory teoría de Lucas
luciferin luciferina
ludwigite ludwigita
Lugol solution solución de Lugol
luminance luminancia
luminescence luminiscencia
luminescent luminiscente
luminiferous luminífero
luminol luminol

luminometry luminometría
luminosity luminosidad
luminous luminoso
Lundergardh vaporizer vaporizador de
    Lundergardh
lupinidine lupinidina
luster brillo
lutein luteína
luteotropin luteotropina
lutetia lutecia
lutetium lutecio
lutetium chloride cloruro de lutecio
lutetium fluoride fluoruro de lutecio
lutetium nitrate nitrato de lutecio
lutetium oxide óxido de lutecio
lutetium sulfate sulfato de lutecio
lutidine lutidina
lycopene licopeno
lycopodium licopodio

lye lejía
Lyman continuum continuo de Lyman
Lyman series serie de Lyman
lyogel liogel
lyophilic liofílico
lyophilization liofilización
lyophobic liofóbico
lyotopic liotópico
lyotrope liótropo
lyotropic series serie liotrópica
Lyovac process proceso de Lyovac
lysergic acid ácido lisérgico
lysergic acid diethylamide dietilamida
    de ácido lisérgico
lysidine lisidina
lysine lisina
lysis lisis
lysozyme lisozima
lyxose lixosa

M

Macquer salt sal de Macquer
macro macro
macroanalysis macroanálisis
macrochemistry macroquímica
macrocyclic macrocíclico
macrolide macrólido
macromolecular macromolecular
macromolecular crystal cristal
    macromolecular
macromolecule macromolécula
macroscopic macroscópico
macrose macrosa
Maddrell salt sal de Maddrell
Madelung constant constante de
    Madelung
Madelung synthesis síntesis de
    Madelung
magenta magenta
magic numbers números mágicos
magnalite magnalita
magnalium magnalio
magnesia magnesia
magnesioferrite magnesioferrita
magnesite magnesita
magnesium magnesio
magnesium acetate acetato magnésico,
    acetato de magnesio
magnesium acetylacetonate
    acetilacetonato magnésico
magnesium alloy aleación magnésica
magnesium aluminate aluminato
    magnésico
magnesium amide amida magnésica
magnesium ammonium phosphate
    fosfato amónico magnésico
magnesium arsenate arseniato
    magnésico
magnesium benzoate benzoato
    magnésico
magnesium biphosphate bifosfato
    magnésico
magnesium borate borato magnésico
magnesium bromate bromato magnésico

magnesium bromide bromuro magnésico
magnesium carbonate carbonato
    magnésico
magnesium chlorate clorato magnésico
magnesium chloride cloruro magnésico
magnesium chromate cromato
    magnésico
magnesium citrate citrato magnésico
magnesium dichromate dicromato
    magnésico
magnesium dihydrogensulfate
    dihidrogenosulfato magnésico
magnesium dioxide dióxido magnésico
magnesium fluoride fluoruro magnésico
magnesium fluosilicate fluosilicato
    magnésico
magnesium flux fundente magnésico
magnesium formate formiato magnésico
magnesium gluconate gluconato
    magnésico
magnesium halide haluro magnésico
magnesium hexafluorosilicate
    hexafluorosilicato magnésico
magnesium hydride hidruro magnésico
magnesium hydrogen carbonate
    carbonato de hidrógeno magnésico
magnesium hydroxide hidróxido
    magnésico
magnesium hyposulfite hiposulfito
    magnésico
magnesium iodate yodato magnésico
magnesium iodide yoduro magnésico
magnesium lactate lactato magnésico
magnesium laurate laurato magnésico
magnesium lauryl sulfate laurilsulfato
    magnésico
magnesium limestone caliza magnésica
magnesium methoxide metóxido
    magnésico
magnesium methylate metilato
    magnésico
magnesium molybdate molibdato
    magnésico

magnesium myristate miristato magnésico

magnesium nitrate nitrato magnésico

magnesium nitride nitruro magnésico

magnesium nitrite nitrito magnésico

magnesium oleate oleato magnésico

magnesium orthophosphate ortofosfato magnésico

magnesium oxalate oxalato magnésico

magnesium oxide óxido magnésico

magnesium palmitate palmitato magnésico

magnesium perborate perborato magnésico

magnesium perchlorate perclorato magnésico

magnesium permanganate permanganato magnésico

magnesium peroxide peróxido magnésico

magnesium phosphate fosfato magnésico

magnesium phosphide fosfuro magnésico

magnesium phosphinate fosfinato magnésico

magnesium phosphonate fosfonato magnésico

magnesium propionate propionato magnésico

magnesium pyrophosphate pirofosfato magnésico

magnesium salicylate salicilato magnésico

magnesium silicate silicato magnésico

magnesium silicide siliciuro magnésico

magnesium silicofluoride silicofluoruro magnésico

magnesium stannate estannato magnésico

magnesium stannide estanniuro magnésico

magnesium stearate estearato magnésico

magnesium succinate succinato magnésico

magnesium sulfate sulfato magnésico

magnesium sulfide sulfuro magnésico

magnesium sulfite sulfito magnésico

magnesium tartrate tartrato magnésico

magnesium thiocyanate tiocianato magnésico

magnesium thiosulfate tiosulfato magnésico

magnesium trisilicate trisilicato magnésico

magnesium tungstate tungstato magnésico

magnesium urate urato magnésico

magnesium wolframate wolframato magnésico

magnesium zirconate zirconato magnésico

magnesyl megnesilo

magnetic magnético

magnetic activity actividad magnética

magnetic deflection deflexión magnética

magnetic element elemento magnético

magnetic moment momento magnético

magnetic quantum number número cuántico magnético

magnetic resonance resonancia magnética

magnetic resonance imaging formación de imágenes por resonancia magnética

magnetic separation separación magnética

magnetic spectrum espectro magnético

magnetic stirrer agitador magnético

magnetic susceptibility susceptibilidad magnética

magnetism magnetismo

magnetite magnetita

magnetization magnetización

magnetochemistry magnetoquímica

magnetohydrodynamics magnetohidrodinámica

magnetometric titration titulación magnetométrica, valoración magnetométrica

magneton magnetón

magnolite magnolita

magnolium magnolio

Magnus rule regla de Magnus

Magnus salt sal de Magnus

Maillard reaction reacción de Maillard

Majorana force fuerza de Majorana

malachite malaquita

malachite green verde malaquita

malamide malamida

Malaprade reaction reacción de
Malaprade
malate malato
malathion malatión
malchite malquita
maldonite maldonita
maleamic acid ácido maleámico
maleate maleato
maleic acid ácido maleico
maleic anhydride anhídrido maleico
maleic hydrazide hidrazida maleica
malenoid malenoide
maleoyl maleoilo
malic acid ácido málico
malic amide amida málica
mallardite mallardita
malleability maleabilidad
malleable maleable
malonaldehydic acid ácido
malonaldehídico
malonamide malonamida
malonate malonato
malonic acid ácido malónico
malonic amide amida malónica
malonic dinitrile dinitrilo malónico
malonic ester éster malónico
malonic ester synthesis síntesis de
ésteres malónicos
malonic mononitrile mononitrilo
malónico
malononitrile malononitrilo
malonyl malonilo
malonyl urea malonilurea
maloyl maloilo
malt malta
malt extract extracto de malta
maltase maltasa
maltha malta
malthene malteno
maltol maltol
maltose maltosa
mandarin oil aceite de mandarina
mandelic acid ácido mandélico
Mandelin reagent reactivo de Mandelin
mandelonitrile mandelonitrilo
maneb maneb
manganate manganato
manganese manganeso
manganese acetate acetato de manganeso

manganese arsenate arseniato de
manganeso
manganese binoxide binóxido de
manganeso
manganese black negro de manganeso
manganese borate borato de manganeso
manganese boride boruro de manganeso
manganese-boron manganeso-boro
manganese bromide bromuro de
manganeso
manganese bronze bronce de manganeso
manganese carbide carburo de
manganeso
manganese carbonate carbonato de
manganeso
manganese carbonyl carbonilo de
manganeso
manganese chloride cloruro de
manganeso
manganese chromate cromato de
manganeso
manganese citrate citrato de manganeso
manganese dioxide dióxido de
manganeso
manganese dithionate ditionato de
manganeso
manganese fluoride fluoruro de
manganeso
manganese gluconate gluconato de
manganeso
manganese glycerophosphate
glicerofosfato de manganeso
manganese halide haluro de manganeso
manganese heptoxide heptóxido de
manganeso
manganese hexacyanoferrate
hexacianoferrato de manganeso
manganese hexafluorosilicate
hexafluorosilicato de manganeso
manganese hydrate hidrato de
manganeso
manganese hydroxide hidróxido de
manganeso
manganese hypophosphite hipofosfito de
manganeso
manganese iodide yoduro de manganeso
manganese lactate lactato de manganeso
manganese linoleate linoleato de
manganeso

manganese monoxide monóxido de manganeso

manganese naphthenate naftenato de manganeso

manganese nitrate nitrato de manganeso

manganese octoate octoato de manganeso

manganese oleate oleato de manganeso

manganese oxalate oxalato de manganeso

manganese oxide óxido de manganeso

manganese peroxide peróxido de manganeso

manganese phosphate fosfato de manganeso

manganese phosphinate fosfinato de manganeso

manganese potassium sulfate sulfato potásico de manganeso

manganese protoxide protóxido de manganeso

manganese resinate resinato de manganeso

manganese sesquioxide sesquióxido de manganeso

manganese silicate silicato de manganeso

manganese sulfate sulfato de manganeso

manganese sulfide sulfuro de manganeso

manganese sulfite sulfito de manganeso

manganese tallate talato de manganeso

manganese tetrachloride tetracloruro de manganeso

manganese tetrafluoride tetrafluoruro de manganeso

manganese tetraoxide tetróxido de manganeso

manganese-titanium manganeso-titanio

manganese trioxide trióxido de manganeso

manganese tungstate tungstato de manganeso

manganic mangánico

manganic acetylacetonate acetilacetonato mangánico

manganic acid ácido mangánico

manganic fluoride fluoruro mangánico

manganic hydroxide hidróxido mangánico

manganic oxide óxido mangánico

manganin manganina

manganite manganita

manganosite manganosita

manganostilbite manganoestilbita

manganous manganoso

manganous acetate acetato manganoso

manganous arsenate arseniato manganoso

manganous bromide bromuro manganoso

manganous carbonate carbonato manganoso

manganous chloride cloruro manganoso

manganous chromate cromato manganoso

manganous citrate citrato manganoso

manganous fluoride fluoruro manganoso

manganous formate formiato manganoso

manganous hydroxide hidróxido manganoso

manganous iodide yoduro manganoso

manganous nitrate nitrato manganoso

manganous orthophosphate ortofosfato manganoso

manganous oxide óxido manganoso

manganous phosphate fosfato manganoso

manganous pyrophosphate pirofosfato manganoso

manganous silicate silicato manganoso

manganous sulfate sulfato manganoso

manganous sulfide sulfuro manganoso

manganous sulfite sulfito manganoso

manganous tartrate tartrato manganoso

Manila resin resina de Manila

manna maná

Mannich condensation reaction reacción de condensación de Mannich

Mannich reaction reacción de Mannich

mannite manita

mannitol manitol

mannitol hexanitrate hexanitrato de manitol

mannitose manitosa

mannose manosa

manometer manómetro

Marathon-Howard process proceso de Marathon-Howard

marble mármol

marcasite marcasita
margaric acid ácido margárico
margarine margarina
margarite margarita
marialite marialita
marihuana mariguana
marijuana mariguana
Mark-Houwink equation ecuación de
  Mark-Houwink
Markovnikoff rule regla de
  Markovnikoff
Markovnikov rule regla de Markovnikov
Marme reagent reactivo de Marme
marsh gas gas de pantano
Marsh test prueba de Marsh
martensite martensita
Martinet synthesis síntesis de Martinet
martite martita
martonite martonita
mascagnite mascagnita
maser máser
mash templa
masked element elemento enmascarado
masked radical radical enmascarado
masking enmascaramiento
mass masa
mass action law ley de acción de masas
mass conservation law ley de
  conservación de masa
mass defect defecto de masa
mass fraction fracción de masa
mass number número de masa
mass spectrograph espectrógrafo de
  masa
mass spectrometer espectrómetro de
  masa
mass spectrometry espectrometría de
  masa
mass spectroscope espectroscopio de
  masa
mass spectroscopy espectroscopia de
  masa
mass spectrum espectro de masa
mass-to-charge ratio razón de masa a
  carga
massicot masicote
master batch lote matriz
master equation ecuación maestra
mastication masticación

material material
materialization materialización
materials handling manejo de materiales
matildite matildita
matlockite matlockita
matrix matriz
matrix spectrophotometry
  espectrofotometría de matriz
matte mata
matter materia
maxivalence maxivalencia
Maxwell-Boltzmann distribution
  distribución de Maxwell-Boltzmann
Mayer reagent reactivo de Mayer
McFayden-Stevens reaction reacción de
  McFayden-Stevens
McLafferty rearrangement
  reordenamiento de McLafferty
McMillan-Mayer theory teoría de
  McMillan-Mayer
mean free path recorrido libre medio
mean free time tiempo libre medio
mean life vida media
measurement medida
measurement error error de medida
measuring beaker vaso de laboratorio de
  medida
measuring flask matraz de medida
mechanical analysis análisis mecánico
mechanism mecanismo
mechanism of reaction mecanismo de
  reacción
mechanochemistry mecanoquímica
meclizine hydrochloride clorhidrato de
  meclizina
meconin meconina
medicinal chemistry química medicinal
medium medio
Meisenheimer complex complejo de
  Meisenheimer
Meisenheimer rearrangement
  reordenamiento de Meisenheimer
melamine melamina
melamine resin resina de melamina
melaniline melanilina
melanin melanina
melibiose melibiosa
melissic acid ácido melísico
melissyl alcohol alcohol melisílico

mellitate melitato
mellitic acid ácido melítico
melonite melonita
melphalan melfalano
melt fundir, derretir
melting point punto de fusión
melting-point depression depresión de punto de fusión
membrane membrana
membrane hydrolysis hidrólisis por membrana
memtetrahydrophthalic anhydride anhídrido memtetrahidroftálico
menachanite menacanita
menadione menadiona
Mendeleev law ley de Mendeleev
mendelevium mendelevio
Mendius reaction reacción de Mendius
mendozite mendocita
meneghinite meneghinita
menhaden oil aceite de menhaden
meniscus menisco
Menschutkin reaction reacción de Menschutkin
mensuration mensuración
menthacamphor mentalcanfor
menthadiene mentadieno
menthane mentano
menthanediamine mentanodiamina
menthanol mentanol
menthanone mentanona
menthene menteno
menthenol mentenol
menthenyl mentenilo
menthol mentol
menthol valerate valerato de mentol
menthonaphthene mentonafteno
menthone mentona
menthyl mentilo
menthyl acetate acetato de mentilo
menthyl salicylate salicilato de mentilo
meperidine hydrochloride clorhidrato de meperidina
meprobamate meprobamato
mercaptal mercaptal
mercaptamine mercaptamina
mercaptan mercaptano
mercapto mercapto
mercaptoacetic acid ácido mercaptoacético
mercaptobenzoic acid ácido mercaptobenzoico
mercaptobenzothiazole mercaptobenzotiazol
mercaptoethanol mercaptoetanol
mercaptol mercaptol
mercaptophenyl mercaptofenilo
mercaptopropionic acid ácido mercaptopropiónico
mercaptosuccinic acid ácido mercaptosuccínico
mercaptothiazoline mercaptotiazolina
mercapturic acid ácido mercaptúrico
mercuration mercuración
mercurial mercurial
mercuric mercúrico
mercuric acetate acetato mercúrico
mercuric arsenate arseniato mercúrico
mercuric barium bromide bromuro bárico mercúrico
mercuric barium iodide yoduro bárico mercúrico
mercuric benzoate benzoato mercúrico
mercuric borate borato mercúrico
mercuric bromide bromuro mercúrico
mercuric carbonate carbonato mercúrico
mercuric chloride cloruro mercúrico
mercuric chromate cromato mercúrico
mercuric cyanate cianato mercúrico
mercuric cyanide cianuro mercúrico
mercuric dichromate dicromato mercúrico
mercuric fluoride fluoruro mercúrico
mercuric hydroxide hidróxido mercúrico
mercuric iodide yoduro mercúrico
mercuric lactate lactato mercúrico
mercuric nitrate nitrato mercúrico
mercuric oleate oleato mercúrico
mercuric oxalate oxalato mercúrico
mercuric oxide óxido mercúrico
mercuric oxycyanide oxicianuro mercúrico
mercuric phosphate fosfato mercúrico
mercuric potassium cyanide cianuro potásico mercúrico
mercuric salicylate salicilato mercúrico
mercuric sodium phenolsulfate fenolsulfato sódico mercúrico

mercuric stearate estearato mercúrico

mercuric succinate succinato mercúrico

mercuric sulfate sulfato mercúrico

mercuric sulfide sulfuro mercúrico

mercuric sulfocyanate sulfocianato mercúrico

mercuric thiocyanate tiocianato mercúrico

mercuride mercururo

mercurification mercurificación

mercurization mercurización

mercurous mercurioso

mercurous acetate acetato mercurioso

mercurous acetylide acetiluro mercurioso

mercurous bromide bromuro mercurioso

mercurous chlorate clorato mercurioso

mercurous chloride cloruro mercurioso

mercurous chromate cromato mercurioso

mercurous formate formiato mercurioso

mercurous iodide yoduro mercurioso

mercurous nitrate nitrato mercurioso

mercurous oxalate oxalato mercurioso

mercurous oxide óxido mercurioso

mercurous phosphate fosfato mercurioso

mercurous sulfate sulfato mercurioso

mercurous sulfide sulfuro mercurioso

mercurous tartrate tartrato mercurioso

mercury mercurio

mercury acetate acetato de mercurio

mercury alloy aleación de mercurio

mercury amalgam amalgama de mercurio

mercury arsenate arseniato de mercurio

mercury benzoate benzoato de mercurio

mercury bichloride bicloruro de mercurio

mercury bromide bromuro de mercurio

mercury cell pila de mercurio

mercury chloride cloruro de mercurio

mercury chromate cromato de mercurio

mercury cyanide cianuro de mercurio

mercury dichromate dicromato de mercurio

mercury dithiocarbonate ditiocarbonato de mercurio

mercury fulminate fulminato de mercurio

mercury hexacyanoferrate hexacianoferrato de mercurio

mercury hexafluorosilicate hexafluorosilicato de mercurio

mercury nitrate nitrato de mercurio

mercury oxide óxido de mercurio

mercury peroxide peróxido de mercurio

mercury phosphate fosfato de mercurio

mercury selenide seleniuro de mercurio

mercury telluride telururo de mercurio

mercury tetraborate tetraborato de mercurio

meridional meridional

merwinite merwinita

mesaconic acid ácido mesacónico

mescaline mescalina

mesitite mesitita

mesitol mesitol

mesityl mesitilo

mesityl oxide óxido de mesitilo

mesitylene mesitileno

mesitylenesulfonyl chloride cloruro de mesitilensulfonilo

meso meso

mesocolloid mesocoloide

mesoionic mesoiónico

mesomer mesómero

mesomeric mesomérico

mesomerism mesomería

mesomorphic mesomórfico

mesomorphic state estado mesomórfico

meson mesón

meson mass masa mesónica

mesonic mesónico

mesoscopic mesoscópico

mesothorium mesotorio

mesotomy mesotomía

mesoxalic acid ácido mesoxálico

mesoxalo mesoxalo

mesoxalyl mesoxalilo

mesyl mesilo

mesyl chloride cloruro de mesilo

meta meta

metabolic metabólico

metabolism metabolismo

metabolite metabolito

metaborate metaborato

metaboric acid ácido metabórico

metachromasia metacromasia
metachromatism metacromatismo
metaformaldehyde meaformaldehído
metal metal
metal-clad blindado
metal-coated metalizado
metal deposition deposición metálica
metal electrode electrodo metálico
metal film película metálica
metal foil hoja metálica
metal glass vidrio metálico
metal-metal bond enlace metal-metal
metalation metalación
metaldehyde metaldehído
metallic metálico
metallic binding forces fuerzas ligantes
      metálicas
metallic bond enlace metálico
metallic bonding enlazado metálico
metallic conduction conducción metálica
metallic crystal cristal metálico
metallic soap jabón metálico
metallify metalificar
metallization metalización
metallize metalizar
metallized metalizado
metallizing metalización
metallocene metaloceno
metallochromic metalocrómico
metallochromic indicator indicador
      metalocrómico
metallography metalografía
metalloid metaloide
metallurgy metalurgia
metalorganic metalorgánico
metamer metámero
metamerism metamerismo
metanilic acid ácido metalínico
metaphenylene metafenileno
metaphosphate metafosfato
metaphosphoric acid ácido
      metafosfórico
metasilicic acid ácido metasilícico
metasomatosis metasomatosis
metastable metaestable
metastable electron electrón metaestable
metastable ion ion metaestable
metastable phase fase metaestable
metastable state estado metaestable

metastasic electron electrón metastásico
metastasis metástasis
metastructure metaestructura
metathesis descomposición doble
metathetical reaction reacción
      metatética
metatitanic acid ácido metatitánico
meteoric meteórico
metepa metepa
meter metro, medidor
methacetin metacetina
methacrolein metacroleína
methacrylamide metacrilamida
methacrylate metacrilato
methacrylic acid ácido metacrílico
methacrylonitrile metacrilonitrilo
methacryloyl chloride cloruro de
      metacriloilo
methadone metadona
methadone hydrochloride clorhidrato de
      metadona
methallydine diacetate diacetato de
      metalidina
methallyl metalilo
methallyl alcohol alcohol metalílico
methanal metanal
methanamide metanamida
methanation metanación
methane metano
methanecarboxylic acid ácido
      metanocarboxílico
methanedicarboxylic acid ácido
      metanodicarboxílico
methanesulfonate metanosulfonato
methanesulfonic acid ácido
      metanosulfónico
methanesulfonyl chloride cloruro de
      metanosulfonilo
methanethiol metanotiol
methano metano
methanoate metanoato
methanoic acid ácido metanoico
methanol metanol
methanolate metanolato
methanoyl metanoilo
methenamine metenamina
methene meteno
methenyl metenilo
methicillin meticilina

methide metida
methine metina
methionic acid ácido metiónico
methionine metionina
methionyl metionilo
methomyl metomilo
methose metosa
methotrexate metotrexato
methoxalyl metoxalilo
methoxide metóxido
methoxy metoxi
methoxyacetaldehyde metoxiacetaldehído
methoxyacetanilide metoxiacetanilida
methoxyacetic acid ácido metoxiacético
methoxyacetophenone metoxiacetofenona
methoxyamine metoxiamina
methoxybenzaldehyde metoxibenzaldehído
methoxybenzene metoxibenceno
methoxybenzoic acid ácido metoxibenzoico
methoxybenzyl alcohol alcohol metoxibencílico
methoxybutanol metoxibutanol
methoxyethyl metoxietilo
methoxyl metoxilo
methoxynaphthalene metoxinaftaleno
methoxyphenol metoxifenol
methoxyphenyl metoxifenilo
methoxypropanol metoxipropanol
methoxypropylamine metoxipropilamina
methyl metilo
methyl abietate abietato de metilo
methyl acetate acetato de metilo
methyl acetoacetate acetoacetato de metilo
methyl acetone metilacetona
methyl acetylricinoleate acetilricinoleato de metilo
methyl acrylate acrilato de metilo
methyl alcohol alcohol metílico
methyl arachidate araquidato de metilo
methyl behenate behenato de metilo
methyl benzoate benzoato de metilo
methyl blue azul de metilo
methyl borate borato de metilo
methyl bromide bromuro de metilo

methyl butyrate butirato de metilo
methyl caprate caprato de metilo
methyl caproate caproato de metilo
methyl caprylate caprilato de metilo
methyl carbonate carbonato de metilo
methyl ceroate ceroato de metilo
methyl chloride cloruro de metilo
methyl chloroacetate cloroacetato de metilo
methyl chloroformate cloroformiato de metilo
methyl chlorosilane clorosilano de metilo
methyl chlorosulfonate clorosulfonato de metilo
methyl cinnamate cinamato de metilo
methyl cyanate cianato de metilo
methyl cyanide cianuro de metilo
methyl cyanoacetate cianoacetato de metilo
methyl cyanoformate cianoformiato de metilo
methyl dichloroacetate dicloroacetato de metilo
methyl dichlorostearate dicloroestearato de metilo
methyl elaidate elaidato de metilo
methyl ester éster metílico
methyl ether éter metílico
methyl fluoride fluoruro de metilo
methyl fluorosulfonate fluorosulfonato de metilo
methyl formate formiato de metilo
methyl gallate galato de metilo
methyl glycol metilglicol
methyl heneicosanoate heneicosanoato de metilo
methyl heptadecanoate heptadecanoato de metilo
methyl hydrazine hidrazina de metilo
methyl hydride hidruro de metilo
methyl iodide yoduro de metilo
methyl isocyanate isocianato de metilo
methyl isonicotinate isonicotinato de metilo
methyl isothiocyanate isotiocianato de metilo
methyl lactate lactato de metilo
methyl laurate laurato de metilo

methyl lauroleate lauroleato de metilo
methyl lignocerate lignocerato de metilo
methyl linoleate linoleato de metilo
methyl linolenate linolenato de metilo
methyl methacrylate metacrilato de
    metilo
methyl myristate miristato de metilo
methyl nitrate nitrato de metilo
methyl nitrite nitrito de metilo
methyl oleate oleato de metilo
methyl orange anaranjado de metilo
methyl oxide óxido de metilo
methyl palmitate palmitato de metilo
methyl pentadecanoate pentadecanoato
    de metilo
methyl propionate propionato de metilo
methyl pyruvate piruvato de metilo
methyl red rojo de metilo
methyl ricinoleate ricinoleato de metilo
methyl salicylate salicilato de metilo
methyl stearate estearato de metilo
methyl sulfate sulfato de metilo
methyl sulfide sulfuro de metilo
methyl tridecanoate tridecanoato de
    metilo
methyl violet violeta de metilo
methyl yellow amarillo de metilo
methylacetic acid ácido metilacético
methylacetophenone metilacetofenona
methylacetylene metilacetileno
methylacrylamide metilacrilamida
methylal metilal
methylalanine metilalanina
methylallyl acetate acetato de metilalilo
methylallyl alcohol alcohol metilalílico
methylallyl chloride cloruro de
    metilalilo
methylaluminum metilaluminio
methylamine metilamina
methylamino metilamino
methylaminoacetic acid ácido
    metilaminoacético
methylaminophenol metilaminofenol
methylamyl metilamilo
methylamyl acetate acetato de
    metilamilo
methylamyl alcohol alcohol metilamílico
methylamyl ketone metilamilcetona
methylaniline metilanilina

methylanisole metilanisol
methylanthracene metilantraceno
methylanthranilate metilantranilato
methylanthraquinone metilantraquinona
methylate metilar
methylated metilado
methylbenzene metilbenceno
methylbenzoic acid ácido metilbenzoico
methylbenzyl metilbencilo
methylbenzyl alcohol alcohol
    metilbencílico
methylbenzyl ether éter metilbencílico
methylbenzylamine metilbencilamina
methylbutane metilbutano
methylbutanol metilbutanol
methylbutene metilbuteno
methylbutyl ketone metilbutilcetona
methylbutylamine metilbutilamina
methylbutynol metilbutinol
methylcellulose metilcelulosa
methylcholanthrene metilcolantreno
methylcyclohexane metilciclohexano
methylcyclohexanol metilciclohexanol
methylcyclohexanone
    metilciclohexanona
methylcyclohexanyl metilciclohexanilo
methylcyclohexylamine
    metilciclohexilamina
methylcyclopentane metilciclopentano
methyldichloroarsine metildicloroarsina
methyldichlorosilane metildiclorosilano
methyldiethanolamine
    metildietanolamina
methyldioxolane metildioxolano
methyldiphenylamine metildifenilamina
methyldipropylmethane
    metildipropilmetano
methylene metileno
methylene blue azul de metileno
methylene bromide bromuro de metileno
methylene chloride cloruro de metileno
methylene dichloride dicloruro de
    metileno
methylene diiodide diyoduro de metileno
methylene iodide yoduro de metileno
methylene oxide óxido de metileno
methylenecyclopentadiene
    metilenociclopentadieno
methylenedisalicylic acid ácido

metilenodisalicílico

methylethyl ketone metiletilcetona

methylethylcellulose metiletilcelulosa

methylformanilide metilformanilida

methylfuran metilfurano

methylfurfurylamine metilfurfulrilamina

methylheptane metilheptano

methylheptenone metilheptenona

methylheptylamine metilheptilamina

methylhexane metilhexano

methylhexyl ketone metilhexilcetona

methylhydrazine metilhidrazina

methylhydrazone metilhidrazona

methylhydroxybutanone
metilhidroxibutanona

methylic metílico

methylidyne metilidino

methylindole metilindol

methylisoamyl ketone
metilisoamilcetona

methylisobutyl carbinol
metilisobutilcarbinol

methylisobutyl ketone
metilisobutilcetona

methylmagnesium bromide bromuro de
metilmagnesio

methylmagnesium chloride cloruro de
metilmagnesio

methylmagnesium iodide yoduro de
metilmagnesio

methylmaleic acid ácido metilmaleico

methylmercury cyanide cianuro de
metilmercurio

methylmethane metilmetano

methylnaphthalene metilnaftaleno

methylnitrobenzene metilnitrobenceno

methylol metilol

methylolurea metilolurea

methylparaben metilparabeno

methylpentadiene metilpentadieno

methylpentaldehyde metilpentaldehído

methylpentane metilpentano

methylpentanoic acid ácido
metilpentanoico

methylpentanol metilpentanol

methylpentene metilpenteno

methylpentose metilpentosa

methylphenyl ether éter metilfenílico

methylphenyl ketone metilfenilcetona

methylphenyldichlorosilane
metilfenildiclorosilano

methylphosphonic acid ácido
metilfosfónico

methylphosphoric acid ácido
metilfosfórico

methylpiperazine metilpiperazina

methylpropane metilpropano

methylpropene metilpropeno

methylpropyl ketone metilpropilcetona

methylpyrrole metilpirrol

methylpyrrolidine metilpirrolidina

methylquinoline metilquinolina

methylstyrene metilestireno

methylsulfonic acid ácido metilsulfónico

methylsulfonyl metilsulfonilo

methylsulfuric acid ácido metilsulfúrico

methyltaurine metiltaurina

methyltrichlorosilane metiltriclorosilano

methyltrinitrobenzene
metiltrinitrobenceno

methylundecanoate metilundecanoato

methyne metino

metopon hydrochloride clorhidrato de
metopón

metre metro

mevalonic acid ácido mevalónico

Meyer formula fórmula de Meyer

Meyer law ley de Meyer

Meyer reaction reacción de Meyer

Meyer synthesis síntesis de Meyer

meymacite meimacita

miargyrite miargirita

miazine miazina

mica mica

micelle micela

Michael condensation condensación de
Michael

Michael reaction reacción de Michael

Michaelis-Arbuzov reaction reacción de
Michaelis-Arbuzov

Michaelis constant constante de
Michaelis

miconazole nitrate nitrato de miconazol

micril micrilo

micro micro

microanalysis microanálisis

microanalyzer microanalizador

microbalance microbalanza

microbe microbio
microbicide microbicida
microbody microcuerpo
microcapsule microcápsula
microchemical microquímico
microchemistry microquímica
microcrystal microcristal
microcrystalline microcristalino
microcrystalline wax cera
    microcristalina
microcurie microcurie
microdensitometer microdensitómetro
microdiffusion microdifusión
microelectrode microelectrodo
microelectrolysis microelectrólisis
microelectrophoresis microelectroforesis
microencapsulation microencapsulación
microgalvanometer microgalvanómetro
microgammil microgamilo
microgram microgramo
micrography micrografía
microincineration microincineración
microline microlina
micrometer micrómetro
micron micrón
micronutrient micronutriente
microorganism microorganismo
microphotometer microfotómetro
micropipet micropipeta
microporous microporoso
microprobe microsonda
microreaction microrreacción
microscope microscopio
microscopic microscópico
microscopic examination examen
    microscópico
microscopy microscopia
microsecond microsegundo
microspectrophotometer
    microespectrofotómetro
microspectroscope microespectroscopio
microwave microonda
microwave absorber absorbedor de
    microondas
microwave spectrometer espectrómetro
    de microondas
microwave spectroscope espectroscopio
    de microondas
microwave spectroscopy espectroscopia

de microondas
microwave spectrum espectro de
    microondas
miemite miemita
Miescher degradation degradación de
    Miescher
Mignonac reaction reacción de
    Mignonac
migration migración
migration current corriente de
    migración
migration velocity velocidad de
    migración
mildew moho
milk of magnesia leche de magnesia
milk sugar azúcar de leche
milky nublado
mill molino
milli mili
millibar milibarra
millicurie milicurie
milliequivalent miliequivalente
milligram miligramo
milliliter mililitro
millimeter milímetro
millimeter of mercury milímetro de
    mercurio
millimole milimol
millinormal milinormal
Millon base base de Millon
Millon reagent reactivo de Millon
Millon test prueba de Millon
mimetite mimetita
mineral mineral
mineral acid ácido mineral
mineral alkali álcali mineral
mineral carbon carbón mineral
mineral fat grasa mineral
mineral jelly jalea mineral
mineral oil aceite mineral
mineral pigment pigmento mineral
mineral pitch brea mineral
mineral tar alquitrán mineral
mineral wax cera mineral
mineralization mineralización
mineralogy mineralogía
minimum ionizing speed velocidad
    mínima de ionización
minium minio

minivalence minivalencia
minulite minulita
mirabilite mirabilita
mirror-image imagen de espejo
mirror nuclei núcleos espejos
mischmetal mischmetal
miscibility miscibilidad
miscible miscible
misfire falla de encendido
mispickel mispiquel
mitomycin mitomicina
Mitsunobu reaction reacción de
    Mitsunobu
mixed acid ácido mixto
mixed crystal cristal mixto
mixed indicator indicador mixto
mixed melting point punto de fusión
    mixto
mixed salt sal mixta
mixing mezcla
mixite mixita
mixo mixo
mixture mezcla
mobility movilidad
mode modo
model modelo
moderator moderador
modification modificación
modulation spectroscopy espectroscopia
    de modulación
modulus of elasticity módulo de
    elasticidad
Mohr method método de Mohr
Mohr salt sal de Mohr
Mohr titration titulación de Mohr,
    valoración de Mohr
Mohs scale escala de Mohs
moiety porción, mitad
moistness humedad
moisture humedad
moisture-proof a prueba de humedad
mol mol
molal molal
molal concentration concentración molal
molal conductivity conductividad molal
molal solution solución molal
molal specific heat calor específico
    molal
molal volume volumen molal

molality molalidad
molar molar
molar conductivity conductividad molar
molar heat calor molar
molar heat capacity capacidad térmica
    molar
molar latent heat calor latente molar
molar solution solución molar
molar specific heat calor específico
    molar
molar surface superficie molar
molar susceptibility susceptibilidad
    molar
molar volume volumen molar
molar weight peso molar
molarity molaridad
molasses melaza
mold moho
mole mol
mole fraction fracción molar
mole volume volumen molar
molecular molecular
molecular adhesion adhesión molecular
molecular amplitude amplitud molecular
molecular association asociación
    molecular
molecular attraction atracción molecular
molecular beam haz molecular
molecular biology biología molecular
molecular colloid coloide molecular
molecular combination combinación
    molecular
molecular concentration concentración
    molecular
molecular conductivity conductividad
    molecular
molecular density densidad molecular
molecular depression depresión
    molecular
molecular diagram diagrama molecular
molecular diamagnetism diamagnetismo
    molecular
molecular diameter diámetro molecular
molecular dipole dipolo molecular
molecular dispersion dispersión
    molecular
molecular dissociation disociación
    molecular
molecular distillation destilación

molecular

molecular elevation　elevación molecular

molecular energy level　nivel de energía
　　molecular

molecular equation　ecuación molecular

molecular field　campo molecular

molecular film　película molecular

molecular flow　flujo molecular

molecular flux　flujo molecular

molecular formula　fórmula molecular

molecular free path　trayectoria libre
　　molecular

molecular frequency　frecuencia
　　molecular

molecular gas　gas molecular

molecular heat capacity　capacidad
　　térmica molecular

molecular ion beam　haz iónico
　　molecular

molecular ion collision　colisión iónica
　　molecular

molecular lattice　red molecular

molecular level　nivel molecular

molecular mass　masa molecular

molecular number　número molecular

molecular orbital　orbital molecular

molecular paramagnetism
　　paramagnetismo molecular

molecular polarizability　polarizabilidad
　　molecular

molecular rays　rayos moleculares

molecular rearrangement
　　reordenamiento molecular

molecular relaxation　relajación
　　molecular

molecular repulsion　repulsión molecular

molecular rotation　rotación molecular

molecular sieve　cedazo molecular

molecular solution　solución molecular

molecular specific heat　calor específico
　　molecular

molecular spectroscopy　espectroscopia
　　molecular

molecular spectrum　espectro molecular

molecular still　destilador molecular

molecular structure　estructura molecular

molecular velocity　velocidad molecular

molecular vibration　vibración molecular

molecular volume　volumen molecular

molecular weight　peso molecular

molecularity　molecularidad

molecule　molécula

Molisch test　prueba de Molisch

molten　fundido

molten salt　sal fundida

molybdate　molibdato

molybdenite　molibdenita

molybdenum　molibdeno

molybdenum acetylacetonate
　　acetilacetonato de molibdeno

molybdenum blue　azul de molibdeno

molybdenum boride　boruro de
　　molibdeno

molybdenum chloride　cloruro de
　　molibdeno

molybdenum dichloride　dicloruro de
　　molibdeno

molybdenum dioxide　dióxido de
　　molibdeno

molybdenum disilicide　disiliciuro de
　　molibdeno

molybdenum disulfide　disulfuro de
　　molibdeno

molybdenum hexacarbonyl
　　hexacarbonilo de molibdeno

molybdenum hydroxide　hidróxido de
　　molibdeno

molybdenum orange　anaranjado de
　　molibdeno

molybdenum oxide　óxido de molibdeno

molybdenum pentachloride　pentacloruro
　　de molibdeno

molybdenum sesquioxide　sesquióxido de
　　molibdeno

molybdenum sulfide　sulfuro de
　　molibdeno

molybdenum tetrasulfide　tetrasulfuro de
　　molibdeno

molybdenum trioxide　trióxido de
　　molibdeno

molybdenum trisulfide　trisulfuro de
　　molibdeno

molybdenyl　molibdenilo

molybdic　molíbdico

molybdic acid　ácido molíbdico

molybdic anhydride　anhídrido molíbdico

molybdite　molibdita

molybdosilicic acid　ácido molibdosilícico

molysite molisita
monad mónada
monatomic monoatómico
monazite monacita
Mond process proceso de Mond
monetite monetita
monistic monístico
monistic compound compuesto monístico
mono mono
monoacetate monoacetato
monoacetin monoacetina
monoacid monoácido
monoamine monoamina
monoamino acid monoaminoácido
monoammonium monoamonio
monoatomic monoatómico
monoatomic gas gas monoatómico
monoatomic molecule molécula monoatómica
monobasic monobásico
monobasic calcium phosphate fosfato cálcico monobásico
monobromated monobromado
monobromo monobromo
monocalcium fluoride fluoruro monocálcico
monocalcium phosphate fosfato monocálcico
monochloro monocloro
monochloroacetone monocloroacetona
monochlorobenzene monoclorobenceno
monochloroethane monocloroetano
monochloromethane monoclorometano
monochlorophenol monoclorofenol
monochromatic monocromático
monodisperse monodisperso
monodisperse system sistema monodisperso
monoester monoéster
monoethanolamine monoetanolamina
monoethylamine monoetilamina
monofunctional monofuncional
monofunctional compound compuesto monofuncional
monoglyceride monoglicérido
monoglyme monoglime
monohydrate monohidrato
monohydric monohídrico

monohydric alcohol alcohol monohídrico
monolayer monocapa
monomer monómero
monomethylamine monometilamina
monometric monométrico
monomolecular monomolecular
monomolecular film película monomolecular
monomorphous monomorfo
mononuclear mononuclear
monophosphate monofosfato
monoprotic acid ácido monoprótico
monorefringent monorrefringente
monosaccharide monosacárido
monosodium glutamate glutamato monosódico
monosulfide monosulfuro
monoterpene monoterpeno
monotropic monotrópico
monotropy monotropía
monovalent monovalente
monoxide monóxido
monoxychlor monoxicloro
montan wax cera montana
montanite montanita
monticellite monticelita
montmorillonite montmorillonita
montroydite montroidita
mordant mordiente
morenosite morenosita
morin morina
morphine morfina
morpholine morfolina
morphology morfología
morphosan morfosán
Morse equation ecuación de Morse
Morse potential potencial de Morse
mortar mortero
mosaic gold oro musivo, oro de mosaico
Moseley diagram diagrama de Moseley
Moseley law ley de Moseley
Mössbauer effect efecto de Mössbauer
Mössbauer spectroscopy espectroscopia de Mössbauer
mu mu
mucic acid ácido múcico
mucopeptide mucopéptido
mucopolysaccharide mucopolisacárido
mullite mullita

multi multi
multiple bond enlace múltiple
multiple bonding enlazado múltiple
multiple decay desintegración múltiple
multiple disintegration desintegración
    múltiple
multiple proportions law ley de
    proporciones múltiples
multiplet multiplete
multivalent multivalente
Muntz metal metal Muntz
muon muon
murexide murexida
muriatic acid ácido muriático
muscarine muscarina
muscovite muscovita
musk almizcle
musk xylene xileno de almizcle
mustard gas gas mostaza
mustard oil aceite de mostaza
mutagenic agent agente mutagénico

mutamer mutámero
mutamerism mutamerismo
mutarotation mutarrotación
mutase mutasa
mutual exclusion rule regla de exclusión
    mutua
mutuality of phases mutualidad de fases
mycotoxin micotoxina
myoglobin mioglobina
myokinase mioquinasa
myosin miosina
myrcene mirceno
myricyl miricilo
myristic acid ácido mirístico
myristoyl miristoilo
myristoyl peroxide peróxido de
    miristoilo
myristyl alcohol alcohol miristílico
myristyl chloride cloruro de miristilo
myrrh mirra
myxin mixina

N

nabam nabam
nacre nácar
nacrite nacrita
nadorite nadorita
naftolen naftolén
nakrite nacrita
naled naled
nalidixic acid ácido nalidíxico
nalorphine nalorfina
naloxone naloxona
nanometer nanómetro
nanotechnology nanotecnología
nantokite nantoquita
napalin napalina
napalm napalm
naphtha nafta
naphthacene naftaceno
naphthacetol naftecetol
naphthacridine naftacridina
naphthal naftal
naphthaldehyde naftaldehído
naphthalene naftaleno
naphthaleneacetic acid ácido
　　naftalenacético
naphthalenecarboxylic acid ácido
　　naftalencarboxílico
naphthalenediamine naftalendiamina
naphthalenedicarboxylic acid ácido
　　naftalendicarboxílico
naphthalenediol naftalendiol
naphthalenedisulfonic acid ácido
　　naftalendisulfónico
naphthalenesulfinic acid ácido
　　naftalensulfínico
naphthalenesulfonic acid ácido
　　naftalensulfónico
naphthalenethiol naftalentiol
naphthalenyl naftalenilo
naphthalic acid ácido naftálico
naphthalide naftalida
naphthalimido naftalimido
naphthalin naftalina
naphthamide naftamida

naphthamine naftamina
naphthane naftano
naphthene nafteno
naphthenic acid ácido nafténico
naphthenyl naftenilo
naphthieno naftieno
naphthindene naftindeno
naphthionic acid ácido naftiónico
naphtho nafto
naphthodianthrene naftodiantreno
naphthoic acid ácido naftoico
naphthoic aldehyde aldehído naftoico
naphthol naftol
naphthol blue azul de naftol
naphthol green verde de naftol
naphthol orange anaranjado de naftol
naphthol yellow amarillo de naftol
naphtholate naftolato
naphtholsulfonic acid ácido
　　naftolsulfónico
naphthonitrile naftonitrilo
naphthoquinone naftoquinona
naphthoresorcinol naftorresorcina
naphthothiazole naftotiazol
naphthoxy naftoxi
naphthoxyacetic acid ácido
　　naftoxiacético
naphthoyl naftoilo
naphthyl naftilo
naphthyl alcohol alcohol naftílico
naphthyl aldehyde aldehído naftílico
naphthylacetic acid ácido naftilacético
naphthylamine naftilamina
naphthylamine hydrochloride
　　clorhidrato de naftilamina
naphthylaminesulfonic acid ácido
　　naftilaminosulfónico
naphthylbenzene naftilbenceno
naphthylbenzoate naftilbenzoato
naphthylene naftileno
naphthylenediamine naftilendiamina
naphthylethyl ether naftiletiléter
naphthylmercuric naftilmercúrico

naphthylmethanol naftilmetanol
naphthylmethyl naftilmetilo
naphthylmethylene naftilmetileno
naphthylnaphthyl naftilnaftilo
naphthyloxy naftiloxi
naphthylthiourea naftiltiourea
naphthyridine naftiridina
naproxen naproxén
naptalam naptalam
narceine narceína
narcotic narcótico
naringin naringina
nascent nasciente
nascent hydrogen hidrógeno nasciente
native compound compuesto nativo
native element elemento nativo
native metal metal nativo
natrolite natrolita
natron natrón
natural natural
natural abundance abundancia natural
natural base base natural
natural gas gas natural
Nazarov cyclization reaction reacción de ciclización de Nazarov
neatsfoot oil aceite de pata de vaca
Neber rearrangement reordenamiento de Neber
nectar nectar
Néel temperature temperatura de Néel
Nef reaction reacción de Nef
Nef synthesis síntesis de Nef
negative negativo
negative adsorption adsorción negativa
negative catalysis catálisis negativa
negative element elemento negativo
negative group grupo negativo
negative ion ion negativo
negative radical radical negativo
nematic nemático
nematic crystal cristal nemático
nematic phase fase nemática
Nencki reaction reacción de Nencki
Nenitzescu indole synthesis síntesis de indoles de Nenitzescu
neo neo
neocianite neocianita
neodecanoic acid ácido neodecanoico
neodymium neodimio

neodymium acetate acetato de neodimio
neodymium acetylacetate acetilacetato de neodimio
neodymium ammonium nitrate nitrato amónico de neodimio
neodymium carbonate carbonato de neodimio
neodymium chloride cloruro de neodimio
neodymium fluoride fluoruro de neodimio
neodymium hydroxide hidróxido de neodimio
neodymium iodide yoduro de neodimio
neodymium nitrate nitrato de neodimio
neodymium oxalate oxalato de neodimio
neodymium oxide óxido de neodimio
neodymium phosphate fosfato de neodimio
neodymium sulfate sulfato de neodimio
neodymium sulfide sulfuro de neodimio
neohexane neohexano
neolite neolita
neomycin neomicina
neon neón
neon lamp lámpara de neón
neopentane neopentano
neopentanoic acid ácido neopentanoico
neopentyl neopentilo
neopentyl glycol neopentilglicol
neoprene neopreno
neotridecanoic acid ácido neotridecanoico
nephelite nefelita
nephelometer nefelómtero
nephelometry nefelometría
nephrite nefrita
neptunium neptunio
neptunium dioxide dióxido de neptunio
neptunyl neptunilo
Nernst-Einstein equation ecuación de Nernst-Einstein
Nernst equation ecuación de Nernst
Nernst law ley de Nernst
Nernst potential potencial de Nernst
Nernst-Thomson rule regla de Nernst-Thomson
nerol nerol
nerolidol nerolidol

nerve gas gas neurotóxico
nervone nervona
nervonic acid ácido nervónico
neryl nerilo
nesquehonite nesquehonita
Nessler reagent reactivo de Nessler
Nessler tubes tubos de Nessler
neurine neurina
neutral neutro
neutral atom átomo neutro
neutral compound compuesto neutro
neutral element elemento neutro
neutral molecule molécula neutra
neutral oil aceite neutro
neutral reaction reacción neutra
neutral red rojo neutro
neutral salt sal neutra
neutral solution solución neutra
neutral violet violeta neutra
neutrality neutralidad
neutralization neutralización
neutralization equivalent equivalente de
    neutralización
neutralization number número de
    neutralización
neutralization ratio razón de
    neutralización
neutralize neutralizar
neutralized neutralizado
neutralizing neutralizante
neutrino neutrino
neutron neutrón
neutron absorber absorbedor
    neutrónico, absorbedor de neutrones
neutron absorption absorción neutrónica
neutron activation analysis análisis de
    activación neutrónica
neutron beam haz neutrónico
neutron binding energy energía de
    separación de neutrón
neutron bombardment bombardeo
    neutrónico
neutron capture captura neutrónica
neutron converter convertidor de
    neutrones
neutron current corriente neutrónica
neutron cycle ciclo neutrónico
neutron density densidad de neutrones
neutron diffraction difracción neutrónica

neutron energy energía neutrónica
neutron excess exceso de neutrones
neutron flux flujo neutrónico
neutron-induced inducido por neutrones
neutron-induced reaction reacción
    inducida por neutrones
neutron magnetic moment momento
    magnético neutrónico
neutron multiplication multiplicación
    neutrónica
neutron number número de neutrones
neutron producer productor de
    neutrones
neutron rest mass masa en reposo de
    neutrón
neutron source fuente de neutrones
neutron spectrometer espectrómetro
    neutrónico
neutron spectrometry espectrometría
    neutrónica
neutron spectroscopy espectroscopia
    neutrónica
neutron spectrum espectro neutrónico
neutronic neutrónico
neutronics neutrónica
Neville-Winter acid ácido de Neville-
    Winter
Newman projection proyección de
    Newman
Newtonian fluid fluido newtoniano
niacin niacina
niacinamide niacinamida
niacinamide ascorbate ascorbato de
    niacinamida
nialamide nialamida
niccolite niccolita
nichrome nicromo
nickel níquel
nickel acetate acetato de níquel
nickel acetylacetonate acetilacetonato de
    níquel
nickel ammonium chloride cloruro
    amónico de níquel
nickel ammonium sulfate sulfato
    amónico de níquel
nickel arsenate arseniato de níquel
nickel benzoate benzoato de níquel
nickel bromide bromuro de níquel
nickel-cadmium níquel-cadmio

nickel carbonate carbonato de níquel
nickel carbonyl carbonilo de níquel
nickel chloride cloruro de níquel
nickel cyanide cianuro de níquel
nickel dimethylglyoxime dimetilglioxima
 de níquel
nickel fluoride fluoruro de níquel
nickel formate formiato de níquel
nickel hexafluorosilicate
 hexafluorosilicato de níquel
nickel iodide yoduro de níquel
nickel-iron alloy aleación de níquel-
 hierro
nickel layer capa de níquel
nickel nitrate nitrato de níquel
nickel oxalate oxalato de níquel
nickel oxide óxido de níquel
nickel perchlorate perclorato de níquel
nickel peroxide peróxido de níquel
nickel phosphate fosfato de níquel
nickel potassium sulfate sulfato potásico
 de níquel
nickel protoxide protóxido de níquel
nickel-rhodium níquel-rodio
nickel sesquioxide sesquióxido de níquel
nickel-silver níquel-plata
nickel stannate estannato de níquel
nickel steel acero al níquel
nickel sulfate sulfato de níquel
nickel tetracarbonyl tetracarbonilo de
 níquel
nickel titanate titanato de níquel
nickelic niquélico
nickelic hydroxide hidróxido niquélico
nickelic oxide óxido niquélico
nickelic sulfide sulfuro niquélico
nickeline niquelina
nickelocene niqueloceno
nickelous niqueloso
nickelous acetate acetato niqueloso
nickelous arsenate arseniato niqueloso
nickelous bromide bromuro niqueloso
nickelous carbonate carbonato niqueloso
nickelous chloride cloruro niqueloso
nickelous formate formiato niqueloso
nickelous hydroxide hidróxido niqueloso
nickelous iodide yoduro niqueloso
nickelous nitrate nitrato niqueloso
nickelous oxalate oxalato niqueloso

nickelous oxide óxdo niqueloso
nickelous phosphate fosfato niqueloso
nickelous sulfate sulfato niqueloso
Nicol prism prisma de Nicol
nicotinamide nicotinamida
nicotine nicotina
nicotine hydrochloride clorhidrato de
 nicotina
nicotinic acid ácido nicotínico
nicotinic acid amide amida de ácido
 nicotínico
nicotinoyl nicotinoilo
nido nido
Niementowski quinazoline synthesis
 síntesis de quinazolinas de
 Niementowski
Niementowski quinoline synthesis
 síntesis de quinolinas de
 Niementowski
Nierenstein reaction reacción de
 Nierenstein
ninhydrin ninhidrina
niobate niobato
niobe oil aceite de niobe
niobic nióbico
niobic acid ácido nióbico
niobite niobita
niobium niobio
niobium bromide bromuro de niobio
niobium carbide carburo de niobio
niobium chloride cloruro de niobio
niobium diselenide diseleniuro de niobio
niobium fluoride fluoruro de niobio
niobium oxalate oxalato de niobio
niobium oxide óxido de niobio
niobium pentachloride pentacloruro de
 niobio
niobium pentoxide pentóxido de niobio
niobium potassium oxyfluoride
 oxifluoruro potásico de niobio
niobium silicide siliciuro de niobio
niobium-tin niobio-estaño
niobium-titanium niobio-titanio
niobium-uranium niobio-uranio
nioboxy nioboxi
niobus niobo
niobyl niobilo
nionel nionel
niranium niranio

nisin nisina
nisinic acid ácido nisínico
niter nitro
nitracidium nitracidio
nitralin nitralina
nitramide nitramida
nitramine nitramina
nitranilic acid ácido nitranílico
nitranilide nitranilida
nitraniline nitranilina
nitrate nitrato
nitrating acid ácido nitrante
nitration nitración
nitrene nitreno
nitric acid ácido nítrico
nitric ether éter nítrico
nitric oxide óxido nítrico
nitridation nitridación
nitride nitruro
nitrido nitrido
nitrification nitrificación
nitrilase nitrilasa
nitrile nitrilo
nitrile rubber caucho nitrilo
nitrilo nitrilo
nitrilotriacetic acid ácido
     nitrilotriacético
nitrilotriacetonitrile nitrilotriacetonitrilo
nitrite nitrito
nitrito nitrito
nitro nitro
nitro dye colorante nitro
nitroacetanilide nitroacetanilida
nitroamine nitroamina
nitroamino nitroamino
nitroaniline nitroanilina
nitroanisole nitroanisol
nitroanthracene nitroantraceno
nitroanthraquinone nitroantraquinona
nitroaromatic nitroaromático
nitrobarbituric acid ácido
     nitrobarbitúrico
nitrobarite nitrobarita
nitrobenzaldehyde nitrobenzaldehído
nitrobenzamide nitrobenzamida
nitrobenzene nitrobenceno
nitrobenzenesulfonic acid ácido
     nitrobencenosulfónico
nitrobenzoic acid ácido nitrobenzoico

nitrobenzonitrile nitrobenzonitrilo
nitrobenzotrifluoride
     nitrobenzotrifluoruro
nitrobenzoyl nitrobenzoilo
nitrobenzoyl chloride cloruro de
     nitrobenzoilo
nitrobenzyl nitrobencilo
nitrobenzyl alcohol alcohol
     nitrobencílico
nitrobenzyl bromide bromuro de
     nitrobencilo
nitrobenzyl cyanide cianuro de
     nitrobencilo
nitrobiphenyl nitobifenilo
nitrobromoform nitrobromoformo
nitrobutanol nitrobutanol
nitrocellulose nitrocelulosa
nitrochlorobenzene nitroclorobenceno
nitrochloroform nitrocloroformo
nitrocinnamic acid ácido nitrocinámico
nitrocobalamin nitrocobalamina
nitrocresol nitrocresol
nitrodiphenyl nitrodifenilo
nitrodiphenylamine nitrodifenilamina
nitroethane nitroetano
nitroethylpropanediol
     nitroetilpropanodiol
nitrofuran nitrofurano
nitrofurantoin nitrofurantoína
nitrogen nitrógeno
nitrogen chloride cloruro de nitrógeno
nitrogen cycle ciclo de nitrógeno
nitrogen dioxide dióxido de nitrógeno
nitrogen equivalent equivalente de
     nitrógeno
nitrogen fixation fijación de nitrógeno
nitrogen fluoride fluoruro de nitrógeno
nitrogen hydride hidruro de nitrógeno
nitrogen iodide yoduro de nitrógeno
nitrogen monoxide monóxido de
     nitrógeno
nitrogen mustard mostaza de nitrógeno
nitrogen oxide óxido de nitrógeno
nitrogen pentoxide pentóxido de
     nitrógeno
nitrogen peroxide peróxido de nitrógeno
nitrogen sulfide sulfuro de nitrógeno
nitrogen tetraoxide tetróxido de
     nitrógeno

nitrogen tribromide tribromuro de
    nitrógeno
nitrogen trichloride tricloruro de
    nitrógeno
nitrogen trifluoride trifluoruro de
    nitrógeno
nitrogen triiodide triyoduro de nitrógeno
nitrogen trioxide trióxido de nitrógeno
nitrogenase nitrogenasa
nitrogenated nitrogenado
nitrogenous base base nitrogenosa
nitroglycerin nitroglicerina
nitroguanidine nitroguanidina
nitrohydrochloric acid ácido
    nitroclorhídrico
nitrol nitrol
nitromagnesite nitromagnesita
nitromannite nitromanita
nitromannitol nitromanitol
nitromersol nitromersol
nitromethane nitrometano
nitromethoxyaniline nitrometoxianilina
nitromethylpropanediol
    nitrometilpropanodiol
nitron nitrón
nitronaphthalene nitronaftaleno
nitronaphthol nitronaftol
nitronaphthylamine nitronaftilamina
nitrone nitrona
nitronic acid ácido nitrónico
nitronium nitronio
nitronium perchlorate perclorato de
    nitronio
nitroparaffin nitroparafina
nitrophenetole nitrofenetol
nitrophenide nitrofenida
nitrophenol nitrofenol
nitrophenylacetic acid ácido
    nitrofenilacético
nitrophenylhydrazine nitrofenilhidrazina
nitrophenylpropiolic acid ácido
    nitrofenilpropiólico
nitrophosphate nitrofosfato
nitrophthalic acid ácido nitroftálico
nitropropane nitropropano
nitroquinoline nitroquinolina
nitrosalicylic acid ácido nitrosalicílico
nitroso nitroso
nitroso dye colorante nitroso

nitroso polymer polímero nitroso
nitrosoamine nitrosoamina
nitrosobenzene nitrosobenceno
nitrosobenzoic acid ácido
    nitrosobenzoico
nitrosodimethylamine
    nitrosodimetilamina
nitrosodimethylaniline
    nitrosodimetilanilina
nitrosoethane nitrosoetano
nitrosoguanidine nitrosoguanidina
nitrosomethylurea nitrosometilurea
nitrosonaphthol nitrosonaftol
nitrosonaphthylamine nitrosonaftilamina
nitrosonium nitrosonio
nitrosooxy nitrosooxi
nitrosophenol nitrosofenol
nitrosotoluene nitrosotolueno
nitrostarch nitroalmidón
nitrostyrene nitroestireno
nitrosulfathiazole nitrosulfatiazol
nitrosyl nitrosilo
nitrosyl bromide bromuro de nitrosilo
nitrosyl chloride cloruro de nitrosilo
nitrosyl fluoride fluoruro de nitrosilo
nitrosyl perchlorate perclorato de
    nitrosilo
nitrosylsulfuric acid ácido
    nitrosilsulfúrico
nitrothiophene nitrotiofeno
nitrotoluene nitrotolueno
nitrotoluidine nitrotoluidina
nitrotrichloromethane
    nitrotriclorometano
nitrotrifluoromethyl chlorobenzene
    nitrotrifluorometilclorobenceno
nitrourea nitrourea
nitrous acid ácido nitroso
nitrous oxide óxido nitroso
nitroxanthic acid ácido nitroxántico
nitroxyl nitroxilo
nitroxylene nitroxileno
nitroxylic acid ácido nitroxílico
nitryl nitrilo
nitryl chloride cloruro de nitrilo
nitryl fluoride fluoruro de nitrilo
nivenite nivenita
Nobel explosive explosivo de Nobel
nobelium nobelio

noble noble
noble gas gas noble
noble metal metal noble
noble potential potencial noble
nodule nódulo
nomenclature nomenclatura
non non
nona nona
nonacosane nonacosano
nonacosanol nonacosanol
nonacyclic nonacíclico
nonadecane nonadecano
nonadecanoic acid ácido nonadecanoico
nonadecanol nonadecanol
nonadecanone nonadecanona
nonahydrate nonahidrato
nonalactone nonalactona
nonanal nonanal
nonane nonano
nonanedioic acid ácido nonanodioico
nonanediol nonanodiol
nonanenitrile nonanonitrilo
nonanoate nonanoato
nonanoic acid ácido nonanoico
nonanol nonanol
nonanone nonanona
nonanoyl nonanoilo
nonanoyl chloride cloruro de nonanoilo
nonaqueous no acuoso
nonaqueous solution solución no acuosa
nonaqueous solvent solvente no acuosa
noncombustible no combustible
noncombustible material material no
    combustible
noncompetitive inhibition inhibición no
    competitiva
nondestructive no destructivo
nondestructive test prueba no destructiva
nonelectrolyte no electrólito
nonene noneno
nonferrous no ferroso
nonflammable no inflamable
nonideal no ideal
nonionic no iónico
nonionizing no ionizante
nonionizing solvent solvente no ionizante
nonlinear molecule molécula no lineal
nonmetal no metal
nonmetallic no metálico

nonnutritive sweetner endulcorante no
    nutritivo
nonoic acid ácido nonoico
nonpolar no polar
nonpolar compound compuesto no polar
nonpolar crystal cristal no polar
nonpolar molecule molécula no polar
nonpolar solvent solvente no polar
nonreducing sugar azúcar no reductor
nonstoichiometric no estequiométrico
nonyl nonilo
nonyl acetate acetato de nonilo
nonyl alcohol alcohol nonílico
nonyl aldehyde aldehído nonílico
nonyl bromide bromuro de nonilo
nonyl chloride cloruro de nonilo
nonyl hydride hidruro de nonilo
nonyl lactone lactona de nonilo
nonyl nonanoate nonanoato de nonilo
nonyl thiocyanate tiocianato de nonilo
nonylamine nonilamina
nonylbenzene nonilbenceno
nonylcarbinol nonilcarbinol
nonylene nonileno
nonylic acid ácido nonílico
nonylphenol nonilfenol
nonylphenoxyacetic acid ácido
    nonilfenoxiacético
nonyltrichlorosilane noniltriclorosilano
nonyne nonino
nopinene nopineno
nor nor
noradrenaline noradrenalina
norbornadiene norbornadieno
norbornene norborneno
norbornenemethanol norbornenometanol
norepinephrine norepinefrina
norgestrel norgestrel
norleucine norleucina
normal normal
normal conditions condiciones normales
normal hydrocarbon hidrocarburo
    normal
normal pressure presión normal
normal salt sal normal
normal solution solución normal
normal state estado normal
normal temperature temperatura normal
normality normalidad

Normant reagent reactivo de Normant
nornicotine nornicotina
noscapine noscapina
noselite noselita
notation notación
novobiocin novobiocina
novolak novolaca
noxious nocivo
noxious chemical producto químico nocivo
noxious gas gas nocivo
noxious substance substancia nociva
nuclear nuclear
nuclear absorption absorción nuclear
nuclear angular momentum momento angular nuclear
nuclear atom átomo nuclear
nuclear attraction atracción nuclear
nuclear binding energy energía de unión nuclear
nuclear bombardment bombardeo nuclear
nuclear capture captura nuclear
nuclear chain reaction reacción en cadena nuclear
nuclear charge carga nuclear
nuclear chemistry química nuclear
nuclear collision colisión nuclear
nuclear cross-section sección transversal nuclear
nuclear decay desintegración nuclear
nuclear-decay mode modo de desintegración nuclear
nuclear density densidad nuclear
nuclear disintegration desintegración nuclear
nuclear energy energía nuclear
nuclear energy level nivel de energía nuclear
nuclear equation ecuación nuclear
nuclear excitation excitación nuclear
nuclear fission fisión nuclear
nuclear force fuerza nuclear
nuclear fragment fragmento nuclear
nuclear fuel combustible nuclear
nuclear fusion fusión nuclear
nuclear ground state estado fundamental nuclear
nuclear heat calor nuclear

nuclear interaction interacción nuclear
nuclear isomer isómero nuclear
nuclear isomerism isomerismo nuclear
nuclear magnetic moment momento magnético nuclear
nuclear magnetic resonance resonancia magnética nuclear
nuclear magneton magnetón nuclear
nuclear mass masa nuclear
nuclear moment momento nuclear
nuclear neutron neutrón nuclear
nuclear number número nuclear
nuclear paramagnetism paramagnetismo nuclear
nuclear particle partícula nuclear
nuclear physics física nuclear
nuclear polarization polarización nuclear
nuclear potential energy energía potencial nuclear
nuclear power energía nuclear
nuclear process proceso nuclear
nuclear proton protón nuclear
nuclear quadripole moment momento cuadripolar nuclear
nuclear quadripole resonance resonancia cuadripolar nuclear
nuclear radiation radiación nuclear
nuclear radius radio nuclear
nuclear reaction reacción nuclear
nuclear reactor reactor nuclear
nuclear repulsion repulsión nuclear
nuclear resonance resonancia nuclear
nuclear scattering dispersión nuclear
nuclear species especie nuclear
nuclear spectroscopy espectroscopia nuclear
nuclear spectrum espectro nuclear
nuclear spin espín nuclear
nuclear stability estabilidad nuclear
nuclear structure estructura nuclear
nuclear theory teoría nuclear
nuclear transformation transformación nuclear
nuclear volume volumen nuclear
nuclear waste desperdicios nucleares
nuclease nucleasa
nucleate nucleado
nucleation nucleación

nucleic acid ácido nucleico
nucleogenesis nucleogénesis
nucleon nucleón
nucleonic nucleónico
nucleophile nucleófilo
nucleophilic nucleofílico
nucleophilic addition adición
    nucleofílica
nucleophilic substitution substitución
    nucleofílica
nucleoprotein nucleoproteína
nucleoside nucleósido

nucleotide nucleótido
nucleus núcleo
nuclide nucleido
nutrient nutriente
nutrient solution solución de nutrientes
nutrification nutrificación
nutrition nutrición
Nylander reagent reactivo de Nylander
nylidrin hydrochloride clorhidrato de
    nilidrina
nylon nilón

O

Obermayer reagent reactivo de
  Obermayer
occlusion oclusión
occupied molecular orbital orbital
  molecular ocupado
ocher ocre
ocimene ocimeno
octa octa
octabenzone octabenzona
octachloronaphthalene
  octacloronaftaleno
octacosane octacosano
octacosanoic acid ácido octacosanoico
octadeca octadeca
octadecanal octadecanal
octadecane octadecano
octadecanoic acid ácido octadecanoico
octadecanol octadecanol
octadecatrienoic acid ácido
  octadecatrienoico
octadecene octadeceno
octadecenoic acid ácido octadecenoico
octadecenol octadecenol
octadecenyl aldehyde aldehído
  octadecenílico
octadecyl octadecilo
octadecyl alcohol alcohol octadecílico
octadecyl isocyanate isocianato de
  octadecilo
octadecyltrichlorosilane
  octadeciltriclorosilano
octadiene octadieno
octafluorobutene octafluorobuteno
octafluorocyclobutane
  octafluorociclobutano
octafluoropropane octafluoropropano
octahedral octahedral
octahedro octahedro
octahedron octahedro
octahydrate octahidrato
octamethyltrisiloxane
  octametiltrisiloxano
octanal octanal

octane octano
octane number índice de octano
octane rating índice de octano
octanedicarboxylic acid ácido
  octanodicarboxílico
octanedioic acid ácido octanodioico
octanoate octanoato
octanoic acid ácido octanoico
octanol octanol
octanone octanona
octanoyl octanoilo
octanoyl chloride cloruro de octanoilo
octavalent octavalente
octene octeno
octet octeto
octet theory teoría de octeto
octoic acid ácido octoico
octomethylene octometileno
octyl octilo
octyl acetate acetato de octilo
octyl alcohol alcohol octílico
octyl aldehyde aldehído octílico
octyl bromide bromuro de octilo
octyl carbinol octilcarbinol
octyl chloride cloruro de octilo
octyl formate formiato de octilo
octyl gallate galato de octilo
octyl iodide yoduro de octilo
octyl mercaptan octilmercaptano
octyl peroxide peróxido de octilo
octyl phosphate fosfato de octilo
octylacetic acid ácido octilacético
octylamine octilamina
octylene octileno
octylene oxide óxido de octileno
octylmagnesium chloride cloruro de
  octilmagnesio
octylphenol octilfenol
octylphenyl octilfenilo
octylphenyl salicylate salicilato de
  octilfenilo
octyltrichlorosilane octiltriclorosilano
octyne octino

odd-even nucleus núcleo impar-par
odd-odd nucleus núcleo impar-impar
odor olor
odor intensity intensidad de olor
odor permanence permanencia de olor
odorant odorante
odoriferous odorífero
odorimetry odorimetría
odorometer odorómetro
oic oico
oid oide
oil aceite, petróleo
oil black negro de petróleo
oil gas gas de petróleo
oil of almond aceite de almendra
oil of anise aceite de anís
oil of apple aceite de manzana
oil of avocado aceite de aguacate
oil of banana aceite de plátano
oil of calamus aceite de cálamo
oil of camphor aceite de alcanfor
oil of cananga aceite de cananga
oil of caraway aceite de alcaravea
oil of cardamom aceite de cardamomo
oil of cascarilla aceite de cascarilla
oil of cashew aceite de anacardo
oil of cassia aceite de casia
oil of chamomile aceite de manzanilla
oil of chenopodium aceite de quenopodio
oil of cinnamon aceite de canela
oil of citronella aceite de citronela
oil of clove aceite de clavero
oil of coconut aceite de nuez de coco
oil of cod-liver aceite de hígado de
     bacalao
oil of copaiba aceite de copaiba
oil of cottonseed aceite de semilla de
     algodón
oil of croton aceite de crotón
oil of fennel aceite de hinojo
oil of flaxseed aceite de linaza
oil of garlic aceite de ajo
oil of ginger aceite de jengibre
oil of hempseed aceite de cañamón
oil of jasmine aceite de jazmín
oil of jojoba aceite de jojoba
oil of laurel leaves aceite de hojas de
     laurel
oil of lavender aceite de lavanda

oil of lemongrass aceite de luisa
oil of linseed aceite de linaza
oil of mustard aceite de mostaza
oil of neatsfoot aceite de pata de vaca
oil of orange-peel aceite de corteza de
     naranja
oil of palm aceite de palma
oil of palmarosa aceite de palmarosa
oil of parsley aceite de semilla de perejil
oil of patchouli aceite de pachulí
oil of pennyroyal aceite de poleo
oil of peppermint aceite de menta
oil of perilla aceite de perilla
oil of pine aceite de pino
oil of rapeseed aceite de colza
oil of ricinus aceite de ricino
oil of safflower aceite de cártamo
oil of sage aceite de salvia
oil of sandalwood aceite de sándalo
oil of sassafras aceite de sasafrás
oil of sesame aceite de sésamo
oil of shale aceite de pizarra
oil of spearmint aceite de menta verde
oil of sunflower aceite de girasol
oil of tangerine aceite de tangerina
oil of tarragon aceite de estragón
oil of thuja aceite de tuya
oil of tiglium aceite de tiglio
oil of verbena aceite de verbena
oil of vitriol aceite de vitriolo
oil of wintergreen aceite de pirola
oil of wormseed aceite de quenopondio
ointment ungüento
okonite okonita
ol ol
ole ol
oleamide oleamida
oleate oleato
olefin olefina
olefin copolymer copolímero olefínico
olefin fiber fibra olefínica
olefin resin resina olefínica
oleic acid ácido oleico
olein oleína
oleo oleo
oleoresin oleorresina
oleoyl oleoilo
oleoyl chloride cloruro de oleoilo
oleoylsarcosine oleoilsarcosina

oleyl oleilo
oleyl alcohol alcohol oleílico
oleyl aldehyde aldehído oleílico
oligo oligo
oligomer oligómero
oligonite oligonita
oligonucleotide oligonucleótido
oligopeptide oligopéptido
oligosaccharide oligosacárido
olive oil aceite de oliva
olivenite olivenita
olivetol olivetol
olivine olivina
omega omega
oncogen oncógeno
one ona
one-way tap espita unidireccional
onium onio
Onsager equation ecuación de Onsager
opacity opacidad
opal ópalo
opaque opaco
open chain cadena abierta
open shell capa abierta
operation operación
opiate opiato
opioid opioide
opium opio
Oppenauer oxidation oxidación de
    Oppenauer
optical activity actividad óptica
optical-activity index índice de actividad
    óptica
optical axis eje óptico
optical brightener abrillantador óptico
optical crystal cristal óptico
optical density densidad óptica
optical electrons electrones ópticos
optical fiber fibra óptica
optical glass vidrio óptico
optical isomer isómero óptico
optical microscope microscopio óptico
optical purity pureza óptica
optical rotation rotación óptica
optical rotatory dispersion dispersión
    rotatoria óptica
optical spectrograph espectrógrafo
    óptico
optical spectrometer espectrómetro

óptico
optical spectroscope espectroscopio
    óptico
optical spectroscopy espectroscopia
    óptica
optical spectrum espectro óptico
optically ópticamente
optically active ópticamente activo
optimum conditions condiciones óptimas
orange cadmium cadmio anaranjado
orange mineral mineral anaranjado
orange oil aceite de corteza de naranja
orange-peel oil aceite de corteza de
    naranja
orbit órbita
orbital orbital
orbital angular momentum momento
    angular orbital
orbital electron electrón orbital
orbital moment momento orbital
orbital quantum number número
    cuántico orbital
orbital state estado orbital
orbital symmetry simetría orbital
orbital theory teoría orbital
orcin orcina
order orden
ordinal test prueba ordinal
ore mena, mineral
organic orgánico
organic acid ácido orgánico
organic analysis análisis orgánico
organic base base orgánica
organic chemistry química orgánica
organic combustion combustión orgánica
organic compound compuesto orgánico
organic pigment pigmento orgánico
organic radical radical orgánico
organic reaction mechanism mecanismo
    de reacción orgánica
organic salt sal orgánica
organic solvent solvente orgánico
organic synthesis síntesis orgánica
organoaluminum compound compuesto
    organoalumínico
organoborane organoborano
organoclay organoarcilla
organogel organogel
organolithium compound compuesto

organolítico
organomagnesium compound
    compuesto organomagnesiano
organomercury compound compuesto
    organomercúrico
organometallic organometálico
organometallic compound compuesto
    organometálico
organometalloid organometaloide
organophosphorus compound
    compuesto organofosforado
organosilane organosilano
organosilicon organosilíceo
organosol organosol
organozinc compound compuesto
    organocíncico
Orgel diagram diagrama de Orgel
orientation effect efecto de orientación
ornithine ornitina
ornithuric acid ácido ornitúrico
orotic acid ácido orótico
orpiment oropimente
Orsat analyzer analizador de Orsat
Orsat apparatus aparato de Orsat
orthamine ortamina
orthite ortita
ortho orto
ortho acid ácido orto
orthoacetic acid ácido ortoacético
orthoarsenic acid ácido ortoarsénico
orthoboric acid ácido ortobórico
orthocarbonic acid ácido ortocarbónico
orthoclase ortoclasa
orthoformic ester éster ortofórmico
orthohydrogen ortohidrógeno
orthonitrogen ortonitrógeno
orthophosphate ortofosfato
orthophosphoric acid ácido
    ortofosfórico
orthorhombic system sistema
    ortorrómbico
orthosilicate ortosilicato
orthosilicic acid ácido ortosilícico
orthotungstic acid ácido ortotúngstico
osamine osamina
osazone osazona
oscillating crystal cristal oscilante
oscillating reaction reacción oscilante
oscillation oscilación

oscillometric titration titulación
    oscilométrica, valoración
    oscilométrica
ose osa
osmate osmiato
osmic acid ácido ósmico
osmic anhydride anhídrido ósmico
osmics ósmica
osmiridium osmiridio
osmium osmio
osmium chloride cloruro de osmio
osmium dichloride dicloruro de osmio
osmium dioxide dióxido de osmio
osmium disulfide disulfuro de osmio
osmium monoxide monóxido de osmio
osmium oxide óxido de osmio
osmium potassium chloride cloruro
    potásico de osmio
osmium tetrachloride tetracloruro de
    osmio
osmium tetraoxide tetróxido de osmio
osmocene osmoceno
osmolality osmolalidad
osmolarity osmolaridad
osmole osmol
osmometer osmómetro
osmometry osmometría
osmosis ósmosis
osmotic osmótico
osmotic cell célula osmótica
osmotic coefficient coeficiente osmótico
osmotic equivalent equivalente osmótico
osmotic gradient gradiente osmótico
osmotic pressure presión osmótica
osmous osmoso
osmous chloride cloruro osmoso
osone osono
osotriazole osotriazol
Ostwald dilution law ley de dilución de
    Ostwald
Ostwald ripening maduración de
    Ostwald
Ostwald rule regla de Ostwald
Oudeman law ley de Oudeman
ous oso, o
outer electron electrón exterior
outer orbital orbital exterior
outer shell capa exterior
outgassing desgaseamiento

ovalene ovaleno
Overhauser effect efecto de Overhauser
overlapping orbitals orbitales solapados
overpotential sobrepotencial
overvoltage sobretensión
ovicide ovicida
oxa oxa
oxalacetic acid ácido oxalacético
oxalate oxalato
oxaldehydic acid ácido oxaldehídico
oxalic acid ácido oxálico
oxalo oxalo
oxalyl oxalilo
oxalyl chloride cloruro de oxalilo
oxalyl dichloride dicloruro de oxalilo
oxalylurea oxalilurea
oxamic acid ácido oxámico
oxamide oxamida
oxamido oxamido
oxamoyl oxamoilo
oxanthrol oxantrol
oxatyl oxatilo
oxazole oxazol
oxetane oxetano
oxethyl oxetilo
oxetone oxetona
oxidant oxidante
oxidase oxidasa
oxidate oxidar
oxidation oxidación
oxidation number número de oxidación
oxidation potential potencial de
    oxidación
oxidation process proceso de oxidación
oxidation reaction reacción de oxidación
oxidation-reduction oxidación-reducción
oxidation-reduction cycle ciclo de
    oxidación-reducción
oxidation-reduction indicator indicador
    de oxidación-reducción
oxidation-reduction potential potencial
    de oxidación-reducción
oxidation-reduction reaction reacción de
    oxidación-reducción
oxidation state estado de oxidación
oxidative oxidativo
oxidative addition adición oxidativa
oxidative coupling acoplamiento
    oxidativo

oxide óxido
oxidimetry oxidimetría
oxidize oxidar
oxidizer oxidante
oxidizing oxidante
oxidizing acid ácido oxidante
oxidizing agent agente oxidante
oxidizing atmosphere atmósfera oxidante
oxidizing flame llama oxidante
oxidizing material material oxidante
oxidizing reaction reacción oxidante
oxidoreductase oxidorreductasa
oxime oxima
oximido oximido
oxine oxina
oxirane oxirano
oxirene oxireno
oxo oxo
oxo acid oxoácido
oxo process oxoproceso
oxo reaction oxorreacción
oxoethanoic acid ácido oxoetanoico
oxonium ion ion oxonio
oxosilane oxosilano
oxy oxi
oxyacetylene flame llama oxiacetilénica
oxyacid oxiácido
oxyamide oxiamida
oxyanion oxianión
oxyazo dye colorante oxiazoico
oxybenzoic acid ácido oxibezoico
oxycarbonyl oxicarbonilo
oxycyanogen oxicianógeno
oxydipropionitrile oxidipropionitrilo
oxygen oxígeno
oxygen absorbent absorbente de oxígeno
oxygen balance equilibrio de oxígeno
oxygen carrier portador de oxígeno
oxygen consumed oxígeno consumido
oxygen electrode electrodo de oxígeno
oxygen fluoride fluoruro de oxígeno
oxygen inlet entrada de oxígeno
oxygenate oxigenar
oxygenated oxigenado
oxygenation oxigenación
oxyhydrogen oxihidrógeno
oxyluminescence oxiluminiscencia
oxymethurea oximeturea
oxymethylene oximetileno

oxynaphthoic acid ácido oxinaftoico
oxyphosphorane oxifosforano
oxyquinoline oxiquinolina
oxyquinone oxiquinona
oxytetracycline oxitetraciclina
oxytoxin oxitoxina
ozocerite ozocerita
ozokerite ozoquerita
ozonation ozonación

ozonator ozonizador
ozone ozono
ozone layer capa de ozono
ozonidation ozonidación
ozonide ozónido
ozonization ozonización
ozonizer ozonizador
ozonolysis ozonólisis

P

Paal-Knorr synthesis síntesis de Paal-Knorr

pachnolite pacnolita

packaging empaque, empaquetado

packed empaquetado

packed column columna empaquetada

packed tower torre empaquetada

packing empaquetamiento

packing effect efecto de empaquetamiento

packing fraction fracción de empaquetamiento

pagodite pagodita

paint pintura

paint remover quitapintura

paired electrons electrones emparejados

paligorskite paligorskita

palladate paladato

palladic paládico

palladium paladio

palladium amine amina de paladio

palladium bichloride bicloruro de paladio

palladium black negro de paladio

palladium chloride cloruro de paladio

palladium diacetate diacetato de paladio

palladium dibromide dibromuro de paladio

palladium dicyanide dicianuro de paladio

palladium diiodide diyoduro de paladio

palladium dinitrate dinitrato de paladio

palladium dioxide dióxido de paladio

palladium disulfide disulfuro de paladio

palladium hydride hidruro de paladio

palladium hydroxide hidróxido de paladio

palladium iodide yoduro de paladio

palladium monosulfite monosulfito de paladio

palladium monoxide monóxido de paladio

palladium nitrate nitrato de paladio

palladium oxide óxido de paladio

palladium potassium chloride cloruro potásico de paladio

palladium sodium chloride cloruro sódico de paladio

palladium subsulfate subsulfato de paladio

palladium sulfate sulfato de paladio

palladous paladioso

palladous chloride cloruro paladioso

palladous iodide yoduro paladioso

palladous nitrate nitrato paladioso

palladous potassium chloride cloruro potásico paladioso

palladous sodium chloride cloruro sódico paladioso

pallamine palamina

pallas palas

palm oil aceite de palma

palmarosa oil aceite de palmarosa

palmic pálmico

palmierite palmierita

palmitate palmitato

palmitic acid ácido palmítico

palmitic cyanide cianuro palmítico

palmitin palmitina

palmitinic acid ácido palmitínico

palmitoleic acid ácido palmitoleico

palmitolic acid ácido palmitólico

palmitonitrile palmitonitrilo

palmitoyl palmitoilo

palmitoyl chloride cloruro de palmitoilo

palmityl alcohol alcohol palmitílico

palustric acid ácido palústrico

palygorskite paligorskita

pamoic acid ácido pamoico

pan pan

panchromatic pancromático

panclastite panclastita

pancreatin pancreatina

pandermite pandermita

Paneth rule regla de Paneth

Paneth technique técnica de Paneth

pannic acid ácido pánico
pantachromatic pantacromático
pantethine pantetina
panthenol pantenol
pantocaine pantocaína
pantolactone pantolactona
pantomorphism pantomorfismo
pantothenic acid ácido pantoténico
pantothenol pantotenol
papain papaína
papaverine papaverina
papaverine hydrochloride clorhidrato de papaverina
paper papel
paper chromatography cromatografía sobre papel
paper electrochromatography electrocromatografía sobre papel
paper electrophoresis electroforesis sobre papel
para para
para red rojo para
parabanic acid ácido parabánico
paraben parabeno
parabituminous parabituminoso
paracasein paracaseína
paracetaldehyde paracetaldehído
paracetamol paracetamol
paraconic acid ácido paracónico
paraconine paraconina
paracyanic acid ácido paraciánico
paracyanogen paracianógeno
paradiazine paradiazina
paraffin parafina
paraffin oil aceite de parafina
paraffin wax cera de parafina
paraformaldehyde paraformaldehído
paragonite paragonita
parahelium parahelio
parahydrogen parahidrógeno
paralactic acid ácido paraláctido
paraldehyde paraldehído
paraldol paraldol
parallax paralaje
parallax error error de paralaje
parallel spins espines paralelos
parallelosterism paralelosterismo
paralysol paralisol
paramagnetic paramagnético

paramagnetic resonance resonancia paramagnética
paramagnetic spectrum espectro paramagnético
paramagnetism paramagnetismo
parameter parámetro
paramorph paramorfo
paramorphism paramorfismo
paranitraniline paranitranilina
paranitraniline red rojo de paranitranilina
parapectic acid ácido parapéctico
paraquat paraquat
pararosaniline pararrosanilina
pararosolic acid ácido pararosólico
parasite parásito
parasiticide parasiticida
parasorbic acid ácido parasórbico
parathion paratión
parawolframate parawolframato
paraxanthine paraxantina
paraxylene paraxileno
parent compound compuesto base
parent molecule molécula base
parent name nombre base
parietic acid ácido pariético
parinaric acid ácido parinárico
Paris green verde de París
Paris red rojo de París
Paris violet violeta de París
parisite parisita
Parkes process proceso de Parkes
paromomycin paromomicina
paromomycin sulfate sulfato de paromomicina
paroxazine paroxazina
parsley oil aceite de semilla de perejil
parthenine partenina
partial parcial
partial condenser condensador parcial
partial molal quantity cantidad molal parcial
partial molar volume volumen molar parcial
partial pressure presión parcial
partial racemization racemización parcial
partially permeable membrane membrana parcialmente permeable

particle partícula

particle accelerator acelerador de partículas

particle counting conteo de partículas

particle distribution distribución de partículas

particle emission emisión de partículas

particle energy energía de partícula

particle mass masa de partícula

particle scattering dispersión de partículas

particle size tamaño de partícula

particulate particulado

partinium partinio

partition partición

partition chromatography cromatografía de partición

partition coeffcient coeficiente de partición

partition function función de partición

partition law ley de partición

parvoline parvolina

parvuline parvulina

parylene parileno

Pascal law ley de Pascal

Pascal rules reglas de Pascal

Paschen-Back effect efecto de Paschen-Back

Paschen series serie de Paschen

Passerini reaction reacción de Passerini

passivation pasivación

passivator pasivador

passive pasivo

passive iron hierro pasivo

passive metal metal pasivo

passivity pasividad

paste pasta

Pasteur effect efecto de Pasteur

Pasteur filter filtro de Pasteur

Pasteur reaction reacción de Pasteur

Pasteur salt solution solución de sal de Pasteur

pasteurization pasteurización

patchouli oil aceite de pachulí

patentability patentabilidad

Paterno-Buchi reaction reacción de Paterno-Buchi

pathochemistry patoquímica

pathogenic patogénico

pathway vía

patronite patronita

Patterson function función de Patterson

Patterson synthesis síntesis de Patterson

Pattinson process proceso de Pattinson

paucine paucina

Pauli exclusion principle principio de exclusión de Pauli

Pauli-Fermi exclusion principle principio de exclusión de Pauli-Fermi

Pauli principle principio de Pauli

Pauli rule regla de Pauli

Pavy solution solución de Pavy

peacock blue azul pavo real

peak pico, punta

peanut oil aceite de maní

peanut oil meal harina de aceite de maní

pearl ash cenizas de perla

pearl essence esencia de perla

pearl white blanco perla

Pearson solution solución de Pearson

peat turba

peat tar alquitrán de turba

pebulate pebulato

Pechmann pyrazole synthesis síntesis de pirazoles de Pechmann

pectate pectato

pectic péctico

pectic acid ácido péctico

pectin pectina

pectin sugar azúcar de pectina

pectinase pectinasa

pectization pectización

pectolite pectolita

pectose pectosa

pedesis pedesis

pegmatite pegmatita

pelargonic acid ácido pelargónico

pelargonic alcohol alcohol pelargónico

pelargonic aldehyde aldehído pelargónico

pelargonin pelargonina

pelargonitrile pelargonitrilo

pelargonyl pelargonilo

pelargonyl chloride cloruro de pelargonilo

pelargonyl peroxide peroxido de pelargonilo

Peligot salt sal de Pelilgot

pellet pastilla, comprimido
pelletierine peletierina
Pellizzari reaction reacción de Pellizzari
pellotine pelotina
Pelouze synthesis síntesis de Pelouze
penetrant penetrante
penetration penetración
penicillamine penicilamina
penicillanase penicilanasa
penicillanic acid ácido penicilánico
penicillic acid ácido penicílico
penicillin penicilina
Penning ionization ionización de Penning
pennyroyal oil aceite de poleo
Pensky-Martens apparatus aparato de Pensky-Martens
penta penta
pentaamino pentaamino
pentabasic pentabásico
pentaborane pentaborano
pentabromide pentabromuro
pentabromo pentabromo
pentabromobenzene pentabromobenceno
pentacarbonyl pentacarbonilo
pentacarboxylic pentacarboxílico
pentacene pentaceno
pentacetate pentacetato
pentachloride pentacloruro
pentachloro pentacloro
pentachloroaniline pentacloroanilina
pentachlorobenzene pentaclorobenceno
pentachloroethane pentacloroetano
pentachloronaphthalene pentacloronaftaleno
pentachloronitrobenzene pentacloronitrobenceno
pentachlorophenol pentaclorofenol
pentachlorothiophenol pentaclorotiofenol
pentacosamic acid ácido pentacosámico
pentacosane pentacosano
pentacosanoic acid ácido pentacosanoico
pentacyclic pentacíclico
pentad pentada
pentadecane pentadecano
pentadecanoic acid ácido pentadecanoico
pentadecanol pentadecanol
pentadecanolide pentadecanolida

pentadecanone pentadecanona
pentadecenyl pentadecenilo
pentadecyl pentadecilo
pentadecyl alcohol alcohol pentadecílico
pentadecylic acid ácido pentadecílico
pentadiene pentadieno
pentadienedioic acid ácido pentadienodioico
pentaerythritol pentaeritritol
pentaerythritol tetraacetate tetraacetato de pentaeritritol
pentaerythritol tetranitrate tetranitrato de pentaeritritol
pentaethylbenzene pentaetilbenceno
pentafluoride pentafluoruro
pentagalloyl pentagaloilo
pentaglucose pentaglucosa
pentaglycerin pentaglicerina
pentaglycol pentaglicol
pentahydrate pentahidrato
pentahydro pentahidro
pentahydroxy pentahidroxi
pentahydroxycyclohexane pentahidroxiciclohexano
pentaiodo pentayodo
pentalin pentalina
pentamethyl pentametilo
pentamethylbenzene pentametilbenceno
pentamethylbenzoic acid ácido pentametilbenzoico
pentamethylene pentametileno
pentamethylene bromide bromuro de pentametileno
pentamethylene dibromide dibromuro de pentametileno
pentamethylene glycol pentametilenglicol
pentamethylene oxide óxido de pentametileno
pentamethyleneamine pentametilenamina
pentamethylenediamine pentametilendiamina
pentamethylphenol pentametilfenol
pentamino pentaamino
pentanal pentanal
pentane pentano
pentane thermometer termómetro de pentano

pentanecarboxylic acid ácido pentanocarboxílico

pentanedicarboxylic acid ácido pentanodicarboxílico

pentanedioic acid ácido pentanodioico

pentanediol pentanodiol

pentanedione pentanodiona

pentanethiol pentanotiol

pentanoic acid ácido pentanoico

pentanol pentanol

pentanone pentanona

pentaoxide pentaóxido

pentaprismo pentaprismo

pentasodium triphosphate trifosfato pentasódico

pentasulfide pentasulfuro

pentatriacontane pentatriacontano

pentavalent pentavalente

pentazdiene pentazdieno

pentazocine pentazocina

pentazolyl pentazolilo

pentel pentel

pentene penteno

pentenedioic acid ácido pentenodioico

pentenic acid ácido penténico

pentenoic acid ácido pentenoico

pentenol pentenol

pentenyl pentenilo

penthrite pentrita

pentinic acid ácido pentínico

pentite pentita

pentitol pentitol

pentlandite pentlandita

pentobarbital pentobarbital

pentoic acid ácido pentoico

pentol pentol

pentolite pentolita

pentonic acid ácido pentónico

pentosan pentosana

pentose pentosa

pentoside pentósido

pentoxide pentóxido

pentoxyphenol pentoxifenol

pentrite pentrita

pentyl pentilo

pentyl acetate acetato de pentilo

pentyl alcohol alcohol pentílico

pentyl aldehyde aldehído pentílico

pentyl bromide bromuro de pentilo

pentyl butanoate butanoato de pentilo

pentyl carbamate carbamato de pentilo

pentyl chloride cloruro de pentilo

pentyl chlorocarbonate clorocarbonato de pentilo

pentyl cyanide cianuro de pentilo

pentyl ether éter pentílico

pentyl formate formiato de pentilo

pentyl furoate furoato de pentilo

pentyl hydrate hidrato de pentilo

pentyl hydride hidruro de pentilo

pentyl iodide yoduro de pentilo

pentyl isocyanide isocianuro de pentilo

pentyl ketone pentilcetona

pentyl nitrate nitrato de pentilo

pentyl oxide óxido de pentilo

pentyl phenyl ketone pentilfenilcetona

pentyl phenylhydrazine fenilhidrazina de pentilo

pentyl phthalate ftalato de pentilo

pentyl propionate propionato de pentilo

pentyl salicylate salicilato de pentilo

pentyl sulfate sulfato de pentilo

pentyl thiocyanate tiocianato de pentilo

pentyl valerate valerato de pentilo

pentyl xanthate xantato de pentilo

pentylacetic ether éter pentilacético

pentylamine pentilamina

pentylbenzene pentilbenceno

pentylbenzoate pentilbenzoato

pentylcinnamaldehyde pentilcinamaldehído

pentylene pentileno

pentylformic acid ácido pentilfórmico

pentylidene pentilideno

pentylidyne pentilidino

pentyloxy pentiloxi

pentyloxyhydrate pentiloxihidrato

pentyloxyphenol pentiloxifenol

pentylphenol pentilfenol

pentylurea pentilurea

pentylurethane pentiluretano

pentyne pentino

peppermint camphor alcanfor de menta

peppermint oil aceite de menta

pepsin pepsina

pepsinogen pepsinógeno

peptidase peptidasa

peptide péptido

peptide bond enlace de péptido
peptide linkage enlace de péptido
peptization peptización
peptone peptona
per per
peracetic acid ácido peracético
peracid perácido
peracidity peracidez
peralcohol peralcohol
perbenzoic acid ácido perbenzoico
perborate perborato
perbromate perbromato
perbromo perbromo
percarbamide percarbamida
percarbide percarburo
percarbonate percarbonato
percarbonic acid ácido percarbónico
percentage porcentaje
perchlorate perclorato
perchloric acid ácido perclórico
perchloric acid anhydride anhídrido de
    ácido perclórico
perchloride percloruro
perchloro percloro
perchlorobenzene perclorobenceno
perchlorocyclopentadiene
    perclorociclopentadieno
perchloroethane percloroetano
perchloroether percloroéter
perchloroethylene percloroetileno
perchloromethane perclorometano
perchloromethyl perclorometilo
perchloromethyl mercaptan
    perclorometilmercaptano
perchloropropylene percloropropileno
perchloryl perclorilo
perchloryl fluoride fluoruro de
    perclorilo
perchromate percromato
perchromic acid ácido percrómico
percolate percolar
percolation percolación
percolator percolador
percrystallization percristalización
perdistillation perdestilación
perfect gas gas perfecto
perfect solution solución perfecta
perfluoro perfluoro
perfluorobutene perfluorobuteno

perfluorobutyric acid ácido
    perfluorobutírico
perfluorocarbon perfluorocarburo
perfluorocyclobutane
    perfluorociclobutano
perfluoroethylene perfluoroetileno
perfluoropropane perfluoropropano
perfluoropropene perfluoropropeno
perfluorosulfonic acid ácido
    perfluorosulfónico
perforated plate placa perforada
performic acid ácido perfórmico
perfume perfume
perfusion perfusión
pergenol pergenol
perhydrate perhidrato
perhydro perhidro
perhydrol perhidrol
perhydronaphthalene perhidronaftaleno
peri peri
peri acid ácido peri
periclase periclasa
pericline periclina
pericyclic pericíclico
pericyclic reaction reacción pericíclica
pericyclo periciclo
perilla oil aceite de perilla
perilogic series serie perilógica
perimidine perimidina
perimidinyl perimidinilo
perimorph perimorfo
period periodo
periodate peryodato
periodic acid ácido peryódico
periodic chain cadena periódica
periodic classification clasificación
    periódica
periodic law ley periódica
periodic property propiedad periódica
periodic series serie periódica
periodic system sistema periódico
periodic table tabla periódica
periodicity periodicidad
peritectic peritéctico
peritectic point punto peritéctico
peritectic system sistema peritéctico
peritectic temperature temperatura
    peritéctica
peritectoid peritectoide

Perkin alicyclic synthesis síntesis alicíclica de Perkin
Perkin reaction reacción de Perkin
Perkin rearrangement reordenamiento de Perkin
perkinetic pericinético
Perkow reaction reacción de Perkow
perlite perlita
permalloy permaleación
permanent gas gas permanente
permanent hardness dureza permanente
permanent white blanco permanente
permanganate permanganato
permanganate titration titulación con permanganato, valoración con permanganato
permanganic acid ácido permangánico
permanganyl permanganilo
permeability permeabilidad
permeable permeable
permeable membrane membrana permeable
permeametry permeametría
permeation impregnación
permethrin permetrina
permissible explosive explosivo permisible
permonosulfuric acid ácido permonosulfúrico
permutation permutación
permutite permutita
pernitric acid ácido pernítrico
perovskite perovskita
peroxidase peroxidasa
peroxidation peroxidación
peroxide peróxido
peroxide number número de peróxido
peroxide value valor de peróxido
peroxo peroxo
peroxomonosulfuric acid ácido peroxomonosulfúrico
peroxonitric acid ácido peroxonítrico
peroxy peroxi
peroxyacetic acid ácido peroxiacético
peroxybenzoic acid ácido peroxibenzoico
peroxybenzoyl nitrate nitrato de peroxibenzoilo
peroxychromium peroxicromo
peroxydol peroxidol

peroxyethanoic acid ácido peroxietanoico
peroxyformic acid ácido peroxifórmico
peroxymonosulfuric acid ácido peroximonosulfúrico
peroxysulfuric acid ácido peroxisulfúrico
perphenazine perfenazina
perphosphoric acid ácido perfosfórico
perrhenate perrenato
perrhenic acid ácido perrénico
Perrin equation ecuación de Perrin
Perrin rule regla de Perrin
Persian red rojo de Persia
persorption persorción
persulfate persulfato
persulfide persulfuro
persulfuric acid ácido persulfúrico
perthio pertio
perthiocarbonate pertiocarbonato
Peru balsam bálsamo de Perú
pervaporation pervaporación
perylene perileno
pesticide pesticida
pestle mano de mortero
petalite petalita
Peterson reaction reacción de Peterson
petri dish caja de Petri
petrochemical petroquímico
Petroff reagent reactivo de Petroff
petrolatum petrolato
petrolatum wax cera de petrolato
petroleum petróleo
petroleum asphalt asfalto de petróleo
petroleum benzin bencina de petróleo
petroleum coke coque de petróleo
petroleum ether éter de petróleo
petroleum gas gas de petróleo
petroleum jelly petrolato
petroleum naphtha nafta de petróleo
petroleum refining refinación de petróleo
petroleum wax cera de petróleo
petroxolin petroxolina
petzite petzita
pewter peltre
Pfeiffer effect efecto de Pfeiffer
Pfitzinger reaction reacción de Pfitzinger

Pfitzner-Moffatt oxidation   oxidación de Pfitzner-Moffatt

pH   pH

pH indicator   indicador de pH

pH meter   medidor de pH

pH recorder   registrador de pH

pH scale   escala de pH

pH value   valor de pH

pharmaceutical   farmacéutico

pharmacokinetics   farmacocinética

pharmacolite   farmacolita

pharmacology   farmacología

phase   fase

phase diagram   diagrama de fases

phase equilibrium   equilibrio de fases

phase reversal   inversión de fase

phase rule   regla de fases

phase transfer   transferencia de fase

phase transition   transición de fase

phasotropy   fasotropía

phellandrene   felandreno

phen   fen

phenacaine hydrochloride   clorhidrato de fenacaína

phenacetin   fenacetina

phenacetol   fenacetol

phenacyl   fenacilo

phenacyl alcohol   alcohol fenacílico

phenacyl chloride   cloruro de fenacilo

phenacylidene   fenacilideno

phenanthraquinone   fenantraquinona

phenanthrene   fenantreno

phenanthrenequinone   fenantrenoquinona

phenanthrenyl   fenantrenilo

phenanthridine   fenantridina

phenanthrol   fenantrol

phenanthroline   fenantrolina

phenanthroline indicator   indicador de fenantrolina

phenanthrone   fenantrona

phenanthrophenazine   fenantrofenacina

phenanthryl   fenantrilo

phenanthrylene   fenantrileno

phenarsazine chloride   cloruro de fenarsazina

phenate   fenato

phenazine   fenazina

phenazone   fenazona

phenazonium   fenazonio

phenenyl   fenenilo

phenethyl   fenetilo

phenethyl acetate   acetato de fenetilo

phenethyl alcohol   alcohol fenetílico

phenethyl isobutyrate   isobutirato de fenetilo

phenethylamine   fenetilamina

phenethylene   fenetileno

phenetidine   fenetidina

phenetidino   fenetidino

phenetole   fenetol

phenetyl   fenetilo

phenic acid   ácido fénico

phenidone   fenidona

phenindione   fenindiona

pheniramine maleate   maleato de feniramina

phenixin   fenixina

phenmethyl   fenmetilo

phenobarbital   fenobarbital

phenobarbitone   fenobarbitona

phenodiazine   fenodiazina

phenol   fenol

phenol acid   ácido fenólico

phenol aldehyde   aldehído fenólico

phenol bismuth   bismuto de fenol

phenol coefficient   coeficiente de fenol

phenol ether   éter fenólico

phenol-formaldehyde resin   resina de fenol-formaldehído

phenol-furfural resin   resina de fenol-furfural

phenol red   rojo fenol

phenol trinitrate   trinitrato de fenol

phenolate   fenolato

phenolate process   proceso de fenolato

phenoldisulfonic acid   ácido fenoldisulfónico

phenolic   fenólico

phenolic acid   ácido fenólico

phenolic resin   resina fenólica

phenolphthalein   fenolftaleína

phenolphthalide   fenolftalida

phenolsafranin   fenolsafranina

phenolsulfonate   fenolsulfonato

phenolsulfonephthalein   fenolsulfonaftaleína

phenolsulfonic acid   ácido fenolsulfónico

phenone   fenona

phenothiazine fenotiazina
phenothrin fenotrina
phenotole fenotol
phenoxazine fenoxazina
phenoxide fenóxido
phenoxy fenoxi
phenoxy resin resina fenoxi
phenoxyacetic acid ácido fenoxiacético
phenoxybenzene fenoxibenceno
phenoxydihydroxypropane
    fenoxidihidroxipropano
phenoxyethanol fenoxietanol
phenoxyethylpenicillin
    fenoxietilpenicilina
phenoxypropanediol fenoxipropanodiol
phenoxypropylene oxide óxido de
    fenoxipropileno
phenyl fenilo
phenyl acetate acetato de fenilo
phenyl azide azida de fenilo
phenyl bromide bromuro de fenilo
phenyl carbamate carbamato de fenilo
phenyl carbonate carbonato de fenilo
phenyl chloride cloruro de fenilo
phenyl cyanide cianuro de fenilo
phenyl ether éter fenílico
phenyl fluoride fluoruro de fenilo
phenyl hydrate hidrato de fenilo
phenyl hydroxide hidróxido de fenilo
phenyl isocyanate isocianato de fenilo
phenyl isocyanide isocianuro de fenilo
phenyl isothiocyanate isotiocianato de
    fenilo
phenyl ketone fenilcetona
phenyl mercaptan fenilmercaptano
phenyl methyl ketone fenilmetilcetona
phenyl mustard oil aceite de
    fenilmostasa
phenyl phthalate ftalato de fenilo
phenyl salicylate salicilato de fenilo
phenyl sulfate sulfato de fenilo
phenyl valerate valerato de fenilo
phenyl vinyl ketone fenilvinilcetona
phenylacetaldehyde fenilacetaldehído
phenylacetaldehyde dimethylacetal
    dimetilacetal de fenilacetaldehído
phenylacetamide fenilacetamida
phenylacetic acid ácido fenilacético
phenylacetonitrile fenilacetonitrilo

phenylacetyl fenilacetilo
phenylacetyl chloride cloruro de
    fenilacetilo
phenylalanine fenilalanina
phenylamine fenilamina
phenylaniline fenilanilina
phenylarsonic acid ácido fenilarsónico
phenylation fenilación
phenylazo fenilazo
phenylbarbital fenilbarbital
phenylbenzamide fenilbenzamida
phenylbenzene fenilbenceno
phenylbenzoate fenilbenzoato
phenylbenzoyl fenilbenzoilo
phenylbenzyl fenilbencilo
phenylbutane fenilbutano
phenylbutanoic acid ácido fenilbutanoico
phenylbutazone fenilbutazona
phenylbutene fenilbuteno
phenylbutynol fenilbutinol
phenylbutyric acid ácido fenilbutírico
phenylcarbamoyl fenilcarbamoilo
phenylcarbethoxypyrazolone
    fenilcarbetoxipirazolona
phenylcarbimide fenilcarbimida
phenylcarbinol fenilcarbinol
phenylcarbylamine chloride cloruro de
    fenilcarbilamina
phenylchloroform fenilcloroformo
phenylcyclidene hydrochloride
    clorhidrato de fenilciclideno
phenylcyclohexane fenilciclohexano
phenylcyclohexanol fenilciclohexanol
phenyldichloroarsine fenildicloroarsina
phenyldidecyl phosphite fosfito de
    fenildidecilo
phenyldiethanolamine
    fenildietanolamina
phenyldimethylurea fenildimetilurea
phenylene fenileno
phenylene blue azul fenileno
phenylenediamine fenilendiamina
phenylethane feniletano
phenylethanoic acid ácido feniletanoico
phenylethyl feniletilo
phenylethyl acetate acetato de feniletilo
phenylethyl alcohol alcohol feniletílico
phenylethyl mercaptan
    feniletilmercaptano

phenylethyl salicylate salicilato de feniletilo

phenylethylacetic acid ácido feniletilacético

phenylethylamine feniletilamina

phenylethylethanolamine feniletiletanolamina

phenylethylene feniletileno

phenylethylene glycol feniletilenglicol

phenylethylether feniletiléter

phenylformamide fenilformamida

phenylformic acid ácido fenilfórmico

phenylglycine fenilglicina

phenylglycol fenilglicol

phenylglycolic acid ácido fenilglicólico

phenylhydrazine fenilhidrazina

phenylhydrazone fenilhidrazona

phenylhydroxylamine fenilhidroxilamina

phenylic fenílico

phenylic acid ácido fenílico

phenylidene fenilideno

phenyllithium fenil-litio

phenylmagnesium fenilmagnesio

phenylmagnesium bromide bromuro de fenilmagnesio

phenylmagnesium chloride cloruro de fenilmagnesio

phenylmercuric fenilmercúrico

phenylmercuric acetate acetato fenilmercúrico

phenylmercuric benzoate benzoato fenilmercúrico

phenylmercuric borate borato fenilmercúrico

phenylmercuric chloride cloruro fenilmercúrico

phenylmercuric ethanoate etanoato fenilmercúrico

phenylmercuric hydroxide hidroxido fenilmercúrico

phenylmercuric lactate lactato fenilmercúrico

phenylmercuric naphthenate naftenato fenilmercúrico

phenylmercuric nitrate nitrato fenilmercúrico

phenylmercuric oleate oleato fenilmercúrico

phenylmercuric propionate propionato fenilmercúrico

phenylmercuric salicylate salicilato fenilmercúrico

phenylmethane fenilmetano

phenylmethanol fenilmetanol

phenylmethyl fenilmetilo

phenylmethyl acetate acetato de fenilmetilo

phenylmethylcarbinol fenilmetilcarbinol

phenylmethylene fenilmetileno

phenylmethylethanolamine fenilmetiletanolamina

phenylmorpholine fenilmorfolina

phenylnaphthalene fenilnaftaleno

phenylnaphthylamine fenilnaftilamina

phenylneopentyl phosphite fosfito de fenilneopentilo

phenylnitroamine fenilnitroamina

phenylnonane fenilnonano

phenylpentane fenilpentano

phenylphenol fenilfenol

phenylphosphine fenilfosfina

phenylphosphinic acid ácido fenilfosfínico

phenylphosphonic acid ácido fenilfosfónico

phenylpiperazine fenilpiperazina

phenylpropane fenilpropano

phenylpropanol fenilpropanol

phenylpropanone fenilpropanona

phenylpropenal fenilpropenal

phenylpropenoic acid ácido fenilpropenoico

phenylpropenol fenilpropenol

phenylpropionic acid ácido fenilpropiónico

phenylpropyl fenilpropilo

phenylpropyl acetate acetato de fenilpropilo

phenylpropyl alcohol alcohol fenilpropílico

phenylpropyl aldehyde aldehído fenilpropílico

phenylpropyl chloride cloruro de fenilpropilo

phenylpropyl ketone fenilpropilcetona

phenylpropylpyridine fenilpropilpiridina

phenylpyrazolide fenilpirazolida

phenylpyrazolidone fenilpirazolidona

phenylpyridine fenilpiridina
phenylsalicylic acid ácido fenilsalicílico
phenylstearic acid ácido fenilesteárico
phenylsulfanilic acid ácido
    fenilsulfanílico
phenylsulfonic acid ácido fenilsulfónico
phenylsulfonyl fenilsulfonilo
phenylthiourea feniltiourea
phenyltoluene feniltolueno
phenyltrichlorosilane feniltriclorosilano
phenyltridecane feniltridecano
phenylurethane feniluretano
phenytoin fenitoína
pheromone feromona
phi phi
phillipite filipita
phloroglucinol floroglucinol
phonochemistry fonoquímica
phorate forato
phorbol forbol
phorone forona
phoryl forilo
phosalone fosalona
phosgene fosgeno
phosmet fosmet
phospha fosfa
phosphamic acid ácido fosfámico
phosphamide fosfamida
phosphamidon fosfamidón
phosphane fosfano
phosphatase fosfatasa
phosphate fosfato
phosphate buffer amortiguador de
    fosfato
phosphate glass vidrio de fosfato
phosphate rock roca de fosfato
phosphate slag escoria de fosfato
phosphatide fosfátido
phosphazene fosfaceno
phosphenyl fosfenilo
phosphide fosfuro
phosphinate fosfinato
phosphine fosfina
phosphinic acid ácido fosfínico
phosphinico fosfinico
phosphinite fosfinita
phosphino fosfino
phosphinoso fosfinoso
phosphinous acid ácido fosfinoso

phosphinoyl fosfinoilo
phosphite fosfito
phospho fosfo
phosphobenzene fosfobenceno
phosphocreatine fosfocreatina
phosphodiester fosfodiéster
phosphodiesterase fosfodiesterasa
phosphoglyceride fosfoglicérido
phospholipase fosfolipasa
phospholipid fosfolípido
phosphomolybdate fosfomolibdato
phosphomolybdic acid ácido
    fosfomolíbdico
phosphonate fosfonato
phosphonic acid ácido fosfónico
phosphonite fosfonita
phosphonitrile fosfonitrilo
phosphonium fosfonio
phosphonium bromide bromuro de
    fosfonio
phosphonium chloride cloruro de
    fosfonio
phosphonium hydroxide hidróxido de
    fosfonio
phosphonium iodide yoduro de fosfonio
phosphonium salt sal de fosfonio
phosphono fosfono
phosphonoso fosfonoso
phosphonous acid ácido fosfonoso
phosphonoyl fosfonoilo
phosphoprotein fosfoproteína
phosphor fósforo
phosphor bronze bronce fosforado
phosphorane fosforano
phosphorate fosforato
phosphorated fosforado
phosphorescence fosforescencia
phosphorescent fosforescente
phosphoric fosfórico
phosphoric acid ácido fosfórico
phosphoric anhydride anhídrido
    fosfórico
phosphoric bromide bromuro fosfórico
phosphoric chloride cloruro fosfórico
phosphoric oxide óxido fosfórico
phosphoric perbromide perbromuro
    fosfórico
phosphoric perchloride percloruro
    fosfórico

phosphoric sulfide sulfuro fosfórico
phosphorimetry fosforimetría
phosphorite fosforita
phosphoroso fosforoso
phosphorous acid ácido fosforoso
phosphorus fósforo
phosphorus bromide bromuro de fósforo
phosphorus chloride cloruro de fósforo
phosphorus cycle ciclo de fósforo
phosphorus heptasulfide heptasulfuro de fósforo
phosphorus hydride hidruro de fósforo
phosphorus nitride nitruro de fósforo
phosphorus oxide óxido de fósforo
phosphorus oxyacid oxiácido de fósforo
phosphorus oxybromide oxibromuro de fósforo
phosphorus oxychloride oxicloruro de fósforo
phosphorus pentabromide pentabromuro de fósforo
phosphorus pentachloride pentacloruro de fósforo
phosphorus pentafluoride pentafluoruro de fósforo
phosphorus pentaselenide pentaseleniuro de fósforo
phosphorus pentasulfide pentasulfuro de fósforo
phosphorus pentoxide pentóxido de fósforo
phosphorus persulfide persulfuro de fósforo
phosphorus sesquisulfide sesquisulfuro de fósforo
phosphorus sulfide sulfuro de fósforo
phosphorus thiochloride tiocloruro de fósforo
phosphorus tribromide tribromuro de fósforo
phosphorus trichloride tricloruro de fósforo
phosphorus triiodide triyoduro de fósforo
phosphorus trisulfide trisulfuro de fósforo
phosphoryl fosforilo
phosphoryl bromide bromuro de fosforilo

phosphoryl chloride cloruro de fosforilo
phosphoryl fluoride fluoruro de fosforilo
phosphoryl nitride nitruro de fosforilo
phosphorylase fosforilasa
phosphorylation fosforilación
phosphotungstate fosfotungstato
phosphotungstic acid ácido fosfotúngstico
photoacoustic spectroscopy espectroscopia fotoacústica
photocatalysis fotocatálisis
photocatalyst fotocatalizador
photocathode fotocátodo
photochemical fotoquímico
photochemical activation activación fotoquímica
photochemical catalysis catálisis fotoquímica
photochemical dissociation disociación fotoquímica
photochemical effect efecto fotoquímico
photochemical equivalent equivalente fotoquímico
photochemical equivalent law ley de equivalente fotoquímico
photochemical excitation excitación fotoquímica
photochemical induction inducción fotoquímica
photochemical oxidant oxidante fotoquímico
photochemical process proceso fotoquímico
photochemical reaction reacción fotoquímica
photochemical yield rendimiento fotoquímico
photochemistry fotoquímica
photochromic fotocrómico
photochromic compound compuesto fotocrómico
photochromism fotocromismo
photoconduction fotoconducción
photodecomposition fotodescomposición
photodegradation fotodegradación
photodeposition fotodeposición
photodimerization fotodimerización
photodisintegration fotodesintegración
photodissociation fotodisociación

photoelectric fotoeléctrico
photoelectric absorption absorción fotoeléctrica
photoelectric cell célula fotoeléctrica
photoelectric colorimeter colorímetro fotoeléctrico
photoelectron fotoelectrón
photoelectron spectromicroscopy espectromicroscopia de fotoelectrones
photoelectron spectroscopy espectroscopia de fotoelectrones
photoemission fotoemisión
photoemissive fotoemisivo
photoemitter fotoemisor
photographic fotográfico
photoionization fotoionización
photoluminescence fotoluminiscencia
photolysis fotólisis
photolyte fotólito
photolytic fotolítico
photometer fotómetro
photometric fotométrico
photometric analysis análisis fotométrico
photometric titration titulación fotométrica, valoración fotométrica
photon fotón
photon absorption absorción de fotones
photon emission emisión de fotones
photoneutron fotoneutrón
photonuclear fotonuclear
photonuclear reaction reacción fotonuclear
photooxidation fotooxidación
photopolymer fotopolímero
photopolymerization fotopolimerización
photoproton fotoprotón
photoreaction fotorreacción
photoreduction fotorreducción
photosensitive fotosensible
photosensitive substance substancia fotosensible
photosensitivity fotosensibilidad
photosensitization fotosensibilización
photosensitizer fotosensibilizador
photostabilization fotoestabilización
photosynthesis fotosíntesis
phototropism fototropismo
phototropy fototropia

photovoltaic fotovoltaico
photovoltaic cell célula fotovoltaica
phthalamic acid ácido ftalámico
phthalamide ftalamida
phthalate ftalato
phthalazine ftalazina
phthaldiamide ftaldiamida
phthalein ftaleína
phthalic acid ácido ftálico
phthalic anhydride anhídrido ftálico
phthalic ester éster ftálico
phthalide ftalida
phthalidyl ftalidilo
phthalidylidene ftalidilideno
phthalimide ftalimida
phthalimido ftalimido
phthalocyanine ftalocianina
phthalodinitrile ftalodinitrilo
phthalonitrile ftalonitrilo
phthaloyl ftaloilo
phthaloyl chloride cloruro de ftaloilo
phthalyl ftalilo
phthioic acid ácido ftioico
phycocolloid ficocoloide
physical analysis análisis físico
physical chemistry química física, fisicoquímica
physical constant constante física
physical phenomenon fenómeno físico
physical property propiedad física
physical quantity cantidad física
physical solution solución física
physical solvent solvente físico
physical testing pruebas físicas
physicochemical fisicoquímico
physostigmine fisostigmina
physostigmine sulfate sulfato de fisostigmina
phytane fitano
phytanic acid ácido fitánico
phytic acid ácido fítico
phytin fitina
phytochemistry fitoquímica
phytol fitol
phytotoxic fitotóxico
phytyl fitilo
pi pi
pi bond enlace pi
pi bonding enlazado pi

pi complex complejo pi
pi electron electrón pi
picene piceno
pickling decapado
picocurie picocurie
picoline picolina
picolinic acid ácido picolínico
picolyl picolilo
picolylamine picolilamina
picotite picotita
picramic acid ácido picrámico
picramide picramida
picraminic acid ácido picramínico
picrate picrato
picric acid ácido pícrico
picro picro
picrolonic acid ácido picrolónico
picromerite picromerita
picronitric acid ácido picronítrico
picrotoxin picrotoxina
picryl picrilo
picryl chloride cloruro de picrilo
Pictet-Hubert reaction reacción de
    Pictet-Hubert
piedmontite piedmontita
piezochemistry piezoquímica
piezocrystallization piezocristalización
piezoelectric effect efecto piezoeléctrico
piezoelectricity piezoelectricidad
pig iron hierro en lingotes
pigment pigmento
pigment volume concentration
    concentración de volmen de pigmento
pilocarpine pilocarpina
pilot plant planta piloto
Piloty-Robinson synthesis síntesis de
    Piloty-Robinson
pimaric acid ácido pimárico
pimaricin pimaricina
pimelic acid ácido pimélico
pimelite pimelita
pinacoid pinacoide
pinacol pinacol
pinacol condensation condensación de
    pinacol
pinacol rearrangement reordenamiento
    de pinacol
pinacolone pinacolona
pinacone pinacona

pinane pinano
pinanyl pinanilo
pine oil aceite de pino
pine tar alquitrán de pino
pine-tar oil aceite de alquitrán de pino
pine-tar pitch brea de alquitrán de pino
pinene pineno
pinene hydrochloride clorhidrato de
    pineno
pinic acid ácido pínico
Pinner reaction reacción de Pinner
Pinner synthesis síntesis de Pinner
pinocarveol pinocarveol
pipe clay tierra de pipas
pipecoline pipecolina
piperaline piperalina
piperazine piperazina
piperazine dihydrochloride diclorhidrato
    de piperazina
piperazine hexahydrate hexahidrato de
    piperazina
piperidine piperidina
piperidinium piperidinio
piperidino piperidino
piperidinoethanol piperidinoetanol
piperidyl piperidilo
piperine piperina
piperitol piperitol
piperitone piperitona
piperonal piperonal
piperonyl piperonilo
piperonyl butoxide butóxido de
    piperonilo
piperonylic acid ácido piperonílico
piperonylidene piperonilideno
piperonyloyl piperoniloilo
piperyl piperilo
piperylene piperileno
pipet pipeta
pipette pipeta
Piria reaction reacción de Piria
pitch pez, brea
pitchblende pecblenda
Pitzer equation ecuación de Pitzer
pivaldehyde pivaldehído
pivalic acid ácido piválico
Planck constant constante de Planck
plane of polarization plano de
    polarización

plane of symmetry plano de simetría
plane-polarized polarizado en un plano
plane-polarized light luz polarizada en
   un plano
planetary electrons electrones
   planetarios
plant planta
plasma plasma
plasmid plásmido
plasmin plasmina
plasmogen plasmógeno
plaster of Paris yeso de París
plastic plástico
plastic film película de plástico
plastic flow flujo plástico
plastic foam espuma de plástico
plastic pipe tubo de plástico
plasticity plasticidad
plasticizer plasticizador
plastisol plastisol
plastoquinone plastoquinona
plate placa
plate column columna de placas
plate efficiency eficiencia de placas
plated depositado, electrodepositado,
   metalizado, galvanoplasteado
plating depósito, electrodeposición,
   metalización, galvanoplastia
platinic platínico
platinic ammonium chloride cloruro
   amónico platínico
platinic chloride cloruro platínico
platinic oxide óxido platínico
platinic sodium chloride cloruro sódico
   platínico
platinic sulfate sulfato platínico
platinization platinado
platinochloride platinocloruro
platinocyanide platinocianuro
platinous platinoso
platinous chloride cloruro platinoso
platinous iodide yoduro platinoso
platinum platino
platinum amine amina de platino
platinum ammonium chloride cloruro
   amónico de platino
platinum bichloride bicloruro de platino
platinum black negro de platino
platinum chloride cloruro de platino

platinum-cobalt alloy aleación de
   platino-cobalto
platinum dibromide dibromuro de
   platino
platinum dichloride dicloruro de platino
platinum diiodide diyoduro de platino
platinum dioxide dióxido de platino
platinum iodide yoduro de platino
platinum-iridium alloy aleación de
   platino-iridio
platinum-lithium platino-litio
platinum monosulfide monosulfuro de
   platino
platinum oxide óxido de platino
platinum potassium chloride cloruro
   potásico de platino
platinum-rhodium alloy aleación de
   platino-rodio
platinum sodium chloride cloruro
   sódico de platino
platinum sponge esponja de platino
platinum sulfate sulfato de platino
platinum tetrachloride tetracloruro de
   platino
plazolite plazolita
pleochroism pleocroismo
plumbago plumbago
plumbate plumbato
plumbic plúmbico
plumbic acid ácido plúmbico
plumbous plumboso
plumbous oxide óxido plumboso
plumbous sulfide sulfuro plumboso
plutonium plutonio
plutonium oxide óxido de plutonio
Podbielniak analysis análisis de
   Podbielniak
poison veneno
poison gas gas venenoso
poison vapor vapor venenoso
poisonous venenoso
polar polar
polar compound compuesto polar
polar covalent bond enlace covalente
   polar
polar liquid líquido polar
polar molecule molécula polar
polar solvent solvente polar
polarimeter polarímetro

polarimetric analysis análisis polarimétrico

polarimetry polarimetría

polarity polaridad

polarizability polarizabilidad

polarizable polarizable

polarization polarización

polarization current corriente de polarización

polarization potential potencial de polarización

polarize polarizar

polarized polarizado

polarized electrode electrodo polarizado

polarized light luz polarizada

polarizer polarizador

polarizing polarizante

polarogram polarograma

polarographic analysis análisis polarográfico

polarography polarografía

Polenske number número de Polenske

pollucite polucita

pollution contaminación

polonium polonio

Polonovski reaction reacción de Polonovski

poly poli

polyacetal poliacetal

polyacetaldehyde poliacetaldehído

polyacetylene poliacetileno

polyacrylamide poliacrilamida

polyacrylate poliacrilato

polyacrylic acid ácido poliacrílico

polyacrylic fiber fibra poliacrílica

polyacrylonitrile poliacrilonitrilo

polyalcohol polialcohol

polyalkane polialcano

polyalkene polialqueno

polyalkylidene polialquilideno

polyallomer polialómero

polyamide poliamida

polyamide resin resina de poliamida

polyamine poliamina

polyamine-methylene resin resina de poliamina-metileno

polyaminotriazole poliaminotriazol

polyatomic poliatómico

polyatomic gas gas poliatómico

polyatomic molecule molécula poliatómica

polybasic polibásico

polybasic acid ácido polibásico

polybenzimidazole polibencimidazol

polyblend polimezcla

polybutadiene polibutadieno

polybutene polibuteno

polybutylene polibutileno

polycarbonate policarbonato

polycarboxylic policarboxílico

polycarboxylic acid ácido policarboxílico

polychlor policloro

polychloral policloral

polychlorinated biphenyl bifenilo policlorado

polychloroprene policloropreno

polychlorotrifluoroethane policlorotrifluoroetano

polychromatic policromático

polycondensation policondensación

polycyclic policíclico

polycyclic hydrocarbon hidrocarburo policíclico

polydimethylsiloxane polidimetilsiloxano

polydispersion polidispersión

polyelectrolyte polielectrólito

polyene polieno

polyester poliéster

polyester fiber fibra poliéster

polyester film película poliéster

polyester resin resina poliéster

polyester rubber caucho poliéster

polyethenoid polietenoide

polyether poliéter

polyether foam espuma poliéter

polyether glycol polieterglicól

polyether resin resina poliéter

polyethylene polietileno

polyethylene glycol polietilenglicol

polyethylene glycol chloride cloruro de polietilenglicol

polyethylene oxide óxido de polietileno

polyethylene resin resina de polietileno

polyethylene tetraphthalate tetraftalato de polietileno

polyethyleneimine polietilenimina

polyformaldehyde poliformaldehído

polyforming poliformación

polyfunctional polifuncional

polyfunctional system sistema polifuncional

polyfurfuryl alcohol alcohol polifurfurílico

polygen polígeno

polyglycerol poliglicerol

polyglycerol ester éster de poliglicerol

polyglycol poliglicol

polyglycol distearate diestearato de poliglicol

polyhalite polihalita

polyhexafluoropropene polihexafluoropropeno

polyhexamethylene polixematileno

polyhydrate polihidrato

polyhydric polihídrico

polyhydric alcohol alcohol polihídrico

polyhydric phenol fenol polihídrico

polyimide poliimida

polyimide resin resina de poliimida

polyiodide poliyoduro

polyisobutene poliisobuteno

polyisobutylene poliisobutileno

polyisocyanurate poliisocianurato

polyisoprene poliisopreno

polymer polímero

polymeric polimérico

polymeric reagent reactivo polimérico

polymerism polimerismo

polymerization polimerización

polymerize polimerizar

polymethacrylate resin resina de polimetacrilato

polymethanal polimetanal

polymethylbenzene polimetilbenceno

polymethylene glycol polimetilenglicol

polymethylpentene polimetilpenteno

polymorphism polimorfismo

polymyxin polimixina

polynuclear polinuclear

polynuclear hydrocarbon hidrocarburo polinuclear

polynuclidic polinuclídico

polyol poliol

polyolefin poliolefina

polyoxadiazole polioxadiazol

polyoxamide polioxamida

polyoxy polioxi

polyoxyalkylene resin resina de polioxialquileno

polyoxyethylene polioxietileno

polyoxyethylene monostearate monoestearato de polioxietileno

polyoxyethyleneoxypropylene polioxietilenoxipropileno

polyoxymethylene polioximetileno

polyoxypropylenediamine polioxipropilendiamina

polypeptide polipéptido

polyphase polifásico

polyphenyl polifenilo

polyphenylene oxide óxido de polifenileno

polyphosphate polifosfato

polyphosphoric acid ácido polifosfórico

polyprenol poliprenol

polypropylene polipropileno

polypropylene glycol polipropilenglicol

polypropylene oxide óxido de polipropileno

polypropylenebenzene polipropilenbenceno

polypyrrolidine polipirrolidina

polyrotaxame polirrotaxama

polysaccharide polisacárido

polysiloxane polisiloxano

polysorbate polisorbato

polyspiro poliespiro

polystyrene poliestireno

polysulfide polisulfuro

polysulfone polisulfona

polyterpene resin resina de politerpeno

polytetrafluoroethene politetrafluoroeteno

polytetrafluoroethylene politetrafluoroetileno

polytetrafluoroethylene resin resina de politetrafluoroetileno

polythene politeno

polythiadazole politiadazol

polythiazyl politiacilo

polythionate politionato

polythionic acid ácido politiónico

polyunsaturated poliinsaturado

polyunsaturated acid ácido poliinsaturado

polyunsaturated fat grasa poliinsaturada

polyurethane poliuretano
polyurethane resin resina de poliuretano
polyurethane rubber caucho de
    poliuretano
polyvalence polivalencia
polyvalent polivalente
polyvinyl polivinilo
polyvinyl acetal polivinilacetal
polyvinyl acetate acetato de polivinilo
polyvinyl alcohol alcohol polivinílico
polyvinyl butyral polivinilbutiral
polyvinyl carbazole polivinilcarbazol
polyvinyl chloride cloruro de polivinilo
polyvinyl chloride acetate acetato de
    cloruro de polivinilo
polyvinyl dichloride dicloruro de
    polivinilo
polyvinyl ether éter polivinílico
polyvinyl ethyl ether éter etílico
    polivinílico
polyvinyl fluoride fluoruro de polivinilo
polyvinyl isobutyl ether éter isobutílico
    polivinílico
polyvinyl methyl ether éter metílico
    polivinílico
polyvinyl resin resina de polivinilo
polyvinylidene polivinilideno
polyvinylidene chloride cloruro de
    polivinilideno
polyvinylidene fluoride fluoruro de
    polivinilideno
polyvinylidene resin resina de
    polivinilideno
polyvinylpyrrolidine polivinilpirrolidina
Pomeranz-Fritsch reaction reacción de
    Pomeranz-Fritsch
Ponzio reaction reacción de Ponzio
porcelain porcelana
porcelain enamel esmalte de porcelana
pore poro
porometric porométrico
porosimeter porosímetro
porosity porosidad
porous poroso
porphin porfina
porphyrin porfirina
porphyrinogen porfirinógeno
portland cement cemento portland
positive positivo

positive crystal cristal positivo
positive electron electrón positivo
positive element elemento positivo
positive group grupo positivo
positive ion ion positivo
positive radical radical positivo
positive-ray analysis análisis de rayos
    positivos
positron positrón
positron emission emisión de positrón
positronium positronio
postprecipitation posprecipitación
potasan potasán
potash potasa
potash alum alumbre potásico
potassic potásico
potassium potasio
potassium abietate abietato potásico
potassium acetate acetato potásico,
    acetato de potasio
potassium acetylsalicylate acetilsalicilato
    potásico
potassium acid carbonate carbonato
    ácido potásico
potassium acid fluoride fluoruro ácido
    potásico
potassium acid oxalate oxalato ácido
    potásico
potassium acid phosphate fosfato ácido
    potásico
potassium acid saccharate sacarato
    ácido potásico
potassium acid sulfate sulfato ácido
    potásico
potassium acid sulfite sulfito ácido
    potásico
potassium acid tartrate tartrato ácido
    potásico
potassium alginate alginato potásico
potassium alum alumbre potásico
potassium aluminate aluminato potásico
potassium aluminosilicate
    aluminosilicato potásico
potassium aluminum fluoride fluoruro
    alumínico potásico
potassium aluminum sulfate sulfato
    alumínico potásico
potassium aminobenzoate
    aminobenzoato potásico

potassium anthranilate antranilato potásico

potassium antimonate antimoniato potásico

potassium argentocyanide argentocianuro potásico

potassium arsenate arseniato potásico

potassium arsenite arsenito potásico

potassium aurate aurato potásico

potassium benzenedisulfonate bencenodisulfonato potásico

potassium benzoate benzoato potásico

potassium bicarbonate bicarbonato potásico

potassium bichromate bicromato potásico

potassium bifluoride bifluoruro potásico

potassium binoxalate binoxalato potásico

potassium biphthalate biftalato potásico

potassium bisulfate bisulfato potásico

potassium bisulfite bisulfito potásico

potassium bitartrate bitartrato potásico

potassium borate borato potásico

potassium borofluoride borofluoruro potásico

potassium borohydride borohidruro potásico

potassium bromate bromato potásico

potassium bromide bromuro potásico

potassium butanoate butanoato potásico

potassium butoxide butóxido potásico

potassium carbonate carbonato potásico

potassium chlorate clorato potásico

potassium chloride cloruro potásico

potassium chlorochromate clorocromato potásico

potassium chloroplatinate cloroplatinato potásico

potassium chromate cromato potásico

potassium cinnamate cinamato potásico

potassium citrate citrato potásico

potassium cuprocyanide cuprocianuro potásico

potassium cyanate cianato potásico

potassium cyanide cianuro potásico

potassium cyclamate ciclamato potásico

potassium dichloroisocyanurate dicloroisocianurato potásico

potassium dichromate dicromato potásico

potásico

potassium dicyanoargenate dicianoargenato potásico

potassium dicyanoaurate dicianoaurato potásico

potassium diphosphate difosfato potásico

potassium disulfate disulfato potásico

potassium disulfite disulfito potásico

potassium dithionate ditionato potásico

potassium ethanoate etanoato potásico

potassium feldspar feldespato potásico

potassium ferrate ferrato potásico

potassium ferricyanide ferricianuro potásico

potassium ferrite ferrita potásica

potassium ferrocyanide ferrocianuro potásico

potassium fluoride fluoruro potásico

potassium fluoroborate fluoroborato potásico

potassium fluosilicate fluosilicato potásico

potassium fluozirconate fluozirconato potásico

potassium formate formiato potásico

potassium gluconate gluconato potásico

potassium glutamate glutamato potásico

potassium glycerophosphate glicerofosfato potásico

potassium hexacarbonyl hexacarbonilo potásico

potassium hexachloroplatinate hexacloroplatinato potásico

potassium hexacyanoferrate hexacianoferrato potásico

potassium hexafluorophosphate hexafluorofosfato potásico

potassium hexanitrocobaltate hexanitrocobaltato potásico

potassium hippurate hipurato potásico

potassium hydrate hidrato potásico

potassium hydride hidruro potásico

potassium hydrogen carbonate carbonato de hidrógeno potásico

potassium hydrogen fluoride fluoruro de hidrógeno potásico

potassium hydrogen phosphate fosfato de hidrógeno potásico

potassium hydrogen phthalate ftalato de hidrógeno potásico

potassium hydrogen sulfate sulfato de hidrógeno potásico

potassium hydrogen tartrate tartrato de hidrógeno potásico

potassium hydrosulfide hidrosulfuro potásico

potassium hydroxide hidróxido potásico

potassium hypochlorite hipoclorito potásico

potassium hypophosphite hipofosfito potásico

potassium hyposulfate hiposulfato potásico

potassium hyposulfite hiposulfito potásico

potassium iodate yodato potásico

potassium iodide yoduro potásico

potassium laurate laurato potásico

potassium linoleate linoleato potásico

potassium manganate manganato potásico

potassium mercuric iodide yoduro mercúrico potásico

potassium metaarsenite metaarsenito potásico

potassium metabisulfite metabisulfito potásico

potassium metaphosphate metafosfato potásico

potassium methoxide metóxido potásico

potassium molybdate molibdato potásico

potassium monophosphate monofosfato potásico

potassium monosulfide monosulfuro potásico

potassium naphthenate naftenato potásico

potassium nitrate nitrato potásico

potassium nitrite nitrito potásico

potassium oleate oleato potásico

potassium orthophosphate ortofosfato potásico

potassium osmate osmiato potásico

potassium oxalate oxalato potásico

potassium oxide óxido potásico

potassium penicillin penicilina potásica

potassium pentasulfide pentasulfuro potásico

potassium percarbonate percarbonato potásico

potassium perchlorate perclorato potásico

potassium periodate peryodato potásico

potassium permanganate permanganato potásico

potassium peroxide peróxido potásico

potassium peroxoborate peroxoborato potásico

potassium peroxosulfate peroxosulfato potásico

potassium peroxydisulfate peroxidisulfato potásico

potassium persulfate persulfato potásico

potassium phosphate fosfato potásico

potassium phosphide fosfuro potásico

potassium phosphinate fosfinato potásico

potassium phosphite fosfito potásico

potassium phosphonate fosfonato potásico

potassium picrate picrato potásico

potassium polymetaphosphate polimetafosfato potásico

potassium polysulfide polisulfuro potásico

potassium prussiate prusiato potásico

potassium pyroborate piroborato potásico

potassium pyrophosphate pirofosfato potásico

potassium pyrosulfate pirosulfato potásico

potassium pyrosulfite pirosulfito potásico

potassium ricinoleate ricinoleato potásico

potassium salicylate salicilato potásico

potassium silicate silicato potásico

potassium silicofluoride silicofluoruro potásico

potassium sodium carbonate carbonato sódico potásico

potassium sodium tartrate tartrato sódico potásico

potassium sorbate sorbato potásico

potassium stannate estannato potásico

potassium stearate estearato potásico
potassium sulfate sulfato potásico
potassium sulfide sulfuro potásico
potassium sulfite sulfito potásico
potassium sulfocarbonate
  sulfocarbonato potásico
potassium sulfocyanate sulfocianato
  potásico
potassium sulfocyanide sulfocianuro
  potásico
potassium superoxide superóxido
  potásico
potassium tartrate tartrato potásico
potassium tellurite telurita potásica
potassium tetraborate tetraborato
  potásico
potassium tetrachloropalladate
  tetracloropaladato potásico
potassium tetracyanoaurate
  tetracianoaurato potásico
potassium tetrafluoroborate
  tetrafluoroborato potásico
potassium tetrahydroborate
  tetrahidroborato potásico
potassium tetraiodoaurate
  tetrayodoaurato potásico
potassium tetroxalate tetroxalato
  potásico
potassium thiocyanate tiocianato
  potásico
potassium thiosulfate tiosulfato potásico
potassium titanate titanato potásico
potassium titanium fluoride fluoruro de
  titanio potásico
potassium trichlorophenate
  triclorofenato potásico
potassium triiodide triyoduro potásico
potassium tripolyphosphate
  tripolifosfato potásico
potassium tungstate tungstato potásico
potassium undecylenate undecilenato
  potásico
potassium wolframate wolframato
  potásico
potassium xanthate xantato potásico
potassium zinc iodide yoduro de cinc
  potásico
potassium zirconium chloride cloruro
  de zirconio potásico

potential potencial
potentiometric potenciométrico
potentiometric titration titulación
  potenciométrica, valoración
  potenciométrica
pour point punto de flujo, punto de
  fluidez
powder polvo
powdered en polvo
powdered resin resina en polvo
Prandtl number número de Prandtl
praseodymium praseodimio
praseodymium acetate acetato de
  praseodimio
praseodymium chloride cloruro de
  praseodimio
praseodymium oxalate oxalato de
  praseodimio
praseodymium oxide óxido de
  praseodimio
praseodymium sulfate sulfato de
  praseodimio
precious metal metal precioso
precious stone piedra preciosa
precipitability precipitabilidad
precipitable precipitable
precipitant precipitante
precipitate precipitado
precipitated precipitado
precipitation precipitación
precipitation indicator indicador de
  precipitación
precipitation number número de
  precipitación
precipitator precipitador
precursor precursor
predissociation predisociación
prednisolone prednisolona
prednisone prednisona
preferential preferencial
Pregl procedure procedimento de Pregl
pregnanediol pregananodiol
pregnenedione pregnenodiona
pregnenolone pregnenolona
prehnite prehnita
prenyl prenilo
preparation preparación
prephenic acid ácido prefénico
prepolymer prepolímero

preservative preservativo
pressure presión
pressure effect efecto de presión
pressure filter filtro de presión
Prevost reaction reacción de Prevost
primary primario
primary alcohol alcohol primario
primary amine amina primaria
primary azo dye colorante azoico primario
primary electron electrón primario
primary emission emisión primaria
primary ion ion primario
primary ionization ionización primaria
primary nucleus núcleo primario
primary radiation radiación primaria
primary reaction reacción primaria
principal principal
principal axis eje principal
principal quantum number número cuántico principal
principal series serie principal
principal valency valencia principal
Prins reaction reacción de Prins
prism prisma
pristane pristano
probe sonda
procaine hydrochloride clorhidrato de procaína
procedure procedimiento
process proceso
processing procesamiento
prochiral proquiral
producer gas gas pobre
product producto
production producción
production reactor reactor de producción
progesterone progesterona
proguanil proguanil
projection proyección
prolactin prolactina
prolamine prolamina
proline prolina
prolyl prolilo
promazine promazina
promazine hydrochloride clorhidrato de promazina
promethazine prometazina

promethium prometio
promoter promotor
proof prueba, graduación normal
propadiene propadieno
propagation propagación
propagation rate velocidad de propagación
propanal propanal
propanamide propanamida
propane propano
propane hydrate hidrato de propano
propane sultone propanosultona
propanecarboxylic acid ácido propanocarboxílico
propanediamine propanodiamina
propanediol propanodiol
propanenitrile propanonitrilo
propanethiol propanotiol
propanil propanil
propano propano
propanoic acid ácido propanoico
propanol propanol
propanol nitrate nitrato de propanol
propanolamine propanolamina
propanolpyridine propanolpiridina
propanone propanona
propanoyl propanoilo
propanoyl chloride cloruro de propanoilo
propargite propargita
propargyl propargilo
propargyl alcohol alcohol propargílico
propargyl bromide bromuro de propargilo
propargyl chloride cloruro de propargilo
propellant propelente
propenal propenal
propene propeno
propene polymer polímero de propeno
propenenitrile propenonitrilo
propenethiol propenotiol
propenetricarboxylic acid ácido propenotricarboxílico
propenoic acid ácido propenoico
propenol propenol
propenyl propenilo
propenyl alcohol alcohol propenílico
propenyl guaethol propenilguaetol
propenyl hexanoate hexanoato de

propenilo
propenyl isothiocyanate isotiocianato de
  propenilo
propenylamine propenilamina
propenylanisole propenilanisol
propenylene propenileno
propenylidene propenilideno
propenylthiourea propeniltiourea
properties propiedades
propinyl propinilo
propiolactone propiolactona
propiolic acid ácido propiólico
propioloyl propioloilo
propionaldehyde propionaldehído
propionate propionato
propione propiona
propionic acid ácido propiónico
propionic aldehyde aldehído propiónico
propionic anhydride anhídrido
  propiónico
propionic ether éter propiónico
propionitrile propionitrilo
propiono propiono
propionyl propionilo
propionyl chloride cloruro de propionilo
propionyl peroxide peróxido de
  propionilo
propionylbenzene propionilbenceno
propiophenone propiofenona
proportion proporción
proportional proporcional
proportional ionization ionización
  proporcional
propoxy propoxi
propoxyphene propoxifeno
propoxypropanol propoxipropanol
propranonol propranonol
propyl propilo
propyl acetate acetato de propilo
propyl alcohol alcohol propílico
propyl aldehyde aldehído propílico
propyl benzoate benzoato de propilo
propyl bromide bromuro de propilo
propyl butanoate butanoato de propilo
propyl butyrate butirato de propilo
propyl carbamate carbamato de propilo
propyl chloride cloruro de propilo
propyl chlorosulfonate clorosulfonato de
  propilo

propyl cyanide cianuro de propilo
propyl formate formiato de propilo
propyl furoate furoato de propilo
propyl gallate galato de propilo
propyl iodide yoduro de propilo
propyl mercaptan propilmercaptano
propyl nitrate nitrato de propilo
propyl nitrite nitrito de propilo
propyl pelargonate pelargonato de
  propilo
propyl propionate propionato de propilo
propyl xanthate xantato de propilo
propylacetone propilacetona
propylamine propilamina
propylaniline propilanilina
propylbenzene propilbenceno
propylene propileno
propylene carbonate carbonato de
  propileno
propylene chloride cloruro de propileno
propylene chlorohydrin clorohidrina de
  propileno
propylene dichloride dicloruro de
  propileno
propylene glycol propilenglicol
propylene glycol dinitrate dinitrato de
  propilenglicol
propylene glycol distearate diestearato
  de propilenglicol
propylene glycol monoricinoleate
  monorricinoleato de propilenglicol
propylene glycol monostearate
  monoestearato de propilenglicol
propylene glycol phenyl ether éter
  fenílico de propilenglicol
propylene oxide óxido de propileno
propylenediamine propilendiamina
propyleneimine propilenimina
propylformic acid ácido propilfórmico
propylhydroxylamine
  propilhidroxilamina
propylidene propilideno
propylmagnesium bromide bromuro de
  propilmagnesio
propylmalonic acid ácido propilmalónico
propylparaben propilparabeno
propylpiperidine propilpiperidina
propylpyridine propilpiridina
propyltrichlorosilane propiltriclorosilano

propylurea propilurea
propynal propinal
propyne propino
propynol propinol
propynyl propinilo
prostaglandin prostaglandina
prosthetic group grupo prostético
protactinium protactinio
protamine protamina
protease proteasa
protecting group grupo protector
protective coating revestimiento protector
protective glass vidrio protector
protective gloves guantes protectores
protein proteína
proteolysis proteólisis
proteolytic enzyme enzima proteolítica
prothrombin protrombina
protium protio
proto proto
protogenic protogénico
protolysis protólisis
protolytic catalysis catálisis protolítica
proton protón
proton accelerator acelerador protónico, acelerador de protones
proton acceptor aceptor protónico
proton bombardment bombardeo protónico
proton capture captura protónica
proton cycle ciclo protónico
proton donor donador protónico
proton microscope microscopio protónico
proton number número protónico
protonation protonación
protonic acid ácido protónico
protonic bridge puente protónico
protophilic protofílico
protoplasm protoplasma
protosulfate protosulfato
prototropic prototrópico
protyl protilo
provitamin provitamina
Prussian blue azul de Prusia
prussic acid ácido prúsico
Pschorr reaction reacción de Pschorr
pseudo seudo, pseudo

pseudo acid seudoácido
pseudo base seudobase
pseudoaromatic seudoaromático
pseudoasymmetric seudoasimétrico
pseudocritical properties propiedades seudocríticas
pseudocumene seudocumeno
pseudocumidine seudocumidina
pseudocumol seudocumol
pseudocumyl seudocumilo
pseudohalogen seudohalógeno
pseudohexyl alcohol alcohol seudohexílico
pseudoisomerism seudoisomerismo
pseudoisotope seudoisótopo
pseudomorphic seudomórfico
pseudoreduced properties propiedades seudorreducidas
psi psi
psilocin psilocina
psilocybin psilocibina
psilomelane psilomelano
psychotropic drug droga psicotrópica
pteridine pteridina
pterin pterina
ptomaine ptomaína
ptyalin ptialina
pulegone pulegona
pullulanase pululanasa
pulp pulpa
pulse impulso
pulverization pulverización
pumice pómez
Pummerer rearrangement reordenamiento de Pummerer
pump bomba
pungent acre
Purdie methylation metilación de Purdie
pure puro
purification purificación
purified water agua purificada
purine purina
purity pureza
puromycin puromicina
purpurin purpurina
putrefaction putrefacción
putrescine putrescina
putty masilla
pyran pirano

pyranone piranona
pyranose piranosa
pyranoside piranosida
pyranthrene pirantreno
pyranyl piranilo
pyrazine pirazina
pyrazole pirazol
pyrazolidine pirazolidina
pyrazoline pirazolina
pyrazolone pirazolona
pyrazolyl pirazolilo
pyrene pireno
pyreno pireno
pyrenyl pirenilo
pyrethrin piretrina
pyrethroid piretroide
pyrethrolone piretrolona
pyrethrum piretro
pyridazine piridazina
pyridine piridina
pyridine oxide óxido de piridina
pyridine polymer polímero de piridina
pyridinecarboxylic acid ácido piridinocarboxílico
pyridinedicarboxylic acid ácido piridinodicarboxílico
pyridinepentacarboxylic acid ácido piridinopentacarboxílico
pyridinesulfonic acid ácido piridinosulfónico
pyridinium piridinio
pyridinyl piridinilo
pyridone piridona
pyridoxal piridoxal
pyridoxal hydrochloride clorhidrato de piridoxal
pyridoxal phosphate fosfato de piridoxal
pyridoxine piridoxina
pyridyl piridilo
pyridylamine piridilamina
pyridylcarbinol piridilcarbinol
pyrimidine pirimidina
pyrimidinyl pirimidinilo
pyrimithate pirimitato
pyrite pirita
pyrithiamine piritiamina
pyro piro
pyroacemic acid ácido piroacémico
pyroboric acid ácido pirobórico

pyrocatechol pirocatecol
pyrocellulose pirocelulosa
pyrochlore pirocloro
pyrochroite pirocroíta
pyrogallate pirogalato
pyrogallol pirogalol
pyrolan pirolán
pyroligneous acid ácido pirolígneo
pyrolusite pirolusita
pyrolysis pirólisis
pyromellitic acid ácido piromelítico
pyromellitic dianhydride dianhídrido piromelítico
pyrometer pirómetro
pyrometric cone cono pirométrico
pyrometry pirometría
pyromorphite piromorfita
pyromucic acid ácido piromúcico
pyrone pirona
pyrophoric pirofórico
pyrophoric material material pirofórico
pyrophosphate pirofosfato
pyrophosphoric acid ácido pirofosfórico
pyrophosphoryl pirofosforilo
pyrophyllite pirofilita
pyrosin pirosina
pyrosulfate pirosulfato
pyrosulfite pirosulfito
pyrosulfuric acid ácido pirosulfúrico
pyrosulfurous acid ácido pirosulfuroso
pyrosulfuryl pirosulfurilo
pyrosulfuryl chloride cloruro de pirosulfurilo
pyrotartaric acid ácido pirotartárico
pyrotechnics pirotecnia
pyrovanadic acid ácido pirovanádico
pyroxene piroxeno
pyroxylin piroxilina
pyrrhotite pirrotita
pyrrocoline pirrocolina
pyrrole pirrol
pyrrolidine pirrolidina
pyrrolidinyl pirrolidinilo
pyrrolidone pirrolidona
pyrroline pirrolina
pyrrolyl pirrolilo
pyrrone pirrona
pyrroyl pirroilo
pyrryl pirrilo

pyruvaldehyde piruvaldehído
pyruvic acid ácido pirúvico
pyruvic alcohol alcohol pirúvico

pyruvic aldehyde aldehído pirúvico
pyruvonitrile piruvonitrilo

Q

quadri cuadri

quadridentate ligand ligando cuadridentado

quadrimolecular cuadrimolecular

quadrivalence cuadrivalencia

quadrivalent cuadrivalente

quadruple point punto cuádruple

quadruple resonance resonancia cuádruple

quadrupole cuadripolo

qualitative cualitativo

qualitative analysis análisis cualitativo

qualitative reaction reacción cualitativa

qualitative test prueba cualitativa

quality assurance comprobación de calidad

quality control control de calidad

quanta cuantos

quantitative cuantitativo

quantitative analysis análisis cuantitativo

quantitative measurement medida cuantitativa

quantitative microanalysis microanálisis cuantitativo

quantitative reaction reacción cuantitativa

quantitative test prueba cuantitativa

quantity cantidad

quantivalence cuantivalencia

quantization cuantificación

quantize cuantificar

quantized cuantificado

quantized system sistema cuantificado

quantum cuanto

quantum chemistry química cuántica

quantum efficiency eficiencia cuántica

quantum equivalence equivalencia cuántica

quantum group grupo cuántico

quantum jump salto cuántico

quantum level nivel cuántico

quantum mechanics mecánica cuántica

quantum number número cuántico

quantum of energy cuanto de energía

quantum of light cuanto de luz

quantum of radiation cuanto de radiación

quantum state estado cuántico

quantum statistics estadística cuántica

quantum theory teoría cuántica

quantum transition transición cuántica

quantum unit unidad cuántica

quantum yield rendimiento cuántico

quark quark

quartet cuarteto

quartz cuarzo

quartz-controlled controlado por cuarzo

quartz crystal cristal de cuarzo

quartz rock cuarcita

quartzite cuarcita

quasicrystal cuasicristal

quassia cuasia

quaternary ammonium salt sal amónica cuaternaria

quaternary carbon carbono cuaternario

quaternary carbon atom átomo de carbono cuaternario

quaternary salt sal cuaternaria

quaternary system sistema cuaternario

quebracho quebracho

Quelet reaction reacción de Quelet

quench extinguir, apagar, enfriar rápidamente

quenching extinción, apagado, enfriamiento rápido

quercinic acid ácido quercínico

quercitol quercitol

quercitrin quercitrina

quicklime cal viva

quicksilver azogue

quinacrine quinacrina

quinaldic acid ácido quináldico

quinaldine quinaldina

quinamine quinamina

quinaphthol quinaftol

quinazoline quinazolina

quinhydrone quinhidrona

quinhydrone electrode electrodo de
    quinhidrona

quinic acid ácido quínico

quinidine quinidina

quinine quinina

quinine sulfate sulfato de quinina

quininic acid ácido quinínico

quininone quininona

quinitol quinitol

quinizarin quinizarina

quinogen quinógeno

quinol quinol

quinoline quinolina

quinoline blue azul de quinolina

quinoline dye colorante de quinolina

quinoline red rojo de quinolina

quinolinic acid ácido quinolínico

quinolinium quinolinio

quinolinol quinolinol

quinolinone quinolinona

quinolizine quinolizina

quinolone quinolona

quinolyl quinolilo

quinone quinona

quinonedioxime quinonadioxima

quinonyl quinonilo

quinosol quinosol

quinoxaline quinoxalina

quinoxalinyl quinoxalinilo

quinoyl quinoilo

quinquevalent quinquevalente

quintet quinteto

quintozene quintoceno

quinuclidine quinuclidina

R

Racah parameters parámetros de Racah
racemase racemasa
racemate racemato
racemation racemación
raceme racema
racemic racémico
racemic acid ácido racémico
racemic mixture mezcla racémica
racemic substance substancia racémica
racemization racemización
rad rad
radial chromatography cromatografía radial
radiance radiancia
radiant radiante
radiate radiar
radiated radiado
radiating radiante
radiation radiación
radiation absorber absorbedor de radiación
radiation biochemistry bioquímica de radiación
radiation catalysis catálisis de radiación
radiation chemistry química de radiación
radiation counter contador de radiación
radiation danger peligro de radiación
radiation detector detector de radiación
radiation dose dosis de radiación
radiation effect efecto de radiación
radiation-induced inducido por radiación
radiation intensity intensidad de radiación
radiation ionization ionización por radiación
radiation-sensitive sensible a radiación
radiation source fuente de radiación
radiation unit unidad de radiación
radiative radiativo
radiative capture captura radiativa
radiative correction corrección radiativa
radiative neutron capture captura neutrónica radiativa
radiative transfer transferencia radiativa
radiator radiador
radical radical
radicofunctional nomenclature nomenclatura radicofuncional
radio radio
radioactinium radioactinio
radioactive radiactivo
radioactive atom átomo radiactivo
radioactive chain cadena radiactiva
radioactive clock reloj radiactivo
radioactive constant constante radiactiva
radioactive contamination contaminación radiactiva
radioactive dating datación radiactiva
radioactive decay desintegración radiactiva
radioactive decay constant constante de desintegración radiactiva
radioactive decay product producto de desintegración radiactiva
radioactive decay series serie de desintegración radiactiva
radioactive deposit depósito radiactivo
radioactive disintegration desintegración radiactiva
radioactive displacement law ley de desplazamiento radiactivo
radioactive element elemento radiactivo
radioactive emanation emanación radiactiva
radioactive emission emisión radiactiva
radioactive equilibrium equilibrio radiactivo
radioactive family familia radiactiva
radioactive fission fisión radiactiva
radioactive half-life media vida radiactiva
radioactive indicator indicador radiactivo
radioactive isotope isótopo radiactivo
radioactive material material radiactivo

radioactive metal metal radiactivo
radioactive mineral mineral radiactivo
radioactive nucleus núcleo radiactivo
radioactive nuclide nucleido radiactivo
radioactive product producto radiactivo
radioactive series serie radiactiva
radioactive source fuente radiactiva
radioactive tracer trazador radiactivo
radioactive transformation transformación radiactiva
radioactive waste desperdicios radiactivos
radioactivity radiactividad
radioactivity standard patrón de radiactividad
radioassay radioanálisis
radiobarium radiobario
radiobiology radiobiología
radiocarbon radiocarbono
radiocarbon dating datación por radiocarbono
radiocesium radiocesio
radiochemical radioquímico
radiochemical process proceso radioquímico
radiochemistry radioquímica
radiochromatography radiocromatografía
radiocrystallography radiocristalografía
radioelement radioelemento
radiofrequency spectrometer espectrómetro de radiofrecuencia
radiofrequency spectroscopy espectroscopia de radiofrecuencia
radiogenic radiogénico
radiography radiografía
radioimmunoassay radioinmunoanálisis
radioiodine radioyodo
radioisotope radioisótopo
radiolead radioplomo
radiolite radiolita
radiological radiológico
radiology radiología
radioluminescence radioluminiscencia
radiolysis radiólisis
radiometallography radiometalografía
radiometer radiómetro
radiometric radiométrico
radiometric analysis análisis
radiométrico
radiometric dating datación radiométrica
radiometric titration titulación radiométrica, valoración radiométrica
radiometry radiometría
radiomimetic radiomimético
radiomimetic substance substancia radiomimética
radion radión
radionitrogen radionitrógeno
radionuclide radionucleido
radiosodium radiosodio
radiotellurium radiotelurio
radiothor radiotor
radiothorium radiotorio
radium radio
radium bromide bromuro de radio
radium carbonate carbonato de radio
radium chloride cloruro de radio
radium sulfate sulfato de radio
radium unit unidad de radio
radius radio
radius ratio razón de radios
radon radón
raffinate rafinato
raffinose rafinosa
ralstonite ralstonita
Raman effect efecto de Raman
Raman lines líneas de Raman
Raman spectrophotometry espectrofotometría de Raman
Raman spectroscopy espectroscopia de Raman
Raman spectrum espectro de Raman
Ramberg-Backlund reaction reacción de Ramberg-Backlund
ramie ramio
Ramsauer effect efecto de Ramsauer
Ramsay-Shields equation ecuación de Ramsay-Shields
Ramsay-Young equation ecuación de Ramsay-Young
Ramsay-Young law ley de Ramsay-Young
rancid rancio
random copolymer copolímero aleatorio
Raney catalyst catalizador de Raney
Raney nickel níquel de Raney
Rankine scale escala Rankine

rankinite rankinita
ranksite ranksita
Raoult law ley de Raoult
rapeseed oil aceite de colza
rapid reaction reacción rápida
rare earth tierra rara
rare-earth element elemento de tierras raras
rare-earth metal metal de tierras raras
rare-earth salts sales de tierras raras
rare gas gas raro
rare metal metal raro
rarefaction rarefacción
rarefy rarefícar
Raschig process proceso de Raschig
Raschig rings anillos de Raschig
rasorite rasorita
Rast method método de Rast
rate velocidad, tasa
rate constant constante de velocidad de reacción
rate-determining step paso determinante de velocidad
rate of decay velocidad de desintegración
rate of disintegration velocidad de desintegración
rate of formation velocidad de formación
rate of reaction velocidad de reacción
rate of sedimentation velocidad de sedimentación
ratio razón, relación
ratio of specific heats razón de calores específicos
rational formula fórmula racional
rauwolfia rauwolfia
raw material materia prima
ray rayo
Rayleigh line línea de Rayleigh
rayon rayón
reactant reactante
reaction reacción
reaction constant constante de reacción
reaction control control de reacción
reaction energy energía de reacción
reaction enthalpy number número de entalpía de reacción
reaction equation ecuación de reacción
reaction formula fórmula de reacción

reaction heat calor de reacción
reaction indicator indicador de reacción
reaction inhibitor inhibidor de reacción
reaction intermediate producto intermedio de reacción
reaction kinetics cinética de reacciones
reaction law ley de reacción
reaction mechanism mecanismo de reacción
reaction of first order reacción de primer orden
reaction of second order reacción de segundo orden
reaction order orden de reacción
reaction product producto de reacción
reaction promotor promotor de reacción
reaction rate velocidad de reacción
reaction step paso de reacción
reaction time tiempo de reacción
reaction velocity velocidad de reacción
reactivation reactivación
reactive reactivo
reactive anode ánodo reactivo
reactive atmosphere atmósfera reactiva
reactive bond enlace reactivo
reactive dye colorante reactivo
reactivity reactividad
reactor reactor
reagent reactivo
reagent solution solución de reactivo
real gas gas real
realgar rejalgar
rearrangement reordenamiento, redisposición, transposición
rearrangement reaction reacción de reordenamiento
recalescence recalescencia
receiver receptor
receptor receptor
reciprocal proportions law ley de proporciones recíprocas
recombinant deoxyribonucleic acid ácido desoxirribonucleico recombinante
recombination process proceso de recombinación
reconstitution reconstitución
recording balance balanza de registro
recovery recuperación

recrystallization recristalización
recrystallization nucleus núcleo de recristalización
recrystallize recristalizar
rectification rectificación
rectifying column columna rectificante
rectorite rectorita
recycling reciclado
red acetate acetato rojo
red arsenic arsénico rojo
red brass latón rojo
red glass vidrio rojo
red lead plomo rojo
red manganese manganeso rojo
red mercury oxide óxido de mercurio rojo
red metal metal rojo
red ocher ocre rojo
red oil aceite rojo
red oxide óxido rojo
red phosphorus fósforo rojo
red potassium chromate cromato potásico rojo
red precipitate precipitado rojo
reddingite redingita
redistillation redestilación
redox redox
redox catalyst catalizador redox
redox cell célula de redox
redox equilibrium equilibrio de redox
redox indicator indicador redox
redox potential potencial de redox
redox potentiometry potenciometría redox
redox reaction reacción de redox
redox series serie de redox
redox system sistema de redox
redox titration titulación redox, valoración redox
reduce reducir
reduced reducido
reduced mass masa reducida
reduced state estado reducido
reducer reductor
reducing reductor
reducing agent agente reductor
reducing atmosphere atmósfera reductora
reducing flame llama reductora

reducing material material reductor
reducing reaction reacción reductora
reducing sugar azúcar reductor
reductant reductor
reductase reductasa
reduction reducción
reduction potential potencial de reducción
reductone reductona
reductor reductor
Reed reaction reacción de Reed
reference referencia
reference electrode electrodo de referencia
refine refinar
refined refinado
refinement refinamiento
refinery refinería
refinery gas gas de refinería
refining refinación
reflux reflujo
reflux condenser condensador de reflujo
refluxing reflujo
Reformatsky reaction reacción de Reformatsky
reforming reformación
refraction refracción
refractive refractivo
refractive index índice de refracción
refractivity refractividad
refractometer refractómetro
refractory refractario
refrigerant refrigerante
refrigeration refrigeración
regelation regelación
regeneration regeneración
regenerative regenerativo
regioselectivity regioselectividad
Regnault method método de Regnault
regular polymer polímero regular
regular system sistema regular
regulator regulador
rehydration rehidratación
Reich process proceso de Reich
Reimer reaction reacción de Reimer
Reimer-Tiemann reaction reacción de Reimer-Tiemann
Reinecke acid ácido de Reinecke
Reinecke salt sal de Reinecke

reinforced plastic plástico reforzado
reinite reinita
Reinsch test prueba de Reinsch
Reissert indole synthesis síntesis de indoles de Reissert
Reissert reaction reacción de Reissert
relative relativo
relative atomic mass masa atómica relativa
relative density densidad relativa
relative error error relativo
relative humidity humedad relativa
relative molecular mass masa molecular relativa
relative stereochemistry estereoquímica relativa
relative weight peso relativo
relativistic relativista
relativistic particle partícula relativista
relativistic quantum theory teoría cuántica relativista
relativity relatividad
relaxation relajación
relaxation effect efecto de relajación
relaxation time tiempo de relajación
release liberación
rem rem
remote handling manejo remoto
remote-handling equipment equipo de manejo remoto
renaturation renaturación
renin renina
rennet cuajo
rennin renina
repeatable repetible
repeating unit unidad repetitiva
repellency repelencia
repellent repelente
repercolation repercolación
replacement reemplazo
replicate duplicar
replication duplicación
Reppe process proceso de Reppe
reprocessing reprocesamiento
reproducibility reproducibilidad
reproducible reproducible
reptation reptación
repulsion repulsión
resbenzophenone resbenzofenona

research investigación
research and development investigación y desarrollo
research laboratory laboratorio de investigación
resene reseno
reserpine reserpina
reserve reserva
residual residual
residual oil aceite residual
residue residuo
resin resina
resinamine resinamina
resinate resinato
resinification resinificación
resinography resinografía
resinoid resinoide
resinol resinol
resinous resinoso
resist reserva
resite resita
resitol resitol
resocyanin resocianina
resol resol
resolution resolución
resolving power poder de resolución
resonance resonancia
resonance energy energía de resonancia
resonance fluorescence fluorescencia de resonancia
resonance ionization spectroscopy espectroscopia de ionización en resonancia
resonance luminescence luminiscencia de resonancia
resonance neutron neutrón de resonancia
resonance radiation radiación de resonancia
resonance spectrum espectro de resonancia
resonant resonante
resorcin resorcina
resorcinol resorcina, resorcinol
resorcinol acetate acetato de resorcina
resorcinol blue azul de resorcina
resorcinol dimethyl ether éter dimetílico de resorcina
resorcinol-formaldehyde resin resina de resorcina-formaldehído

resorcinol monoacetate monoacetato de resorcina
resorcinol monobenzoate monobenzoato de resorcina
resorcinol phenolate fenolato de resorcina
resorcyl resorcilo
resorcylic acid ácido resorcílico
restitution restitución
restricted internal rotation rotación interna restringida
resultant resultante
retardation retardo
retarder retardador
retene reteno
retention retención
retention index índice de retención
retention time tiempo de retención
retention volume volumen de retención
Retger law ley de Retger
reticulated reticulado
retinal retinal
retinene retineno
retinite retinita
retinol retinol
retort retorta
retro retro
retrogradation retrogradación
retrograde condensation condensación retrógrada
retrograde evaporation evaporación retrógrada
retropinacol rearrangement reordenamiento de retropinacol
retrorsine retrorcina
retrosynthesis retrosíntesis
Reverdin reaction reacción de Reverdin
reversal inversión
reverse osmosis ósmosis inversa
reversible reversible
reversible action acción reversible
reversible chemical reaction reacción química reversible
reversible process proceso reversible
reversible reaction reacción reversible
reversion reversión
revertose revertosa
revive revivir
rhamnitol ramnitol

rhamnose ramnosa
rhenate renato
rhenic acid ácido rénico
rhenium renio
rhenium black negro de renio
rhenium chloride cloruro de renio
rhenium fluoride fluoruro de renio
rhenium heptasulfide heptasulfuro de renio
rhenium heptoxide heptóxido de renio
rhenium oxide óxido de renio
rhenium pentachloride pentacloruro de renio
rhenium sulfide sulfuro de renio
rhenium trichloride tricloruro de renio
rhenium trioxychloride trioxicloruro de renio
rheology reología
rheometer reómetro
rhigolene rigoleno
rhizonic acid ácido rizónico
rho rho
rho acid ácido rho
rhodalline rodalina
rhodamine rodamina
rhodanate rodanato
rhodanic acid ácido rodánico
rhodanide rodanida
rhodanine rodanina
rhodanometry rodanometría
rhodinol rodinol
rhodinyl rodinilo
rhodinyl acetate acetato de rodinilo
rhodite rodita
rhodium rodio
rhodium black negro de rodio
rhodium carbonyl chloride cloruro de rodiocarbonilo
rhodium chloride cloruro de rodio
rhodium hydroxide hidróxido de rodio
rhodium nitrate nitrato de rodio
rhodium oxide óxido de rodio
rhodium sulfate sulfato de rodio
rhodium sulfide sulfuro de rodio
rhodium trichloride tricloruro de rodio
rhodizite rodizita
rhodizonic acid ácido rodizónico
rhodochrosite rodocrosita
rhodolite rodolita

rhodonite rodonita
rhodopsin rodopsina
rhodoxanthin rodoxantina
rhombic rómbico
rhombic dodecahedron dodecahedro rómbico
rhombic sulfur azufre rómbico
rhombic system sistema rómbico
rhombohedral rombohedral
rhombohedral system sistema rombohedral
rhombohedron rombohedro
rhotanium rotanio
rhythmic deposition deposición rítmica
rhythmic precipitation precipitación rítmica
riboflavin riboflavina
riboflavin phosphate fosfato de riboflavina
ribofuranosyladenine ribofuranosiladenina
ribonic acid ácido ribónico
ribonuclease ribonucleasa
ribonucleic acid ácido ribonucleico
ribonucleoprotein ribonucleoproteína
ribose ribosa
ribosephosphoric acid ácido ribosfosfórico
riboside ribosida
ribosome ribosoma
ribosyl ribosilo
ribulose ribulosa
rich gas gas rico
rich mixture mezcla rica
richmondite richmondita
Richter law ley de Richter
ricin ricina
ricinic acid ácido ricínico
ricinine ricinina
ricinoleate ricinoleato
ricinoleic acid ácido ricinoleico
ricinolein ricinoleína
ricinoleyl alcohol alcohol ricinoleílico
ricinus oil aceite de ricino
rickardite rickardita
Riegler test prueba de Riegler
Riehm quinoline synthesis síntesis de quinolinas de Riehm
Riemschneider synthesis síntesis de Riemschneider
Riley oxidation oxidación de Riley
ring anillo
ring breakage rompimiento de anillo
ring closure cierre de anillo
ring compound compuesto en anillo
ring formation formación de anillo
ring isomerism isomerismo de anillo
ring reaction reacción de anillo
ring structure estructura de anillo
ring system sistema de anillo
ring test prueba de anillo
Ringer solution solución de Ringer
ripening maduración
ripidolite ripidolita
rissic acid ácido rísico
ristocetin ristocetina
Ritter reaction reacción de Ritter
roast cocer, calcinar
roasting cocido, calcinación
Robinson reaction reacción de Robinson
Rochelle salt sal de Rochelle
rock crystal cristal de roca
rock salt sal de roca
rocket fuel cobustible de cohete
rodenticide rodenticida
rodinol rodinol
roentgen roentgen
roentgenogram roentgenograma
Roesler process proceso de Roesler
Rohrbach solution solución de Rohrbach
röntgen roentgen
room temperature temperatura de sala, temperatura ambiente
rosaniline rosanilina
roscoelite roscoelita
Rose Bengal rosa de Bengala
roselite roselita
Rosenheim color test prueba de color de Rosenheim
Rosenmund reaction reacción de Rosenmund
Rosenmund reduction reducción de Rosenmund
Rosenmund-von Braun synthesis síntesis de Rosenmund-von Braun
Rosenstein process proceso de Rosenstein
rosin colofonia, resina

rosin oil aceite de colofonia
rosinol rosinol
rosolic acid ácido rosólico
rotamer rotámero
rotameter rotámetro
rotation rotación
rotational rotacional
rotational constant constante rotacional
rotational energy energía rotacional
rotational spectrum espectro rotacional
rotatory rotatorio
rotatory dispersion dispersión rotatoria
rotatory power poder rotatorio
rotaversion rotaversión
rotenone rotenona
Rothemund reaction reacción de Rothemund
Rowe rearrangement reordenamiento de Rowe
Rowland ghost fantasma de Rowland
rowlandite rowlandita
rubber caucho
rubber accelerator acelerador de caucho
rubber hydrochloride clorhidrato de caucho
rubber sponge esponja de caucho
rubber sulfide sulfuro de caucho
rubene rubeno
ruberite ruberita
ruberythric acid ácido ruberítrico
rubicene rubiceno
rubidine rubidina
rubidium rubidio
rubidium acetate acetato de rubidio
rubidium alum alumbre de rubidio
rubidium bromide bromuro de rubidio
rubidium carbonate carbonato de rubidio
rubidium chloride cloruro de rubidio
rubidium chromate cromato de rubidio
rubidium dichromate dicromato de rubidio
rubidium fluoride fluoruro de rubidio
rubidium hexachloroplatinate hexacloroplatinato de rubidio
rubidium hexafluorosilicate hexafluorosilicato de rubidio
rubidium hydride hidruro de rubidio
rubidium hydroxide hidróxido de rubidio

rubidium iodate yodato de rubidio
rubidium iodide yoduro de rubidio
rubidium nitrate nitrato de rubidio
rubidium oxide óxido de rubidio
rubidium perchlorate perclorato de rubidio
rubidium peroxide peróxido de rubidio
rubidium sulfate sulfato de rubidio
rubidium sulfide sulfuro de rubidio
rubidium sulfite sulfito de rubidio
rubidium tartrate tartrato de rubidio
rubitannic acid ácido rubitánico
rubrene rubreno
ruby rubí
ruby spinel rubí espinela
Ruff-Fenton degradation degradación de Ruff-Fenton
rust herrumbre, moho
ruthenate rutenato
ruthenic ruténico
ruthenic chloride cloruro ruténico
ruthenious rutenioso
ruthenium rutenio
ruthenium bromide bromuro de rutenio
ruthenium carbonyl carbonilo de rutenio
ruthenium chloride cloruro de rutenio
ruthenium dichloride dicloruro de rutenio
ruthenium fluoride fluoruro de rutenio
ruthenium hydroxide hidróxido de rutenio
ruthenium nitrosonitrate nitrosonitrato de rutenio
ruthenium oxide óxido de rutenio
ruthenium oxychloride oxicloruro de rutenio
ruthenium red rojo de rutenio
ruthenium sesquichloride sesquicloruro de rutenio
ruthenium sesquioxide sesquióxido de rutenio
ruthenium silicide siliciuro de rutenio
ruthenium sulfide sulfuro de rutenio
ruthenium tetraoxide tetróxido de rutenio
ruthenium trichloride tricloruro de rutenio
ruthenium trifluoride trifluoruro de

rutenio

ruthenocene rutenoceno

Rutherford nuclear atom átomo nuclear de Rutherford

Rutherford scattering dispersión de Rutherford

rutherfordite rutefordita

rutile rutilo

rutin rutina

rydberg rydberg

Rydberg constant constante de Rydberg

Rydberg correction corrección de Rydberg

Rydberg formula fórmula de Rydberg

Rydberg spectrum espectro de Rydberg

S

sabadilla cebadilla
Sabatier-Senderens reduction reducción de Sabatier-Senderens
sabinene sabineno
sabinic acid ácido sabínico
sabinol sabinol
saccharase sacarasa
saccharate sacarato
saccharic sacárico
saccharic acid ácido sacárico
saccharide sacárido
saccharification sacarificación
saccharify sacarificar
saccharimeter sacarímetro
saccharin sacarina
saccharinic acid ácido sacarínico
saccharonic acid ácido sacarónico
saccharose sacarosa
saccharose unit unidad de sacarosa
Sachse reaction reacción de Sachse
Sackur-Tetrode equation ecuación de Sackur-Tetrode
sacrificial protection protección sacrificial
safe storage almacenamiento seguro
safety seguridad
safety device dispositivo de seguridad
safety engineering ingeniería de seguridad
safety glass vidrio de seguridad
safety goggles anteojos de seguridad, gafas protectoras
safflower oil aceite de cártamo
safranine safranina
safrole safrol
sage oil aceite de salvia
sahlite sahlita
sal ammoniac sal amoníaco
sal soda sal sosa
salamide salamida
salic sálico
salicin salicina
salicoyl salicoilo

salicyl salicilo
salicyl alcohol alcohol salicílico
salicylal salicilal
salicylaldehyde salicilaldehído
salicylamide salicilamida
salicylanilide salicilanilida
salicylate salicilato
salicylated salicilado
salicylic acid ácido salicílico
salicylic aldehyde aldehído salicílico
salicylic amide amida salicílica
salicylidene salicilideno
salicylonitrile salicilonitrilo
salicyloyl saliciloilo
saligenin saligenina
saligenol saligenol
salimeter salímetro
saline salino
saline water agua salina
salinity salinidad
salinometer salinómetro
salmine salmina
salol salol
salt sal
salt bridge puente salino
salt cake torta de sal
salt of Lemery sal de Lemery
salt of tartar sal de tártaro
salting out precipitación asistida por sal
saltpeter salitre
samaric samárico
samaric bromide bromuro samárico
samaric chloride cloruro samárico
samaric hydroxide hidróxido samárico
samaric nitrate nitrato samárico
samaric oxalate oxalato samárico
samaric sulfate sulfato samárico
samaric sulfide sulfuro samárico
samarium samario
samarium chloride cloruro de samario
samarium oxide óxido de samario
samarium trioxide trióxido de samario
samarous samaroso

samarous chloride cloruro samaroso
samarous sulfate sulfato samaroso
samarskite samarskita
sample muestra
sampling muestreo, muestra
sand arena
sand bath baño de arena
sandalwood oil aceite de sándalo
sandarac sandaraca
Sandmeyer reaction reacción de Sandmeyer
Sandmeyer synthesis síntesis de Sandmeyer
sanguinaria sanguinaria
santalol santalol
santalyl santalilo
santalyl acetate acetato de santalilo
santalyl chloride cloruro de santalilo
saponification saponificación
saponification equivalent equivalente de saponificación
saponification number número de saponificación
saponification value valor de saponificación
saponin saponina
saponite saponita
sapphire zafiro
sarcolactate sarcolactato
sarcolactic acid ácido sarcoláctico
sarcolite sarcolita
sarcolysin sarcolisina
sarcosine sarcosina
sardinianite sardinianita
sardonyx sardónica
Sarett oxidation oxidación de Sarett
Sargent curve curva de Sargent
sarin sarina
sassafras oil aceite de sasafrás
satin spar espato satinado
saturable saturable
saturant saturante
saturate saturar
saturated saturado
saturated ammonia amoníaco saturado
saturated compound compuesto saturado
saturated fat grasa saturada
saturated hydrocarbon hidrocarburo saturado

saturated liquid líquido saturado
saturated solution solución saturada
saturated vapor vapor saturado
saturating saturante
saturation saturación
saturation effect efecto de saturación
saturation isomerism isomerismo de saturación
saturation level nivel de saturación
saturation point punto de saturación
saturation spectroscopy espectroscopia de saturación
saturator saturador
sawhorse projection proyección en cabrilla
saxitoxin saxitoxina
scale escala, balanza, costra, capa de óxido
scan exploración, barrido
scandium escandio
scandium chloride cloruro de escandio
scandium fluoride fluoruro de escandio
scandium hydroxide hidróxido de escandio
scandium oxalate oxalato de escandio
scandium oxide óxido de escandio
scandium sulfate sulfato de escandio
scandium sulfite sulfito de escandio
scanning exploración, barrido
scanning electron microscope microscopio electrónico de barrido
scanning electron microscopy microscopia electrónica de barrido
scapolite escapolita
scatter dispersión, esparcimiento
scattered disperso, difuso
scattered electrons electrones dispersos
scattering dispersión, esparcimiento
scavenger depurador
Schäffer acid ácido de Schäffer
Schäffer salt sal de Schäffer
Scheele green verde Scheele
scheelite scheelita
Scheibler reagent reactivo de Scheibler
Schiemann reaction reacción de Schiemann
Schiff base base de Schiff
Schiff reagent reactivo de Schiff
Schiff solution solución de Schiff

Schiff test prueba de Schiff
Schmidlin ketene synthesis síntesis de cetenas de Schmidlin
Schmidt lines líneas de Schmidt
Schmidt reaction reacción de Schmidt
Schoenflies system sistema de Schoenflies
Scholl reaction reacción de Scholl
Schomaker-Stevenson equation ecuación de Schomaker-Stevenson
schönite schonita
Schorigin reaction reacción de Schorigin
Schotten-Baumann reaction reacción de Schotten-Baumann
Schottky defect defecto de Schottky
Schrödinger equation ecuación de Schrödinger
Schweitzer reagent reactivo de Schweitzer
scientific method método científico
scintillating centelleante
scintillation centelleo
scintillation counter contador de centelleos
scintillation spectrometer espectrómetro de centelleos
scintillator centelleador
scission escisión
scleroscope escleroscopio
scolecite escolecita
scopolamine escopolamina
scopoline escopolina
scorodite escorodita
screen cedazo, tamiz, criba, pantalla
screening cribado, tamizado, apantallamiento
screening constant constante de apantallamiento
screening effect efecto de apantallamiento
scrubber depurador
scrubbing depuración
sealant sellador
sealed sellado
sebacic acid ácido sebácico
sebaconitrile sebaconitrilo
sebacoyl sebacoilo
sebacoyl chloride cloruro de sebacoilo
sebacylic acid ácido sebacílico

sec sec
seco seco
secobarbital secobarbital
second segundo
second law of thermodynamics segunda ley de termodinámica
second-order reaction reacción de segundo orden
secondary secundario
secondary alcohol alcohol secundario
secondary amine amina secundaria
secondary carbon carbono secundario
secondary carbon atom átomo de carbono secundario
secondary cell pila secundaria
secondary emission emisión secundaria
secondary-ion mass spectrometry espectrometría de masa de iones secundarios
secondary radiation radiación secundaria
secondary reaction reacción secundaria
sedanolic acid ácido sedanólico
sedative sedante
sediment sedimento
sedimentation sedimentación
sedimentation coefficient coeficiente de sedimentación
sedimentation constant constante de sedimentación
sedimentation equilibrium equilibrio de sedimentación
sedimentation potential potencial de sedimentación
sedimentation rate velocidad de sedimentación
sedimentation velocity velocidad de sedimentación
seed semilla
seeding siembra
Seger cone cono de Seger
segregation segregación
Seidlitz salt sal de Seidlitz
selection rules reglas de selección
selective selectivo
selectivity selectividad
selectivity coefficient coeficiente de selectividad
selena selena

selenate seleniato
selenic selénico
selenic acid ácido selénico
selenide seleniuro
seleninic acid ácido selenínico
selenino selenino
seleninyl seleninilo
selenious selenioso
selenious acid ácido selenioso
selenious oxide óxido selenioso
selenite selenita, selenito
selenium selenio
selenium bromide bromuro de selenio
selenium chloride cloruro de selenio
selenium dibromide dibromuro de
    selenio
selenium dichloride dicloruro de selenio
selenium diethyldithiocarbamate
    dietilditiocarbamato de selenio
selenium dioxide dióxido de selenio
selenium monobromide monobromuro
    de selenio
selenium monochloride monocloruro de
    selenio
selenium nitride nitruro de selenio
selenium oxide óxido de selenio
selenium oxyacid oxiácido de selenio
selenium oxychloride oxicloruro de
    selenio
selenium sulfide sulfuro de selenio
selenium tetrabromide tetrabromuro de
    selenio
selenium tetrafluoride tetrafluoruro de
    selenio
seleno seleno
selenocyanate selenocianato
selenofuran selenofurano
selenol selenol
selenomium selenomio
selenone selenona
selenonic acid ácido selenónico
selenonium selenonio
selenono selenono
selenonyl selenonilo
selenourea selenourea
selenous selenioso
selenous acid ácido selenioso
selenous anhydride anhídrido selenioso
selenoxide selenóxido

selenyl selenilo
self-absorption autoabsorción
self-ionization autoionización
self-organization autoorganización
self-reduction autorreducción
sellaite sellaíta
semi semi
semicarbazide semicarbazida
semicarbazide hydrochloride clorhidrato
    de semicarbazida
semicarbazido semicarbazido
semicarbazino semicarbazino
semicarbazone semicarbazona
semicarbazono semicarbazono
semiconductor semiconductor
semidine semidina
semiempirical semiempírico
semimetal semimetal
semimetallic semimetálico
semimicroanalysis semimicroanálisis
semimicrochemistry semimicroquímica
semipermeable semipermeable
semipermeable membrane membrana
    semipermeable
semiprecious semiprecioso
semisolid semisólido
semisynthetic semisintético
semivalence semivalencia
Semmler-Wolff reaction reacción de
    Semmler-Wolff
senarmontite senarmontita
seniority prioridad
sensitive sensible
sensitizer sensibilizador
sensor sensor
separating funnel embudo de separación
separation separación
separation energy energía de separación
separator separador
separatory funnel embudo de separación
sepia sepia
sepiolite sepiolita
septanose ring anillo de septanosa
septiphene septifeno
sequence rules reglas de secuencia
sequestering secuestrante
sequestering agent agente secuestrante
sequestration secuestro
serendipity serendipismo

series serie
series of lines serie de líneas
serine serina
Serini reaction reacción de Serini
serotonin serotonina
serpentine serpentina
serum suero
sesame oil aceite de sésamo
sesamin sesamina
sesamol sesamol
sesamolin sesamolina
sesone sesona
sesqui sesqui
sesquicarbonate sesquicarbonato
sesquichloride sesquicloruro
sesquioxide sesquióxido
sesquisalt sesquisal
sesquisoda sesquisosa
sesquiterpene sesquiterpeno
setting solidificación, endurecimiento
setting point punto de solidificación, punto de endurecimiento
settling asentamiento, estabilización
sexivalent sexivalente
sextet sexteto
shale oil aceite de pizarra
shape-selective catalyst catalizador selectivo de tamaño
Sharpless reaction reacción de Sharpless
shell capa, concha, cáscara
shell structure estructura de capas
shellac goma laca
shielding blindaje
shift deslizamiento, desplazamiento
shifting deslizamiento, desplazamiento
Shoelkopf acid ácido de Shoelkopf
sialic acid ácido siálico
side chain cadena lateral
side reaction reacción secundaria
siderite siderita
siderophile element elemento siderófilo
siderotilate siderotilato
Siemens-Halske process proceso de Siemens-Halske
Siemens process proceso de Siemens
sienna siena
sieve cedazo
sievert sievert
Sievert law ley de Sievert

sigma sigma
sigma bond enlace sigma
sigma electron electrón sigma
sigma function función sigma
sigma particle partícula sigma
sigma phenomenon fenómeno sigma
sigma value valor sigma
sigmatropic sigmatrópico
sigmatropic reaction reacción sigmatrópica
silane silano
silanediyl silanodiilo
silanetriyl silanotriilo
silanol silanol
silatrane silatrano
silazane silazano
silica sílice
silica gel gel de sílice
silica sand arena de sílice
silicane silicano
silicate silicato
silicate of soda silicato de sosa
siliceous silíceo
silicic silícico
silicic acid ácido silícico
silicide siliciuro
silicious silíceo
silico silico
silicobenzoic acid ácido silicobenzoico
silicobromoform silicobromoformo
silicochloroform silicocloroformo
silicoethane silicoetano
silicofluoride silicofluoruro
silicoheptane silicoheptano
silicoiodoform silicoyodoformo
silicol silicol
silicomanganese silicomanganeso
silicomethane silicometano
silicon silicio
silicon alloy aleación de silicio
silicon boride boruro de silicio
silicon bromide bromuro de silicio
silicon bronze bronce de silicio
silicon carbide carburo de silicio
silicon chloride cloruro de silicio
silicon-copper silicio-cobre
silicon-copper alloy aleación de silicio-cobre
silicon dioxide dióxido de silicio

silicon disulfide disulfuro de silicio
silicon fluoride fluoruro de silicio
silicon-gold alloy aleación de silicio-oro
silicon hydride hidruro de silicio
silicon iodide yoduro de silicio
silicon monoxide monóxido de silicio
silicon nitride nitruro de silicio
silicon octachloride octacloruro de silicio
silicon oxide óxido de silicio
silicon oxychloride oxicloruro de silicio
silicon rubber caucho de silicio
silicon sulfide sulfuro de silicio
silicon tetrabromide tetrabromuro de silicio
silicon tetrachloride tetracloruro de silicio
silicon tetrafluoride tetrafluoruro de silicio
silicon tetrahydride tetrahidruro de silicio
silicon tetraiodide tetrayoduro de silicio
silicone silicona
silicone oil aceite de silicona
silicone rubber caucho de silicona
silicotungstic acid ácido silicotúngstico
silicyl silicilo
silicylene silicileno
silk seda
sillimanite silimanita
siloxane siloxano
siloxy siloxi
silumin silumina
silver plata
silver acetate acetato de plata
silver acetylide acetiluro de plata
silver alloy aleación de plata
silver amalgam amalgama de plata
silver antimonide antimoniuro de plata
silver arsenate arseniato de plata
silver arsenide arseniuro de plata
silver arsenite arsenito de plata
silver azide azida de plata
silver benzamide benzamida de plata
silver benzoate benzoato de plata
silver borate borato de plata
silver bromate bromato de plata
silver bromide bromuro de plata
silver bronze bronce de plata

silver carbonate carbonato de plata
silver chlorate clorato de plata
silver chloride cloruro de plata
silver chlorite clorito de plata
silver chromate cromato de plata
silver citrate citrato de plata
silver cyanate cianato de plata
silver cyanide cianuro de plata
silver dichromate dicromato de plata
silver dithionate ditionato de plata
silver ferricyanide ferricianuro de plata
silver fluoride fluoruro de plata
silver fluoroborate fluoroborato de plata
silver fluosilicate fluosilicato de plata
silver fulminate fulminato de plata
silver hexacyanoferrate hexacianoferrato de plata
silver hexafluorosilicate hexafluorosilicato de plata
silver hypochlorite hipoclorito de plata
silver hypophosphate hipofosfato de plata
silver iodate yodato de plata
silver iodide yoduro de plata
silver lactate lacatato de plata
silver laurate laurato de plata
silver mercury iodide yoduro mercúrico de plata
silver migration migración de plata
silver myristate miristato de plata
silver nitrate nitrato de plata
silver nitride nitruro de plata
silver nitrite nitrito de plata
silver ore mineral de plata
silver orthophosphate ortofosfato de plata
silver oxalate oxalato de plata
silver oxide óxido de plata
silver palmitate palmitato de plata
silver perchlorate perclorato de plata
silver permanganate permanganato de plata
silver peroxide peróxido de plata
silver phosphate fosfato de plata
silver phosphide fosfuro de plata
silver picrate picrato de plata
silver-plated plateado
silver-plating plateado
silver potassium cyanide cianuro

potásico de plata

silver salicylate salicilato de plata

silver salt sal de plata

silver selenide seleniuro de plata

silver sodium chloride cloruro sódico de plata

silver sodium cyanide cianuro sódico de plata

silver sodium thiosulfate tiosulfato sódico de plata

silver stearate estearato de plata

silver suboxide subóxido de plata

silver sulfate sulfato de plata

silver sulfide sulfuro de plata

silver sulfite sulfito de plata

silver tartrate tartrato de plata

silver telluride telururo de plata

silver thiocyanate tiocianato de plata

silver thiosulfate tiosulfato de plata

silver vanadate vanadato de plata

silvestrene silvestreno

silvex silvex

silvichemical silviquímico

silylation sililación

silylene silileno

simazine simazina

Simmons-Smith reagent reactivo de Simmons-Smith

Simmons-Smith reaction reacción de Simmons-Smith

Simonini reaction reacción de Simonini

Simons process proceso de Simons

simple distillation destilación simple

simple salt sal simple

simultaneous reaction reacción simultánea

single bond enlace sencillo

singlet singulete

sinigrin sinigrina

sinter sínter, sinterizado, aglutinación

sintered sinterizado, aglutinado

sintering sinterización, aglutinación

sinterization sinterización, aglutinación

sinterize sinterizar, aglutinar

sisal sisal

sitosterol sitosterol

skatole escatol

sklodowskite sklodowskita

Skraup synthesis síntesis de Skraup

skutterudite skutterudita

slack apagado, flojo, lento

slag escoria

slaked lime cal apagada

slate pizarra

slate flour harina de pizarra

slow chemical reaction reacción química lenta

slow combustion combustión lenta

slow neutron neutrón lento

slow-neutron spectrometry espectrometría de neutrones lentos

slow reaction reacción lenta

sludge lodo, sedimento

slurry suspensión, suspensión acuosa, lechada

smalt esmaltín

smectic esméctico

smectic phase fase esméctica

smelt fundir

smelting fundido

Smiles rearrangement reordenamiento de Smiles

smog smog, niebla contaminante

smoke humo

smokeless fuel combustible sin humo

smokeless powder pólvora sin humo

snake venom veneno de serpiente

soap jabón

soapstone esteatita

soda sosa, soda

soda alum alumbre de sosa

soda ash ceniza de sosa

soda lime cal sodada

soda niter nitro de sosa

soda pulp pulpa de sosa

sodalite sodalita

sodamide sodamida

Soddy displacement law ley de desplazamiento de Soddy

sodium sodio

sodium abietate abietato sódico

sodium acetate acetato sódico, acetato de sodio

sodium acetone bisulfate bisulfato de acetona sódico

sodium acetylformate acetilformiato sódico

sodium acetylsalicylate acetilsalicilato

sódico

sodium acid carbonate carbonato ácido
sódico

sodium acid chromate cromato ácido
sódico

sodium acid fluoride fluoruro ácido
sódico

sodium acid phosphate fosfato ácido
sódico

sodium acid pyrophosphate pirofosfato
ácido sódico

sodium acid sulfate sulfato ácido sódico

sodium acid sulfite sulfito ácido sódico

sodium acid tartrate tartrato ácido
sódico

sodium alginate alginato sódico

sodium alum alumbre sódico

sodium aluminate aluminato sódico

sodium aluminosilicate aluminosilicato
sódico

sodium aluminum hydride hidruro
alumínico sódico

sodium aluminum phosphate fosfato
alumínico sódico

sodium aluminum silicofluoride
silicofluoruro alumínico sódico

sodium aluminum sulfate sulfato
alumínico sódico

sodium amalgam amalgama sódica

sodium amide amida sódica

sodium aminohippurate aminohipurato
sódico

sodium aminosalicylate aminosalicilato
sódico

sodium ammonium phosphate fosfato
amónico sódico

sodium ammonium sulfate sulfato
amónico sódico

sodium amytal amital sódico

sodium anoxynaphthonate
anoxinaftonato sódico

sodium antimonate antimoniato sódico

sodium antimonyl nitrate nitrato de
antimonilo sódico

sodium arsanilate arsanilato sódico

sodium arsenate arseniato sódico

sodium arsenide arseniuro sódico

sodium arsenite arsenito sódico

sodium ascorbate ascorbato sódico

sodium aspartate aspartato sódico

sodium auribromide auribromuro sódico

sodium aurichloride auricloruro sódico

sodium aurothiomalate aurotiomalato
sódico

sodium azide azida sódica

sodium barbiturate barbiturato sódico

sodium benzenesulfonate
bencenosulfonato sódico

sodium benzoate benzoato sódico

sodium benzosulfimide benzosulfimida
sódica

sodium benzyl succinate succinato de
bencilo sódico

sodium beryllium fluoride fluoruro de
berilio sódico

sodium bicarbonate bicarbonato sódico

sodium bichromate bicromato sódico

sodium bifluoride bifluoruro sódico

sodium binoxalate binoxalato sódico

sodium binoxide binóxido sódico

sodium biphosphate bifosfato sódico

sodium bismuthate bismutato sódico

sodium bisulfate bisulfato sódico

sodium bisulfide bisulfuro sódico

sodium bisulfite bisulfito sódico

sodium bitartrate bitartrato sódico

sodium bithionolate bitionolato sódico

sodium borate borato sódico

sodium boroformate boroformiato
sódico

sodium borohydride bohidruro sódico

sodium bromate bromato sódico

sodium bromide bromuro sódico

sodium bromite bromito sódico

sodium butyrate butirato sódico

sodium cacodylate cacodilato sódico

sodium caprylate caprilato sódico

sodium carbide carburo sódico

sodium carbolate carbolato sódico

sodium carbonate carbonato sódico

sodium carbonate decahydrate
decahidrato de carbonato sódico

sodium carbonate monohydrate
monohidrato de carbonato sódico

sodium carbonate peroxide peróxido de
carbonato sódico

sodium carbonate peroxohydrate
peroxohidrato de carbonato sódico

sodium carboxymethylcellulose
carboximetilcelulosa sódica
sodium carminate carminato sódico
sodium caseinate caseinato sódico
sodium cellulosate celulosato sódico
sodium chlorate clorato sódico
sodium chloride cloruro sódico
sodium chlorite clorito sódico
sodium chloroacetate cloroacetato
sódico
sodium chloroaluminate cloroaluminato
sódico
sodium chloroaurate cloroaurato sódico
sodium chlorophthalate cloroftalato
sódico
sodium chloroplatinate cloroplatinato
sódico
sodium chloroplatinite cloroplatinita
sódica
sodium chlorosulfonate clorosulfonato
sódico
sodium chromate cromato sódico
sodium chromite cromita sódica
sodium cinnamate cinamato sódico
sodium citrate citrato sódico
sodium copper chloride cloruro de cobre
sódico
sodium copper cyanide cianuro de cobre
sódico
sodium cyanamide cianamida sódica
sodium cyanate cianato sódico
sodium cyanide cianuro sódico
sodium cyanoaurite cianoaurita sódica
sodium cyanoborohydride
cianoborohidruro sódico
sodium cyanocuprate cianocuprato
sódico
sodium cyanoethanesulfonate
cianoetanosulfonato sódico
sodium cyclamate ciclamato sódico
sodium cyclopentadienide
ciclopentadienida sódica
sodium dehydroacetate deshidroacetato
sódico
sodium dextran sulfate sulfato de
dextrano sódico
sodium diacetate diacetato sódico
sodium diatrizoate diatrizoato sódico
sodium dibutyldithiocarbamate

dibutilditiocarbamato sódico
sodium dichloroisocyanurate
dicloroisocianurato sódico
sodium dichloropropionate
dicloropropionato sódico
sodium dichromate dicromato sódico
sodium dicyanoaurate dicianoaurato
sódico
sodium dihydrogenorthoperiodate
dihidrogenoortoperyodato sódico
sodium dihydrogenphosphate
dihidrogenofosfato sódico
sodium dihydroxyethylglycine
dihidroxietilglicina sódica
sodium dimethylarsenate
dimetilarseniato sódico
sodium dimethyldithiocarbamate
dimetilditiocarbamato sódico
sodium dinitrocresolate dinitrocresolato
sódico
sodium dioxide dióxido sódico
sodium diphenylhydantoin
difenilhidantoína sódica
sodium dispersion dispersión sódica
sodium disulfate disulfato sódico
sodium disulfite disulfito sódico
sodium dithionate ditionato sódico
sodium dithionite ditionito sódico
sodium dithiosalicylate ditiosalicilato
sódico
sodium diuranate diuranato sódico
sodium divanadate divanadato sódico
sodium dodecylbenzenesulfonate
dodecilbencenosulfonato sódico
sodium edetate edetato sódico
sodium erythorbate eritorbato sódico
sodium ethanoate etanoato sódico
sodium ethanolate etanolato sódico
sodium ethoxide etoxido sódico
sodium ethylate etilato sódico
sodium ethylxanthate exilxantato sódico
sodium ferricyanide ferricianuro sódico
sodium ferrocyanide ferrocianuro sódico
sodium fluorescein fluoresceína sódica
sodium fluoride fluoruro sódico
sodium fluoroacetate fluoroacetato
sódico
sodium fluoroborate fluoroborato sódico
sodium fluorophosphate fluorofosfato

sódico

sodium fluorosilicate fluorosilicato sódico

sodium folate folato sódico

sodium formaldehyde bisulfite bisulfito de formaldehído sódico

sodium formate formiato sódico

sodium fusidate fusidato sódico

sodium germanate germanato sódico

sodium glucoheptonate glucoheptonato sódico

sodium gluconate gluconato sódico

sodium glutamate glutamato sódico

sodium glycolate glicolato sódico

sodium gold chloride cloruro de oro sódico

sodium gold cyanide cianuro de oro sódico

sodium guanylate guanilato sódico

sodium halide haluro sódico

sodium hexachloroosmate hexacloroosmiato sódico

sodium hexachloroplatinate hexacloroplatinato sódico

sodium hexacyanoferrate hexacianoferrato sódico

sodium hexafluoroaluminate hexafluoroaluminato sódico

sodium hexafluorosilicate hexafluorosilicato sódico

sodium hexametaphosphate hexametafosfato sódico

sodium hexanitrocobaltate hexanitrocobaltato sódico

sodium hydrate hidrato sódico

sodium hydride hidruro sódico

sodium hydrogen carbonate carbonato de hidrógeno sódico

sodium hydrogen difluoride difluoruro de hidrógeno sódico

sodium hydrogen fluoride fluoruro de hidrógeno sódico

sodium hydrogen peroxide peróxido de hidrógeno sódico

sodium hydrogen phosphate fosfato de hidrógeno sódico

sodium hydrogen sulfate sulfato de hidrógeno sódico

sodium hydrogen sulfide sulfuro de hidrógeno sódico

sodium hydrogen sulfite sulfito de hidrógeno sódico

sodium hydrogen tartrate tartrato de hidrógeno sódico

sodium hydrosulfide hidrosulfuro sódico

sodium hydrosulfite hidrosulfito sódico

sodium hydroxide hidróxido sódico

sodium hypobromite hipobromito sódico

sodium hypochlorite hipoclorito sódico

sodium hyponitrite hiponitrito sódico

sodium hypophosphite hipofosfito sódico

sodium hyposulfate hiposulfato sódico

sodium hyposulfite hiposulfito sódico

sodium inosinate inosinato sódico

sodium iodate yodato sódico

sodium iodide yoduro sódico

sodium iodohippurate yodohipurato sódico

sodium iothalamate iotalamato sódico

sodium iron pyrophosphate pirofosfato de hierro sódico

sodium isoascorbate isoascorbato sódico

sodium isobutylxanthate isobutilxantato sódico

sodium isovalerate isovalerato sódico

sodium lactate lactato sódico

sodium lamp lámpara de sodio

sodium lauryl sulfate laurilsulfato sódico

sodium-lead alloy aleación de sodio-plomo

sodium lead hyposulfite hiposulfito de plomo sódico

sodium lead thiosulfate tiosulfato de plomo sódico

sodium malonic ester éster malónico sódico

sodium manganate manganato sódico

sodium mercaptoacetate mercaptoacetato sódico

sodium mercaptobenzothiazole mercaptobenzotiazol sódico

sodium metabisulfite metabisulfito sódico

sodium metaborate metaborato sódico

sodium metanilate metanilato sódico

sodium metaperiodate metaperyodato sódico

sodium metaphosphate metafosfato sódico

sodium metasilicate metasilicato sódico

sodium metavanadate metavanadato sódico

sodium methacrylate metacrilato sódico

sodium methanoate metanoato sódico

sodium methanolate metanloato sódico

sodium methiodal metiodal sódico

sodium methoxide metóxido sódico

sodium methyl carbonate carbonato de metilo sódico

sodium methyl siliconate siliconato de metilo sódico

sodium methylate metilato sódico

sodium molybdate molibdato sódico

sodium molybdophosphate molibdofosfato sódico

sodium molybdosilicate molibdosilicato sódico

sodium monosulfide monosulfuro sódico

sodium monoxide monóxido sódico

sodium naphthalenesulfonate naftalenosulfonato sódico

sodium naphthenate naftenato sódico

sodium naphthionate naftionato sódico

sodium niobate niobato sódico

sodium nitranilate nitranilato sódico

sodium nitrate nitrato sódico

sodium nitride nitruro sódico

sodium nitrilotriacetate nitrilotriacetato sódico

sodium nitrite nitrito sódico

sodium nitroferricyanide nitroferricianuro sódico

sodium nitrophenolate nitrofenolato sódico

sodium novobiocin novobiocina sódica

sodium octyl sulfate octilsulfato sódico

sodium oleate oleato sódico

sodium orthophosphate ortofosfato sódico

sodium orthosilicate ortosilicato sódico

sodium orthovanadate ortovanadato sódico

sodium oxalate oxalato sódico

sodium oxide óxido sódico

sodium palconate palconato sódico

sodium palladium chloride cloruro de paladio sódico

sodium palmitate palmitato sódico

sodium paraperiodate paraperyodato sódico

sodium pentaborate pentaborato sódico

sodium pentaborate decahydrate decahidrato de pentaborato sódico

sodium pentachlorophenate pentaclorofenato sódico

sodium pentobarbital pentobarbital sódico

sodium perborate perborato sódico

sodium perborate monohydrate monohidrato de perborato sódico

sodium perborate tetrahydrate tetrahidrato de perborato sódico

sodium percarbonate percarbonato sódico

sodium perchlorate perclorato sódico

sodium periodate peryodato sódico

sodium permanganate permanganato sódico

sodium peroxide peróxido sódico

sodium peroxodicarbonate peroxodicarbonato sódico

sodium peroxydisulfate peroxidisulfato sódico

sodium persulfate persulfato sódico

sodium phenate fenato sódico

sodium phenoacetate fenoacetato sódico

sodium phenobarbital fenobarbital sódico

sodium phenolate fenolato sódico

sodium phenolsulfonate fenolsulfonato sódico

sodium phenylacetate fenilacetato sódico

sodium phenylphenate fenilfenato sódico

sodium phenylphenolate fenilfenolato sódico

sodium phenylphosphinate fenilfosfinato sódico

sodium phosphate fosfato sódico

sodium phosphide fosfuro sódico

sodium phosphinate fosfinato sódico

sodium phosphite fosfito sódico

sodium phosphoaluminate fosfoaluminato sódico

sodium phosphomolybdate fosfomolibdato sódico

sodium phosphonate fosfonato sódico

sodium phosphotungstate fosfotungstato sódico

sodium phosphovanadate fosfovanadato sódico

sodium phthalate ftalato sódico

sodium phytate fitato sódico

sodium picramate picramato sódico

sodium platinichloride platinicloruro sódico

sodium platinochloride platinocloruro sódico

sodium plumbate plumbato sódico

sodium plumbite plumbito sódico

sodium polyphosphate polifosfato sódico

sodium polysulfide polisulfuro sódico

sodium-potassium alloy aleación de sodio-potasio

sodium potassium tartrate tartrato potásico sódico

sodium propionate propionato sódico

sodium prussiate prusiato sódico

sodium pyroantimonate piroantimoniato sódico

sodium pyroborate piroborato sódico

sodium pyrophosphate pirofosfato sódico

sodium pyrophosphate peroxide peróxido de pirofosfato sódico

sodium pyroracemate pirorracemato sódico

sodium pyrosulfate pirosulfato sódico

sodium pyrosulfite pirosulfito sódico

sodium pyrovanadate pirovanadato sódico

sodium pyruvate piruvato sódico

sodium resinate resinato sódico

sodium rhodanate rodanato sódico

sodium ricinoleate ricinoleato sódico

sodium saccharin saccarina sódica

sodium salicylate salicilato sódico

sodium sarcosinate sarcosinato sódico

sodium secobarbital secobarbital sódico

sodium selenate seleniato sódico

sodium selenite selenito sódico

sodium sesquicarbonate sesquicarbonato sódico

sodium sesquisilicate sesquisilicato sódico

sodium silicate silicato sódico

sodium silicoaluminate silicoaluminato sódico

sodium silicofluoride silicofluoruro sódico

sodium silicomolybdate silicomolibdato sódico

sodium silicotungstate silicotungstato sódico

sodium silver chloride cloruro de plata sódico

sodium silver cyanide cianuro de plata sódico

sodium silver thiosulfate tiosulfato de plata sódico

sodium sodioacetate sodioacetato sódico

sodium sorbate sorbato sódico

sodium stannate estannato sódico

sodium stannite estannita sódica

sodium stearate estearato sódico

sodium stearoyl lactylate estearoilolactilato sódico

sodium styrenesulfonate estirenosulfonato sódico

sodium subsulfite subsulfito sódico

sodium succinate succinato sódico

sodium sulfalizarate sulfalizarato sódico

sodium sulfanilate sulfanilato sódico

sodium sulfantimonate sulfantimoniato sódico

sodium sulfate sulfato sódico

sodium sulfate decahydrate decahidrato de sulfato sódico

sodium sulfate monohydrate monohidrato de sulfato sódico

sodium sulfhydrate sulfhidrato sódico

sodium sulfide sulfuro sódico

sodium sulfite sulfito sódico

sodium sulfobromophthalein sulfobromoftaleína sódica

sodium sulfocarbolate sulfocarbolato sódico

sodium sulfocarbonate sulfocarbonato sódico

sodium sulfocyanate sulfocianato sódico

sodium sulfocyanide sulfocianuro sódico

sodium sulfonate sulfonato sódico

sodium sulfopropionitrile sulfopropionitrilo sódico

sodium sulforicinoleate sulforricinoleato sódico

sodium sulfoxylate sulfoxilato sódico

sodium superoxide superóxido sódico

sodium tartrate tartrato sódico

sodium tellurate telurato sódico

sodium tellurite telurita sódica

sodium tetraborate tetraborato sódico

sodium tetrabromoaurate tetrabromoaurato sódico

sodium tetrachloroaluminate tetracloroaluminato sódico

sodium tetrachloroaurate tetracloroaurato sódico

sodium tetrachloropalladate tetracloropaladato sódico

sodium tetrachlorophenate tetraclorofenato sódico

sodium tetrachloroplatinate tetracloroplatinato sódico

sodium tetradecyl sulfate tetradecilsulfato sódico

sodium tetrahydroborate tetrahidroborato sódico

sodium tetraphenylborate tetrafenilborato sódico

sodium tetraphosphate tetrafosfato sódico

sodium tetrasulfide tetrasulfuro sódico

sodium thioantimonate tioantimoniato sódico

sodium thioaurate tioaurato sódico

sodium thiocarbonate tiocarbonato sódico

sodium thiocyanate tiocianato sódico

sodium thioglycolate tioglicolato sódico

sodium thiosulfate tiosulfato sódico

sodium titanate titanato sódico

sodium toluate toluato sódico

sodium toluenesulfonate toluenosulfonato sódico

sodium trichloroacetate tricloroacetato sódico

sodium trichlorophenate triclorofenato sódico

sodium triphosphate trifosfato sódico

sodium tripolyphosphate tripolifosfato sódico

sodium trithiocarbonate tritiocarbonato sódico

sodium trititanate trititanato sódico

sodium tungstate tungstato sódico

sodium tungstophosphate tungstofosfato sódico

sodium tungstosilicate tungstosilicato sódico

sodium undecylenate undecilenato sódico

sodium uranate uranato sódico

sodium valproate valproato sódico

sodium vanadate vanadato sódico

sodium vanadophosphate vanadofosfato sódico

sodium warfarin warfarina sódica

sodium wolframate wolframato sódico

sodium xanthate xantato sódico

sodium xanthogenate xantogenato sódico

sodium xylenesulfonate xilenosulfonato sódico

sodium zirconium glycolate glicolato de zirconio sódico

sodium zirconium lactate lactato de zirconio sódico

sodium zirconium sulfate sulfato de zirconio sódico

sodyl sodilo

soft acid ácido blando

soft base base blanda

soft water agua blanda

softener ablandador

soil suelo, tierra, tierra negra

soil conditioner acondicionador de suelo

sol sol

solan solana

solanine solanina

solar cell célula solar

solar energy energía solar

solar furnace horno solar

solar pond tanque solar

solate solato

solation solación

solder soldante

solid sólido

solid-liquid equilibrium equilibrio sólido-líquido

solid solution solución sólida

solid state estado sólido

solid-state chemistry química de estado

sólido
solidification solidificación
solidify solidificar
solidus solidus
solubility solubilidad
solubility coefficient coeficiente de solubilidad
solubility curve curva de solubilidad
solubility product producto de solubilidad
solubility-product constant constante de producto de solubilidad
solubility test prueba de solubilidad
solubilize solubilizar
soluble soluble
soluble glass vidrio soluble
soluble oil aceite soluble
soluble starch almidón soluble
solute soluto
solution solución, disolución
solution conductivity conductividad de solución
solution pressure presión de disolución
solutrope solutropo
solvate solvato
solvation solvatación
solvatochromism solvatocromismo
solvent solvente, disolvente
solvent drying secado por solvente
solvent extraction extracción por solvente
solvent refining refinación por solvente
solventless sin solvente
solvolysis solvólisis
solvolytic solvolítico
solvus solvus
Sommelet-Hauser rearrangement reordenamiento de Sommelet-Hauser
Sommelet reaction reacción de Sommelet
Sonn-Muller method método de Sonn-Muller
sonochemistry sonoquímica
sonolysis sonólisis
soot hollín
sorbent sorbente
sorbic acid ácido sórbico
sorbide sórbido
sorbitan sorbitán

sorbitol sorbitol
sorbitol anhydride anhídrido de sorbitol
sorbose sorbosa
Sörensen indicator indicador de Sörensen
Soret effect efecto de Soret
sorghum zahína
sorption sorción
sosoloid sosoloide
sour agrio, rancio
source fuente
Soxhlet extractor extractor de Soxhlet
soybean soja
soybean oil aceite de soja
space espacio
spallation espalación
spandex espandex
Spanish white blanco de España
spar espato
spark chispa
sparking chisporroteo
spearmint menta verde
spearmint oil aceite de menta verde
species especie
specific específico
specific activity actividad específica
specific charge carga específica
specific conductance conductancia específica
specific conductivity conductividad específica
specific energy energía específica
specific gravity gravedad específica
specific heat calor específico
specific heat capacity capacidad térmica específica
specific reaction rate velocidad de reacción específica
specific refractivity refractividad específica
specific rotation rotación específica
specific volume volumen específico
specific weight peso específico
spectra espectros
spectral espectral
spectral analysis análisis espectral
spectral line línea espectral
spectral region región espectral
spectral series serie espectral

spectral shift deslizamiento espectral
spectral tube tubo espectral
spectro espectro
spectroanalysis espectroanálisis
spectroanalytical espectroanalítco
spectrochemical espectroquímico
spectrochemical analysis análsis
  espectroquímico
spectrochemical series serie
  espectroquímica
spectrochemistry espectroquímica
spectrofluorometer espectrofluorómetro
spectrofluorometric
  espectrofluorométrico
spectrofluorometry espectrofluorometría
spectrogram espectrograma
spectrograph espectrógrafo
spectrographic espectrográfico
spectrographically espectrográficamente
spectrography espectrografía
spectrometer espectrómetro
spectrometric espectrométrico
spectrometric analysis análisis
  espectrométrico
spectrometric instrument instrumento
  espectrométrico
spectrometrically espectrométricamente
spectrometry espectrometría
spectromicroscope espectromicroscopio
spectromicroscopic
  espectromicroscópico
spectrophotometer espectrofotómetro
spectrophotometric espectrofotométrico
spectrophotometric analysis análisis
  espectrofotométrico
spectrophotometry espectrofotometría
spectropolarimeter espetropolarímetro
spectropolarimetric espetropolarimétrico
spectroradiometer espectrorradiómetro
spectroscope espectroscopio
spectroscopic espectroscópico
spectroscopic analysis análisis
  espectroscópico
spectroscopically espectroscópicamente
spectroscopy espectroscopia
spectrum espectro
spectrum analysis análisis espectral
spectrum analyzer analizador espectral
spectrum line línea espectral

spectrum series serie espectral
specular especular
spelter cinc industrial
Sperry process proceso de Sperry
sperrylite esperrilita
sphalerite esfalerita
sphingomyelin esfingomielina
sphingosine esfingosina
spin espín
spin magnetic moment momento
  magnético de espín
spin of electron espín de electrón
spin-orbit coupling acoplamiento espín-
  orbital
spin-orbit interaction interacción espín-
  orbital
spin quantum number número cuántico
  de espín
spin resonance resonancia de espín
spin-spin coupling acoplamiento espín-
  espín
spin-spin interaction interacción espín-
  espín
spinel espinela
spinning electron electrón giratorio
spinodal curve curva espinodal
spiran espirano
spirit espíritu
spiro atom átomo espiro
spiro system sistema espiro
spiro union unión espiro
spirochete espiroqueta
spironolactone espironolactona
spiropentane espiropentano
splitting división, partición,
  desdoblamiento
spodumene espodumena
sponge esponja
spontaneous espontáneo
spontaneous combustion combustión
  espontánea
spontaneous decay desintegración
  espontánea
spontaneous disintegration
  desintegración espontánea
spontaneous emission emisión
  espontánea
spontaneous fission fisión espontánea
spontaneous heating calentamiento

espontáneo

spontaneous ignition ignición espontánea

spontaneous reaction reacción
espontánea

sputter evaporación catódica,
chisporroteo, deposición electrónica

sputtering evaporación catódica,
chisporroteo, deposición electrónica

squalane escualano

squalene escualeno

stability estabilidad

stability constant constante de
estabilidad

stabilization estabilización

stabilization energy energía de
estabilización

stabilizer estabilizador

stable estable

stable equilibrium equilibrio estable

stable isobar isóbaro estable

stable isotope isótopo estable

stable nucleus núcleo estable

stable system sistema estable

stacking apilamiento

Staedel-Rugheimer pyrazine synthesis
síntesis de pirazinas de Staedel-
Rugheimer

staggered conformation conformación
alternada

stain tintura, colorante

stainless steel acero inoxidable

stalagmometer estalagmómetro

stalagmometry estalagmometría

standard patrón, estándar, norma

standard cell pila patrón

standard conditions condiciones
normales

standard pressure presión normal

standard solution solución normal

standard temperature temperatura
normal

standard temperature and pressure
temperatura y presión normales

standardization estandarización,
normalización

standardize estandarizar, normalizar

standardized estandarizado, normalizado

stannane estannano

stannate estannato

stannic estánnico

stannic acid ácido estánnico

stannic anhydride anhídrido estánnico

stannic bromide bromuro estánnico

stannic chloride cloruro estánnico

stannic chromate cromato estánnico

stannic fluoride fluoruro estánnico

stannic hydroxide hidróxido estánnico

stannic iodide yoduro estánnico

stannic oxide óxido estánnico

stannic phosphide fosfuro estánnico

stannic sulfate sulfato estánnico

stannic sulfide sulfuro estánnico

stannite estannita

stannous estannoso

stannous acetate acetato estannoso

stannous bromide bromuro estannoso

stannous chloride cloruro estannoso

stannous chromate cromato estannoso

stannous citrate citrato estannoso

stannous ethylhexoate etilhexoato
estannoso

stannous fluoride fluoruro estannoso

stannous hydroxide hidróxido estannoso

stannous iodide yoduro estannoso

stannous octoate octoato estannoso

stannous oleate oleato estannoso

stannous oxalate oxalato estannoso

stannous oxide óxido estannoso

stannous pyrophosphate pirofosfato
estannoso

stannous sulfate sulfato estannoso

stannous sulfide sulfuro estannoso

stannous tartrate tartrato estannoso

stannum estaño

stannyl estannilo

starch almidón

starch dialdehyde dialdehído de almidón

starch phosphate fosfato de almidón

starch syrup jarabe de almidón

starch xanthate xantato de almidón

Stark effect efecto de Stark

Stark-Einstein law ley de Stark-Einstein

Stark-Luneland effect efecto de Stark-
Luneland

state estado

static estático

stationary phase fase estacionaria

stationary state estado estacionario

statistical mechanics mecánica estadística

Staudinger reaction reacción de Staudinger

Stead reagent reactivo de Stead

steady state estado estacionario

steam vapor de agua

steam bath baño de vapor

steam distillation destilación al vapor

steam point punto de vapor

steam reforming reformación al vapor

stearaldehyde estearaldehído

stearamide estearamida

stearate estearato

stearic acid ácido esteárico

stearin estearina

stearone estearona

stearonitrile estearonitrilo

stearoyl estearoilo

stearoyl chloride cloruro de estearoilo

stearoylaminophenol estearoilaminofenol

stearyl estearilo

stearyl alcohol alcohol estearílico

stearyl mercaptan estearilmercaptano

stearyl methacrylate metacrilato de estearilo

steatite esteatita

steel acero

Steffen process proceso de Steffen

Stengel process proceso de Stengel

step paso, etapa

stereoblock polymer polímero de estereobloques

stereochemistry estereoquímica

stereoisomer estereoisómero

stereoisomerism estereoisomerismo

stereoregular estereorregular

stereoregular polymer polímero estereorregular

stereoscope estereoscopio

stereoselectivity estereoselectividad

stereospecific estereoespecífico

stereospecific catalyst catalizador estereoespecífico

stereospecific polymer polímero estereoespecífico

stereospecific reaction reacción estereoespecífica

stereospecific synthesis síntesis estereoespecífica

steric estérico

steric effect efecto estérico

steric geometry geometría estérica

steric hindrance impedimento estérico

sterile esteril

sterilization esterilización

sterling silver plata esterlina

steroid esteriode

sterol esterol

Stevens rearrangement reordenamiento de Stevens

stibic anhydride anhídrido estíbico

stibine estibina

stibino estibino

stibinoso estibinoso

stibnite estibnita

stibo estibo

stibonium estibonio

stibono estibono

stibonoso estibonoso

stiboso estiboso

Stieglitz rearrangement reordenamiento de Stieglitz

stigmasterol estigmasterol

stilbene estilbeno

stilbene dye colorante de estilbeno

stilbestrol estilbestrol

stilbite estilbita

still destilador, alambique

stimulation estimulación

stirrer agitador

stirring rod varilla agitadora

Stobbe condensation condensación de Stobbe

Stobbe reaction reacción de Stobbe

stochastic estocástico

stochastic process proceso estocástico

Stock system sistema de Stock

Stoddard solvent solvente de Stoddard

stoichiometric estequiométrico

stoichiometric coefficient coeficiente estequiométrico

stoichiometric compound compuesto estequiométrico

stoichiometric mixture mezcla estequiométrica

stoichiometry estequiometría

Stokes law ley de Stokes
Stokes line línea de Stokes
Stokes shift deslizamiento de Stokes
Stolle synthesis síntesis de Stolle
stopper tapón
stopper with tap tapón con llave
storage almacenamiento
storage battery batería de acumuladores
storage guidelines normas de
    almacenamiento
storage precautions precauciones de
    almacenamiento
Stork enamine reaction reacción de
    enaminas de Stork
straight chain cadena recta
straight-chain hydrocarbon
    hidrocarburo de cadena recta
strain deformación, tensión
strain theory teoría de deformación
streaming potential potencial de
    circulación
streamline filtration filtración por
    bordes
Strecker degradation degradación de
    Strecker
Strecker reaction reacción de Strecker
Strecker sulfite alkylation alquilación de
    sulfitos de Strecker
Strecker synthesis síntesis de Strecker
strength fuerza, intensidad
streptolin estreptolina
streptomycin estreptomicina
stress estrés, tensión, deformación
stripped atom átomo despojado de
    electrones
stripping retroextracción, extracción,
    eliminación, lavado
strong fuerte
strong acid ácido fuerte
strong base base fuerte
strong electrolyte electrólito fuerte
strontia estroncia
strontianite estroncianita
strontium estroncio
strontium acetate acetato de estroncio
strontium arsenite arsenito de estroncio
strontium bicarbonate bicarbonato de
    estroncio
strontium bromate bromato de estroncio

strontium bromide bromuro de estroncio
strontium carbonate carbonato de
    estroncio
strontium chlorate clorato de estroncio
strontium chloride cloruro de estroncio
strontium chromate cromato de
    estroncio
strontium dioxide dióxido de estroncio
strontium dithionate ditionato de
    estroncio
strontium fluoride fluoruro de estroncio
strontium formate formiato de estroncio
strontium hexafluorosilicate
    hexafluorosilicato de estroncio
strontium hydrogen carbonate
    carbonato de hidrógeno de estroncio
strontium hydroxide hidróxido de
    estroncio
strontium hyposulfite hiposulfito de
    estroncio
strontium iodide yoduro de estroncio
strontium lactate lactato de estroncio
strontium molybdate molibdato de
    estroncio
strontium monosulfide monosulfuro de
    estroncio
strontium monoxide monóxido de
    estroncio
strontium nitrate nitrato de estroncio
strontium nitrite nitrito de estroncio
strontium oxalate oxalato de estroncio
strontium oxide óxido de estroncio
strontium perchlorate perclorato de
    estroncio
strontium peroxide peróxido de
    estroncio
strontium phosphate fosfato de estroncio
strontium-potassium chlorate clorato de
    estroncio-potasio
strontium saccharate sacarato de
    estroncio
strontium salicylate salicilato de
    estroncio
strontium sulfate sulfato de estroncio
strontium sulfide sulfuro de estroncio
strontium sulfite sulfito de estroncio
strontium tartrate tartrato de estroncio
strontium thiosulfate tiosulfato de
    estroncio

strontium titanate titanato de estroncio
strontium zirconate zirconato de
    estroncio
structural estructural
structural antagonist antagonista
    estructural
structural formula fórmula estructural
structural isomerism isomerismo
    estructural
structure estructura
strychnidine estricnidina
strychnine estricnina
Stuffer rule regla de Stuffer
styphnic acid ácido estífnico
styracin estiracina
styralyl acetate acetato estiralílico
styralyl alcohol alcohol estiralílico
styrenated oil aceite estirenado
styrene estireno
styrene-acrylonitrile resin resina de
    estireno-acrilonitrilo
styrene-butadiene rubber caucho de
    estireno-butadieno
styrene glycol estirenglicol
styrene monomer monómero de estireno
styrene nitrosite nitrosita de estireno
styrene oxide óxido de estireno
styrene plastic plástico de estireno
styrene polymer polímero de estireno
styryl estirilo
sub sub
subacetate subacetato
subatomic subatómico
subatomic decomposition
    descomposición subatómica
subatomic particle partícula subatómica
subatomic reaction reacción subatómica
subatomics subatómica
subcarbonate subcarbonato
suberane suberano
suberic acid ácido subérico
suberone suberona
suberyl suberilo
sublation sublación
sublethal subletal
sublevel subnivel
sublimate sublimar
sublimation sublimación
sublimation point punto de sublimación

sublimator sublimador
subnitrate subnitrato
subnuclear subnuclear
subnuclear particle partícula subnuclear
suboxide subóxido
subshell subcapa
substance substancia
substance concentration concentración
    de substancia
substantive dye colorante substantivo
substituent átomo substituyente, grupo
    substituyente
substitute substituir
substitution substitución
substitution reaction reacción de
    substitución
substrate substrato
substructure subestructura
subtilin subtilina
succinaldehyde succinaldehído
succinamic acid ácido succinámico
succinamoyl succinamoilo
succinamyl succinamilo
succinate succinato
succinic acid ácido succínico
succinic acid peroxide peróxido de ácido
    succínico
succinic anhydride anhídrido succínico
succinimide succinimida
succinimido succinimido
succiniodimide succiniodimida
succinonitrile succinonitrilo
succinyl succinilo
succinyl chloride cloruro de succinilo
succinyl oxide óxido de succinilo
sucrase sucrasa
sucrate sucrato
sucrol sucrol
sucrose sucrosa, sacarosa
sucrose acetate isobutyrate
    acetatoisobutirato de sucrosa
sucrose monostearate monoestearato de
    sucrosa
sucrose octaacetate octaacetato de
    sucrosa
sucrose polyester poliéster de sucrosa
suction pump bomba de succión
sugar azúcar
sugar of lead azúcar de plomo

sugar of milk azúcar de leche
sugar substitute substituto de azúcar
sugarcane wax cera de caña de azúcar
sulfa drug droga sulfa, sulfonamida
sulfacetamide sulfacetamida
sulfadiazine sulfadiazina
sulfadimidine sulfadimidina
sulfaguanidine sulfaguanidina
sulfaldehyde sulfaldehído
sulfamate sulfamato
sulfamerazine sulfamerazina
sulfamethizole sulfametizol
sulfamic sulfámico
sulfamic acid ácido sulfámico
sulfamide sulfamida
sulfamidic acid ácido sulfamídico
sulfaminic acid ácido sulfamínico
sulfamoyl sulfamoilo
sulfamyl sulfamilo
sulfane sulfano
sulfanilamide sulfanilamida
sulfanilate sulfanilato
sulfanilic acid ácido sulfanílico
sulfarsenate sulfarseniato
sulfatase sulfatasa
sulfate sulfato
sulfate mineral mineral de sulfato
sulfathiazole sulfatiazol
sulfation sulfatación
sulfazide sulfazida
sulfenamide sulfenamida
sulfenic sulfénico
sulfeno sulfeno
sulfenone sulfenona
sulfhydryl sulfhidrilo
sulfide sulfuro
sulfide dye colorante al sulfuro
sulfime sulfima
sulfimide sulfimida
sulfinate sulfinato
sulfine sulfina
sulfino sulfino
sulfinoxide sulfonóxido
sulfinyl sulfinilo
sulfinyl bromide bromuro de sulfinilo
sulfinyl chloride cloruro de sulfinilo
sulfinyl fluoride fluoruro de sulfinilo
sulfinylimide sulfinilimida
sulfite sulfito

sulfite pulp pulpa de sulfito
sulfo sulfo
sulfoacetic acid ácido sulfoacético
sulfoamino sulfoamino
sulfobenzoic acid ácido sulfobenzoico
sulfobenzoic anhydride anhídrido sulfobenzoico
sulfocarbanilide sulfocarbanilida
sulfocarbimide sulfocarbimida
sulfocarbolic acid ácido sulfocarbólico
sulfocarbolide sulfocarbolida
sulfocarbonic acid ácido sulfocarbónico
sulfocyanate sulfocianato
sulfocyanic acid ácido sulfociánico
sulfocyanide sulfocianuro
sulfohydrate sulfohidrato
sulfolane sulfolano
sulfoleic acid ácido sulfoleico
sulfonamide sulfonamida
sulfonate sulfonato
sulfonation sulfonación
sulfone sulfona
sulfonic acid ácido sulfónico
sulfonic anhydride anhídrido sulfónico
sulfonium sulfonio
sulfonohydrazonic acid ácido sulfonohidrazónico
sulfonyl sulfonilo
sulfonyl chloride cloruro de sulfonilo
sulfonyl fluoride fluoruro de sulfonilo
sulfonyldiphenol sulfonildifenol
sulfophthalic anhydride anhídrido sulfoftálico
sulfosalicylic acid ácido sulfosalicílico
sulfoselenide sulfoseleniuro
sulfoxide sulfóxido
sulfoxylate sulfoxilato
sulfoxylic acid ácido sulfoxílico
sulfur azufre
sulfur bichloride bicloruro de azufre
sulfur bromide bromuro de azufre
sulfur chloride cloruro de azufre
sulfur cycle ciclo de azufre
sulfur dichloride dicloruro de azufre
sulfur dioxide dióxido de azufre
sulfur dye colorante al azufre
sulfur fluoride fluoruro de azufre
sulfur hexafluoride hexafluoruro de azufre

sulfur hexaiodide hexayoduro de azufre
sulfur iodide yoduro de azufre
sulfur monobromide bromuro de azufre
sulfur monochloride monocloruro de azufre
sulfur monofluoride monofluoruro de azufre
sulfur monoxide monóxido de azufre
sulfur oxide óxido de azufre
sulfur oxyacid oxiácido de azufre
sulfur oxychloride oxicloruro de azufre
sulfur pentachloride pentacloruro de azufre
sulfur subchloride subcloruro de azufre
sulfur test prueba de azufre
sulfur tetrachloride tetracloruro de azufre
sulfur tetrafluoride tetrafluoruro de azufre
sulfur trioxide trióxido de azufre
sulfurated sulfurado
sulfurated lime cal sulfurada
sulfuration sulfuración
sulfuretted sulfurado
sulfuric sulfúrico
sulfuric acid ácido sulfúrico
sulfuric anhydride anhídrido sulfúrico
sulfuric chloride cloruro sulfúrico
sulfuric monobromide monobromuro sulfúrico
sulfuric monochloride monocloruro sulfúrico
sulfuric oxychloride oxicloruro sulfúrico
sulfurous sulfuroso
sulfurous acid ácido sulfuroso
sulfurous oxychloride oxicloruro sulfuroso
sulfuryl sulfurilo
sulfuryl chloride cloruro de sulfurilo
sulfuryl fluoride fluoruro de sulfurilo
Sullivan reaction reacción de Sullivan
sulphur azufre
sultam sultama
sultone sultona
sulvanite sulvanita
sunflower oil aceite de girasol
superacid superácido
superactinide superactínido
superalloy superaleación

supercarbonate supercarbonato
superconductivity superconductividad
superconductor superconductor
supercooled subenfriado
supercooled liquid líquido subenfriado
supercooling subenfriado
supercritical supercrítico
supercritical fluid fluido supercrítico
superfluid superfluido
superfluidity superfluidez
superheating supercalentamiento
superheavy superpesado
superheavy element elemento superpesado
supernatant sobrenadante
supernormal supernormal
superoxide superóxido
superoxide dismutase dismutasa superóxida
superpalite superpalita
superphosphate superfosfato
superphosphoric acid ácido superfosfórico
superplasticity superplasticidad
superpolymer superpolímero
supersaturated supersaturado
supersaturated solution solución supersaturada
supersaturation supersaturación
supersolubility supersolubilidad
supra supra
supramolecular chemistry química supramolecular
surface superficie
surface-active agent agente tensoactivo
surface activity actividad superficial
surface area área superficial
surface chemistry química superficial
surface combustion combustión superficial
surface compound compuesto superficial
surface energy energía superficial
surface orientation orientación superficial
surface pressure presión superficial
surface reaction reacción superficial
surface tension tensión superficial
surface viscocity viscocidad superficial
surfactant surfactante

surfusion surfusión
surrosion surrosión
suspended solids sólidos suspendidos
suspending agent agente de suspensión
suspension suspensión
suspensoid suspensoide
suxamethonium chloride cloruro de
    suxametonio
Swarts reaction reacción de Swarts
sweet oil aceite dulce
sweet water agua dulce
sweeten endulzar
sweetener endulcorante, endulzador
sydnone sidnona
sylvan silvano
sylvanite silvanita
sylvic acid ácido sílvico
sylvine silvina
sylvinite silvinita
sylvite silvita
sylyl sililo
symbol símbolo
symmetric simétrico
symmetrical simétrico
symmetrical carbon atom átomo de
    carbono simétrico
symmetrical compound compuesto

    simétrico
symmetry simetría
syndiotactic sindiotáctico
syneresis sinéresis
synergism sinergismo
synergist sinergista
syntactic sintáctico
synthesis síntesis
synthesis gas gas de síntesis
synthesize sintetizar
synthetic sintético
synthetic detergent detergente sintético
synthetic fiber fibra sintética
synthetic fuel combustible sintético
synthetic gas gas sintético
synthetic natural gas gas natural
    sintético
synthetic oil aceite sintético
synthetic resin resina sintética
synthetic rubber caucho sintético
synthol sintol
syrosingopine sirosingopina
syrup jarabe
systematic name nombre sistemático
systemic sistémico
Szilard-Chalmers effect efecto de
    Szilard-Chalmers

T

table salt sal de mesa
tabun tabún
tachiol taquiol
taconite taconita
tactic polymer polímero táctico
tacticity tacticidad
tactosol tactosol
taenite taenita
Tafel rearrangement reordenamiento de
    Tafel
tag marcador
tagged atom átomo marcado
tagged molecule molécula marcada
tailings residuos, desechos, colas
talc talco
tall oil resina líquida
tallow sebo
tallow oil aceite de sebo
talose talosa
Tanabe-Sugano diagram diagrama de
    Tanabe-Sugano
tanacetyl alcohol alcohol tanacetílico
tangerine oil aceite de tangerina
tannase tanasa
tannic acid ácido tánico
tannin tanino
tanning curtido
tannyl tanilo
tantalate tantalato
tantalic acid ácido tantálico
tantalic chloride cloruro tantálico
tantalite tantalita
tantalous tantaloso
tantalous chloride cloruro tantaloso
tantalum tántalo, tantalio
tantalum alcoholate alcoholato de
    tántalo
tantalum bromide bromuro de tántalo
tantalum carbide carburo de tántalo
tantalum chloride cloruro de tántalo
tantalum disulfide disulfuro de tántalo
tantalum fluoride fluoruro de tántalo
tantalum nitride nitruro de tántalo

tantalum oxide óxido de tántalo
tantalum pentabromide pentabromuro
    de tántalo
tantalum pentachloride pentacloruro de
    tántalo
tantalum pentafluoride pentafluoruro de
    tántalo
tantalum pentoxide pentóxido de tántalo
tantalum potassium fluoride fluoruro
    potásico de tántalo
tantiron tantirón
tapioca tapioca
tapiolite tapolita
tar alquitrán, brea
tar acid ácido de alquitrán
tar base base de alquitrán
tar camphor alcanfor de alquitrán
tar oil aceite de alquitrán
tar soap jabón de alquitrán
tare tara
target blanco
tarnish deslustre, descoloramiento
tarragon oil aceite de estragón
tartar emetic tártaro emético
tartaric acid ácido tartárico
tartaroyl tartaroilo
tartrate tartrato
tartrazine tartrazina
tartronic acid ácido tartrónico
tartronoyl tartronoilo
taste sabor, gusto
tau tau
tau particle partícula tau
tau value valor tau
Tauber test prueba de Tauber
taurine taurina
taurocholic acid ácido taurocólico
tauryl taurilo
tautomer tautómero
tautomerase tautomerasa
tautomeric tautomérico
tautomerism tautomería
tautourea tautourea

taxol taxol
technetium tecnecio
technical técnico
Teclu burner mechero de Teclu
teichoic acid ácido teicoico
tektite tectita
Teller-Redlich rule regla de Teller-Redlich
tellurate telurato
telluric acid ácido telúrico
telluric bromide bromuro telúrico
telluric lead plomo telúrico
telluride telururo
tellurinic acid ácido telurínico
tellurinyl telurinilo
tellurite telurita
tellurium telurio
tellurium bromide bromuro de telurio
tellurium chloride cloruro de telurio
tellurium dibromide dibromuro de telurio
tellurium dichloride dicloruro de telurio
tellurium dioxide dióxido de telurio
tellurium disulfide disulfuro de telurio
tellurium fluoride fluoruro de telurio
tellurium hexafluoride hexafluoruro de telurio
tellurium hydroxide hidróxido de telurio
tellurium iodide yoduro de telurio
tellurium monoxide monóxido de telurio
tellurium nitrate nitrato de telurio
tellurium oxide óxido de telurio
tellurium oxychloride oxicloruro de telurio
tellurium sulfate sulfato de telurio
tellurium sulfide sulfuro de telurio
tellurium tetrabromide tetrabromuro de telurio
tellurium trioxide trióxido de telurio
telluroketone telurocetona
telluronium teluronio
tellurous teluroso
tellurous acid ácido teluroso
tellurous bromide bromuro teluroso
tellurous chloride cloruro teluroso
telluryl telurilo
telodrin telodrina
telomerization telomerización
telomerization reaction reacción de

telomerización
temper templar
temperature temperatura
temperature coefficient coeficiente de temperatura
temperature control control de temperatura
temperature degree grado de temperatura
temperature regulator regulador de temperatura
temperature scale escala de temperatura
temperature-sensitive sensible a la temperatura
tenderization ablandamiento
tenderize ablandar
tennantite tennantita
tenorite tenorita
tenth-normal solution solución decinormal
tephroite tefroíta
ter ter
teratogen teratógeno
terbia terbia
terbium terbio
terbium chloride cloruro de terbio
terbium hydroxide hidróxido de terbio
terbium nitrate nitrato de terbio
terbium oxide óxido de terbio
terbium sulfate sulfato de terbio
terebene terebeno
terephthalaldehyde tereftalaldehído
terephthalic acid ácido tereftálico
terephthalonitrile tereftalonitrilo
terephthaloyl chloride cloruro de tereftaloilo
terephthaloylbenzoic acid ácido tereftaloilbenzoico
terminology terminología
ternary ternario
ternary compound compuesto ternario
ternary system sistema ternario
teroxide teróxido
terpadiene terpadieno
terpene terpeno
terpenoid terpenoide
terpenol terpenol
terphenyl terfenilo
terpilenol terpilenol

terpin hydrate hidrato de terpina
terpinene terpineno
terpineol terpineol
terpinolene terpinoleno
terpinyl acetate acetato de terpinilo
terpolymer terpolímero
tert terc
tertiary terciario
tertiary alcohol alcohol terciario
tertiary amine amina terciaria
tertiary carbon carbono terciario
tertiary carbon atom átomo de carbono
    terciario
tervalent tervalente
test prueba
test paper papel de prueba
test tube probeta, tubo de ensayo
testosterone testosterona
tetra tetra
tetrabasic tetrabásico
tetrabenzyl tetrabencilo
tetraborane tetraborano
tetrabromo tetrabromo
tetrabromobenzene tetrabromobenceno
tetrabromoethane tetrabromoetano
tetrabromoethylene tetrabromoetileno
tetrabromofluorescein
    tetrabromofluoresceína
tetrabromomethane tetrabromometano
tetrabromosilane tetrabromosilano
tetrabutyl titanate titanato de tetrabutilo
tetrabutyl zirconate zirconato de
    tetrabutilo
tetrabutylammonium chloride cloruro
    de tetrabutilamonio
tetrabutyltin tetrabutilestaño
tetracaine tetracaína
tetracaine hydrochloride clorhidrato de
    tetracaína
tetracarboxybutane tetracarboxibutano
tetracarboxylic acid ácido
    tetracarboxílico
tetracene tetraceno
tetrachloride tetracloruro
tetrachloro tetracloro
tetrachloroaniline tetracloroanilina
tetrachlorobenzene tetraclorobenceno
tetrachlorodiphenylethane
    tetraclorodifeniletano

tetrachloroethane tetracloroetano
tetrachloroethylene tetracloroetileno
tetrachloromethane tetraclorometano
tetrachloronaphthalene
    tetracloronaftaleno
tetrachlorophenol tetraclorofenol
tetrachlorophthalic acid ácido
    tetracloroftálico
tetrachlorophthalic anhydride anhídrido
    tetracloroftálico
tetrachlorosalicylanilide
    tetraclorosalicilanilida
tetrachlorosilane tetraclorosilano
tetracosane tetracosano
tetracosanoic acid ácido tetracosanoico
tetracosyl tetracosilo
tetracyano tetraciano
tetracyanoethylene tetracianoetileno
tetracyanoquinonedimethane
    tetracianoquinonadimetano
tetracycline tetraciclina
tetracyclone tetraciclona
tetrad tetrada
tetradecane tetradecano
tetradecanoic acid ácido tetradecanoico
tetradecanol tetradecanol
tetradecene tetradeceno
tetradecyl tetradecilo
tetradecyl chloride cloruro de tetradecilo
tetradecyl thiol tetradeciltiol
tetradecylamine tetradecilamina
tetradecylene tetradecileno
tetradifon tetradifón
tetraethanolammonium hydroxide
    hidróxido de tetraetanolamonio
tetraethoxypropane tetraetoxipropano
tetraethyl tetraetilo
tetraethyl phosphate fosfato de tetraetilo
tetraethylbenzene tetraetilbenceno
tetraethylene glycol tetraetilenglicol
tetraethylene glycol dimethacrylate
    dimetacrilato de tetraetilenglicol
tetraethylene glycol monostearate
    monoestearato de tetraetilenglicol
tetraethylenepentamine
    tetraetilenopentamina
tetraethylgermanium tetraetilgermanio
tetraethylhexyl titanate titanato de
    tetraetilhexilo

tetraethyllead tetraetilplomo
tetraethyltin tetraetilestaño
tetrafluoroborate tetrafluoroborato
tetrafluorodichloroethane tetrafluorodicloroetano
tetrafluoroethylene tetrafluoroetileno
tetrafluoroethylene epoxide epóxido de tetrafluoroetileno
tetrafluorohydrazine tetrafluorohidrazina
tetrafluoromethane tetrafluorometano
tetrafluorosilane tetrafluorosilano
tetragalloyl tetragaloilo
tetraglycol dichloride dicloruro de tetraglicol
tetragonal tetragonal
tetragonal system sistema tetragonal
tetrahedral tetraédrico
tetrahedral angle ángulo tetraédrico
tetrahedral atom átomo tetraédrico
tetrahedral carbon carbono tetraédrico
tetrahedral compound compuesto tetraédrico
tetrahedral coordination coordinación tetraédrica
tetrahedron tetraedro
tetrahedronal tetraédrico
tetrahydrate tetrahidrato
tetrahydro tetrahidro
tetrahydrobenzene tetrahidrobenceno
tetrahydrocannibol tetrahidrocannibol
tetrahydrofuran tetrahidrofurano
tetrahydrofuran polymer polímero de tetrahidrofurano
tetrahydrofurandimethanol tetrahidrofuranodimetanol
tetrahydrofurfuryl alcohol alcohol tetrahidrofurfurílico
tetrahydrofurfuryl benzoate benzoato de tetrahidrofurfurilo
tetrahydrofurfuryl laurate laurato de tetrahidrofurfurilo
tetrahydrofurfurylamine tetrahidrofurfurilamina
tetrahydronaphthalene tetrahidronaftaleno
tetrahydrophenol tetrahidrofenol
tetrahydrophthalic anhydride anhídrido tetrahidroftálico

tetrahydropyranmethanol tetrahidropiranometanol
tetrahydropyridine tetrahidropiridina
tetrahydrothiophene tetrahidrotiofeno
tetrahydroxy tetrahidroxi
tetrahydroxybenzene tetrahidroxibenceno
tetrahydroxybenzoic acid ácido tetrahidroxibenzoico
tetrahydroxybutane tetrahidroxibutano
tetraiodo tetrayodo
tetraisopropyl titanate titanato de tetraisopropilo
tetraisopropyl zirconate zirconato de tetraisopropilo
tetraketone tetracetona
tetrakis tetrakis
tetralin tetralina
tetralite tetralita
tetralone tetralona
tetramer tetrámero
tetramethoxypropane tetrametoxipropano
tetramethyl tetrametilo
tetramethylammonium chloride cloruro de tetrametilamonio
tetramethylbenzene tetrametilbenceno
tetramethylbenzidine tetrametilbencidina
tetramethylcyclobutanediol tetrametilciclobutanodiol
tetramethyldiaminobenzhydrol tetrametildiaminobenzhidrol
tetramethyldiaminobenzophenone tetrametildiaminobenzofenona
tetramethyldiaminodiphenylsulfone tetrametildiaminodifenilsulfona
tetramethylene tetrametileno
tetramethylenediamine tetrametilendiamina
tetramethylethylenediamine tetrametiletilendiamina
tetramethyllead tetrametilplomo
tetramethylmethane tetrametilmetano
tetramethylsilane tetrametilsilano
tetramethyltin tetrametilestaño
tetramethylurea tetrametilurea
tetramine tetramina
tetramolecular tetramolecular
tetramorphism tetramorfismo

tetranitrate tetranitrato
tetranitro tetranitro
tetranitroaniline tetranitroanilina
tetranitrobiphenyl tetranitrobifenilo
tetranitromethane tetranitrometano
tetranitrophenol tetranitrofenol
tetraoxide tetróxido, tetraóxido
tetraoxophosphoric acid ácido
     tetraoxofosfórico
tetraoxosulfuric acid ácido
     tetraoxosulfúrico
tetraphenyl tetrafenilo
tetraphenylbutadiene tetrafenilbutadieno
tetraphenylcyclopentadienone
     tetrafenilciclopentadienona
tetraphenylene tetrafenileno
tetraphenylethane tetrafeniletano
tetraphenylmethane tetrafenilmetano
tetraphenylsilane tetrafenilsilano
tetraphenyltin tetrafenilestaño
tetraphenylurea tetrafenilurea
tetraphosphoric acid ácido tetrafosfórico
tetraphosphorus selenide seleniuro de
     tetrafósforo
tetrapotassium pyrophosphate
     pirofosfato tetrapotásico
tetrapropylene tetrapropileno
tetrasilane tetrasilano
tetrasodium diphosphate difosfato
     tetrasódico
**tetrasodium ethylenediaminetetraacetic
     acid** ácido etilendiaminotetraacético
     tetrasódico
tetrasodium pyrophosphate pirofosfato
     tetrasódico
tetrathioarsenate tetratioarseniato
tetrathioarsenite tetratioarsenito
tetrathionate tetrationato
tetrathionic acid ácido tetratiónico
tetravalent tetravalente
tetrazene tetraceno
tetrazole tetrazol
tetrazolium tetrazolio
tetrazolyl tetrazolilo
tetrazone tetrazona
tetrel tetrel
tetritol tetritol
tetrol tetrol
tetrolic acid ácido tetrólico

tetrone tetrona
tetrose tetrosa
tetroxide tetróxido
tetryl tetrilo
texaphyrin texafirina
texture textura
thalenite thalenita
thalidomide talidomida
thallate talato
thallic tálico
thallium talio
thallium acetate acetato de talio
thallium alum alumbre de talio
thallium amalgam amalgama de talio
thallium bromide bromuro de talio
thallium carbonate carbonato de talio
thallium chloride cloruro de talio
thallium chromate cromato de talio
thallium ethanolate etanolato de talio
thallium fluoride fluoruro de talio
thallium formate formiato de talio
thallium hexafluorosilicate
     hexafluorosilicato de talio
thallium hydroxide hidróxido de talio
thallium iodide yoduro de talio
thallium monoxide monóxido de talio
thallium nitrate nitrato de talio
thallium oxide óxido de talio
thallium oxysulfide oxisulfuro de talio
thallium sesquichloride sesquicloruro de
     talio
thallium sulfate sulfato de talio
thallium sulfide sulfuro de talio
thallium sulfoxide sulfóxido de talio
thallium trifluoroacetate trifluoroacetato
     de talio
thallous talioso
thallous bromide bromuro talioso
thallous carbonate carbonato talioso
thallous chloride cloruro talioso
thallous hydroxide hidróxido talioso
thallous iodide yoduro talioso
thallous monoxide monóxido talioso
thallous nitrate nitrato talioso
thallous oxide óxido talioso
thallous sesquichloride sesquicloruro
     talioso
thallous sulfate sulfato talioso
thallous sulfide sulfuro talioso

thebaine tebaína
thebenidine tebenidina
theine teína
Thenard blue azul de Thenard
thenoic acid ácido tenoico
thenyl tenilo
thenyl alcohol alcohol tenílico
thenyldiamine tenildiamina
theobroma oil aceite de teobroma
theobromine teobromina
theophylline teofilina
theoretical teorético
theoretical plate placa teorética
thermal térmico, termal
thermal activation activación térmica
thermal analysis análisis térmico
thermal capacity capacidad térmica
thermal collision colisión térmica
thermal column columna térmica
thermal conductance conductancia
    térmica
thermal conductivity conductividad
    térmica
thermal constant constante térmica
thermal cracking craqueo térmico
thermal decomposition descomposición
    térmica
thermal degradation degradación
    térmica
thermal diffusion difusión térmica
thermal effect efecto térmico
thermal energy energía térmica
thermal equilibrium equilibrio térmico
thermal excitation excitación térmica
thermal expansion expansión térmica
thermal expansion coefficient
    coeficiente de expansión térmica
thermal fission fisión térmica
thermal fragmentation fragmentación
    térmica
thermal insulation aislamiento térmico
thermal intensity intensidad térmica
thermal ionization ionización térmica
thermal neutron neutrón térmico
thermal pollution contaminación térmica
thermal reactor reactor térmico
thermal reforming reformación térmica
thermal stability estabilidad térmica
thermal titration titulación térmica,

valoración térmica
thermalize termalizar
thermatomic process proceso
    termatómico
thermic térmico, termal
thermionic effect efecto termiónico
thermite termita
thermo termo
thermoanalysis termoanálisis
thermobalance termobalanza
thermocatalytic termocatalítico
thermochemical termoquímico
thermochemical standard patrón
    termoquímico
thermochemistry termoquímica
thermocouple termopar
thermocroic termocroico
thermodiffusion termodifusión
thermodynamic termodinámico
thermodynamic data datos
    termodinámicos
thermodynamic potential potencial
    termodinámico
thermodynamic stability estabilidad
    termodinámica
thermodynamics termodinámica
thermodynamics law ley de
    termodinámica
thermoelectric termoeléctrico
thermoelectricity termoelectricidad
thermofission termofisión
Thermofor process proceso de
    Thermofor
thermofusion termofusión
thermogenic termogénico
thermograph termógrafo
thermogravimetric analysis análisis
    termogravimétrico
thermokinetic analysis análisis
    termocinético
thermoluminescence termoluminiscencia
thermoluminescent termoluminiscente
thermolysis termólisis
thermolytic termolítico
thermomagnetic effect efecto
    termomagnético
thermometamorphism
    termometamorfismo
thermometer termómetro

thermometer scale escala termométrica
thermometric termométrico
thermometric analysis análisis termométrico
thermometric titration titulación termométrica, valoración termométrica
thermonatrite termonatrita
thermoneutrality termoneutralidad
thermonuclear termonuclear
thermonuclear energy energía termonuclear
thermonuclear fusion fusión termonuclear
thermonuclear neutron neutron termonuclear
thermonuclear reaction reacción termonuclear
thermonuclear transformation transformación termonuclear
thermoplastic termoplástico
thermoplastic resin resina termoplástica
thermoregulator termorregulador
thermostable termoestable
thermostat termostato
thermostatic termostático
thermotropic termotrópico
theta theta
thetine tetina
thexyl texilo
thia tia
thiabendazole tiabendazol
thial tial
thiamine tiamina
thiamine hydrochloride clorhidrato de tiamina
thiamine pyrophosphate pirofosfato de tiamina
thianthrene tiantreno
thiazine tiazina
thiazole tiazol
thiazole dye colorante de tiazol
thiazolyl tiazolilo
thiazyl tiacilo
thick film película gruesa
thickener espesador
thickening espesamiento
thickening agent agente de espesamiento
Thiele apparatus aparato de Thiele

Thiele reaction reacción de Thiele
thienyl tienilo
thimerosal timerosal
thin film película fina
thin-film chromatography cromatografía de película fina
thin layer capa fina
thin-layer chromatography cromatografía de capa fina
thinner diluente
thio tio
thioacetal tioacetal
thioacetaldehyde tioacetaldehído
thioacetamide tioacetamida
thioacetanilide tioacetanilida
thioacetic acid ácido tioacético
thioacid tioácido
thioalcohol tioalcohol
thioaldehyde tioaldehído
thioanhydride tioanhídrido
thioarsenate tioarseniato
thioate tioato
thiobarbituric acid ácido tiobarbitúrico
thiobenzaldehyde tiobenzaldehído
thiobenzamide tiobenzamida
thiobenzoic acid ácido tiobenzoico
thiobenzophenone tiobenzofenona
thiocarbamoyl tiocarbamoilo
thiocarbanilide tiocarbanilida
thiocarbonate tiocarbonato
thiocarbonic acid ácido tiocarbónico
thiocarbonyl tiocarbonilo
thiocarboxy tiocarboxi
thiocarboxylic acid ácido tiocarboxílico
thiochrome tiocromo
thiocyanate tiocianato
thiocyanato tiocianato
thiocyanic acid ácido tiociánico
thiocyanide tiocianuro
thiocyanogen tiocianógeno
thiodiglycol tiodiglicol
thiodiglycolic acid ácido tiodiglicólico
thiodiphenol tiodifenol
thiodipropionic acid ácido tiodipropiónico
thiodipropionitrile tiodipropionitrilo
thioester tioéster
thioether tioéter
thioethyl alcohol alcohol tioetílico

thioflavine tioflavina
thiofuran tiofurano
thioglycerol tioglicerol
thioglycolic acid ácido tioglicólico
thiohydantoin tiohidantoína
thiohydroxy tiohidroxi
thioic acid ácido tioico
thioketone tiocetona
thiol tiol
thiolactic acid ácido tioláctico
thiolate tiolato
thiomalic acid ácido tiomálico
thiomolybdate tiomolibdato
thionate tionato
thionic acid ácido tiónico
thionyl tionilo
thionyl bromide bromuro de tionilo
thionyl chloride cloruro de tionilo
thiopental sodium tiopental sódico
thiophane tiofano
thiophene tiofeno
thiophenealdehyde tiofenaldehído
thiophenol tiofenol
thiophenyl tiofenilo
thiophosgene tiofosgeno
thiophosphoric acid ácido tiofosfórico
thiophosphoryl tiofosforilo
thioridazine tioridazina
thiosalicylic acid ácido tiosalicílico
thiosemicarbazide tiosemicarbazida
thiosorbitol tiosorbitol
thiosulfate tiosulfato
thiosulfonic acid ácido tiosulfónico
thiosulfuric acid ácido tiosulfúrico
thiotungstate tiotungstato
thiouracil tiouracil
thiourea tiourea
thioxanthene tioxanteno
thioxo tioxo
thioxylenol tioxilenol
thiram tiram
third law of thermodynamics tercera ley de termodinámica
third-order reaction reacción de tercer orden
thixotropy tixotropía
Thomas-Fermi model modelo de Thomas-Fermi
thomsonite thomsonita

thoria toria
thorianite torianita
thoriated toriado
thorin torina
thorite torita
thorium torio
thorium anhydride anhídrido de torio
thorium carbide carburo de torio
thorium chloride cloruro de torio
thorium dioxide dióxido de torio
thorium disulfide disulfuro de torio
thorium fluoride fluoruro de torio
thorium hydroxide hidróxido de torio
thorium nitrate nitrato de torio
thorium oxalate oxalato de torio
thorium oxide óxido de torio
thorium picrate picrato de torio
thorium reactor reactor de torio
thorium series serie de torio
thorium sulfate sulfato de torio
thorium tetrachloride tetracloruro de torio
thoron torón
Thorpe reaction reacción de Thorpe
thortveitite thortveitita
three-necked flask matraz de tres cuellos
threnardite trenardita
threo treo
threonine treonina
threose treosa
threshold umbral
threshold detector detector de umbral
threshold energy energía umbral
threshold limit value valor umbral límite
threshold of detection umbral de detección
threshold value valor umbral
thrombin trombina
thuja oil aceite de tuya
thujene tuyeno
thujone tuyona
thulia tulia
thulium tulio
thulium chloride cloruro de tulio
thulium oxalate oxalato de tulio
thulium oxide óxido de tulio
thymic acid ácido tímico
thymidine timidina
thymidylic acid ácido timidílico

thymine timina

thymol timol

thymol blue azul de timol

thymol iodide yoduro de timol

thymolphthalein timolftaleína

thymolsulfonephthalein timolsulfonaftaleína

thymyl timilo

thyronine tironina

thyrothricin tirotricina

thyrotropic hormone hormona tirotrópica

thyroxine tiroxina

Tiemann rearrangement reordenamiento de Tiemann

tiglic acid ácido tíglico

tiglium oil aceite de tiglio

timolol timolol

tin estaño, hojalata

tin anhydride anhídrido de estaño

tin ash ceniza de estaño

tin bisulfide bisulfuro de estaño

tin bromide bromuro de estaño

tin bronze bronce de estaño

tin chloride cloruro de estaño

tin chromate cromato de estaño

tin dichloride dicloruro de estaño

tin difluoride difluoruro de estaño

tin dioxide dióxido de estaño

tin fluoride fluoruro de estaño

tin hexafluorosilicate hexafluorosilicato de estaño

tin hydride hidruro de estaño

tin hydroxide hidróxido de estaño

tin iodide yoduro de estaño

tin monosulfide monosulfuro de estaño

tin oxide óxido de estaño

tin perchloride percloruro de estaño

tin peroxide peróxido de estaño

tin-plate estañar

tin-plated estañado

tin-plating estañado

tin protochloride protocloruro de estaño

tin protosulfide protosulfuro de estaño

tin protoxide protóxido de estaño

tin resinate resinato de estaño

tin salt sal de estaño

tin sulfate sulfato de estaño

tin sulfide sulfuro de estaño

tin tetrabromide tetrabromuro de estaño

tin tetrachloride tetracloruro de estaño

tin tetraiodide tetrayoduro de estaño

tincal tincal

tincture tintura

tinned estañado

tinning estañado

tinplate hojalata

tinstone casiterita

tintometer tintómetro

Tishchenko reaction reacción de Tishchenko

titanate titanato

titanellow amarillo de titanio

titania titania

titanic titánico

titanic acid ácido titánico

titanic chloride cloruro titánico

titanic hydroxide hidróxido titánico

titanic sulfate sulfato titánico

titanium titanio

titanium acetylacetonate acetilacetonato de titanio

titanium alloy aleación de titanio

titanium ammonium oxalate oxalato amónico de titanio

titanium boride boruro de titanio

titanium bromide bromuro de titanio

titanium carbide carburo de titanio

titanium chelate quelato de titanio

titanium chloride cloruro de titanio

titanium diboride diboruro de titanio

titanium dichloride dicloruro de titanio

titanium dioxide dióxido de titanio

titanium disilicide disiliciuro de titanio

titanium disulfide disulfuro de titanio

titanium ester éster de titanio

titanium ferrocene ferroceno de titanio

titanium fluoride fluoruro de titanio

titanium hydride hidruro de titanio

titanium hydroxide hidróxido de titanio

titanium iodide yoduro de titanio

titanium isopropylate isopropilato de titanio

titanium monoxide monóxido de titanio

titanium nitrate nitrato de titanio

titanium nitride nitruro de titanio

titanium ore mineral de titanio

titanium oxalate oxalato de titanio

titanium oxide óxido de titanio
titanium peroxide peróxido de titanio
titanium potassium fluoride fluoruro
  potásico de titanio
titanium sesquisulfate sesquisulfato de
  titanio
titanium sponge esponja de titanio
titanium sulfate sulfato de titanio
titanium tetrachloride tetracloruro de
  titanio
titanium trichloride tricloruro de titanio
titanium trioxide trióxido de titanio
titanium white blanco de titanio
titanocene dichloride dicloruro de
  titanoceno
titanous titanoso
titanous chloride cloruro titanoso
titanous oxalate oxalato titanoso
titanous sulfate sulfato titanoso
titanyl titanilo
titanyl sulfate sulfato de titanilo
titer título
titrant solución de titulación, solución de
  valoración
titration titulación, valoración,
  volumetría
tobacco tabaco
Tobias acid ácido de Tobias
tocopherol tocoferol
tocophersolan tocofersolán
tolan tolano
tolidine tolidina
Tollens reagent reactivo de Tollens
Tollens test prueba de Tollens
tolnaftate tolnaftato
tolualdehyde tolualdehído
toluene tolueno
toluene trichloride tricloruro de tolueno
toluene trifluoride trifluoruro de tolueno
toluenediamine toluendiamina
toluenediisocyanate toluendiisocianato
toluenesulfamine toluensulfamina
toluenesulfanilide toluensulfanilida
toluenesulfonamide toluensulfonamida
toluenesulfonic acid ácido
  toluensulfónico
toluenesulfonyl toluensulfonilo
toluenesulfonyl chloride cloruro de
  toluensulfonilo

toluenethiol toluentiol
toluenyl toluenilo
toluic acid ácido toluico
toluic aldehyde aldehído toluico
toluidine toluidina
toluidine red rojo de toluidina
toluidino toluidino
toluoyl toluoilo
toluquinone toluquinona
toluyl toluilo
toluylene toluileno
toluylenediamine toluilendiamina
toluylic acid ácido toluílico
tolyl tolilo
tolyl chloride cloruro de tolilo
tolyl isobutyrate isobutirato de tolilo
tolylacetic acid ácido tolilacético
tolylaldehyde tolilaldehído
tolyldiethanolamine tolildietanolamina
tolylene tolileno
tolylenediamine tolilendiamina
tolylidene tolilideno
tomatine tomatina
tongs tenazas
tonka tonca
topaz topacio
topochemical reaction reacción
  topoquímica
topochemistry topoquímica
topotactic topotáctico
toroid toriode
torr torr
torsion balance balanza de torsión
torula yeast levadura de tórula
tosyl tosilo
tosylation tosilación
total heat calor total
total reflux reflujo total
Toth process proceso de Toth
touchstone piedra de toque
tourmaline turmalina
tower torre
toxaphene toxafeno
toxic tóxico
toxic properties propiedades tóxicas
toxic substance substancia tóxica
toxication intoxicación
toxicity toxicidad
toxicology toxicología

toxin toxina
trace traza
trace analysis análisis de trazas
trace element microelemento,
    oligoelemento
tracer trazador, indicador radiactivo
tracer atom átomo trazador
tracer compound compuesto trazador
tracer element elemento trazador
tragacanth gum goma de tragacanto
tranexamic acid ácido tranexámico
trans trans
trans effect efecto trans
trans-isomer isómero trans
transactinide transactínido
transactinide element elemento
    transactínido
transalkylation transalquilación
transaminase transaminasa
transamination transaminación
transducer transductor
transesterification transesterificación
transfer coefficient coeficiente de
    transferencia
transfer ribonucleic acid ácido
    ribonucleico de transferencia
transferase transferasa
transference number número de
    transferencia
transformation transformación
transformation series serie de
    transformaciones
transformer oil aceite de transformador
transition transición
transition effect efecto de transición
transition element elemento de
    transición
transition interval intervalo de
    transición
transition point punto de transición
transition state estado de transición
transition temperature temperatura de
    transición
transition time tiempo de transición
transitional transicional
transmission diffraction difracción por
    transmisión
transmittance transmitancia
transmutation transmutación

transoid transoide
transparency transparencia
transport transporte
transport coefficient coeficiente de
    transporte
transport number número de transporte
transposition transposición
transuranic transuránico
transuranic element elemento
    transuránico
Traube purine synthesis síntesis de
    purinas de Traube
traumatic acid ácido traumático
travertine travertino
tremolite tremolita
tretamine tretamina
tri tri
triacetate triacetato
triacetin triacetina
triacetoneamine triacetonamina
triacontane triacontano
triacontanoic acid ácido triacontanoico
triacontanol triacontanol
triacontyl triacontilo
triad tríada
trialkylsilanol trialquilsilanol
triallyl trialilo
triallyl cyanurate cianurato de trialilo
triallyl phosphate fosfato de trialilo
triallylamine trialilamina
triamcinolone triamcinolona
triamine triamina
triaminobenzene triaminobenceno
triaminotoluene triaminotolueno
triamylamine triamilamina
triamylbenzene triamilbenceno
triangular diagram diagrama triangular
triangulo triangulo
triarylmethane dye colorante de
    triarilmetano
triatomic triatómico
triatomic molecule molécula triatómica
triazane triazano
triazano triazano
triazene triaceno
triazeno triaceno
triazine triazina
triazinyl triazinilo
triazole triazol

triazone resin resina de triazona
tribasic tribásico
tribasic acid ácido tribásico
tribasic calcium phosphate fosfato
   cálcico tribásico
tribenzyl tribencilo
tribenzylamine tribencilamina
triboluminescence triboluminiscencia
tribromide tribromuro
tribromo tribromo
tribromoacetaldehyde
   tribromoacetaldehído
tribromoacetic acid ácido
   tribromoacético
tribromoaniline tribromoanilina
tribromobenzene tribromobenceno
tribromoethanal tribromoetanal
tribromoethane tribromoetano
tribromoethanol tribromoetanol
tribromomethane tribromometano
tribromophenol tribromofenol
tribromopropane tribromopropano
tribromosalicylanilide
   tribromosalicilanilida
tributyl tributilo
tributyl borate borato de tributilo
tributyl citrate citrato de tributilo
tributyl phosphate fosfato de tributilo
tributyl phosphine fosfina de tributilo
tributyl phosphite fosfito de tributilo
tributyl phosphorothiolate fosforotiolato
   de tributilo
tributylaluminum tributilaluminio
tributylamine tributilamina
tributylborane tributilborano
tributyltin tributilestaño
tributyltin chloride cloruro de
   tributilestaño
tributyltin oxide óxido de tributilestaño
tricalcium citrate citrato tricálcico
tricalcium orthophosphate ortofosfato
   tricálcico
tricalcium phosphate fosfato tricálcico
tricalcium silicate silicato tricálcico
tricarbimide tricarbimida
tricarbon dioxide dióxido de tricarbono
tricarboxylic acid cycle ciclo de ácido
   tricarboxílico
trichlorfon triclorfón

trichloride tricloruro
trichloro tricloro
trichloroacetaldehyde
   tricloroacetaldehído
trichloroacetamide tricloroacetamida
trichloroacetic acid ácido tricloroacético
trichloroanisole tricloroanisol
trichlorobenzene triclorobenceno
trichloroborazole tricloroborazol
trichlorobromomethane
   triclorobromometano
trichlorobutylene oxide óxido de
   triclorobutileno
trichloroethanal tricloroetanal
trichloroethane tricloroetano
trichloroethanoic acid ácido
   tricloroetanoico
trichloroethanol tricloroetanol
trichloroethene tricloroeteno
trichloroethylene tricloroetileno
trichlorofluoromethane
   triclorofluorometano
trichloroisocyanuric acid ácido
   tricloroisocianúrico
trichloromethane triclorometano
trichloromethyl triclorometilo
trichloromethyl chloroformate
   cloroformiato de triclorometilo
trichloromethylphosphonic acid ácido
   triclorometilfosfónico
trichloromethylsulfenyl chloride cloruro
   de triclorometilsulfenilo
trichloronaphthalene tricloronaftaleno
trichloronitrosomethane
   tricloronitrosometano
trichlorophenol triclorofenol
trichloropropane tricloropropano
trichlorosilane triclorosilano
trichlorotoluene triclorotolueno
trichlorotrifluoroacetone
   triclorotrifluoroacetona
trichlorotrifluoroethane
   triclorotrifluoroetano
triclinic system sistema triclínico
tricosane tricosano
tricosanoic acid ácido tricosanoico
tricosanol tricosanol
tricresyl tricresilo
tricresyl phosphate fosfato de tricresilo

tricresyl phosphite fosfito de tricresilo
tricyanic acid ácido triciánico
tricyano triciano
tricyclic tricíclico
tricyclodecane triciclodecano
tridecane tridecano
tridecanoic acid ácido tridecanoico
tridecanol tridecanol
tridecene tridecano
tridecyl tridecilo
tridecyl alcohol alcohol tridecílico
tridecylbenzene tridecilbenceno
tridecylic acid ácido tridecílico
tridodecyl amine tridodecilamina
tridymite tridimita
triethanolamine trietanolamina
triethanolamine lauryl sulfate
    laurilsulfato de trietanolamina
triethoxy trietoxi
triethoxyhexane trietoxihexano
triethoxymethane trietoximetano
triethoxymethoxypropane
    trietoximetoxipropano
triethyl trietilo
triethyl borate borato de trietilo
triethyl citrate citrato de trietilo
triethyl phosphate fosfato de trietilo
triethyl phosphite fosfito de trietilo
triethyl phosphorothioate fosforotioato
    de trietilo
triethylaluminum trietilaluminio
triethylamine trietilamina
triethylbenzene trietilbenceno
triethylborane trietilborano
triethylene trietileno
triethylene glycol trietilenglicol
triethylene glycol dichloride dicloruro
    de trietilenglicol
triethylene glycol didecanoate
    didecanoato de trietilenglicol
triethylene glycol dihydroabietate
    dihidoabietato de trietilenglicol
triethylene glycol dimethyl ether éter
    dimetílico de trietilenglicol
triethylene glycol dioctoate dioctoato de
    trietilenglicol
triethylene glycol dipelargonate
    dipelargonato de trietilenglicol
triethylene glycol dipropionate

    dipropionato de trietilenglicol
triethylenediamine trietilendiamina
triethylenemelamine trietilenmelamina
triethylenephosphoramide
    trietilenfosforamida
triethylenetetramine trietilentetramina
triethylenetriamine trietilentriamina
triethylmethane trietilmetano
triethylorthoformate trietilortoformiato
triflate triflato
trifluoro trifluoro
trifluoroacetic acid ácido
    trifluoroacético
trifluoroamine oxide óxido de
    trifluoroamina
trifluorobromomethane
    trifluorobromometano
trifluorochloromethane
    trifluoroclorometano
trifluoroethanoic acid ácido
    trifluoroetanoico
trifluoroiodomethane
    trifluoroyodometano
trifluoromethyl trifluorometilo
trifluoronitrosomethane
    trifluoronitrosometano
trifluorostyrene trifluoroestireno
trifluorotrichloroethane
    trifluorotricloroetano
trigalloyl trigaloilo
triglyceride triglicérido
triglycerol triglicerol
triglycine triglicina
triglycol dichloride dicloruro de triglicol
triglyme triglime
trigol trigol
trigonal trigonal
trigonal bipyramid bipirámide trigonal
trigonal bipyramidal coordination
    coordinación bipiramidal trigonal
trigonal system sistema trigonal
trigonelline trigonelina
trihexyl trihexilo
trihexylene glycol biborate biborato de
    trihexilenglicol
trihydrate trihidrato
trihydric trihídrico
trihydric alcohol alcohol trihídrico
trihydroxy trihidroxi

trihydroxyanthraquinone
trihidroxiantraquinona
trihydroxybenzene trihidroxibenceno
trihydroxybenzoic acid ácido
trihidroxibenzoico
trihydroxyethylamine stearate estearato
de trihidroxietilamina
trihydroxypropane trihidroxipropano
triiodo triyodo
triiodomethane triyodometano
triiodothyronine triyodotironina
triisobutylaluminum triisobutilaluminio
triisobutylene triisobutileno
triisooctyl phosphite fosfito de
triisooctilo
triisopropanolamine triisopropanolamina
triisopropyl phosphite fosfito de
triisopropilo
triketone tricetona
trilauryl trilaurilo
trilauryl phosphite trilaurilfosfito
trilauryl trithiophosphite
trilauriltritiofosfito
trilaurylamine trilaurilamina
trimagnesium phosphate fosfato de
trimagnesio
trimellitic acid ácido trimelítico
trimellitic anhydride anhídrido
trimelítico
trimer trímero
trimercuric orthophosphate ortofosfato
trimercúrico
trimercurous orthophosphate
ortofosfato trimercurioso
trimesoyl trichloride tricloruro de
trimesoilo
trimetallic trimetálico
trimethadione trimetadiona
trimethano trimetano
trimethoxy trimetoxi
trimethoxyboroxine trimetoxiboroxina
trimethoxymethane trimetoximetano
trimethyl trimetilo
trimethyl borate borato de trimetilo
trimethyl phosphate fosfato de trimetilo
trimethyl phosphite fosfito de trimetilo
trimethylacetic acid ácido trimetilacético
trimethyladipic acid ácido
trimetiladípico

trimethylaluminum trimetilaluminio
trimethylamine trimetilamina
trimethylamine oxide óxido de
trimetilamina
trimethylaniline trimetilanilina
trimethylbenzene trimetilbenceno
trimethylbenzoic acid ácido
trimetilbenzoico
trimethylbutane trimetilbutano
trimethylchlorosilane trimetilclorosilano
trimethylcyclododecatriene
trimetilciclododecatrieno
trimethylcyclohexane
trimetilciclohexano
trimethylcyclohexanol
trimetilciclohexanol
trimethylene trimetileno
trimethylene bromide bromuro de
trimetileno
trimethylene chlorohydrin clorohidrina
de trimetileno
trimethylene glycol trimetilenglicol
trimethylene oxide óxido de trimetileno
trimethylhexane trimetilhexano
trimethylhexanol trimetilhexanol
trimethylmethane trimetilmetano
trimethylnonanone trimetilnonanona
trimethylolethane trimetiloletano
trimethylolpropane trimetilolpropano
trimethylpentane trimetilpentano
trimethylpentanediol trimetilpentanodiol
trimethylpentene trimetilpenteno
trimethylpyridine trimetilpiridina
trimethylsilyl trimetilsililo
trimethyltin trimetilestaño
trimolecular trimolecular
trimorphism trimorfismo
trinitrate trinitrato
trinitro trinitro
trinitroaniline trinitroanilina
trinitroanisole trinitroanisol
trinitrobenzene trinitrobenceno
trinitrobenzoic acid ácido
trinitrobenzoico
trinitroglycerin trinitroglicerina
trinitromethane trinitrometano
trinitronaphthalene trinitronaftaleno
trinitrophenol trinitrofenol
trinitroresorcinol trinitrorresorcina

trinitrotoluene trinitrotolueno
trinor trinor
trioctadecyl phosphite fosfito de trioctadecilo
trioctyl phosphate fosfato de trioctilo
trioctylphosphinic oxide óxido trioctilfosfínico
triol triol
trione triona
triose triosa
trioxa trioxa
trioxane trioxano
trioxide trióxido
trioxime trioxima
trioxin trioxina
trioxoboric acid ácido trioxobórico
trioxosulfuric acid ácido trioxosulfúrico
trioxygen trioxígeno
trioxymethylene trioximetileno
tripalmitin tripalmitina
triphasic trifásico
triphenol trifenol
triphenyl trifenilo
triphenyl phosphate fosfato de trifenilo
triphenyl phosphite fosfito de trifenilo
triphenylacetic acid ácido trifenilacético
triphenylantimony trifenilantimonio
triphenylbenzene trifenilbenceno
triphenylboron trifenilboro
triphenylcarbinol trifenilcarbinol
triphenylene trifenileno
triphenylguanidine trifenilguanidina
triphenylmethane dye colorante de trifenilmetano
triphenylmethyl trifenilmetilo
triphenylphosphine trifenilfosfina
triphenylphosphorus trifenilfósforo
triphenyltin trifenilestaño
triphenyltin chloride cloruro de trifenilestaño
triphenyltin hydroxide hidróxido de trifenilestaño
triphosgene trifosgeno
triphosphoric acid ácido trifosfórico
triple bond enlace triple
triple point punto triple
triple superphosphate superfosfato triple
triplet triplete
triplite triplita

tripod trípode
tripoli trípoli
tripolite tripolita
tripotassium orthophosphate ortofosfato tripotásico
tripotassium phosphate fosfato tripotásico
tripropyl tripropilo
tripropylaluminum tripropilaluminio
tripropylamine tripropilamina
tripropylene tripropileno
tripropylene glycol tripropilenglicol
triptane triptano
triptycene tripticeno
tris tris
trischloroethyl phosphite fosfito de triscloroetilo
trisethylhexyl phosphate fosfato de trisetilhexilo
trisethylhexyl phosphite fosfito de trisetilhexilo
trishydroxyphenylpropane trishidroxifenilpropano
trisilane trisilano
trisodium citrate citrato trisódico
trisodium ethylenediaminetetraacetic acid ácido etilendiaminotetraacético trisódico
trisodium orthophosphate ortofosfato trisódico
trisodium phosphate fosfato trisódico
trisulfide trisulfuro
triterpene triterpeno
trithioacetaldehyde tritioacetaldehído
trithiocarbonate tritiocarbonato
trithionic acid ácido tritiónico
tritio tritio
tritium tritio
tritolyl phosphate fosfato de tritolilo
triton tritón
triturate triturar
trityl tritilo
triuranium octoxide octóxido de triuranio
trivalent trivalente
trivial name nombre trivial
trona trona
Trona process proceso de Trona
tropanol tropanol

tropic acid ácido trópico
tropilidene tropilideno
tropine tropina
tropolone tropolona
tropoyl tropoilo
tropylium tropilio
truth serum suero de la verdad
trypsin tripsina
trypsinogen tripsinógeno
tryptophan triptófano
tryptophyl triptofila
Tscherniac-Einhorn reaction reacción
    de Tscherniac-Einhorn
tung oil aceite de tung
tungstate tungstato
tungsten tungsteno
tungsten alloy aleación de tungsteno
tungsten boride boruro de tungsteno
tungsten bronze bronce de tungsteno
tungsten carbide carburo de tungsteno
tungsten carbonyl carbonilo de
    tungsteno
tungsten chloride cloruro de tungsteno
tungsten dichloride dicloruro de
    tungsteno
tungsten diselenide diseleniuro de
    tungsteno
tungsten disulfide disulfuro de tungsteno
tungsten fluoride fluoruro de tungsteno
tungsten hexacarbonyl hexacarbonilo de
    tungsteno
tungsten hexachloride hexacloruro de
    tungsteno
tungsten lake laca de tungsteno
tungsten oxide óxido de tungsteno
tungsten oxychloride oxicloruro de
    tungsteno
tungsten pentachloride pentacloruro de
    tungsteno
tungsten pentoxide pentóxido de

    tungsteno
tungsten silicide siliciuro de tungsteno
tungsten steel acero al tungsteno
tungsten trioxide trióxido de tungsteno
tungstic túngstico
tungstic acid ácido túngstico
tungstic anhydride anhídrido túngstico
tungstic oxide óxido túngstico
tungstite tungstita
tungstophosphate tungstofosfato
tungstosilicate tungstosilicato
tunnel effect efecto túnel
turbidimetric analysis análisis
    turbidimétrico
turbidimetry turbidimetría
turbidity turbidez
turbidity point punto de turbidez
Turkey red rojo turco
Turnbull blue azul de Turnbull
turpentine trementina
turpentine camphor alcanfor de
    trementina
turpentine oil aceite de trementina
turquoise turquesa
twinning maclado
Twitchell process proceso de Twitchell
Twitchell reagent reactivo de Twitchell
two-dimensional chromatography
    cromatografía bidimensional
Tyndall effect efecto de Tyndall
type tipo
typical típico
tyramine tiramina
Tyrian purple púrpura de Tiro
tyrocidine tirocidina
tyrosinase tirosinasa
tyrosine tirosina
tyrosyl tirosilo
tysonite tisonita

U

U-tube tubo en U
ubiquinone ubiquinona
ulexine ulexina
ulexite ulexita
Ullman reaction reacción de Ullman
ulmin ulmina
ultimate analysis análisis elemental
ultra ultra
ultracentrifuge ultracentrífuga
ultrafiltration ultrafiltración
ultramarine blue azul ultramarino
ultramarine green verde ultramarino
ultramicroscope ultramicroscopio
ultramicroscopic ultramicroscópico
ultrapure ultrapuro
ultrarapid ultrarrápido
ultrasensitive ultrasensible
ultrasonic ultrasónico
ultrasonics ultrasónica
ultraviolet ultravioleta
ultraviolet absorber absorbedor
    ultravioleta
ultraviolet absorption absorción
    ultravioleta
ultraviolet-absorption
    spectrophotometry
    espectrofotometría de absorción
    ultravioleta
ultraviolet densitometry densitometría
    ultravioleta
ultraviolet light luz ultravioleta
ultraviolet radiation radiación
    ultravioleta
ultraviolet spectrometer espectrómetro
    ultravioleta
ultraviolet spectrophotometry
    espectrofotometría ultravioleta
ultraviolet spectroscopy espectroscopia
    ultravioleta
ultraviolet spectrum espectro
    ultravioleta
ultraviolet stabilizer estabilizador
    ultravioleta

ultraviolet-visible spectroscopy
    espectroscopia ultravioleta-visible
umangite umangita
umbellic acid ácido umbélico
umbelliferone umbeliferona
umbellulone umbelulona
umber tierra de sombra
unary unario
uncertainty principle principio de
    incertidumbre
uncoupling phenomena fenómenos de
    desacoplamiento
undeca undeca
undecalactone undecalactona
undecanal undecanal
undecane undecano
undecanoic acid ácido undecanoico
undecanol undecanol
undecanone undecanona
undecene undeceno
undecenoic acid ácido undecenoico
undecenol undecenol
undecenyl undecenilo
undecyl undecilo
undecyl alcohol alcohol undecílico
undecylamine undecilamina
undecylene undecileno
undecylenic acid ácido undecilénico
undecylenic alcohol alcohol undecilénico
undecylenyl acetate acetato de
    undecilenilo
undecylic acid ácido undecílico
undercooling subenfriado
uni uni
uniaxial uniaxial
unified atomic mass constant constante
    de masa atómica unificada
unimolecular unimolecular
unimolecular film película unimolecular
unimolecular reaction reacción
    unimolecular
union unión
unit unidad

univalent univalente
universal indicator indicador universal
unoccupied molecular orbital orbital
    molecular desocupado
unpaired electrons electrones no
    emparejados
unpolarized light luz no polarizada
unsaturated insaturado, no saturado
unsaturated compound compuesto
    insaturado
unsaturated hydrocarbon hidrocarburo
    insaturado
unsaturated solution solución no
    saturada
unsaturation insaturación, no saturación
unstable inestable
unstable element elemento inestable
unstable isotope isótopo inestable
unstable nucleus núcleo inestable
unsym asim
unsymmetrical asimétrico
ur ur
uracil uracil
uralite uralita
uramido uramido
uramine uramina
uramino uramino
uranate uranato
urania urania
uranic uránico
uranic acid ácido uránico
uranic chloride cloruro uránico
uranic oxide óxido uránico
uranin uranina
uranine uranina
uranine yellow amarillo de uranina
uraninite uraninita
uranite uranita
uranium uranio
uranium acetate acetato de uranio
uranium bromide bromuro de uranio
uranium carbide carburo de uranio
uranium chloride cloruro de uranio
uranium content contenido de uranio
uranium decay series serie de
    desintegración de uranio
uranium dicarbide dicarburo de uranio
uranium dioxide dióxido de uranio
uranium enrichment enriquecimiento de

uranio
uranium fission fisión de uranio
uranium fluoride fluoruro de uranio
uranium glass vidrio de uranio
uranium hexafluoride hexafluoruro de
    uranio
uranium hydride hidruro de uranio
uranium hydroxide hidróxido de uranio
uranium iodide yoduro de uranio
uranium monocarbide monocarburo de
    uranio
uranium nitrate nitrato de uranio
uranium ore mineral de uranio
uranium oxide óxido de uranio
uranium oxychloride oxicloruro de
    uranio
uranium-radium series serie de uranio-
    radio
uranium reactor reactor de uranio
uranium series serie de uranio
uranium sulfate sulfato dc uranio
uranium tetrabromide tetrabromuro de
    uranio
uranium tetrachloride tetracloruro de
    uranio
uranium tetrafluoride tetrafluoruro de
    uranio
uranium trioxide trióxido de uranio
uranocene uranoceno
uranocircite uranocircita
uranophane uranofano
uranospherite uranosferita
uranothallite uranotalita
uranous uranoso
uranous chloride cloruro uranoso
uranous oxide óxido uranoso
uranyl uranilo
uranyl acetate acetato de uranilo
uranyl benzoate benzoato de uranilo
uranyl chloride cloruro de uranilo
uranyl fluoride fluoruro de uranilo
uranyl formate formiato de uranilo
uranyl hexacyanoferrate
    hexacianoferrato de uranilo
uranyl hydroxide hidróxido de uranilo
uranyl nitrate nitrato de uranilo
uranyl oxide óxido de uranilo
uranyl sulfate sulfato de uranilo
uranyl sulfide sulfuro de uranilo

urea urea
urea acetate acetato de urea
urea adduct aducto de urea
urea-ammonium orthophosphate
    ortofosfato de urea-amonio
urea-ammonium phosphate fosfato de
    urea-amonio
urea anhydride anhídrido de urea
urea cycle ciclo de urea
urea-formaldehyde resin resina de urea-
    formaldehído
urea nitrate nitrato de urea
urea oxalate oxalato de urea
urea peroxide peróxido de urea
urease ureasa
Urech synthesis síntesis de Urech
ureide ureido
ureido ureido

urethane uretano
urethane resin resina de uretano
urethano uretano
uric acid ácido úrico
uridine uridina
uridine monophosphate monofosfato de
    uridina
uridine phosphate fosfato de uridina
uridylic acid ácido uridílico
urobenzoic acid ácido urobenzoico
uronic acid ácido urónico
uronium uronio
urotropin urotropina
ursin ursina
urylon urilón
usnic acid ácido úsnico
uwarowite uvarovita

V

vacancy   vacante
vacant   vacante
vaccenic acid   ácido vaccénico
vacuum   vacío
vacuum condensing point   punto de
    condensación al vacío
vacuum crystallizer   cristalizador al
    vacío
vacuum deposition   deposición al vacío
vacuum distillation   destilación al vacío
vacuum filtration   filtración al vacío
vacuum forming   formación al vacío
vacuum metalizing   metalización al vacío
vacuum pump   bomba de vacío
valacidin   valacidina
valence   valencia
valence angle   ángulo de valencia
valence band   banda de valencia
valence bond   enlace de valencia
valence electrons   electrones de valencia
valence isomerization   isomerización de
    valencia
valence number   número de valencia
valence shell   capa de valencia
valence tautomerism   tautomerismo de
    valencia
valence theory   teoría de valencia
valency   valencia
valentinite   valentinita
valeral   valeral
valeraldehyde   valeraldehído
valeramide   valeramida
valerate   valerato
valerian oil   aceite de valeriana
valerianic acid   ácido valeriánico
valeric acid   ácido valérico
valeric aldehyde   aldehído valérico
valeric anhydride   anhídrido valérico
valerolactone   valerolactona
valeryl   valerilo
valine   valina
valyl   valilo
van der Waals adsorption   adsorción de

van der Waals
van der Waals attraction   atracción de
    van der Waals
van der Waals constant   constante de van
    der Waals
van der Waals covolume   covolumen de
    van der Waals
van der Waals equation   ecuación de van
    der Waals
van der Waals forces   fuerzas de van der
    Waals
van der Waals molecule   molécula de
    van der Waals
van der Waals radius   radio de van der
    Waals
van der Waals structure   estructura de
    van der Waals
Van Slyke determination   determinación
    de Van Slyke
Van Slyke method   método de Van Slyke
van't Hoff equation   ecuación de van't
    Hoff
van't Hoff factor   factor de van't Hoff
van't Hoff formula   fórmula de van't
    Hoff
van't Hoff isochore   isócora de van't
    Hoff
van't Hoff isotherm   isoterma de van't
    Hoff
van't Hoff law   ley de van't Hoff
van't Hoff theory   teoría de van't Hoff
vanadate   vanadato
vanadic acid   ácido vanádico
vanadic acid anhydride   anhídrido de
    ácido vanádico
vanadic anhydride   anhídrido vanádico
vanadic sulfate   sulfato vanádico
vanadic sulfide   sulfuro vanádico
vanadinite   vanadinita
vanadite   vanadita
vanadium   vanadio
vanadium acetylacetonate
    acetilacetonato de vanadio

256

vanadium bromide bromuro de vanadio

vanadium carbide carburo de vanadio

vanadium chloride cloruro de vanadio

vanadium dichloride dicloruro de vanadio

vanadium difluoride difluoruro de vanadio

vanadium dioxide dióxido de vanadio

vanadium disulfide disulfuro de vanadio

vanadium ethylate etilato de vanadio

vanadium fluoride fluoruro de vanadio

vanadium hexacarbonyl hexacarbonilo de vanadio

vanadium hydroxide hidróxido de vanadio

vanadium monoxide monóxido de vanadio

vanadium nitride nitruro de vanadio

vanadium oxide óxido de vanadio

vanadium oxychloride oxicloruro de vanadio

vanadium oxydichloride oxidicloruro de vanadio

vanadium oxytrichloride oxitricloruro de vanadio

vanadium pentasulfide pentasulfuro de vanadio

vanadium pentoxide pentóxido de vanadio

vanadium sesquioxide sesquióxido de vanadio

vanadium suboxide subóxido de vanadio

vanadium sulfate sulfato de vanadio

vanadium sulfide sulfuro de vanadio

vanadium tetrachloride tetracloruro de vanadio

vanadium tetraoxide tetróxido de vanadio

vanadium trichloride tricloruro de vanadio

vanadium trioxide trióxido de vanadio

vanadium trisulfide trisulfuro de vanadio

vanadol vanadol

vanadous vanadioso

vanadous acid ácido vanadioso

vanadous chloride cloruro vanadioso

vanadous hydroxide hidróxido vanadioso

vanadyl vanadilo

vanadyl bromide bromuro de vanadilo

vanadyl chloride cloruro de vanadilo

vanadyl sulfate sulfato de vanadilo

vanadylic vanadílico

vanadylic chloride cloruro vanadílico

vanadylous vanadiloso

vancomycin hydrochloride clorhidrato de vancomicina

vanillic acid ácido vainíllico

vanillic aldehyde aldehído vainíllico

vanillin vainillina

vanilloyl vainilloilo

vanillyl vainillilo

vanillylidene vainillilideno

vapor vapor

vapor bath baño de vapor

vapor density densidad de vapor

vapor-liquid equilibrium equilibrio vapor-líquido

vapor-phase chromatography cromatografía en fase vapor

vapor pressure presión de vapor

vapor-pressure osmometer osmómetro de presión de vapor

vapor tension tensión de vapor

vaporization vaporización

vaporize vaporizar

variscite variscita

varnish barniz

Varrentrapp reaction reacción de Varrentrapp

Vaska compound compuesto de Vaska

vasopressin vasopresina

vat dye colorante de tina

Vauquelin salt sal de Vauquelin

vauquelinite vauquelinita

vector model of atomic structure modelo vectorial de estructura atómica

vegetable dye colorante vegetal

vegetable gum goma vegetal

vegetable oil aceite vegetal

vehicle vehículo

velocity velocidad

velocity constant constante de velocidad

velocity of reaction velocidad de reacción

Venetian red rojo veneciano

venturi venturi

veratraldehyde veratraldehído

veratric acid ácido verátrico
veratrole veratrol
veratroyl veratroilo
veratryl veratrilo
veratrylidene veratrilideno
verbena oil aceite de verbena
verbenone verbenona
verdigris cardenillo
vermiculite vermiculita
vermillion bermellón
vernolepin vernolepina
verxite verxita
vessel recipiente
vetivone vetivona
vibration vibración
vibrational energy energía vibracional
vibrational level nivel vibracional
vibrational spectrum espectro
    vibracional
vic vec
vicinal vecinal
Victoria blue azul Victoria
Victoria green verde Victoria
villiaumite villiaumita
Vilsmeier-Haack reaction reacción de
    Vilsmeier-Haack
Vilsmeier reagent reactivo de Vilsmeier
vinegar vinagre
vinic acid ácido vínico
vinyl vinilo
vinyl acetal resin resina de vinilacetal
vinyl acetate acetato de vinilo
vinyl acetate resin resina de acetato de
    vinilo
vinyl alcohol alcohol vinílico
vinyl bromide bromuro de vinilo
vinyl butyl ether vinilbutiléter
vinyl butyrate butirato de vinilo
vinyl chloride cloruro de vinilo
vinyl chloride resin resina de cloruro de
    vinilo
vinyl cyanide cianuro de vinilo
vinyl ether viniléter
vinyl ether resin resina de viniléter
vinyl ethyl ether viniletiléter
vinyl fluoride fluoruro de vinilo
vinyl iodide yoduro de vinilo
vinyl isobutyl ether vinilisobutiléter
vinyl ketone vinilcetona

vinyl methyl ether vinilmetiléter
vinyl methyl ketone vinilmetilcetona
vinyl oxide óxido de vinilo
vinyl polymer polímero de vinilo
vinyl polymerization polimerización de
    vinilo
vinyl propionate propionato de vinilo
vinyl resin resina de vinilo
vinyl stearate estearato de vinilo
vinyl sulfide sulfuro de vinilo
vinyl trichloride tricloruro de vinilo
vinylacetonitrile vinilacetonitrilo
vinylacetylene vinilacetileno
vinylamine vinilamina
vinylation vinilación
vinylbenzene vinilbenceno
vinylcarbazole vinilcarbazol
vinylcyclohexene vinilciclohexeno
vinylcyclohexene dioxide dióxido de
    vinilciclohexeno
vinylcyclohexene monoxide monóxido
    de vinilciclohexeno
vinylene vinileno
vinylethoxyethyl sulfide sulfuro de
    viniletoxietilo
vinylethylene viniletileno
vinylethylhexyl ether viniletilhexiléter
vinylethylpyridine viniletilpiridina
vinylidene vinilideno
vinylidene chloride cloruro de vinilideno
vinylidene chloride polymer polímero de
    cloruro de vinilideno
vinylidene fluoride fluoruro de
    vinilideno
vinylidene resin resina de vinilideno
vinylimine vinilimina
vinylmagnesium chloride cloruro de
    vinilmagnesio
vinylog vinílogo
vinylpyridine vinilpiridina
vinylpyrrolidone vinilpirrolidona
vinylstyrene vinilestireno
vinyltoluene viniltolueno
vinyltrichlorosilane viniltriclorosilano
vioform vioformo
violanthrone violantrona
violuric acid ácido violúrico
viomycin viomicina
virtual level nivel virtual

virtual particle partícula virtual
virtual quantum cuanto virtual
virtual state estado virtual
virus virus
viscid viscoso
viscometer viscosímetro
viscoplastic viscoplástico
viscose viscosa
viscosimeter viscosímetro
viscosimetry viscosimetría
viscosity viscosidad
viscosity index índice de viscosidad
viscous viscoso
visible absorption spectrophotometry
espectrofotometría de absorción
visible
visible spectrophotometry
espectrofotometría visible
visible spectrum espectro visible
visual colorimetry colometría visual
vitamin vitamina
vitamin complex complejo vitamínico
vitellin vitelina
vitreous vítreo
vitrification vitrificación
vitriol vitriolo
vivianite vivianita
Voight amination aminación de Voight
volatile volátil
volatile compound compuesto volátil
volatile fluid fluido volátil
volatile organic compound compuesto
orgánico volátil
volatile solvent solvente volátil

volatility volatilidad
volatility product producto de volatilidad
volatilization volatilización
volatilize volatilizar
Volhard-Erdmann cyclization
ciclización de Volhard-Erdmann
Volhard method método de Volhard
Volhard solution solución de Volhard
Volhard titration titulación de Volhard,
valoración de Volhard
Volta series serie de Volta
voltaic cell pila voltaica
voltaite voltaíta
voltzite voltzina
volume volumen
volumenometer volumenómetro
volumeter volúmetro
volumetric volumétrico
volumetric analysis análisis volumétrico
volumetric factor factor volumétrico
volumetric flask matraz volumétrico
volumetric pipet pipeta volumétrica
volumetric solution solución volumétrica
von Braun reaction reacción de von
Braun
von Richter reaction reacción de von
Richter
von Richter synthesis síntesis de von
Richter
vulcanite vulcanita
vulcanization vulcanización
vulcanize vulcanizar
vulpinite vulpinita

W

Wackenroder reaction reacción de
  Wackenroder
Wacker process proceso de Wacker
Wacker reaction reacción de Wacker
Wagner-Jauregg reaction reacción de
  Wagner-Jauregg
Wagner-Meerwein rearrangement
  reordenamiento de Wagner-Meerwein
Wagner reagent reactivo de Wagner
Wagner solution solución de Wagner
wagnerite wagnerita
Walden inversion inversión de Walden
Wallach rearrangement reordenamiento
  de Wallach
warfarin warfarina
warning odor olor de advertencia
wash bottle frasco lavador
washing lavado
washing bottle frasco lavador
washing soda sosa de lavar
waste desperdicios
waste control control de desperdicios
water agua
water analysis análisis de agua
water bath baño de agua, baño María
water constants constantes de agua
water detection detección de agua
water distillation destilación de agua
water gas gas de agua
water glass vidrio de agua
water of constitution agua de
  constitución
water of crystallization agua de
  cristalización
water of hydration agua de hidratación
water pollution contaminación de agua
water purification purificación de agua
water saturation saturación con agua
water softening ablandamiento de agua
water soluble soluble en agua,
  hidrosoluble
water-soluble gum goma soluble en agua
water-soluble oil aceite soluble en agua

water-soluble resin resina soluble en
  agua
water sterilization esterilización de agua
water still destilador de agua
waterproof a prueba de agua,
  impermeable
waterproofing agent agente
  impermeabilizante
waterproofing compound compuesto
  impermeabilizante
watertight hermético al agua
Watson equation ecuación de Watson
wave onda
wave function función de onda
wavellite wavellita
wax cera
wax tailings residuos de cera
weak débil
weak acid ácido débil
weak base base débil
weak electrolyte electrólito débil
weak salt sal débil
Weerman degradation degradación de
  Weerman
Weigert effect efecto de Weigert
weighing bottle recipiente de pesar
weight peso
weight buret bureta de pesar
Weissenberg effect efecto de
  Weissenberg
welding soldadura
Weldon process proceso de Weldon
Welter rule regla de Welter
Werner complex complejo de Werner
Werner theory teroría de Werner
Wessely-Moser rearrangement
  reordenamiento de Wessely-Moser
Weston cell pila de Weston
Westphalen-Lettre rearrangement
  reordenamiento de Westphalen-Lettre
wet analysis análisis húmedo
wet-bulb thermometer termómetro de
  ampolleta húmeda

260

wet deposition deposición húmeda
wetting agent agente humectante
Wharton reaction reacción de Wharton
whey suero de la leche
whiskers triquitas
whiskey whisky
white acid ácido blanco
white arsenic arsénico blanco
white copperas caparrosa blanca
white gasoline gasolina blanca
white gold oro blanco
white lead plomo blanco
white lotion loción blanca
white metal metal blanco
white oil aceite blanco
white phosphorus fósforo blanco
white precipitate precipitado blanco
white vitriol vitriolo blanco
whitener blanqueador
whitewash lechada de cal
whiting blanco de España
Whiting reaction reacción de Whiting
whitneyite whitneyita
Widman-Stoermer synthesis síntesis de
    Widman-Stoermer
Wien effect efecto de Wien
Wigner nuclides nucleidos de Wigner
Wijs solution solución de Wijs
Wilkinson catalyst catalizador de
    Wilkinson
willemite willemita
Willgerodt reaction reacción de
    Willgerodt
Williamson ether synthesis síntesis de
    éteres de Williamson
Williamson reaction reacción de
    Williamson
Williamson synthesis síntesis de
    Williamson
Wilzbach procedure procedimiento de
    Wilzbach
wine vino
wine ether éter de vino
wintergreen oil aceite de pirola
wire cloth tela de alambre
wire gauze gasa de alambre
Wiswesser line notation notación lineal
    de Wiswesser
Wiswesser notation notación de

Wiswesser
witherite witherita
Wittig ether rearrangement
    reordenamiento de éteres de Wittig
Wittig reaction reacción de Wittig
Wittig rearrangement reordenamiento
    de Wittig
Wohl degradation degradación de Wohl
Wohl reaction reacción de Wohl
Wohl-Ziegler reaction reacción de
    Wohl-Ziegler
Wöhler synthesis síntesis de Wöhler
wöhlerite wohlerita
Wohlwill process proceso de Wohlwill
Wolff-Kishner reaction reacción de
    Wolff-Kishner
Wolff-Kishner reduction reducción de
    Wolff-Kishner
Wolff rearrangement reordenamiento de
    Wolff
Wolffenstein-Boters reaction reacción
    de Wolffenstein-Boters
wolfram wolframio, volframio
wolfram white blanco de wolframio
wolframate wolframato
wolframic acid ácido wolfrámico
wolframite wolframita
wollastonite wollastonita
wood alcohol alcohol de madera
wood ash ceniza de madera
wood cellulose celulosa de madera
wood flour harina de madera
wood meal harina de madera
wood pulp pulpa de madera
wood rosin colofonia de madera
wood sugar azúcar de madera
wood tar alquitrán de madera
wood turpentine trementina de madera
Woodward-Hoffmann rules reglas de
    Woodward-Hoffmann
Woodward hydroxylation hidroxilación
    de Woodward
wool lana
wool fat grasa de lana
work function función trabajo
wormseed oil aceite de quenopondio
wormwood ajenjo, absintio
wulfenite wulfenita
Wulff process proceso de Wulff

Wullner law ley de Wullner
Wurster salt sal de Wurster
Wurtz-Fittig reaction reacción de
    Wurtz-Fittig

Wurtz flask matraz de Wurtz
Wurtz reaction reacción de Wurtz
Wurtz synthesis síntesis de Wurtz
wurtzite wurtzita

X

X-ray analysis análisis con rayos X
X-ray crystallography cristalografía por rayos X
X-ray diffraction difracción de rayos X
X-ray fluorescence fluorescencia de rayos X
X-ray fluorescence analysis análisis de fluorescencia de rayos X
X-ray image spectrography espectrografía de imagen de rayos X
X-ray microanalysis microanálisis con rayos X
X-ray spectrograph espectrógrafo de rayos X
X-ray spectrography espectrografía de rayos X
X-ray spectrometer espectrómetro de rayos X
X-ray spectrometry espectrometría de rayos X
X-ray spectrum espectro de rayos X
X-ray structure estructura por rayos X
X-ray tube tubo de rayos X
xanthan xantano
xanthan gum goma de xantano
xanthate xantato
xanthene xanteno
xanthene dye colorante de xanteno
xanthenol xantenol
xanthenyl xantenilo
xanthic acid ácido xántico
xanthic amide amida xántica
xanthine xantina
xanthine oxidase oxidasa de xantina
xantho xanto
xanthochromium xantocromo
xanthogenic acid ácido xantogénico
xanthone xantona
xanthophyll xantofila
xanthopterin xantopterina

xanthosiderite xantosiderita
xanthosine xantosina
xanthoxylene xantoxileno
xanthyl xantilo
xanthylium xantilio
xenate xenato
xenobiotic xenobiótico
xenol xenol
xenolite xenolita
xenon xenón
xenon effect efecto de xenón
xenon tetrafluoride tetrafluoruro de xenón
xenotime xenotima
xenyl xenilo
xenylamine xenilamina
xerogel xerogel
xerography xerografía
xi xi
xi particle partícula xi
xylan xilano
xylene xileno
xylenediol xilenodiol
xylenesulfonic acid ácido xilenosulfónico
xylenol xilenol
xylidene xilideno
xylidic acid ácido xilídico
xylidine xilidina
xylite xilita
xylitol xilitol
xylocaine xilocaína
xylol xilol
xylonic acid ácido xilónico
xylose xilosa
xyloyl xiloilo
xylyl xililo
xylyl bromide bromuro de xililo
xylyl chloride cloruro de xililo
xylyl dichloride dicloruro de xililo
xylylene xilileno

Y

yacca gum goma de yaca
yeast levadura
yellow acid ácido amarillo
yellow brass latón amarillo
yellow cake torta amarilla
yellow copperas caparrosa amarilla
yellow lake laca amarilla
yellow phosphorus fósforo amarillo
yellow precipitate precipitado amarillo
yellow prussiate of potash prusiato
    amarillo de potasa
yellow prussiate of soda prusiato
    amarillo de sosa
yellow rain lluvia amarilla
yellow resin resina amarilla
yellow salt sal amarilla
yield rendimiento
yl ilo
ylide iluro
ylidene ilideno
ylidyne ilidino
yne ino
yohimbine yohimbina
yperite iperita
ytterbia iterbia
ytterbium iterbio

ytterbium bromide bromuro de iterbio
ytterbium chloride cloruro de iterbio
ytterbium fluoride fluoruro de iterbio
ytterbium oxalate oxalato de iterbio
ytterbium oxide óxido de iterbio
ytterbium sulfate sulfato de iterbio
yttria itria
yttrialite itrialita
yttrium itrio
yttrium acetate acetato de itrio
yttrium antimonide antimoniuro de itrio
yttrium arsenide arseniuro de itrio
yttrium bromide bromuro de itrio
yttrium carbonate carbonato de itrio
yttrium chloride cloruro de itrio
yttrium hydroxide hidróxido de itrio
yttrium nitrate nitrato de itrio
yttrium oxalate oxalato de itrio
yttrium oxide óxido de itrio
yttrium phosphide fosfuro de itrio
yttrium sulfate sulfato de itrio
yttrium sulfide sulfuro de itrio
yttrium vanadate vanadato de itrio
yttrocerite itrocerita
yttrotantalite itrotantalita
Yukawa force fuerza de Yukawa

Z

zaratite zaratita

zeaxanthin zeaxantina

Zeeman displacement desplazamiento de Zeeman

Zeeman effect efecto de Zeeman

Zeeman energy energía de Zeeman

zein zeína

Zeise salt sal de Zeise

Zeisel determination determinación de Zeisel

Zeisel reaction reacción de Zeisel

zeolite zeolita

zeolite catalyst catalizador de zeolita

Zerewitinoff determination determinación de Zerewitinoff

Zerewitinoff reagent reactivo de Zerewitinoff

zero-order reaction reacción de orden cero

zero-point energy energía al cero absoluto

zeroth law of thermodynamics ley cero de termodinámica

zeta potential potencial zeta

Ziegler catalyst catalizador de Ziegler

Ziegler method método de Ziegler

Ziegler-Natta polymerization polimerización de Ziegler-Natta

Ziegler process proceso de Ziegler

Ziehl stain colorante de Ziehl

Ziervogel process proceso de Ziervogel

Ziesel reaction reacción de Ziesel

Zimmermann reaction reacción de Zimmermann

zinc cinc, zinc

zinc abietate abietato de cinc

zinc acetate acetato de cinc

zinc acetylacetonate acetilacetonato de cinc

zinc albuminate albuminato de cinc

zinc amalgam amalgama de cinc

zinc ammonium nitrite nitrito amónico de cinc

zinc antimonide antimoniuro de cinc

zinc arsenate arseniato de cinc

zinc arsenide arseniuro de cinc

zinc arsenite arsenito de cinc

zinc bacitracin bacitracina de cinc

zinc benzoate benzoato de cinc

zinc blende blenda de cinc

zinc borate borato de cinc

zinc bromate bromato de cinc

zinc bromide bromuro de cinc

zinc caprylate caprilato de cinc

zinc carbonate carbonato de cinc

zinc chlorate clorato de cinc

zinc chloride cloruro de cinc

zinc chloroiodide cloroyoduro de cinc

zinc chromate cromato de cinc

zinc citrate citrato de cinc

zinc cyanide cianuro de cinc

zinc dichromate dicromato de cinc

zinc dioxide dióxido de cinc

zinc dithionate ditionato de cinc

zinc dust polvo de cinc

zinc ethanoate etanoato de cinc

zinc ethylhexoate etilhexoato de cinc

zinc ethylsulfate etilsulfato de cinc

zinc ferrocyanide ferrocianuro de cinc

zinc fluoride fluoruro de cinc

zinc fluoroborate fluoroborato de cinc

zinc fluorosilicate fluorosilicato de cinc

zinc formate formiato de cinc

zinc gluconate gluconato de cinc

zinc hexacyanoferrate hexacianoferrato de cinc

zinc hexafluorosilicate hexafluorosilicato de cinc

zinc hydrosulfite hidrosulfito de cinc

zinc hydroxide hidróxido de cinc

zinc hypophosphite hipofosfito de cinc

zinc iodate yodato de cinc

zinc iodide yoduro de cinc

zinc lactate lactato de cinc

zinc laurate laurato de cinc

zinc linoleate linoleato de cinc

zinc malate malato de cinc
zinc metaarsenite metaarsenito de cinc
zinc molybdate molibdato de cinc
zinc naphthenate naftenato de cinc
zinc nitrate nitrato de cinc
zinc nitride nitruro de cinc
zinc octanoate octanoato de cinc
zinc octoate octoato de cinc
zinc oleate oleato de cinc
zinc orthoarsenate ortoarseniato de cinc
zinc orthophosphate ortofosfato de cinc
zinc oxalate oxalato de cinc
zinc oxide óxido de cinc
zinc oxychloride oxicloruro de cinc
zinc palmitate palmitato de cinc
zinc perborate perborato de cinc
zinc perchlorate perclorato de cinc
zinc permanganate permanganato de cinc
zinc peroxide peróxido de cinc
zinc peroxoborate peroxoborato de cinc
zinc phenate fenato de cinc
zinc phenolsulfonate fenolsulfonato de cinc
zinc phosphate fosfato de cinc
zinc phosphide fosfuro de cinc
zinc phosphonate fosfonato de cinc
zinc potassium iodide yoduro potásico de cinc
zinc propionate propionato de cinc
zinc pyrithione piritiona de cinc
zinc pyrophosphate pirofosfato de cinc
zinc resinate resinato de cinc
zinc ricinoleate ricinoleato de cinc
zinc salicylate salicilato de cinc
zinc selenide seleniuro de cinc
zinc silicate silicato de cinc
zinc stearate estearato de cinc
zinc succinate succinato de cinc
zinc sulfate sulfato de cinc
zinc sulfate monohydrate monohidrato de sulfato de cinc
zinc sulfide sulfuro de cinc
zinc sulfite sulfito de cinc
zinc sulfoxylate sulfoxilato de cinc
zinc tartrate tartrato de cinc
zinc telluride telururo de cinc
zinc tetraborate tetraborato de cinc
zinc thiocyanate tiocianato de cinc

zinc thiophenate tiofenato de cinc
zinc undecylenate undecilenato de cinc
zinc vitriol vitriolo de cinc
zinc white blanco de cinc
zinc yellow amarillo de cinc
zincate cincato
zincite cincita
Zincke nitration nitración de Zincke
Zincke-Suhl reaction reacción de Zincke-Suhl
zincous cincoso
zingiberene zingibereno
zinnwaldite zinnwaldita
zippeite zippeíta
ziram ziram
zircon zircón
zirconate zirconato
zirconia zirconia
zirconic zircónico
zirconic acid ácido zircónico
zirconic anhydride anhídrido zircónico
zirconium zirconio
zirconium acetate acetato de zirconio
zirconium boride boruro de zirconio
zirconium bromide bromuro de zirconio
zirconium carbide carburo de zirconio
zirconium carbonate carbonato de zirconio
zirconium chloride cloruro de zirconio
zirconium diboride diboruro de zirconio
zirconium dioxide dióxido de zirconio
zirconium disilicide disiliciuro de zirconio
zirconium disulfide disulfuro de zirconio
zirconium fluoride fluoruro de zirconio
zirconium glycolate glicolato de zirconio
zirconium hydride hidruro de zirconio
zirconium hydroxide hidróxido de zirconio
zirconium iodide yoduro de zirconio
zirconium lactate lactato de zirconio
zirconium naphthenate naftenato de zirconio
zirconium nitrate nitrato de zirconio
zirconium nitride nitruro de zirconio
zirconium orthophosphate ortofosfato de zirconio
zirconium oxide óxido de zirconio

zirconium oxybromide oxibromuro de
  zirconio
zirconium oxychloride oxicloruro de
  zirconio
zirconium phosphate fosfato de zirconio
zirconium potassium fluoride fluoruro
  potásico de zirconio
zirconium pyrophosphate pirofosfato de
  zirconio
zirconium silicate siicato de zirconio
zirconium silicide siliciuro de zirconio
zirconium sulfate sulfato de zirconio
zirconium tetraacetylacetonate
  tetraacetilacetonato de zirconio
zirconium tetrachloride tetracloruro de
  zirconio
zirconium tetrafluoride tetrafluoruro de
  zirconio
zirconocene dichloride dicloruro de
  zirconoceno
zirconyl zirconilo
zirconyl acetate acetato de zirconilo
zirconyl bromide bromuro de zirconilo
zirconyl carbonate carbonato de
  zirconilo

zirconyl chloride cloruro de zirconilo
zirconyl hydroxide hidróxido de
  zirconilo
zirconyl hyroxychloride hidroxicloruro
  de zirconilo
zirconyl nitrate nitrato de zirconilo
zirconyl phosphate fosfato de zirconilo
zirconyl sulfate sulfato de zirconilo
zirlite zirlita
zoalene zoaleno
zoisite zoisita
zone electrophoresis electroforesis por
  zonas
zone refining refinamiento por zonas
zoochemistry zooquímica
Zsigmondy filters filtros de Zsigmondy
zwitterion zwitterion
zymase zimasa
zymochemistry zimoquímica
zymohexase zimohexasa
zymolysis zimólisis
zymose zimosa
zymosis zimosis
zymurgy zimurgia

ESPAÑOL-INGLÉS
SPANISH-ENGLISH

A

a prueba de ácido acidproof

a prueba de agua waterproof

a prueba de fuego fireproof

a prueba de gas gasproof

a prueba de humedad humidity-proof,
moisture-proof

abacá abaca

aberración aberration

aberración cromática chromatic
aberration

abietato cálcico calcium abietate

abietato de bencilo benzyl abietate

abietato de cinc zinc abietate

abietato de cobre copper abietate

abietato de etilo ethyl abietate

abietato de glicerilo glyceryl abietate

abietato de metilo methyl abietate

abietato potásico potassium abietate

abietato sódico sodium abietate

ablación ablation

ablandador softener

ablandamiento de agua water softening

abono orgánico compost

abrasión abrasion

abrasivo abrasive

abrazadera clamp

abrazadera de Bunsen Bunsen clamp

abrazadera de bureta buret clamp

abrillantador brightener

abrillantador óptico optical brightener

abrillantamiento brightening

absintio absinthium, absinthe,
wormwood

absoluto absolute

absorbancia absorbance

absorbedor de microondas microwave
absorber

absorbedor de neutrones neutron
absorber

absorbedor de radiación radiation
absorber

absorbedor neutrónico neutron absorber

absorbedor ultravioleta ultraviolet
absorber

absorbente absorbent

absorbente de oxígeno oxygen absorbent

absorciómetro absorptiometer

absorción absorption

absorción atómica atomic absorption

absorción de calor heat absorption

absorción de fotones photon absorption

absorción de gas gas absorption

absorción de luz absorption of light

absorción fotoeléctrica photoelectric
absorption

absorción neutrónica neutron absorption

absorción nuclear nuclear absorption

absorción térmica heat absorption

absorción ultravioleta ultraviolet
absorption

absortividad absorptivity

abstracción abstraction

abundancia abundance

abundancia isotópica isotope abundance

abundancia natural natural abundance

acacetina acacetin

acacia acacia

acadesina acadesine

acantita acanthite

accesorio accessory

acción action

acción amortiguadora buffer action

acción catalítica catalytic action

acción cíclica cyclic action

acción electrolítica electrolytic action

acción galvánica galvanic action

acción química chemical action

acción reversible reversible action

aceite oil

aceite blanco white oil

aceite carbólico carbolic oil

aceite combustible fuel oil

aceite comestible edible oil

aceite de aguacate avocado oil

aceite de ajo garlic oil

aceite de alcanfor camphor oil

aceite de alcaravea caraway oil
aceite de almendra almond oil
aceite de alquitrán tar oil
aceite de alquitrán de hulla coal-tar oil
aceite de alquitrán de pino pine-tar oil
aceite de anacardo cashew oil
aceite de anís anise oil
aceite de antraceno anthracene oil
aceite de babasú babassu oil
aceite de bergamota bergamot oil
aceite de betula betula oil
aceite de cacao cocoa oil
aceite de cada juniper tar
aceite de cálamo calamus oil
aceite de cananga cananga oil
aceite de canela cinnamon oil
aceite de cañamón hempseed oil
aceite de cardamomo cardamom oil
aceite de cártamo safflower oil
aceite de cascarilla cascarilla oil
aceite de casia cassia oil
aceite de castor castor oil
aceite de citronela citronella oil
aceite de clavero clove oil
aceite de cloronaftaleno
   chloronaphthalene oil
aceite de colofonia rosin oil
aceite de colza rapeseed oil, canola
aceite de copaiba copaiba oil
aceite de corteza de naranja orange-peel
   oil
aceite de crotón croton oil
aceite de estragón tarragon oil
aceite de fenilmostaza phenyl mustard
   oil
aceite de fusel fusel oil
aceite de girasol sunflower oil
aceite de hígado de bacalao cod-liver oil
aceite de hinojo fennel oil
aceite de hojas de laurel laurel leaves oil
aceite de horno furnace oil
aceite de huesos bone oil
aceite de huevo egg oil
aceite de hulla coal oil
aceite de jazmín jasmine oil
aceite de jengibre ginger oil
aceite de jojoba jojoba oil
aceite de lavanda lavender oil
aceite de lima lime oil

aceite de limón lemon oil
aceite de linaza linseed oil, flaxseed oil
aceite de luisa lemongrass oil
aceite de maíz corn oil
aceite de mandarina mandarin oil
aceite de maní peanut oil
aceite de manteca lard oil
aceite de manzana apple oil
aceite de manzanilla chamomile oil
aceite de menhaden menhaden oil
aceite de menta peppermint oil
aceite de menta verde spearmint oil
aceite de mostaza mustard oil
aceite de niobe niobe oil
aceite de nuez de coco coconut oil
aceite de oliva olive oil
aceite de pachulí patchouli oil
aceite de palma palm oil
aceite de palmarosa palmarosa oil
aceite de parafina paraffin oil
aceite de pata de vaca neatsfoot oil
aceite de perilla perilla oil
aceite de pescado fish oil
aceite de pino pine oil
aceite de pirola wintergreen oil
aceite de pizarra shale oil
aceite de plátano banana oil
aceite de poleo pennyroyal oil
aceite de quenopodio chenopodium oil,
   wormseed oil
aceite de ricino ricinus oil
aceite de salvado bran oil
aceite de salvia sage oil
aceite de sándalo sandalwood oil
aceite de sasafrás sassafras oil
aceite de sebo tallow oil
aceite de semilla de algodón cottonseed
   oil
aceite de semilla de perejil parsley oil
aceite de sésamo sesame oil
aceite de silicona silicone oil
aceite de soja soybean oil
aceite de tangerina tangerine oil
aceite de tiglio tiglium oil
aceite de transformador transformer oil
aceite de tung tung oil
aceite de tuya thuja oil
aceite de valeriana valerian oil
aceite de verbena verbena oil

aceite de vitriolo vitriol oil
aceite dulce sweet oil
aceite esencial essential oil
aceite estirenado styrenated oil
aceite fijo fixed oil
aceite hidrogenado hydrogenated oil
aceite ligero light oil
aceite lubricante lubricating oil
aceite mineral mineral oil
aceite neutro neutral oil
aceite pesado heavy oil
aceite residual residual oil
aceite rojo red oil
aceite secante drying oil
aceite sintético synthetic oil
aceite soluble soluble oil
aceite soluble en agua water-soluble oil
aceite soplado blown oil
aceite vegetal vegetable oil
aceite yodado iodized oil
aceleración acceleration
acelerador accelerator
acelerador de caucho rubber accelerator
acelerador de electrones electron
    accelerator
acelerador de partículas particle
    accelerator
acelerador de protones proton
    accelerator
acelerador electrónico electron
    accelerator
acelerador lineal electrónico electron
    linear accelerator
acelerador protónico proton accelerator
acenafteno acenaphthene
acenaftenoquinona acenaphthenequinone
acenaftileno acenaphthylene
aceptor acceptor
aceptor de ácido acid acceptor
aceptor de hidrógeno hydrogen acceptor
aceptor electrónico electron acceptor
aceptor protónico proton acceptor
acero steel
acero al boro boron steel
acero al carbono carbon steel
acero al cromo chromium steel
acero al níquel nickel steel
acero al tungsteno tungsten steel
acero aleado alloy steel

acero de horno eléctrico electric steel
acero inoxidable stainless steel
acetal acetal
acetaldehído acetaldehyde
acetaldol acetaldol
acetamida acetamide
acetamidina acetamidine
acetaminofén acetaminophen
acetaminofeno acetaminophen
acetaminosalol acetaminosalol
acetanilida acetanilide
acetato acetate
acetato alumínico aluminum acetate
acetato amónico ammonium acetate
acetato bárico barium acetate
acetato cálcico calcium acetate
acetato ceroso cerous acetate
acetato cobaltoso cobaltous acetate
acetato crómico chromic acetate
acetato cromoso chromous acetate
acetato cúprico cupric acetate
acetato de alilo allyl acetate
acetato de aluminio aluminum acetate
acetato de amilo amyl acetate
acetato de amonio ammonium acetate
acetato de anilina aniline acetate
acetato de anisilo anisyl acetate
acetato de bario barium acetate
acetato de bencilo benzyl acetate
acetato de benzaldoxima benzaldoxime
    acetate
acetato de benzoilo benzoyl acetate
acetato de berilio beryllium acetate
acetato de bismuto bismuth acetate
acetato de bornilo bornyl acetate
acetato de butanol butanol acetate
acetato de butilo butyl acetate
acetato de cadmio cadmium acetate
acetato de calcio calcium acetate
acetato de cedrilo cedryl acetate
acetato de celulosa cellulose acetate
acetato de cesio cesium acetate
acetato de cianometilo cyanomethyl
    acetate
acetato de ciclohexanol cyclohexanol
    acetate
acetato de cinamilo cinnamyl acetate
acetato de cinc zinc acetate
acetato de cloromadinona

chloromadinone acetate

acetato de cloruro de polivinilo
polyvinyl chloride acetate

acetato de cobalto cobalt acetate

acetato de cobre copper acetate

acetato de cresilo cresyl acetate

acetato de cromo chromium acetate

acetato de decilo decyl acetate

acetato de dicloroetilo dichloroethyl
acetate

acetato de erbio erbium acetate

acetato de estroncio strontium acetate

acetato de etilbutilo ethylbutyl acetate

acetato de etilhexilo ethylhexyl acetate

acetato de etilo ethyl acetate

acetato de fenetilo phenethyl acetate

acetato de feniletilo phenylethyl acetate

acetato de fenilmetilo phenylmethyl
acetate

acetato de fenilo phenyl acetate

acetato de fenilpropilo phenylpropyl
acetate

acetato de furfurilo furfuryl acetate

acetato de galio gallium acetate

acetato de geranilo geranyl acetate

acetato de glicol glycol acetate

acetato de guayacol guaiacol acetate

acetato de heptilo heptyl acetate

acetato de hexilo hexyl acetate

acetato de hidrocinamilo hydrocinnamyl
acetate

acetato de hierro iron acetate

acetato de isoamilo isoamyl acetate

acetato de isobornilo isobornyl acetate

acetato de isobutilo isobutyl acetate

acetato de isopropenilo isopropenyl
acetate

acetato de isopropilo isopropyl acetate

acetato de itrio yttrium acetate

acetato de lantano lanthanum acetate

acetato de linalilo linalyl acetate

acetato de litio lithium acetate

acetato de magnesio magnesium acetate

acetato de manganeso manganese acetate

acetato de mentilo menthyl acetate

acetato de mercurio mercury acetate

acetato de metilalilo methylallyl acetate

acetato de metilamilo methylamyl
acetate

acetato de metilo methyl acetate

acetato de neodimio neodymium acetate

acetato de níquel nickel acetate

acetato de nonilo nonyl acetate

acetato de octilo octyl acetate

acetato de pentilo pentyl acetate

acetato de plata silver acetate

acetato de plomo lead acetate

acetato de polivinilo polyvinyl acetate

acetato de potasio potassium acetate

acetato de praseodimio praseodymium
acetate

acetato de propilo propyl acetate

acetato de resorcina resorcinol acetate

acetato de rodinilo rhodinyl acetate

acetato de rubidio rubidium acetate

acetato de santalilo santalyl acetate

acetato de sodio sodium acetate

acetato de talio thallium acetate

acetato de terpinilo terpinyl acetate

acetato de undecilenilo undecylenyl
acetate

acetato de uranilo uranyl acetate

acetato de uranio uranium acetate

acetato de urea urea acetate

acetato de vinilo vinyl acetate

acetato de yodo iodine acetate

acetato de zirconilo zirconyl acetate

acetato de zirconio zirconium acetate

acetato estannoso stannous acetate

acetato estiralílico styralyl acetate

acetato fenilmercúrico phenylmercuric
acetate

acetato férrico ferric acetate

acetato ferroso ferrous acetate

acetato magnésico magnesium acetate

acetato manganoso manganous acetate

acetato mercúrico mercuric acetate

acetato mercurioso mercurous acetate

acetato niqueloso nickelous acetate

acetato potásico potassium acetate

acetato rojo red acetate

acetato sódico sodium acetate

acetatobutirato de celulosa cellulose
acetatobutyrate

acetatoisobutirato de sucrosa sucrose
acetate isobutyrate

acetazolamida acetazolamide

acetiamina acetiamine

acetilacetato de neodimio neodymium
    acetylacetate
acetilacetona acetylacetone
acetilacetonato acetylacetonate
acetilacetonato de berilio beryllium
    acetylacetonate
acetilacetonato de cinc zinc
    acetylacetonate
acetilacetonato de cromo chromium
    acetylacetonate
acetilacetonato de indio indium
    acetylacetonate
acetilacetonato de molibdeno
    molybdenum acetylacetonate
acetilacetonato de níquel nickel
    acetylacetonate
acetilacetonato de titanio titanium
    acetylacetonate
acetilacetonato de vanadio vanadium
    acetylacetonate
acetilacetonato férrico ferric
    acetylacetonate
acetilacetonato magnésico magnesium
    acetylacetonate
acetilacetonato mangánico manganic
    acetylacetonate
acetilación acetylation
acetilamino acetylamino
acetilbenceno acetylbenzene
acetilcarbromal acetylcarbromal
acetilcetena acetyl ketene
acetilcoenzima acetyl coenzyme
acetilcolina acetylcholine
acetilcolinesterasa acetylcholinesterase
acetilenilo acetylenyl
acetileno acetylene
acetileno de butilo butyl acetylene
acetilenógeno acetylenogen
acetilfenol acetylphenol
acetilferroceno acetylferrocene
acetilformiato sódico sodium
    acetylformate
acetilmetilcarbinol acetylmethylcarbinol
acetilo acetyl
acetilricinoleato de metilo methyl
    acetylricinoleate
acetilsalicilato cálcico calcium
    acetylsalicylate
acetilsalicilato potásico potassium

acetylsalicylate
acetilsalicilato sódico sodium
    acetylsalicylate
acetilurea acetylurea
acetiluro acetylide
acetiluro cálcico calcium acetylide
acetiluro cuproso cuprous acetylide
acetiluro de plata silver acetylide
acetiluro mercurioso mercurous
    acetylide
acetina acetin
acetoacetanilida acetoacetanilide
acetoacetato acetoacetate
acetoacetato de butilo butyl acetoacetate
acetoacetato de etilo ethyl acetoacetate
acetoacetato de metilo methyl
    acetoacetate
acetoarsenito de cobre copper
    acetoarsenite
acetoestearina acetostearin
acetofenetinina acetophenetidine
acetofenona acetophenone
acetoína acetoin
acetol acetol
acetólisis acetolysis
acetona acetone
acetona clorada chlorinated acetone
acetona de acetofenona acetophenone
    acetone
acetonilacetona acetonylacetone
acetonilideno acetonylidene
acetonitrilo acetonitrile
acetotartrato alumínico aluminum
    acetotartrate
acetotoluida acetotoluide
acetotoluidida acetotoluidide
acetoxiestearato de butilo butyl
    acetoxystearate
acetoxilación acetoxylation
acetoxima acetoxime
acetozona acetozone
acíclico acyclic
acicular acicular
acidez acidity
acídico acidic
acidificación acidification
acidificar acidify
acidimetría acidimetry
acidímetro acidimeter

ácido acid
ácido abienínico abieninic acid
ácido abietadienoico abietadienoic acid
ácido abiético abietic acid
ácido abscísico abscisic acid
ácido acético acetic acid
ácido acético glacial glacial acetic acid
ácido acetilacético acetylacetic acid
ácido acetilfórmico acetylformic acid
ácido acetilpropiónico acetylpropionic
    acid
ácido acetilsalicílico acetylsalicylic acid
ácido acetoacético acetoacetic acid
ácido acetonadicarboxílico acetone
    dicarboxylic acid
ácido acetonamonocarboxílico acetone
    monocarboxylic acid
ácido acónico aconic acid
ácido aconítico aconitic acid
ácido acridínico acridinic acid
ácido acrílico acrylic acid
ácido acroleico acroleic acid
ácido adenílico adenylic acid
ácido adípico adipic acid
ácido agárico agaric acid
ácido alánico allanic acid
ácido alántico alantic acid
ácido alantoico allantoic acid
ácido alantoxánico allantoxanic acid
ácido alantúrico allanturic acid
ácido alcanosulfónico alkanesulfonic
    acid
ácido aldónico aldonic acid
ácido aleurítico aleuritic acid
ácido algínico alginic acid
ácido alifático aliphatic acid
ácido alizárico alizaric acid
ácido alocinámico allocinnamic acid
ácido aloerésico aloeresic acid
ácido alofánico allophanic acid
ácido alomaleico allomaleic acid
ácido aloxánico alloxanic acid
ácido alumínico aluminic acid
ácido amalínico amalinic acid
ácido amarillo yellow acid
ácido amigdálico amygdalic acid
ácido amínico aminic acid
ácido aminoacético aminoacetic acid
ácido aminobencenosulfónico
    aminobenzenesulfonic acid
ácido aminobenzoico aminobenzoic acid
ácido aminobutanoico aminobutanoic
    acid
ácido aminobutírico aminobutyric acid
ácido aminocaproico aminocaproic acid
ácido aminoditiofórmico
    aminodithioformic acid
ácido aminoetilsulfúrico
    aminoethylsulfuric acid
ácido aminofenilacético
    aminophenylacetic acid
ácido aminofosfórico aminophosphoric
    acid
ácido aminoglutárico aminoglutaric acid
ácido aminohexanoico aminohexanoic
    acid
ácido aminohipúrico aminohippuric acid
ácido aminoisobutírico aminoisobutyric
    acid
ácido aminonaftolsulfónico
    aminonaphtholsulfonic acid
ácido aminonicotínico aminonicotinic
    acid
ácido aminopropanoico aminopropanoic
    acid
ácido aminosalicílico aminosalicylic acid
ácido amsónico amsonic acid
ácido anacárdico anacardic acid
ácido ancoico anchoic acid
ácido anemónico anemonic acid
ácido angélico angelic acid
ácido anilinosulfónico anilinesulfonic
    acid
ácido anísico anisic acid
ácido antimónico antimonic acid
ácido antranílico anthranilic acid
ácido antroico anthroic acid
ácido apiólico apiolic acid
ácido apocólico apocholic acid
ácido arabónico arabonic acid
ácido araquídico arachidic acid
ácido araquidónico arachidonic acid
ácido aromático aromatic acid
ácido arrénico arrhenic acid
ácido arsanílico arsanilic acid
ácido arsenénico arsenenic acid
ácido arsénico arsenic acid
ácido arsenioso arsenous acid

ácido arsínico arsinic acid
ácido arsónico arsonic acid
ácido asarónico asaronic acid
ácido ascarídico ascaridic acid
ácido ascaridólico ascaridolic acid
ácido ascórbico ascorbic acid
ácido asiático asiatic acid
ácido asparágico asparagic acid
ácido asparagínico asparaginic acid
ácido aspartámico aspartamic acid
ácido aspártico aspartic acid
ácido aspergílico aspergillic acid
ácido atroláctico atrolactic acid
ácido atrónico atronic acid
ácido atrópico atropic acid
ácido auribromhídrico auribromhydric
    acid
ácido áurico auric acid
ácido azelaico azelaic acid
ácido azobenzoico azobenzoic acid
ácido barbitúrico barbituric acid
ácido behénico behenic acid
ácido behenólico behenolic acid
ácido bencenocarbóxílico
    benzenecarboxylic acid
ácido bencenodicarbóxílico
    benzenedicarboxylic acid
ácido bencenodisulfónico
    benzenedisulfonic acid
ácido bencenofosfínico
    benzenephosphinic acid
ácido bencenofosfónico
    benzenephosphonic acid
ácido bencenomonosulfónico
    benzenemonosulfonic acid
ácido bencenosulfínico benzenesulfinic
    acid
ácido bencenosulfónico benzenesulfonic
    acid
ácido bencenotetracarbóxílico
    benzenetetracarboxylic acid
ácido bencenotrisulfónico
    benzenetrisulfonic acid
ácido bencidinodicarbóxílico
    benzidinedicarboxylic acid
ácido bencílico benzilic acid
ácido bencilsulfónico benzylsulfonic acid
ácido benzamidoacético benzamidoacetic
    acid

ácido benzaminoacético benzaminoacetic
    acid
ácido benzocarbólico benzocarbolic acid
ácido benzoglicólico benzoglycolic acid
ácido benzohidroxímico
    benzohydroximic acid
ácido benzoico benzoic acid
ácido benzoilacético benzoylacetic acid
ácido benzoilacrílico benzoylacrylic acid
ácido benzoilaminoacético
    benzoylaminoacetic acid
ácido benzoilbenzoico benzoylbenzoic
    acid
ácido benzoilpropiónico
    benzoylpropionic acid
ácido berberónico berberonic acid
ácido beta beta acid
ácido betulínico betulinic acid
ácido biliar bile acid
ácido bilirrubínico bilirubinic acid
ácido binario binary acid
ácido blanco white acid
ácido blando soft acid
ácido borácico boracic acid
ácido bórico boric acid
ácido borofosfórico borophosphoric acid
ácido borotúngstico borotungstic acid
ácido brasídico brassidic acid
ácido brasílico brassylic acid
ácido bromhídrico hydrobromic acid
ácido brómico bromic acid
ácido bromoacético bromoacetic acid
ácido bromoantrálico bromoanthrallic
    acid
ácido bromoaúrico bromoauric acid
ácido bromobencenosulfónico
    bromobenzenesulfonic acid
ácido bromobenzoico bromobenzoic acid
ácido bromobutírico bromobutyric acid
ácido bromopropiónico bromopropionic
    acid
ácido bromosuccínico bromosuccinic
    acid
ácido butanoico butanoic acid
ácido butenoico butenoic acid
ácido butilestánnico butylstannic acid
ácido butírico butyric acid
ácido cacodílico cacodylic acid
ácido cafeico caffeic acid

ácido cafetánico caffetannic acid
ácido cambopínico cambopinic acid
ácido canfocarboxílico
     camphocarboxylic acid
ácido canfólico campholic acid
ácido canfórico camphoric acid
ácido cáprico capric acid
ácido caprílico caprylic acid
ácido caproico caproic acid
ácido carbámico carbamic acid
ácido carbanílico carbanilic acid
ácido carbázico carbazic acid
ácido carbazótico carbazotic acid
ácido carbílico carbylic acid
ácido carbólico carbolic acid
ácido carbónico carbonic acid
ácido carboxílico carboxylic acid
ácido carcinómico carcinomic acid
ácido cariofílico caryophyllic acid
ácido cárlico carlic acid
ácido carmínico carminic acid
ácido carnáubico carnaubic acid
ácido caróbico carobic acid
ácido carólico carolic acid
ácido cartámico carthamic acid
ácido cerebrónico cerebronic acid
ácido ceritámico ceritamic acid
ácido ceroplástico ceroplastic acid
ácido cerótico cerotic acid
ácido cerotínico cerotinic acid
ácido cetílico cctylic acid
ácido cetoglutárico ketoglutaric acid
ácido cetólico cetolic acid
ácido cetomalónico ketomalonic acid
ácido cetopropiónico ketopropionic acid
ácido cetovalérico ketovaleric acid
ácido cianhídrico hydrocyanic acid
ácido ciánico cyanic acid
ácido cianoacético cyanoacetic acid
ácido cianocarbónico cyanocarbonic acid
ácido cianoetanoico cyanoethanoic acid
ácido cianúrico cyanuric acid
ácido ciclámico cyclamic acid
ácido ciclohexanocarboxílico
     cyclohexanecarboxylic acid
ácido ciclopentilpropiónico
     cyclopentylpropionic acid
ácido cinámico cinnamic acid
ácido cincomerónico cinchomeronic acid

ácido cinconínico cinchoninic acid
ácido citacrónico citraconic acid
ácido citidílico cytidylic acid
ácido citramálico citramalic acid
ácido cítrico citric acid
ácido citrídico citridic acid
ácido citronélico citronellic acid
ácido clorhídrico hydrochloric acid
ácido clórico chloric acid
ácido cloroacético chloroacetic acid
ácido cloroáurico chloroauric acid
ácido cloroazótico chloroazotic acid
ácido clorobenzoico chlorobenzoic acid
ácido clorodifluoroacético
     chlorodifluoroacetic acid
ácido cloroetanoico chloroethanoic acid
ácido cloroetilfosfónico
     chloroethylphosphonic acid
ácido clorofórmico chloroformic acid
ácido clorogénico chlorogenic acid
ácido cloromalónico chloromalonic acid
ácido clorometilfosfónico
     chloromethylphosphonic acid
ácido cloronítrico chloronitric acid
ácido cloronitrobenzoico
     chloronitrobenzoic acid
ácido cloronitroso chloronitrous acid
ácido cloroperbenzoico chloroperbenzoic
     acid
ácido cloroplatínico chloroplatinic acid
ácido cloropropiónico chloropropionic
     acid
ácido clorosalicílico chlorosalicylic acid
ácido cloroso chlorous acid
ácido clorosulfónico chlorosulfonic acid
ácido clorosulfonilbenzoico
     chlorosulfonylbenzoic acid
ácido clorosulfúrico chlorosulfuric acid
ácido clorovalérico chlorovaleric acid
ácido cójico kojic acid
ácido colaico cholaic acid
ácido coleico choleic acid
ácido cólico cholic acid
ácido colofólico colopholic acid
ácido colofónico colophonic acid
ácido coménico comenic acid
ácido complejo complex acid
ácido conjugado conjugate acid
ácido corísmico chorismic acid

ácido cosanoico cosanoic acid
ácido cresílico cresylic acid
ácido cresótico cresotic acid
ácido crisalícico chrysalicic acid
ácido crisanísico chrysanisic acid
ácido crisantemomonocarboxílico
     chrysanthemummonocarboxylic acid
ácido crisofánico chrysophanic acid
ácido crocónico croconic acid
ácido crómico chromic acid
ácido cromoso chromous acid
ácido cromotrópico chromotropic acid
ácido crotónico crotonic acid
ácido cumálico coumalic acid
ácido cumárico coumaric acid
ácido cúmico cumic acid
ácido cumínico cuminic acid
ácido chaulmúgrico chaulmoogric acid
ácido chi chi acid
ácido datúrico daturic acid
ácido de aceite de castor castor oil acid
ácido de alquitrán tar acid
ácido de Armstrong Armstrong acid
ácido de batería battery acid
ácido de Bayer Bayer acid
ácido de Broenner Broenner acid
ácido de Brönner Brönner acid
ácido de Brönsted Brönsted acid
ácido de cámara chamber acid
ácido de Caro Caro acid
ácido de Cassella Cassella acid
ácido de Cleve Cleve acid
ácido de contacto contact acid
ácido de Chicago Chicago acid
ácido de Dahl Dahl acid
ácido de Freund Freund acid
ácido de Koch Koch acid
ácido de Laurent Laurent acid
ácido de Lewis Lewis acid
ácido de manzana apple acid
ácido de Neville-Winter Neville-Winter
     acid
ácido de nuez de coco coconut acid
ácido de Reinecke Reinecke acid
ácido de Schäffer Schäffer acid
ácido de Shoelkopf Shoelkopf acid
ácido de Tobias Tobias acid
ácido débil weak acid
ácido decanocarboxílico

decanecarboxylic acid
ácido decanodioico decanedioic acid
ácido decanoico decanoic acid
ácido decatoico decatoic acid
ácido decíclico decyclic acid
ácido decílico decyclic acid
ácido decoico decoic acid
ácido delfínico delphinic acid
ácido delta delta acid
ácido dérrico derric acid
ácido descarboxirrísico decarboxyrissic
     acid
ácido deshidroabiético dehydroabietic
     acid
ácido deshidroacético dehydroacetic acid
ácido deshidroascórbico
     dehydroascorbic acid
ácido deshidrocólico dehydrocholic acid
ácido desoxicólico deoxycholic acid
ácido desoxirribonucleico
     deoxyribonucleic acid
ácido desoxirribonucleico recombinante
     recombinant deoxyribonucleic acid
ácido deutérico deuteric acid
ácido diacético diacetic acid
ácido dialúrico dialuric acid
ácido diaminocaproico diaminocaproic
     acid
ácido diaminodifénico diaminodiphenic
     acid
ácido diazobencenosulfónico
     diazobenzenesulfonic acid
ácido diazoico diazoic acid
ácido diazosulfónico diazosulfonic acid
ácido dibásico dibasic acid
ácido dibromomalónico dibromomalonic
     acid
ácido dicarboxílico dicarboxylic acid
ácido dicloroacético dichloroacetic acid
ácido diclorobenzoico dichlorobenzoic
     acid
ácido dicloroetanoico dichloroethanoic
     acid
ácido diclorofenoxiacético
     dichlorophenoxyacetic acid
ácido dicloropropiónico
     dichloropropionic acid
ácido dicrómico dichromic acid
ácido dietilacético diethylacetic acid

ácido dietilbarbitúrico diethylbarbituric acid

ácido difénico diphenic acid

ácido difosfoglicérico diphosphoglyceric acid

ácido difosfórico diphosphoric acid

ácido digálico digallic acid

ácido diglicólico diglycolic acid

ácido dihidroxibenzoico dihydroxybenzoic acid

ácido dihidroxibutanodioico dihydroxybutanedioic acid

ácido dihidroxicinámico dihydroxycinnamic acid

ácido dihidroxiesteárico dihydroxystearic acid

ácido dihidroximalónico dihydroxymalonic acid

ácido dihidroxisuccínico dihydroxysuccinic acid

ácido diluido dilute acid

ácido dimetilaminobenzoico dimethylaminobenzoic acid

ácido dimetilarsínico dimethylarsinic acid

ácido dinitrosalicílico dinitrosalicylic acid

ácido dioxonítrico dioxonitric acid

ácido disalicílico disalicylic acid

ácido disulfónico disulfonic acid

ácido disulfúrico disulfuric acid

ácido ditiobenzoico dithiobenzoic acid

ácido ditiocarbámico dithiocarbamic acid

ácido ditiónico dithionic acid

ácido diyodoacético diiodoacetic acid

ácido docosanoico docosanoic acid

ácido docosenoico docosenoic acid

ácido dodecanoico dodecanoic acid

ácido dodecenilsuccinico dodecenylsuccinic acid

ácido durílico durylic acid

ácido duro hard acid

ácido eicosanoico eicosanoic acid

ácido elágico ellagic acid

ácido elaídico elaidic acid

ácido elatérico elateric acid

ácido electrolítico electrolyte acid

ácido elemónico elemonic acid

ácido eleoesteárico eleostearic acid

ácido embélico embelic acid

ácido embónico embonic acid

ácido enántico enanthic acid

ácido epsilón epsilon acid

ácido eritórbico erythorbic acid

ácido erúcico erucic acid

ácido estánnico stannic acid

ácido esteárico stearic acid

ácido estífnico styphnic acid

ácido etacético ethacetic acid

ácido etanodioico ethanedioic acid

ácido etanoico ethanoic acid

ácido etanoico glacial glacial ethanoic acid

ácido etanosulfónico ethanesulfonic acid

ácido etanotiólico ethanethiolic acid

ácido etérico etheric acid

ácido etilacético ethylacetic acid

ácido etilbencenosulfónico ethylbenzenesulfonic acid

ácido etilbutírico ethylbutyric acid

ácido etilendiaminotetraacético ethylenediaminetetraacetic acid

ácido etilendiaminotetraacético tetrasódico tetrasodium ethylenediaminetetraacetic acid

ácido etilendiaminotetraacético trisódico trisodium ethylenediaminetetraacetic acid

ácido etilensulfónico ethylenesulfonic acid

ácido etilfosfórico ethylphosphoric acid

ácido etilhexoico ethylhexoic acid

ácido etinoico ethinoic acid

ácido etisulfúrico ethylsulfuric acid

ácido eugénico eugenic acid

ácido eupitónico eupittonic acid

ácido fencólico fencholic acid

ácido fénico phenic acid

ácido fenilacético phenylacetic acid

ácido fenilarsónico phenylarsonic acid

ácido fenilbutanoico phenylbutanoic acid

ácido fenilbutírico phenylbutyric acid

ácido fenilesteárico phenylstearic acid

ácido feniletanoico phenylethanoic acid

ácido feniletilacético phenylethylacetic acid

ácido fenilfórmico phenylformic acid

ácido fenilfosfínico phenylphosphinic acid

ácido fenilfosfónico phenylphosphonic acid

ácido fenilglicólico phenylglycolic acid

ácido fenílico phenylic acid

ácido fenilpropenoico phenylpropenoic acid

ácido fenilpropiónico phenylpropionic acid

ácido fenilsalicílico phenylsalicylic acid

ácido fenilsulfanílico phenylsulfanilic acid

ácido fenilsulfónico phenylsulfonic acid

ácido fenoldisulfónico phenoldisulfonic acid

ácido fenólico phenolic acid

ácido fenolsulfónico phenolsulfonic acid

ácido fenoxiacético phenoxyacetic acid

ácido ferriciánico ferricyanic acid

ácido ferrociánico ferrocyanic acid

ácido ferúlico ferulic acid

ácido filícico filicic acid

ácido filíxico filixic acid

ácido fitánico phytanic acid

ácido fítico phytic acid

ácido flaviánico flavianic acid

ácido fluobórico fluoboric acid

ácido fluohídrico fluohydric acid

ácido fluorénico fluorenic acid

ácido fluorgermánico fluorgermanic acid

ácido fluorhídrico hydrofluoric acid

ácido fluoroacético fluoroacetic acid

ácido fluorobórico fluoroboric acid

ácido fluoroetanoico fluoroethanoic acid

ácido fluorofosfórico fluorophosphoric acid

ácido fluoroplúmbico fluoroplumbic acid

ácido fluorosilícico fluorosilicic acid

ácido fluorosulfónico fluorosulfonic acid

ácido fluorosulfúrico fluorosulfuric acid

ácido fluosilícico fluosilicic acid

ácido fluosulfónico fluosulfonic acid

ácido fólico folic acid

ácido folínico folinic acid

ácido fórmico formic acid

ácido formilacético formylacetic acid

ácido formílico formylic acid

ácido formonitrólico formonitrolic acid

ácido fosfámico phosphamic acid

ácido fosfínico phosphinic acid

ácido fosfinoso phosphinous acid

ácido fosfomolíbdico phosphomolybdic acid

ácido fosfónico phosphonic acid

ácido fosfonoso phosphonous acid

ácido fosfórico phosphoric acid

ácido fosfórico anhidro anhydrous phosphoric acid

ácido fosforoso phosphorous acid

ácido fosfotúngstico phosphotungstic acid

ácido frangúlico frangulic acid

ácido frangulínico frangulinic acid

ácido ftalámico phthalamic acid

ácido ftálico phthalic acid

ácido ftioico phthioic acid

ácido fuerte strong acid

ácido fulgénico fulgenic acid

ácido fulmínico fulminic acid

ácido fulminúrico fulminuric acid

ácido fumarámico fumaramic acid

ácido fumárico fumaric acid

ácido furacrílico furacrylic acid

ácido furancarboxílico furancarboxylic acid

ácido furoico furoic acid

ácido furónico furonic acid

ácido fusídico fusidic acid

ácido gadoleico gadoleic acid

ácido gaídico gaidic acid

ácido galactárico galactaric acid

ácido galacturónico galacturonic acid

ácido gálico gallic acid

ácido galotánico gallotannic acid

ácido gamma gamma acid

ácido gammaaminobutírico gamma-aminobutyric acid

ácido gentísico gentisic acid

ácido geránico geranic acid

ácido germanofórmico germanoformic acid

ácido giberélico gibberellic acid

ácido gladiólico gladiolic acid

ácido glicérico glyceric acid

ácido glicerofosfórico glycerophosphoric acid

ácido glicídico glycidic acid

ácido glicocólico glycocholic acid
ácido glicogénico glycogenic acid
ácido glicólico glycolic acid
ácido glicónico glyconic acid
ácido glioxálico glyoxalic acid
ácido glioxílico glyoxylic acid
ácido glucónico gluconic acid
ácido glucurónico glucuronic acid
ácido glutacónico glutaconic acid
ácido glutámico glutamic acid
ácido glutárico glutaric acid
ácido glutínico glutinic acid
ácido grafítico graphitic acid
ácido graso fatty acid
ácido graso conjugado conjugated fatty
     acid
ácido guanílico guanylic acid
ácido hemelítico hemellitic acid
ácido heneicosanoico heneicosanoic acid
ácido heptacosanoico heptacosanoic acid
ácido heptadecanoico heptadecanoic acid
ácido heptafluorobutírico
     heptafluorobutyric acid
ácido heptanodicarboxílico
     heptanedicarboxylic acid
ácido heptanodioico heptanedioic acid
ácido heptanoico heptanoic acid
ácido heptílico heptylic acid
ácido heptoico heptoic acid
ácido hexacosanoico hexacosanoic acid
ácido hexadecanoico hexadecanoic acid
ácido hexadecenoico hexadecenoic acid
ácido hexadienoico hexadienoic acid
ácido hexafluorofosfórico
     hexafluorophosphoric acid
ácido hexafluorosilícico hexafluorosilicic
     acid
ácido hexahídrico hexahydric acid
ácido hexahidrobenzoico
     hexahydrobenzoic acid
ácido hexahidroftálico
     hexahydrophthalic acid
ácido hexanodioico hexanedioic acid
ácido hexanoico hexanoic acid
ácido hexenoico hexenoic acid
ácido hexilacético hexylacetic acid
ácido hexílico hexylic acid
ácido hexoico hexoic acid
ácido hialurónico hyaluronic acid

ácido hidantoico hydantoic acid
ácido hidnocárpico hydnocarpic acid
ácido hidracrílico hydracrylic acid
ácido hidratrópico hydratropic acid
ácido hidrazoico hydrazoic acid
ácido hidrocinámico hydrocinnamic acid
ácido hidrofluorosilícico
     hydrofluorosilicic acid
ácido hidrofluosilícico hydrofluosilicic
     acid
ácido hidronítrico hydronitric acid
ácido hidrosulfúrico hydrosulfuric acid
ácido hidrosulfuroso hydrosulfurous
     acid
ácido hidroxámico hydroxamic acid
ácido hidroxiacético hydroxyacetic acid
ácido hidroxiazobencenosulfónico
     hydroxyazobenzenesulfonic acid
ácido hidroxibenzoico hydroxybenzoic
     acid
ácido hidroxibutírico hydroxybutyric
     acid
ácido hidroxiesterárico hydroxystearic
     acid
ácido hidroxietanosulfónico
     hydroxyethanesulfonic acid
ácido hidroxímico hydroximic acid
ácido hidroxinaftoico hydroxynaphthoic
     acid
ácido hidroxioleico hydroxyoleic acid
ácido hidroxipropiónico
     hydroxypropionic acid
ácido hidroxisalicílico hydroxysalicylic
     acid
ácido hidroxisuccinico hydroxysuccinic
     acid
ácido hidroxitoluico hydroxytoluic acid
ácido hígrico hygric acid
ácido hipobromoso hypobromous acid
ácido hipocloroso hypochlorous acid
ácido hipofosfórico hypophosphoric acid
ácido hipofosforoso hypophosphorous
     acid
ácido hiponitroso hyponitrous acid
ácido hiposulfúrico hyposulfuric acid
ácido hiposulfuroso hyposulfurous acid
ácido hipúrico hippuric acid
ácido homoftálico homophthalic acid
ácido homogentísico homogentisic acid

ácido homosalicílico homosalicylic acid
ácido húmico humic acid
ácido icosanoico icosanoic acid
ácido imídico imidic acid
ácido iminodiacético iminodiacetic acid
ácido indolacético indoleacetic acid
ácido indolbutírico indolebutyric acid
ácido inorgánico inorganic acid
ácido inosínico inosinic acid
ácido isetiónico isethionic acid
ácido isoascórbico isoascorbic acid
ácido isobutírico isobutyric acid
ácido isociánico isocyanic acid
ácido isocianúrico isocyanuric acid
ácido isocianúrico clorado chlorinated
     isocyanuric acid
ácido isocincomerónico
     isocinchomeronic acid
ácido isocrotónico isocrotonic acid
ácido isodecanoico isodecanoic acid
ácido isoftálico isophthalic acid
ácido isonicotínico isonicotinic acid
ácido isopentanoico isopentanoic acid
ácido isotiociánico isothiocyanic acid
ácido isovalérico isovaleric acid
ácido itacónico itaconic acid
ácido junipérico juniperic acid
ácido lactámico lactamic acid
ácido láctico lactic acid
ácido lactónico lactonic acid
ácido lanocérico lanoceric acid
ácido láurico lauric acid
ácido lepargílico lepargylic acid
ácido leucónico leuconic acid
ácido levulínico levulinic acid
ácido licánico licanic acid
ácido lignocérico lignoceric acid
ácido linoleico linoleic acid
ácido linolénico linolenic acid
ácido linólico linolic acid
ácido lisérgico lysergic acid
ácido lítico lithic acid
ácido litocólico lithocholic acid
ácido maleámico maleamic acid
ácido maleico maleic acid
ácido málico malic acid
ácido malonaldehídico malonaldehydic
     acid
ácido malónico malonic acid

ácido mandélico mandelic acid
ácido mangánico manganic acid
ácido margárico margaric acid
ácido melísico melissic acid
ácido melítico mellitic acid
ácido mercaptoacético mercaptoacetic
     acid
ácido mercaptobenzoico
     mercaptobenzoic acid
ácido mercaptopropiónico
     mercaptopropionic acid
ácido mercaptosuccínico
     mercaptosuccinic acid
ácido mercaptúrico mercapturic acid
ácido mesacónico mesaconic acid
ácido mesoxálico mesoxalic acid
ácido metabórico metaboric acid
ácido metacrílico methacrylic acid
ácido metafosfórico metaphosphoric acid
ácido metalínico metanilic acid
ácido metanocarboxílico
     methanecarboxylic acid
ácido metanodicarboxílico
     methanedicarboxylic acid
ácido metanoico methanoic acid
ácido metanosulfónico methanesulfonic
     acid
ácido metasilícico metasilicic acid
ácido metatitánico metatitanic acid
ácido metilacético methylacetic acid
ácido metilaminoacético
     methylaminoacetic acid
ácido metilbenzoico methylbenzoic acid
ácido metilenodisalicílico
     methylenedisalicylic acid
ácido metilfosfónico methylphosphonic
     acid
ácido metilfosfórico methylphosphoric
     acid
ácido metilmaleico methylmaleic acid
ácido metilpentanoico methylpentanoic
     acid
ácido metilsulfónico methylsulfonic acid
ácido metilsulfúrico methylsulfuric acid
ácido metiónico methionic acid
ácido metoxiacético methoxyacetic acid
ácido metoxibenzoico methoxybenzoic
     acid
ácido mevalónico mevalonic acid

ácido mineral mineral acid

ácido mirístico myristic acid

ácido mixto mixed acid

ácido molíbdico molybdic acid

ácido molibdosilícico molybdosilicic acid

ácido monoprótico monoprotic acid

ácido múcico mucic acid

ácido muriático muriatic acid

ácido naftalenacético naphthaleneacetic
acid

ácido naftalencarboxílico
naphthalenecarboxylic acid

ácido naftalendicarboxílico
naphthalenedicarboxylic acid

ácido naftalendisulfónico
naphthalenedisulfonic acid

ácido naftalensulfínico
naphthalenesulfinic acid

ácido naftalensulfónico
naphthalenesulfonic acid

ácido naftálico naphthalic acid

ácido nafténico naphthenic acid

ácido naftilacético naphthylacetic acid

ácido naftilaminosulfónico
naphthylaminesulfonic acid

ácido naftiónico naphthionic acid

ácido naftoico naphthoic acid

ácido naftolsulfónico naphtholsulfonic
acid

ácido naftoxiacético naphthoxyacetic
acid

ácido nalidíxico nalidixic acid

ácido neodecanoico neodecanoic acid

ácido neopentanoico neopentanoic acid

ácido neotridecanoico neotridecanoic
acid

ácido nervónico nervonic acid

ácido nicotínico nicotinic acid

ácido nióbico niobic acid

ácido nisínico nisinic acid

ácido nitranílico nitranilic acid

ácido nitrante nitrating acid

ácido nítrico nitric acid

ácido nítrico fumante fuming nitric acid

ácido nitrilotriacético nitrilotriacetic
acid

ácido nitrobarbitúrico nitrobarbituric
acid

ácido nitrobencenosulfónico
nitrobenzenesulfonic acid

ácido nitrobenzoico nitrobenzoic acid

ácido nitrocinámico nitrocinnamic acid

ácido nitroclorhídrico nitrohydrochloric
acid

ácido nitrofenilacético nitrophenylacetic
acid

ácido nitrofenilpropiólico
nitrophenylpropiolic acid

ácido nitroftálico nitrophthalic acid

ácido nitrónico nitronic acid

ácido nitrosalicílico nitrosalicylic acid

ácido nitrosilsulfúrico nitrosylsulfuric
acid

ácido nitroso nitrous acid

ácido nitrosobenzoico nitrosobenzoic
acid

ácido nitroxántico nitroxanthic acid

ácido nitroxílico nitroxylic acid

ácido nonadecanoico nonadecanoic acid

ácido nonanodioico nonanedioic acid

ácido nonanoico nonanoic acid

ácido nonilfenoxiacético
nonylphenoxyacetic acid

ácido nonílico nonylic acid

ácido nonoico nonoic acid

ácido nucleico nucleic acid

ácido octacosanoico octacosanoic acid

ácido octadecanoico octadecanoic acid

ácido octadecatrienoico octadecatrienoic
acid

ácido octadecenoico octadecenoic acid

ácido octanodicarboxílico
octanedicarboxylic acid

ácido octanodioico octanedioic acid

ácido octanoico octanoic acid

ácido octilacético octylacetic acid

ácido octoico octoic acid

ácido oleico oleic acid

ácido orgánico organic acid

ácido ornitúrico ornithuric acid

ácido orótico orotic acid

ácido orto ortho acid

ácido ortoacético orthoacetic acid

ácido ortoarsénico orthoarsenic acid

ácido ortobórico orthoboric acid

ácido ortocarbónico orthocarbonic acid

ácido ortofosfórico orthophosphoric acid

ácido ortosilícico orthosilicic acid

ácido ortotúngstico orthotungstic acid
ácido ósmico osmic acid
ácido oxalacético oxalacetic acid
ácido oxaldehídico oxaldehydic acid
ácido oxálico oxalic acid
ácido oxámico oxamic acid
ácido oxibezoico oxybenzoic acid
ácido oxidante oxidizing acid
ácido oxinaftoico oxynaphthoic acid
ácido oxoetanoico oxoethanoic acid
ácido palmítico palmitic acid
ácido palmitínico palmitinic acid
ácido palmitoleico palmitoleic acid
ácido palmitólico palmitolic acid
ácido palústrico palustric acid
ácido pamoico pamoic acid
ácido pánico pannic acid
ácido pantoténico pantothenic acid
ácido parabánico parabanic acid
ácido paraciánico paracyanic acid
ácido paracónico paraconic acid
ácido paraláctido paralactic acid
ácido parapéctico parapectic acid
ácido pararosólico pararosolic acid
ácido parasórbico parasorbic acid
ácido pariético parietic acid
ácido parinárico parinaric acid
ácido péctico pectic acid
ácido pelargónico pelargonic acid
ácido penicilánico penicillanic acid
ácido penicílico penicillic acid
ácido pentacosámico pentacosamic acid
ácido pentacosanoico pentacosanoic acid
ácido pentadecanoico pentadecanoic acid
ácido pentadecílico pentadecylic acid
ácido pentadienodioico pentadienedioic
   acid
ácido pentametilbenzoico
   pentamethylbenzoic acid
ácido pentanocarboxílico
   pentanecarboxylic acid
ácido pentanodicarboxílico
   pentanedicarboxylic acid
ácido pentanodioico pentanedioic acid
ácido pentanoico pentanoic acid
ácido penténico pentenic acid
ácido pentenodioico pentenedioic acid
ácido pentenoico pentenoic acid
ácido pentilfórmico pentylformic acid

ácido pentínico pentinic acid
ácido pentoico pentoic acid
ácido pentónico pentonic acid
ácido peracético peracetic acid
ácido perbenzoico perbenzoic acid
ácido percarbónico percarbonic acid
ácido perclórico perchloric acid
ácido percrómico perchromic acid
ácido perfluorobutírico perfluorobutyric
   acid
ácido perfluorosulfónico
   perfluorosulfonic acid
ácido perfórmico performic acid
ácido perfosfórico perphosphoric acid
ácido peri peri acid
ácido permangánico permanganic acid
ácido permonosulfúrico permonosulfuric
   acid
ácido pernítrico pernitric acid
ácido peroxiacético peroxyacetic acid
ácido peroxibenzoico peroxybenzoic acid
ácido peroxietanoico peroxyethanoic
   acid
ácido peroxifórmico peroxyformic acid
ácido peroximonosulfúrico
   peroxymonosulfuric acid
ácido peroxisulfúrico peroxysulfuric
   acid
ácido peroxomonosulfúrico
   peroxomonosulfuric acid
ácido peroxonítrico peroxonitric acid
ácido perrénico perrhenic acid
ácido persulfúrico persulfuric acid
ácido peryódico periodic acid
ácido picolínico picolinic acid
ácido picrámico picramic acid
ácido picramínico picraminic acid
ácido pícrico picric acid
ácido picrolónico picrolonic acid
ácido picronítrico picronitric acid
ácido pimárico pimaric acid
ácido pimélico pimelic acid
ácido pínico pinic acid
ácido piperonílico piperonylic acid
ácido piridinocarboxílico
   pyridinecarboxylic acid
ácido piridinodicarboxílico
   pyridinedicarboxylic acid
ácido piridinopentacarboxílico

pyridinepentacarboxylic acid

ácido piridinosulfónico pyridinesulfonic acid

ácido piroacémico pyroacemic acid

ácido pirobórico pyroboric acid

ácido pirofosfórico pyrophosphoric acid

ácido pirolígneo pyroligneous acid

ácido piromelítico pyromellitic acid

ácido piromúcico pyromucic acid

ácido pirosulfúrico pyrosulfuric acid

ácido pirosulfuroso pyrosulfurous acid

ácido pirotartárico pyrotartaric acid

ácido pirovanádico pyrovanadic acid

ácido pirúvico pyruvic acid

ácido piválico pivalic acid

ácido plúmbico plumbic acid

ácido poliacrílico polyacrylic acid

ácido polibásico polybasic acid

ácido policarboxílico polycarboxylic acid

ácido polifosfórico polyphosphoric acid

ácido poliinsaturado polyunsaturated acid

ácido politiónico polythionic acid

ácido prefénico prephenic acid

ácido propanocarboxílico propanecarboxylic acid

ácido propanoico propanoic acid

ácido propenoico propenoic acid

ácido propenotricarboxílico propenetricarboxylic acid

ácido propilfórmico propylformic acid

ácido propilmalónico propylmalonic acid

ácido propiólico propiolic acid

ácido propiónico propionic acid

ácido protónico protonic acid

ácido prúsico prussic acid

ácido quebulínico chebulinic acid

ácido quelidónico chelidonic acid

ácido quercínico quercinic acid

ácido quináldico quinaldic acid

ácido quínico quinic acid

ácido quinínico quininic acid

ácido quinolínico quinolinic acid

ácido racémico racemic acid

ácido rénico rhenic acid

ácido resorcílico resorcylic acid

ácido rho rho acid

ácido ribónico ribonic acid

ácido ribonucleico ribonucleic acid

ácido ribonucleico de transferencia transfer ribonucleic acid

ácido ribosfosfórico ribosephosphoric acid

ácido ricínico ricinic acid

ácido ricinoleico ricinoleic acid

ácido rísico rissic acid

ácido rizónico rhizonic acid

ácido rodánico rhodanic acid

ácido rodizónico rhodizonic acid

ácido rosólico rosolic acid

ácido ruberítrico ruberythric acid

ácido rubitánico rubitannic acid

ácido sabínico sabinic acid

ácido sacárico saccharic acid

ácido sacarínico saccharinic acid

ácido sacarónico saccharonic acid

ácido salicílico salicylic acid

ácido sarcoláctico sarcolactic acid

ácido sebácico sebacic acid

ácido sebacílico sebacylic acid

ácido seco dry acid

ácido sedanólico sedanolic acid

ácido selénico selenic acid

ácido selenínico seleninic acid

ácido selenioso selenious acid

ácido selenónico selenonic acid

ácido siálico sialic acid

ácido silícico silicic acid

ácido silicobenzoico silicobenzoic acid

ácido silicotúngstico silicotungstic acid

ácido sílvico sylvic acid

ácido sórbico sorbic acid

ácido subérico suberic acid

ácido succinámico succinamic acid

ácido succínico succinic acid

ácido sulfámico sulfamic acid

ácido sulfamídico sulfamidic acid

ácido sulfamínico sulfaminic acid

ácido sulfanílico sulfanilic acid

ácido sulfoacético sulfoacetic acid

ácido sulfobenzoico sulfobenzoic acid

ácido sulfocarbólico sulfocarbolic acid

ácido sulfocarbónico sulfocarbonic acid

ácido sulfociánico sulfocyanic acid

ácido sulfoleico sulfoleic acid

ácido sulfónico sulfonic acid

ácido sulfonohidrazónico sulfonohydrazonic acid

ácido sulfosalicílico sulfosalicylic acid
ácido sulfoxílico sulfoxylic acid
ácido sulfúrico sulfuric acid
ácido sulfúrico fumante fuming sulfuric acid
ácido sulfuroso sulfurous acid
ácido superfosfórico superphosphoric acid
ácido tánico tannic acid
ácido tantálico tantalic acid
ácido tartárico tartaric acid
ácido tartrónico tartronic acid
ácido taurocólico taurocholic acid
ácido teicoico teichoic acid
ácido telúrico telluric acid
ácido telurínico tellurinic acid
ácido teluroso tellurous acid
ácido tenoico thenoic acid
ácido tereftálico terephthalic acid
ácido tereftaloilbenzoico
　terephthaloylbenzoic acid
ácido tetracarboxílico tetracarboxylic acid
ácido tetracloroftálico
　tetrachlorophthalic acid
ácido tetracosanoico tetracosanoic acid
ácido tetradecanoico tetradecanoic acid
ácido tetrafosfórico tetraphosphoric acid
ácido tetrahidroxibenzoico
　tetrahydroxybenzoic acid
ácido tetraoxofosfórico
　tetraoxophosphoric acid
ácido tetraoxosulfúrico tetraoxosulfuric acid
ácido tetratiónico tetrathionic acid
ácido tetrólico tetrolic acid
ácido tíglico tiglic acid
ácido tímico thymic acid
ácido timidílico thymidylic acid
ácido tioacético thioacetic acid
ácido tiobarbitúrico thiobarbituric acid
ácido tiobenzoico thiobenzoic acid
ácido tiocarbónico thiocarbonic acid
ácido tiocarboxílico thiocarboxylic acid
ácido tiociánico thiocyanic acid
ácido tiodiglicólico thiodiglycolic acid
ácido tiodipropiónico thiodipropionic acid
ácido tiofosfórico thiophosphoric acid

ácido tioglicólico thioglycolic acid
ácido tioico thioic acid
ácido tioláctico thiolactic acid
ácido tiomálico thiomalic acid
ácido tiónico thionic acid
ácido tiosalicílico thiosalicylic acid
ácido tiosulfónico thiosulfonic acid
ácido tiosulfúrico thiosulfuric acid
ácido titánico titanic acid
ácido tolilacético tolylacetic acid
ácido toluensulfónico toluenesulfonic acid
ácido toluico toluic acid
ácido toluílico toluylic acid
ácido tranexámico tranexamic acid
ácido traumático traumatic acid
ácido triacontanoico triacontanoic acid
ácido tribásico tribasic acid
ácido tribromoacético tribromoacetic acid
ácido triciánico tricyanic acid
ácido tricloroacético trichloroacetic acid
ácido tricloroetanoico trichloroethanoic acid
ácido tricloroisocianúrico
　trichloroisocyanuric acid
ácido triclorometilfosfónico
　trichloromethylphosphonic acid
ácido tricosanoico tricosanoic acid
ácido tridecanoico tridecanoic acid
ácido tridecílico tridecylic acid
ácido trifenilacético triphenylacetic acid
ácido trifluoroacético trifluoroacetic acid
ácido trifluoroetanoico trifluoroethanoic acid
ácido trifosfórico triphosphoric acid
ácido trihidroxibenzoico
　trihydroxybenzoic acid
ácido trimelítico trimellitic acid
ácido trimetilacético trimethylacetic acid
ácido trimetiladípico trimethyladipic acid
ácido trimetilbenzoico trimethylbenzoic acid
ácido trinitrobenzoico trinitrobenzoic acid
ácido trioxobórico trioxoboric acid
ácido trioxosulfúrico trioxosulfuric acid

ácido tritiónico trithionic acid
ácido trópico tropic acid
ácido túngstico tungstic acid
ácido umbélico umbellic acid
ácido undecanoico undecanoic acid
ácido undecenoico undecenoic acid
ácido undecilénico undecylenic acid
ácido undecílico undecylic acid
ácido uránico uranic acid
ácido úrico uric acid
ácido uridílico uridylic acid
ácido urobenzoico urobenzoic acid
ácido urónico uronic acid
ácido úsnico usnic acid
ácido vaccénico vaccenic acid
ácido vainíllico vanillic acid
ácido valeriánico valerianic acid
ácido valérico valeric acid
ácido vanádico vanadic acid
ácido vanadioso vanadous acid
ácido verátrico veratric acid
ácido vínico vinic acid
ácido violúrico violuric acid
ácido wolfrámico wolframic acid
ácido xántico xanthic acid
ácido xantogénico xanthogenic acid
ácido xilenosulfónico xylenesulfonic acid
ácido xilídico xylidic acid
ácido xilónico xylonic acid
ácido yodhídrico hydriodic acid
ácido yódico iodic acid
ácido yodoacético iodoacetic acid
ácido yodogorgoico iodogorgoic acid
ácido zircónico zirconic acid
acidólisis acidolysis
acidulante acidulant
acilación acylation
acilo acyl
aciloína acyloin
acondicionador de suelo soil conditioner
aconitina aconitine
acónito aconite
acoplamiento coupling
acoplamiento de Glaser Glaser coupling
acoplamiento espín-espín spin-spin
    coupling
acoplamiento espín-orbital spin-orbit
    coupling
acoplamiento geminal geminal coupling

acoplamiento oxidativo oxidative
    coupling
acre pungent, acrid
acridina acridine
acriflavina acriflavine
acrilaldehído acrylaldehyde
acrilamida acrylamide
acrilato acrylate
acrilato cálcico calcium acrylate
acrilato de alilo allyl acrylate
acrilato de butilo butyl acrylate
acrilato de cianoetilo cyanoethyl acryate
acrilato de etilhexilo ethylhexyl acrylate
acrilato de etilo ethyl acrylate
acrilato de hidroxipropilo
    hydroxypropyl acrylate
acrilato de isobutilo isobutyl acrylate
acrilato de metilo methyl acrylate
acrílico acrylic
acrilonitrilo acrylonitrile
acrimonioso acrimonious
acroleína acrolein
acromático achromatic
acrómetro acrometer
actínico actinic
actínido actinide
actinio actinium
actinismo actinism
actinoide actinoid
actinometría actinometry
actinómetro actinometer
actinomicina actinomycin
actinón actinon
actinoquímica actinochemistry
actinquinol actinoquinol
activación activation
activación fotoquímica photochemical
    activation
activación térmica thermal activation
activado activated
activador activator
activador de flotación flotation activator
actividad activity
actividad beta beta activity
actividad específica specific activity
actividad iónica ion activity
actividad magnética magnetic activity
actividad óptica optical activity
actividad química chemical activity

actividad superficial surface activity
activo active
actuador actuator
acuación aquation
acumulador accumulator
acumulador de plomo lead accumulator
acuoso aqueous
adamantano adamantane
adamsita adamsite
adaptador adapter
adátomo adatom
adenina adenine
adenosina adenosine
adherencia adherence
adhesión adhesion
adhesión molecular molecular adhesion
adhesivo adhesive
adiabático adiabatic
adición addition
adición anormal de Kharasch Kharasch abnormal addition
adición electrofílica electrophilic addition
adición nucleofílica nucleophilic addition
adición oxidativa oxidative addition
adipato adipate
adipato de dialilo diallyl adipate
adipato de didecilo didecyl adipate
adipato de diisooctilo diisooctyl adipate
adipato de isooctilo isooctyl adipate
adipincetona adipinketone
adiponitrilo adiponitrile
aditamento de mechero burner attachment
aditivo additive
aditivo antigolpeteo antiknock additive
aditivo de alimento food additive
adrenalina adrenaline
adsorbato adsorbate
adsorbente adsorbent
adsorción adsorption
adsorción activada activated adsorption
adsorción apolar apolar adsorption
adsorción cromatográfica chromatographic adsorption
adsorción de gas gas adsorption
adsorción de van der Waals van der Waals adsorption
adsorción diferencial differential

adsorption
adsorción homopolar homopolar adsorption
adsorción negativa negative adsorption
aducto adduct
aducto de urea urea adduct
adulteración adulteration
adulterado adulterated
adulterante adulterant
advección advection
advertencia de peligro hazard warning
adyuvante adjuvant
aeróbico aerobic
aerobio aerobic
aerogel aerogel
aerometría aerometry
aerómetro aerometer
aerosol aerosol
afecto de Auger Auger effect
afinidad affinity
afinidad electrónica electron affinity
afinidad química chemical affinity
aflatoxina aflatoxin
afrodina aphrodine
agar agar
agaricina agaricin
ágata agate
agente agent
agente acetilante acetylating agent
agente anibloqueo antiblock agent
agente antiespumante antifoam agent
agente antiestático antistatic agent
agente antigolpeteo antiknock agent
agente citotóxico cytotoxic agent
agente complejante complexing agent
agente de abrillantamiento brightening agent
agente de adhesión adhesion agent
agente de adición addition agent
agente de curado curing agent
agente de espesamiento thickening agent
agente de flotación flotation agent
agente de suspensión suspending agent
agente decolorante decolorizing agent
agente desespumante defoaming agent
agente disolvente dissolving agent
agente dispersante dispersing agent
agente emulsionante emulsifying agent
agente fijador fixing agent

agente floculante flocculating agent
agente halogenante halogenating agent
agente humectante wetting agent
agente impermeabilizante waterproofing agent
agente lipotrópico lipotropic agent
agente mutagénico mutagenic agent
agente oxidante oxidizing agent
agente quelante chelating agent
agente químico chemical agent
agente reductor reducing agent
agente retardante de teñido dye-retarding agent
agente secante drying agent
agente secuestrante sequestering agent
agente tensoactivo surface-active agent
agitador agitator, stirrer
agitador de vidrio glass stirrer
agitador magnético magnetic stirrer
aglicona aglycone
aglomeración agglomeration
aglucona aglucone
aglutinación agglutination, sinter, sintering
aglutinar agglutinate, sinterize
agmatina agmatine
agregación aggregation
agregación limitada por difusión diffusion limited aggregation
agregado aggregate
agrio sour
agrupamiento electrónico electron bunching
agua water
agua aireada aerated water
agua amoniacal ammonia water
agua blanda soft water
agua de bromo bromine water
agua de cal lime water
agua de cloro chlorine water
agua de constitución water of constitution
agua de cristalización water of crystallization
agua de hidratación water of hydration
agua destilada distilled water
agua dulce sweet water
agua dura hard water
agua ligera light water

agua pesada heavy water
agua purificada purified water
agua regia aqua regia
agua salina saline water
agua salobre brackish water
aire air
aire ambiente ambient air
aire líquido liquid air
airear aerate
aislado isolated
aislador insulator
aislamiento isolation
aislamiento térmico thermal insulation
aislar isolate
ajenjo wormwood, absinthium
alabandina alabandite
alabastro alabaster
alacena protectora hood
alacena protectora contra humo fume cupboard, fume hood
alacreatina alacreatine
alambique still
alanilo alanyl
alanina alanine
alanita allanite
alantoína allantoin
alarma de fuego fire alarm
alarma de llama flame alarm
alazopeptina alazopeptin
albaspidina albaspidin
albata albata
albedo albedo
albertita albertite
albita albite
albomicina albomycin
albumen albumen
albúmina albumin
albuminato albumitate
albuminato de cinc zinc albuminate
albuminoide albuminoid
albutoína albutoin
alcadieno alkadiene
alcalamida alkalamide
alcalescencia alkalescence
alcalescente alkalescent
álcali alkali
álcali mineral mineral alkali
alcalida alkalide
alcalimetría alkalimetry

alcalímetro alkalimeter
alcalinidad alkalinity
alcalino alkaline
alcalizar alkalize
alcaloidal alkaloidal
alcaloide alkaloid
alcalometría alkalometry
alcamina alkamine
alcanal alkanal
alcanfor camphor
alcanfor brominado brominated
    camphor
alcanfor de alquitrán tar camphor
alcanfor de menta peppermint camphor
alcanfor de trementina turpentine
    camphor
alcanina alkannin
alcanización alkanization
alcano alkane
alcanol alkanol
alcanolamina alkanolamine
alcasal alkasal
alcilo alcyl
alclofenac alclophenac
alcogel alcogel
alcohol alcohol
alcohol absoluto absolute alcohol
alcohol ácido acid alcohol
alcohol alílico allyl alcohol
alcohol amilcinámico amylcinnamic
    alcohol
alcohol amílico amyl alcohol
alcohol aminoetílico aminoethyl alcohol
alcohol aminopropílico aminopropyl
    alcohol
alcohol anhidro anhydrous alcohol
alcohol anísico anisic alcohol
alcohol anisílico anisyl alcohol
alcohol aromático aromatic alcohol
alcohol batílico batyl alcohol
alcohol behenílico behenyl alcohol
alcohol bencílico benzyl alcohol
alcohol benzoico benzoic alcohol
alcohol bornílico bornyl alcohol
alcohol bromoetílico bromoethyl alcohol
alcohol butílico butyl alcohol
alcohol butírico butyric alcohol
alcohol cáustico caustic alcohol
alcohol cerílico ceryl alcohol

alcohol cetílico cetyl alcohol
alcohol cinámico cinnamic alcohol
alcohol cinamílico cinnamyl alcohol
alcohol cloroetílico chloroethyl alcohol
alcohol cloroisopropílico
    chloroisopropyl alcohol
alcohol coniferilo coniferyl alcohol
alcohol crotílico crotyl alcohol
alcohol cúmico cumic alcohol
alcohol cumínico cuminic alcohol
alcohol de grano grain alcohol
alcohol de madera wood alcohol
alcohol decatílico decatyl alcohol
alcohol decílico decyl alcohol
alcohol desnaturalizado denatured
    alcohol
alcohol dicloroisopropílico
    dichloroisopropyl alcohol
alcohol dicloropropílico dichloropropyl
    alcohol
alcohol dihídrico dihydric alcohol
alcohol dodecílico dodecyl alcohol
alcohol erucílico erucyl alcohol
alcohol estearílico stearyl alcohol
alcohol estiralílico styralyl alcohol
alcohol etanílico ethanyl alcohol
alcohol etilbencílico ethylbenzyl alcohol
alcohol etilbutílico ethylbutyl alcohol
alcohol etilhexílico ethylhexyl alcohol
alcohol etílico ethyl alcohol
alcohol fenacílico phenacyl alcohol
alcohol fenetílico phenethyl alcohol
alcohol feniletílico phenylethyl alcohol
alcohol fenilpropílico phenylpropyl
    alcohol
alcohol fenquílico fenchyl alcohol
alcohol furfurílico furfuryl alcohol
alcohol glicílico glycyl alcohol
alcohol graso fatty alcohol
alcohol heptílico heptyl alcohol
alcohol hexacosílico hexacosyl alcohol
alcohol hexadecílico hexadecyl alcohol
alcohol hexahídrico hexahydric alcohol
alcohol hexílico hexyl alcohol
alcohol hidroabietílico hydroabietyl
    alcohol
alcohol hidrocinámico hydrocinnamic
    alcohol
alcohol hidroxibencílico hydroxybenzyl

alcohol

alcohol hidroxiestearílico hydroxystearyl alcohol

alcohol industrial industrial alcohol

alcohol isoamílico isoamyl alcohol

alcohol isobutílico isobutyl alcohol

alcohol isononílico isononyl alcohol

alcohol isooctílico isooctyl alcohol

alcohol isopentílico isopentyl alcohol

alcohol isopropílico isopropyl alcohol

alcohol laurílico lauryl alcohol

alcohol linolenílico linolenyl alcohol

alcohol melisílico melissyl alcohol

alcohol metalílico methallyl alcohol

alcohol metilalílico methylallyl alcohol

alcohol metilamílico methylamyl alcohol

alcohol metilbencílico methylbenzyl alcohol

alcohol metílico methyl alcohol

alcohol metoxibencílico methoxybenzyl alcohol

alcohol miristílico myristyl alcohol

alcohol monohídrico monohydric alcohol

alcohol naftílico naphthyl alcohol

alcohol nitrobencílico nitrobenzyl alcohol

alcohol nonílico nonyl alcohol

alcohol octadecílico octadecyl alcohol

alcohol octílico octyl alcohol

alcohol oleílico oleyl alcohol

alcohol palmitílico palmityl alcohol

alcohol pelargónico pelargonic alcohol

alcohol pentadecílico pentadecyl alcohol

alcohol pentílico pentyl alcohol

alcohol pirúvico pyruvic alcohol

alcohol polifurfurílico polyfurfuryl alcohol

alcohol polihídrico polyhydric alcohol

alcohol polivinílico polyvinyl alcohol

alcohol primario primary alcohol

alcohol propargílico propargyl alcohol

alcohol propenílico propenyl alcohol

alcohol propílico propyl alcohol

alcohol ricinoleílico ricinoleyl alcohol

alcohol salicílico salicyl alcohol

alcohol secundario secondary alcohol

alcohol seudohexílico pseudohexyl alcohol

alcohol tanacetílico tanacetyl alcohol

alcohol tenílico thenyl alcohol

alcohol terciario tertiary alcohol

alcohol tetrahidrofurfurílico tetrahydrofurfuryl alcohol

alcohol tioetílico thioethyl alcohol

alcohol tridecílico tridecyl alcohol

alcohol trihídrico trihydric alcohol

alcohol undecilénico undecylenic alcohol

alcohol undecílico undecyl alcohol

alcohol vinílico vinyl alcohol

alcoholato alcoholate

alcoholato alcalino alkali alcoholate

alcoholato de tántalo tantalum alcoholate

alcohólico alcoholic

alcohólisis alcoholysis

alcosol alcosol

alcoxi alkoxy

alcóxido alkoxide

alcóxido alumínico aluminum alkoxide

alcoxilato alkoxylate

aldehidina aldehydine

aldehído aldehyde

aldehído acético acetic aldehyde

aldehído acrílico acrylic aldehyde

aldehído alílico allyl aldehyde

aldehído amilcinámico amylcinnamic aldehyde

aldehído amílico amyl aldehyde

aldehído amónico ammonium aldehyde

aldehído anísico anisic aldehyde

aldehído aromático aromatic aldehyde

aldehído benzoico benzoic aldehyde

aldehído butílico butyl aldehyde

aldehído butírico butyric aldehyde

aldehído cáprico capric aldehyde

aldehído caproico caproic aldehyde

aldehído cinámico cinnamic aldehyde

aldehído cinamílico cinnamyl aldehyde

aldehído crotónico crotonic aldehyde

aldehído cúmico cumic aldehyde

aldehído cumínico cuminic aldehyde

aldehído decílico decyl aldehyde

aldehído fenilpropílico phenylpropyl aldehyde

aldehído fenólico phenol aldehyde

aldehído fórmico formic aldehyde

aldehído glicérico glyceric aldehyde

aldehído glicólico glycolic aldehyde

aldehído hidrocinámico hydrocinnamic

aldehyde

aldehído isobutílico isobutyl aldehyde

aldehído láurico lauric aldehyde

aldehído laurílico lauryl aldehyde

aldehído levulínico levulinic aldehyde

aldehído naftílico naphthyl aldehyde

aldehído naftoico naphthoic aldehyde

aldehído nonílico nonyl aldehyde

aldehído octadecenílico octadecenyl
aldehyde

aldehído octílico octyl aldehyde

aldehído oleílico oleyl aldehyde

aldehído pelargónico pelargonic
aldehyde

aldehído pentílico pentyl aldehyde

aldehído pirúvico pyruvic aldehyde

aldehído propílico propyl aldehyde

aldehído propiónico propionic aldehyde

aldehído salicílico salicylic aldehyde

aldehído toluico toluic aldehyde

aldehído vainíllico vanillic aldehyde

aldehído valérico valeric aldehyde

aldicarb aldicarb

aldocetena aldoketene

aldohexosa aldohexose

aldol aldol

aldolasa aldolase

aldopentosa aldopentose

aldosa aldose

aldosterona aldosterone

aldoxima aldoxime

aldrina aldrin

aleación alloy

aleación alumínica aluminum alloy

aleación binaria binary alloy

aleación cerámica ceramic alloy

aleación de Arndt Arndt alloy

aleación de boro boron alloy

aleación de cobalto cobalt alloy

aleación de cobre copper alloy

aleación de Devarda Devarda alloy

aleación de hierro-níquel iron-nickel
alloy

aleación de mercurio mercury alloy

aleación de níquel-hierro nickel-iron
alloy

aleación de oro gold alloy

aleación de plata silver alloy

aleación de platino-cobalto platinum-

cobalt alloy

aleación de platino-iridio platinum-
iridium alloy

aleación de platino-rodio platinum-
rhodium alloy

aleación de silicio silicon alloy

aleación de silicio-cobre silicon-copper
alloy

aleación de silicio-oro silicon-gold alloy

aleación de sodio-plomo sodium-lead
alloy

aleación de sodio-potasio sodium-
potassium alloy

aleación de titanio titanium alloy

aleación de tungsteno tungsten alloy

aleación eutéctica eutectic alloy

aleación fusible fusible alloy

aleación magnésica magnesium alloy

aleación pesada heavy alloy

alelomorfo allelomorph

aleloquímico allelochemical

alelotropismo allelotropism

alelótropo allelotrope

alemontita allemontite

aleno allene

aletrina allethrin

alexandrita alexandrite

alfa alpha

alfahidroxiácido alphahydroxy acid

alfatópico alphatopic

alfilo alphyl

alfin alfin

algarosa algarose

algas algae

algicida algicide

algina algin

alginato alginate

alginato amónico ammonium alginate

alginato cálcico calcium alginate

alginato potásico potassium alginate

alginato sódico sodium alginate

algodón cotton

algodón polvora guncotton

algulosa algulose

alheña henna

alicíclico alicyclic

alicina allicin

alícuota aliquot

alifático aliphatic

aligación alligation
alilacetona allylacetone
alilacetonitrilo allylacetonitrile
alilamina allylamine
alilanilina allylaniline
alileno allylene
alílico allylic
alilina allylin
alilo allyl
aliltiourea allylthiourea
aliltriclorosilano allyltrichlorosilane
alimentador feeder
alineación alignment
alipita alipite
alita alite
alizaramida alizaramide
alizarato alizarate
alizarina alizarin
almacenamiento storage
almacenamiento seguro safe storage
almandita almandite
almidón starch
almidón de maíz cornstarch
almidón soluble soluble starch
almizcle musk
alneón alneon
álnico alnico
alo allo
aloclasita alloclasite
alocromático allochromatic
áloe aloe
alófana allophane
alofanamida allophanamide
alofanato allophanate
alofánico allophanic
aloína aloin
aloisomerismo alloisomerism
alomérico allomeric
alomerismo allomerism
alomerización allomerization
alomorfismo allomorphism
alonógeno alunogen
alosa allose
alotoxina allotoxin
alotriomorfismo allotriomorphism
alotropía allotropy
alotrópico allotropic
alotropismo allotropism
alótropo allotrope

aloxana alloxan
aloxantina alloxanthin
aloxiprina aloxiprin
alpax alpax
alqueno alkene
alquilación alkylation
alquilación de sulfitos de Strecker
    Strecker sulfite alkylation
alquilamina alkylamine
alquilato alkylate
alquilbenceno alkylbenzene
alquileno alkylene
alquilfenol alkylphenol
alquilideno alkylidene
alquilo alkyl
alquilolamida alkylolamide
alquilolamina alkylolamine
alquiluro alkylide
alquino alkyne
alquitrán tar
alquitrán de hulla coal tar
alquitrán de madera wood tar
alquitrán de pino pine tar
alquitrán de turba peat tar
alquitrán mineral mineral tar
alstonidina alstonidine
alstonina alstonine
alta definición high definition
alta densidad high density
alta energía high energy
alta presión high pressure
alta temperatura high temperature
alteína altheine
alternación alternation
alto horno blast furnace
alto polímero high polymer
alto vacío high vacuum
altrosa altrose
alucinógeno hallucinogen
aludel aludel
alulosa allulose
alumbre alum
alumbre amónico ammonium alum
alumbre amónico férrico ferric
    ammonium alum
alumbre de cesio cesium alum
alumbre de cromo chrome alum
alumbre de filtro filter alum
alumbre de rubidio rubidium alum

alumbre de sosa soda alum
alumbre de talio thallium alum
alumbre negro black alum
alumbre potásico potassium alum, potash alum
alumbre sódico sodium alum
alúmina alumina
alúmina activada activated alumina
alúmina hidratada hydrated alumina
aluminato aluminate
aluminato bárico barium aluminate
aluminato cálcico calcium aluminate
aluminato cobaltoso cobaltous aluminate
aluminato de cobalto cobalt aluminate
aluminato de litio lithium aluminate
aluminato magnésico magnesium aluminate
aluminato potásico potassium aluminate
aluminato sódico sodium aluminate
aluminífero aluminiferous
aluminio aluminum
aluminita aluminite
aluminizado aluminized
aluminoférrico aluminoferric
aluminosilicato aluminosilicate
aluminosilicato potásico potassium aluminosilicate
aluminosilicato sódico sodium aluminosilicate
alumita alumite
alunita alunite
alunogenita alunogenite
alvita alvite
amalgama amalgam
amalgama de cadmio cadmium amalgam
amalgama de cinc zinc amalgam
amalgama de cobre copper amalgam
amalgama de mercurio mercury amalgam
amalgama de plata silver amalgam
amalgama de talio thallium amalgam
amalgama dental dental amalgam
amalgama sódica sodium amalgam
amalgamación amalgamation
amalgamador amalgamator
amanitina amanitine
amaranto amaranth
amargo bitter
amarillo brillante brilliant yellow

amarillo congo congo yellow
amarillo de alizarina alizarin yellow
amarillo de anilina aniline yellow
amarillo de antimonio antimony yellow
amarillo de bismuto bismuth yellow
amarillo de cadmio cadmium yellow
amarillo de cinc zinc yellow
amarillo de cobalto cobalt yellow
amarillo de cromo chrome yellow
amarillo de mantequilla butter yellow
amarillo de metilo methyl yellow
amarillo de naftol naphthol yellow
amarillo de óxido de hierro iron oxide yellow
amarillo de titanio titanellow
amarillo de uranina uranine yellow
amarina amarine
amaroide amaroid
amatista amethyst
amatol amatol
amazonita amazonite
ámbar amber
ambidente ambident
ambiente de laboratorio laboratory environment
ambiente peligroso hazardous environment
ambligonita amblygonite
ambrita ambrite
ameboide ameboid
amelina ammeline
americio americium
amesita amesite
ametocaína amethocaine
amicrón amicron
amíctico amyctic
amida amide
amida cíclica cyclic amide
amida de ácido acid amide
amida de ácido acético acetic acid amide
amida de ácido nicotínico nicotinic acid amide
amida de cloral chloral amide
amida de litio lithium amide
amida magnésica magnesium amide
amida málica malic amide
amida malónica malonic amide
amida salicílica salicylic amide
amida sódica sodium amide

amida xántica xanthic amide
amidación amidation
amidasa amidase
amidina amidine
amidino amidino
amido amido
amidol amidol
amidona amidone
amidrazona amidrazone
amigdalina amygdalin
amilamina amylamine
amilasa amylase
amilato amylate
amilbenceno amylbenzene
amilcarbinol amylcarbinol
amileno amylene
amilfenol amylphenol
amilo amyl
amiloide amyloid
amilólisis amylolysis
amilolítico amylolytic
amilopectina amylopectin
amilosa amylose
amiltriclorosilano amyltrichlorosilane
amina amine
amina aromática aromatic amine
amina de ácido acético acetic acid amine
amina de paladio palladium amine
amina de platino platinum amine
amina grasa fatty amine
amina primaria primary amine
amina secundaria secondary amine
amina terciaria tertiary amine
aminación amination
aminación de Voight Voight amination
aminilo aminyl
aminimida aminimide
amino amino
aminoacetanilida aminoacetanilide
aminoacetofenona aminoacetophenone
aminoacetona aminoacetone
aminoácido amino acid
aminoácido oxidasa amino acid oxidase
aminoalcohol amino alcohol
aminoantipirina aminoantipyrine
aminoantraquinona aminoanthraquinone
aminoazo aminoazo
aminoazobenceno aminoazobenzene
aminoazonaftaleno aminoazonaphthalene

aminoazotolueno aminoazotoluene
aminobenceno aminobenzene
aminobenzoato potásico potassium aminobenzoate
aminobenzotiazol aminobenzothiazole
aminobifenilo aminobiphenyl
aminobutano aminobutane
aminocetona aminoketone
aminociclohexano aminocyclohexane
aminodietilanilina aminodiethylaniline
aminodifenilamina aminodiphenylamine
aminodifenilo aminodiphenyl
aminodimetilanilina aminodimethylaniline
aminodimetilbenceno aminodimethylbenzene
aminoetano aminoethane
aminoetanol aminoethanol
aminofenol aminophenol
aminofilina aminophylline
aminoformo aminoform
aminohexano aminohexane
aminohipurato sódico sodium aminohippurate
aminoide aminoid
aminometano aminomethane
aminometilación aminomethylation
aminonaftaleno aminonaphthalene
aminonaftol aminonaphthol
aminooxidasa amino oxidase
aminopicolina aminopicoline
aminopiridina aminopyridine
aminopirina aminopyrine
aminopropano aminopropane
aminopropanol aminopropanol
aminopropeno aminopropene
aminopropilmorfolina aminopropylmorpholine
aminopterina aminopterin
aminopurina aminopurine
aminoquinolina aminoquinoline
aminorresina amino resin
aminosalicilato sódico sodium aminosalicylate
aminosulfonato amónico ammonium aminosulfonate
aminotiazol aminothiazole
aminotiopeno aminothiopene
aminotiourea aminothiourea

aminotolueno aminotoluene
aminotriazol aminotriazole
aminourea aminourea
aminoxileno aminoxylene
amirina amyrin
amital sódico sodium amytal
amitiozona amithiozone
amitrol amitrole
amobarbital amobarbital
amonación ammonation
amoniacal ammoniacal
amoníaco ammonia
amoníaco anhidro anhydrous ammonia
amoníaco saturado saturated ammonia
amonificación ammonification
amonio ammonium
amonio férrico ferric ammonium
amonio ferroso ferrous ammonium
amonólisis ammonolysis
amorfismo amorphism
amorfo amorphous
amortiguador buffer
amortiguador de fosfato phosphate
    buffer
amortiguamiento buffering
ampelita ampelite
amperometría amperometry
ampicilina ampicillin
amplitud molecular molecular amplitude
ampolla ampule
anabasina anabasine
anabólico anabolic
anabolismo anabolism
anaerobio anaerobic
anafilaxis anaphylaxis
anaforesis anaphoresis
análisis analysis
análisis absorciométrico absorptiometric
    analysis
análisis areamétrico areametric analysis
análisis colorimétrico colorimetric
    analysis
análisis complejométrico
    complexometric analysis
análisis con rayos X X-ray analysis
análisis con sonda electrónica electron
    probe analysis
análisis conductométrico conductometric
    analysis

análisis conformacional conformational
    analysis
análisis cualitativo qualitative analysis
análisis cuantitativo quantitative analysis
análisis de activación neutrónica
    neutron activation analysis
análisis de adsorción adsorption analysis
análisis de agua water analysis
análisis de fluorescencia de rayos X X-
    ray fluorescence analysis
análisis de gases gas analysis
análisis de gases despedidos evolved gas
    analysis
análisis de Podbielniak Podbielniak
    analysis
análisis de rayos positivos positive-ray
    analysis
análisis de trazas trace analysis
análisis electrográfico electrographic
    analysis
análisis electrolítico electrolytic analysis
análisis elemental elementary analysis
análisis espectral spectral analysis,
    spectrum analysis
análisis espectrofotométrico
    spectrophotometric analysis
análisis espectrométrico spectrometric
    analysis
análisis espectroscópico spectroscopic
    analysis
análisis físico physical analysis
análisis fluorométrico fluorometric
    analysis
análisis forense forensic analysis
análisis fotométrico photometric analysis
análisis gravimétrico gravimetric
    analysis
análisis húmedo wet analysis
análisis mecánico mechanical analysis
análisis orgánico organic analysis
análisis polarimétrico polarimetric
    analysis
análisis polarográfico polarographic
    analysis
análisis por activación activation
    analysis
análisis por difusión diffusion analysis
análisis por electrodeposición
    electrodeposition analysis

análisis por fluorescencia fluorescence analysis

análisis químico chemical analysis

análisis radiométrico radiometric analysis

análisis térmico thermal analysis

análisis térmico diferencial differential thermal analysis

análisis termocinético thermokinetic analysis

análisis termogravimétrico thermogravimetric analysis

análisis termométrico thermometric analysis

análisis turbidimétrico turbidimetric analysis

análisis volumétrico volumetric analysis

analítico analytical

analizador analyzer

analizador de Orsat Orsat analyzer

analizador espectral spectrum analyzer

análsis espectroquímico spectrochemical analysis

anamorfismo anamorphism

anaranjado de acridina acridine orange

anaranjado de antimonio antimony orange

anaranjado de bronce bronze orange

anaranjado de cromo chrome orange

anaranjado de metilo methyl orange

anaranjado de molibdeno molybdenum orange

anaranjado de naftol naphthol orange

anarcotina anarcotine

anarmonicidad anharmonicity

anarmónico anharmonic

anatasa anatase

andalucita andalusite

andesina andesine

andrógeno androgen

androstano androstane

androsterona androsterone

anemómetro anemometer

anemosita anemosite

anestesia anesthesia

anestésico anesthetic

anetol anethole

anfetamina amphetamine

anfibol amphibole

anfibolita amphibolite

anficroico amphichroic

anficromático amphichromatic

anfifático amphiphatic

anfifílico amphiphilic

anfífilo amphiphile

anfiprótico amphiprotic

anfolito ampholyte

anfolitoide ampholytoid

anfoterismo amphoterism

anfótero amphoteric

anglesita anglesite

angosturina angosturine

angstrom angstrom

angular angular

ángulo angle

ángulo de Bragg Bragg angle

ángulo de Brewster Brewster angle

ángulo de enlace bond angle

ángulo de valencia valence angle

ángulo dihedral dihedral angle

ángulo tetraédrico tetrahedral angle

anhidrasa carbónica carbonic anhydrase

anhídrido anhydride

anhídrido abiético abietic anhydride

anhídrido acético acetic anhydride

anhídrido antimónico antimonic anhydride

anhídrido arsénico arsenic anhydride

anhídrido básico basic anhydride

anhídrido benzoico benzoic anhydride

anhídrido bórico boric anhydride

anhídrido butírico butyric anhydride

anhídrido caprílico caprylic anhydride

anhídrido cíclico cyclic anhydride

anhídrido cinámico cinnamic anhydride

anhídrido citacrónico citraconic anhydride

anhídrido cloréndico chlorendic anhydride

anhídrido cloroacético chloroacetic anhydride

anhídrido clorocrómico chlorochromic anhydride

anhídrido cloroetanoico chloroethanoic anhydride

anhídrido cloromaleico chloromaleic anhydride

anhídrido crómico chromic anhydride

anhídrido de ácido acid anhydride

anhídrido de ácido benzoico benzoic acid anhydride

anhídrido de ácido perclórico perchloric acid anhydride

anhídrido de ácido vanádico vanadic acid anhydride

anhídrido de ácido yódico iodic acid anhydride

anhídrido de antimonio antimony anhydride

anhídrido de eritritol erythritol anhydride

anhídrido de estaño tin anhydride

anhídrido de sorbitol sorbitol anhydride

anhídrido de torio thorium anhydride

anhídrido de urea urea anhydride

anhídrido decanoico decanoic anhydride

anhídrido estánnico stannic anhydride

anhídrido estíbico stibic anhydride

anhídrido etanoico ethanoic anhydride

anhídrido fosfórico phosphoric anhydride

anhídrido ftálico phthalic anhydride

anhídrido glutárico glutaric anhydride

anhídrido isatoico isatoic anhydride

anhídrido isobutírico isobutyric anhydride

anhídrido láctico lactic anhydride

anhídrido maleico maleic anhydride

anhídrido memtetrahidroftálico memtetrahydrophthalic anhydride

anhídrido molíbdico molybdic anhydride

anhídrido ósmico osmic anhydride

anhídrido propiónico propionic anhydride

anhídrido succínico succinic anhydride

anhídrido sulfobenzoico sulfobenzoic anhydride

anhídrido sulfoftálico sulfophthalic anhydride

anhídrido sulfónico sulfonic anhydride

anhídrido sulfúrico sulfuric anhydride

anhídrido tetracloroftálico tetrachlorophthalic anhydride

anhídrido tetrahidroftálico tetrahydrophthalic anhydride

anhídrido trimelítico trimellitic anhydride

anhídrido túngstico tungstic anhydride

anhídrido valérico valeric anhydride

anhídrido vanádico vanadic anhydride

anhídrido zircónico zirconic anhydride

anhidrita anhydrite

anhidro anhydrous

anhidrona anhydrone

anhidrosíntesis anhydrosynthesis

anibina anibine

anilida anilide

anilida hidroxinaftoica hydroxynaphthoic anilide

anilina aniline

anilina de formaldehído formaldehyde aniline

anilinato anilinate

anilinio anilinium

anilino anilino

anillo ring

anillo bencénico benzene ring

anillo de Kekulé Kekulé ring

anillo de septanosa septanose ring

anillo fundido fused ring

anillo heteroatómico heteroatomic ring

anillo homoatómico homoatomic ring

anillos de Raschig Raschig rings

anión anion

anión ambidente ambident anion

aniónico anionic

anionotropía anionotropy

anisal anisal

anisaldehído anisaldehyde

anísico anisic

anisidina anisidine

anisilacetona anisylacetone

anisilo anisyl

anisoilo anisoyl

anisol anisole

anisomérico anisomeric

anisonitrilo anisonitrile

anisotónico anisotonic

anisotrópico anisotropic

anisótropo anisotropic

anitácido antacid

anitol anytol

anódico anodic

anodización anodizing

anodizar anodize

ánodo anode

ánodo acelerador accelerating anode
ánodo galvánico galvanic anode
ánodo reactivo reactive anode
anolito anolyte
anomalía anomaly
anómalo anomalous
anomérico anomeric
anómero anomer
anorgánico anorganic
anórtico anorthic
anortita anorthite
anoxinaftonato sódico sodium
    anoxynaphthonate
antagonismo antagonism
antagonista antagonist
antagonista estructural structural
    antagonist
antalcalino antalkaline
antemol anthemol
anteojos de seguridad safety goggles
antiadherente abherent
antiaromático antiaromatic
antibacterial antibacterial
antibacteriano antibacterial
antibiótico antibiotic
anticatalizador anticatalyzer
anticátodo anticathode
anticloro antichlor
anticoagulante anticoagulant
anticolinérgico anticholingeric
anticomplemento anticomplement
anticongelante antifreeze
anticorrosivo anticorrosive
anticuerpo antibody
antideflagrante flameproof
antidetonante antiknock
antídoto antidote
antienlazado antibonding
antienzima antienzyme
antiestático antistatic
antiferromagnético antiferromagnetic
antiferromagnetismo
    antiferromagnetism
antiformina antiformin
antígeno antigen
antiglobulina antiglobulin
antigolpeteo antiknock
antigorita antigorite
antihistamina antihistamine

antiisómero antiisomer
antiisomorfismo antiisomorphism
antimateria antimatter
antímero antimer
antimetabolito antimetabolite
antimicina antimycin
antimoniato antimonate
antimoniato amónico ammonium
    antimoniate
antimoniato de plomo lead antimonate
antimoniato potásico potassium
    antimonate
antimoniato sódico sodium antimonate
antimónico antimonic
antimonilo antimonyl
antimonio antimony
antimonio cáustico caustic antimony
antimonioso antimonous
antimonita antimonite
antimoniuro alumínico aluminum
    antimonide
antimoniuro de bismuto bismuth
    antimonide
antimoniuro de cadmio cadmium
    antimonide
antimoniuro de cesio cesium antimonide
antimoniuro de cinc zinc antimonide
antimoniuro de galio gallium antimonide
antimoniuro de indio indium antimonide
antimoniuro de itrio yttrium antimonide
antimoniuro de lantano lanthanum
    antimonide
antimoniuro de manganeso crómico
    chromium manganese antimonide
antimoniuro de plata silver antimonide
antineutrón antineutron
antión anthion
antioxidante antioxidant
antioxígeno antioxygen
antiozonizante antiozonant
antiparasítico antiparasitic
antipartícula antiparticle
antiperspirante antiperspirant
antipireno antipyrene
antipirético antipyretic
antipirilo antipyryl
antipirina antipyrine
antiprotón antiproton
antipsicótico antipsychotic

antiquinol antiquinol
antirradiación antiradiation
antiséptico antiseptic
antitóxico antitoxic
antitoxina antitoxin
antitusivo antitussive
antiviral antiviral
antizimótico antizymotic
antocianidina anthocyanidin
antocianina anthocyanin
antofilita anthophyllite
antozono antozone
antraceno anthracene
antracita anthracite
antradiol anthradiol
antragalol anthragallol
antrahidroquinona anthrahydroquinone
antraldehído anthraldehyde
antralina anthralin
antralinato de butilo butyl anthranilate
antranil anthranil
antranilato potásico potassium
  anthranilate
antranol anthranol
antranona anthranone
antraparaceno anthraparazene
antrapurpurina anthrapurpurin
antraquinona anthraquinone
antrarrufina anthrarufin
antratetrol anthratetrol
antratriol anthratriol
antrileno anthrylene
antrilo anthryl
antrol anthrol
antrona anthrone
anular annular
anuleno annulene
añil indigo
apantallamiento screening
aparato apparatus
aparato de absorción absorption
  apparatus
aparato de análisis de gases gas-analysis
  apparatus
aparato de conductividad conductivity
  apparatus
aparato de destilación distillation
  apparatus
aparato de extracción extraction
  apparatus

aparato de Folin Folin apparatus
aparato de Kipp Kipp apparatus
aparato de laboratorio laboratory
  apparatus
aparato de Orsat Orsat apparatus
aparato de Pensky-Martens Pensky-
  Martens apparatus
aparato de punto de congelación
  freezing-point apparatus
aparato de punto de ebullición boiling-
  point apparatus
aparato de Thiele Thiele apparatus
apatita apatite
apionol apionol
apoatropina apoatropine
apocarotenal apocarotenal
apocodeína apocodeine
apofilita apophyllite
apomorfina apomorphine
aprótico aprotic
aproximación de Born-Oppenheimer
  Born-Oppenheimer approximation
aproximación de Hückel Hückel
  approximation
aproximado approximate
aquiral achiral
arabina arabin
arabinosa arabinose
arabinosida arabinoside
arabita arabite
arabitol arabitol
aragonita aragonite
aralquilo aralkyl
aramida aramid
araquidato de metilo methyl arachidate
arborescente arborescent
arbutina arbutin
arcilla clay
arcilla refractaria fireclay
área peligrosa hazardous area
área superficial surface area
arecaidina arecaidine
arecaína arecaine
arecolina arecoline
arena sand
arena de filtración filter sand
arena de sílice silica sand
areno arene

areómetro areometer
aresniato de cobalto cobalt arsenate
argenato argenate
argentamina argentamine
argentán argentan
argéntico argentic
argentita argentite
argentocianuro argentocyanide
argentocianuro potásico potassium argentocyanide
argentometría argentometry
argentopirita argentopyrite
arginasa arginase
arginina arginine
argirodita argyrodite
argol argol
argón argon
aribina aribine
arilalquilo arylalkyl
arilamina arylamine
arilarsonato arylarsonate
arileno arylene
arilo aryl
ariluro arylide
arino aryne
arnicina arnicin
aroilo aroyl
aromaticidad aromaticity
aromático aromatic
aromatización aromatization
arrurruz arrowroot
arsacetina arsacetin
arsanilato arsanilate
arsanilato sódico sodium arsanilate
arseniato arsenate
arseniato amónico ammonium arsenate
arseniato bárico barium arsenate
arseniato cálcico calcium arsenate
arseniato cobaltoso cobaltous arsenate
arseniato cúprico cupric arsenate
arseniato de cinc zinc arsenate
arseniato de cobre copper arsenate
arseniato de hierro iron arsenate
arseniato de litio lithium arsenate
arseniato de manganeso manganese arsenate
arseniato de mercurio mercury arsenate
arseniato de níquel nickel arsenate
arseniato de plata silver arsenate

arseniato de plomo lead arsenate
arseniato férrico ferric arsenate
arseniato ferroso ferrous arsenate
arseniato magnésico magnesium arsenate
arseniato magnésico amónico ammonium magnesium arsenate
arseniato manganoso manganous arsenate
arseniato mercúrico mercuric arsenate
arseniato niqueloso nickelous arsenate
arseniato potásico potassium arsenate
arseniato sódico sodium arsenate
arsenical arsenical
arsénico arsenic
arsénico blanco white arsenic
arsénico rojo red arsenic
arsenidina arsenidine
arsenilo arsenyl
arsenioso arsenous
arsenito arsenite
arsenito amónico ammonium arsenite
arsenito cálcico calcium arsenite
arsenito de cinc zinc arsenite
arsenito de cobre copper arsenite
arsenito de estroncio strontium arsenite
arsenito de plata silver arsenite
arsenito de plomo lead arsenite
arsenito férrico ferric arsenite
arsenito potásico potassium arsenite
arsenito sódico sodium arsenite
arseniuro arsenide
arseniuro de antimonio antimony arsenide
arseniuro de cesio cesium arsenide
arseniuro de cinc zinc arsenide
arseniuro de cromo chromium arsenide
arseniuro de galio gallium arsenide
arseniuro de indio indium arsenide
arseniuro de itrio yttrium arsenide
arseniuro de lantano lanthanum arsenide
arseniuro de plata silver arsenide
arseniuro sódico sodium arsenide
arseno arseno
arsenobenceno arsenobenzene
arsenolita arsenolite
arsenopirita arsenopyrite
arsina arsine
arsinoso arsinoso
arsonato arsonate

arsonio arsonium
arsono arsono
arsonoilo arsonoyl
arsonoso arsonoso
artesa boat
artesa de combustión combustion boat
artificial artificial
artinita artinite
artranitín arthranitin
asarilo asaryl
asarona asarone
asbestina asbestine
asbesto asbestos
asbolana asbolane
asbolita asbolite
ascaridol ascaridole
ascorbato cálcico calcium ascorbate
ascorbato de niacinamida niacinamide
　　ascorbate
ascorbato sódico sodium ascorbate
asentamiento settling
aseptol aseptol
aseptolina aseptoline
asfalteno asphaltene
asfaltita asphaltite
asfalto asphalt
asfalto de petróleo petroleum asphalt
asimetría asymmetry
asimétrico asymmetrical, unsymmetrical
asimilación assimilation
asimilación de carbono carbon
　　assimilation
asiminina asiminine
asistente de teñido dyeing assistant
askarel askarel
asociación association
asociación molecular molecular
　　association
asparagina asparagine
asparaginasa asparaginase
asparamida asparamide
aspariginilo aspariginyl
aspartama aspartame
aspartamida aspartamide
aspartasa aspartase
aspartato aspartate
aspartato sódico sodium aspartate
aspartilo aspartyl
aspartoilo aspartoyl

aspidinol aspidinol
aspidosamina aspidosamine
aspidospermina aspidospermine
aspirador aspirator
aspirina aspirin
astaceno astacene
astátco astatic
astatino astatine
ástato astatine
asterisco asterisk
astorismo astorism
astrofilita astrophyllite
astroquímica astrochemistry
atacamita atacamite
atáctico atactic
atenolol atenolol
atenuación attenuation
atenuador de haz beam attenuator
atisina atisine
atmólisis atmolysis
atmómetro atmometer
atmósfera atmosphere
atmósfera controlada controlled
　　atmosphere
atmósfera de hidrógeno hydrogen
　　atmosphere
atmósfera iónica ionic atmosphere
atmósfera oxidante oxidizing
　　atmosphere
atmósfera reactiva reactive atmosphere
atmósfera reductora reducing
　　atmosphere
atmosférico atmospheric
ato ate
atómicamente atomically
atomicidad atomicity
atómico atomic
atomización atomization
atomizador atomizer
átomo atom
átomo asimétrico asymmetric atom
átomo clave key atom
átomo de Bohr Bohr atom
átomo de carbono carbon atom
átomo de carbono asimétrico
　　asymmetric carbon atom
átomo de carbono cuaternario
　　quaternary carbon atom
átomo de carbono secundario secondary

carbon atom
átomo de carbono simétrico symmetrical carbon atom
átomo de carbono terciario tertiary carbon atom
átomo de hidrógeno hydrogen atom
átomo de Lewis Lewis atom
átomo donador donor atom
átomo espiro spiro atom
átomo excitado excited atom
átomo exótico exotic atom
átomo-gramo gram-atom
átomo heterocíclico heterocyclic atom
átomo libre free atom
átomo marcado tagged atom
átomo neutro neutral atom
átomo nuclear nuclear atom
átomo nuclear de Rutherford Rutherford nuclear atom
átomo pesado heavy atom
átomo radiactivo radioactive atom
átomo substituyente substituent
átomo tetraédrico tetrahedral atom
átomo trazador tracer atom
atomología atomology
atomsita atomsite
atopita atopite
atracción attraction
atracción atómica atomic attraction
atracción de van der Waals van der Waals attraction
atracción interiónica interionic attraction
atracción molecular molecular attraction
atracción nuclear nuclear attraction
atrazina atrazine
atroneno atronene
atropina atropine
atropoilo atropoyl
attapulgita attapulgite
aubepina aubepine
augita augite
auramina auramine
aurato aurate
aurato potásico potassium aurate
aurelita aurelite
aureolina aureolin
auribromuro auribromide
auribromuro sódico sodium auribromide
auricalcita aurichalcite

auricloruro aurichloride
auricloruro sódico sodium aurichloride
áurico auric
aurífero auriferous
aurina aurin
auro aurous
auroáurico auroauric
aurobromuro aurobromide
aurocianuro aurous cyanide
aurocloruro aurochloride
auromina auromine
aurona aurone
aurosulfuro aurous sulfide
aurotiomalato sódico sodium aurothiomalate
auroyoduro aurous iodide
austenita austenite
autoabsorción self-absorption
autocatálisis autocatalysis
autocatalítico autocatalytic
autoclave autoclave
autodescomposición autodecomposition
autoexcitación autoexcitation
autoionización autoionization
autólisis autolysis
autoluminiscencia autoluminescence
automático automatic
automatización automation
autoorganización self-organization
autooxidación self-oxidation
autoprotólisis autoprotolysis
autorreducción self-reduction
autunita autunite
auxiliar auxiliary
auxina auxin
auxocromo auxochrome
avalancha electrónica electron avalanche
avalancha iónica ion avalanche
avalita avalite
avenina avenine
avidina avidin
axial axial
azaguanina azaguanine
azauridina azauridine
azelato azelate
azeotrópico azeotropic
azeótropo azeotrope
azetidina azetidine
azida azide

azida bárica barium azide
azida de bencilo benzyl azide
azida de benzaldehído benzaldehyde azide
azida de bromo bromine azide
azida de cianógeno cyanogen azide
azida de fenilo phenyl azide
azida de hidrógeno hydrogen azide
azida de plata silver azide
azida de plomo lead azide
azida sódica sodium azide
azido azido
azimida azimide
azimido azimido
azimidobenceno azimidobenzene
azimino azimino
aziminobenceno aziminobenzene
azina azine
azino azino
aziridina aziridine
azlón azlon
azo azo
azobenceno azobenzene
azobisisobutironitrilo azobisisobutyronitrile
azocianuro azocyanide
azocolorante azo dye
azodicabornamida azodicarbonamide
azodina azodine
azofenileno azophenylene
azofenilo azophenyl
azofenol azophenol
azogue quicksilver
azoico azoic
azoimida azoimide
azol azole
azometina azomethine
azonaftaleno azonaphthalene
azonio azonium
azorita azorite
azosulfamida azosulfamide
azotolueno azotoluene
azotómetro azotometer
azoxi azoxy
azoxibenceno azoxybenzene
azoxinaftaleno azoxynaphthalene
azoxitoluidina azoxytoluidine

azúcar sugar
azúcar de caña cane sugar
azúcar de leche milk sugar
azúcar de madera wood sugar
azúcar de pectina pectin sugar
azúcar de plomo sugar of lead
azúcar de remolacha beet sugar
azúcar de uva grape sugar
azúcar invertido invert sugar
azúcar no reductor nonreducing sugar
azúcar reductor reducing sugar
azufre sulfur, sulphur
azufre libre free sulfur
azufre rómbico rhombic sulfur
azul alcalino alkali blue
azul cerúleo cerulean blue
azul congo congo blue
azul de alizarina alizarin blue
azul de anilina aniline blue
azul de antraceno anthracene blue
azul de bromofenol bromophenol blue
azul de bromotimol bromothymol blue
azul de bronce bronze blue
azul de cobalto cobalt blue
azul de cobre copper blue
azul de cresilo cresyl blue
azul de China Chinese blue
azul de hierro iron blue
azul de índigo indigo blue
azul de metileno methylene blue
azul de metilo methyl blue
azul de molibdeno molybdenum blue
azul de naftol naphthol blue
azul de Prusia Prussian blue
azul de quinolina quinoline blue
azul de resorcina resorcinol blue
azul de Thenard Thenard blue
azul de timol thymol blue
azul de Turnbull Turnbull blue
azul fenileno phenylene blue
azul pavo real peacock blue
azul ultramarino ultramarine blue
azul Victoria Victoria blue
azulamiento blueing
azuleno azulene
azulina azulin
azurita azurite

B

bacarina baccarin
bacilo bacillus
bacitracina bacitracin
bacitracina de cinc zinc bacitracin
bacteria bacterium
bacterial bacterial
bacteriano bacterial
bactericida bactericide
bacterina bacterin
bacteriófago bacteriophage
bacteriólisis bacteriolysis
bacteriológico bacteriological
bacteriostático bacteriostatic
badeleyita baddeleyite
badequita baddeckite
bafle baffle
bagazo bagasse
bagrationita bagrationite
baicaleína baicalein
baicalina baicalin
baikalita baikalite
bainita bainite
baja energía low energy
balance balance
balanceado balanced
balanza scale, balance
balanza analítica analytical balance
balanza de astil beam balance
balanza de registro recording balance
balanza de torsión torsion balance
balata balata
baloeléctrico balloelectric
balómetro balometer
bálsamo balsam
bálsamo de Perú balsam of Peru
bálsamo de Tolú balsam of Tolu
banalsita banalsite
banda band
banda de absorción absorption band
banda de conducción conduction band
banda de energía energy band
banda de energía prohibida forbidden
    energy band

banda de valencia valence band
banda electrónica electron band
banda permitida allowed band
banda prohibida forbidden band
banisterina banisterine
baño bath
baño de agua water bath
baño de aire air bath
baño de arena sand bath
baño de vapor vapor bath, steam bath
baño galvánico galvanic bath
baño María water bath
baptisina baptisin
baptitoxina baptitoxine
baquerita bakerite
barbaloína barbaloin
barbana barban
barberita barberite
barbital barbital
barbitona barbitone
barbiturato barbiturate
barbiturato sódico sodium barbiturate
barcenita barcenite
baria barye
bario barium
barión baryon
barisilita barysilite
barita baryte
barita cáustica caustic baryta
barkevita barkevite
barn barn
barniz varnish, glaze
baroforesis barophoresis
baroluminiscencia baroluminescence
barómetro barometer
baroscopio baroscope
barosimina barosimin
barostato barostat
barra bar
barrandita barrandite
barrera barrier
barrera de Coulomb Coulomb barrier
barrera de Gamow Gamow barrier

306

barrido scan, scanning
barril barrel
bartrina barthrin
basanita basanite, bassanite
base base
base blanda soft base
base conjugada conjugate base
base de alquitrán tar base
base de Bandrowski Bandrowski base
base de colina choline base
base de Lewis Lewis base
base de Millon Millon base
base de Schiff Schiff base
base débil weak base
base diluida dilute base
base dura hard base
base fuerte strong base
base natural natural base
base nitrogenosa nitrogenous base
base orgánica organic base
basicidad basicity
básico basic
basisterol bassisterol
basorina bassorin
bastnasita bastnasite
batería battery
batería de acumuladores storage battery
batería de plomo lead battery
batocromático bathochromatic
batocrómico bathochromic
batocuproína bathocuproine
batofenantrolina bathophenanthroline
batracotoxinina batrachotoxinin
batu batu
bauxita bauxite
bauxita calcinada calcined bauxite
bayerita bayerite
becarita beccarite
becquerel becquerel
becquerelita becquerelite
behenato de metilo methyl behenate
behenolilo behenolyl
behenona behenone
beiriquita beyrichite
bel bel
beldongrita beldongrite
belita belite
belonesita belonesite
belladona belladonna

belladonina belladonnine
bemberg bemberg
benazolina benazolin
bencenida benzenide
bencenilo benzenyl
bencenio benzenium
benceno benzene
bencenoazimida benzene azimide
bencenoazoanilida benzeneazoanilide
bencenoazoanilina benzeneazoaniline
bencenoazobenceno benzeneazobenzene
bencenocarbaldehído
    benzenecarbaldehyde
bencenocarbinol benzenecarbinol
bencenocarbonitrilo benzenecarbonitrile
bencenocarboxilato benzenecarboxylate
bencenodicarbinol benzenedicarbinol
bencenodicarbonal benzenedicarbonal
bencenodicarbonitrilo
    benzenedicarbonitrile
bencenodisulfonato potásico potassium
    benzenedisulfonate
bencenodisulfóxido benzenedisulfoxide
bencenoditiol benzenedithiol
bencenoide benzenoid
bencenona benzenone
bencenosulfamida benzenesulfamide
bencenosulfonamida
    benzenesulfonamide
bencenosulfonato amónico ammonium
    benzenesulfonate
bencenosulfonato sódico sodium
    benzenesulfonate
bencenosulfonilo benzenesulfonyl
bencenosulfóxido benzenesulfoxide
bencenosulfuro benzenesulfide
bencenotriol benzenetriol
bencidina benzidine
bencidino benzidino
bencilacetamida benzylacetamide
bencilación benzylation
bencilamina benzylamine
bencilanilina benzylaniline
bencilbenceno benzylbenzene
bencilcarbinol benzyl carbinol
bencilcianamida benzylcyanamide
bencildianilo benzyldianil
bencildimetilamina benzyldimethylamine
bencileno benzylene

benciletanolamina benzylethanolaminc
bencilfenantreno benzylphenanthrene
bencilfenilamina benzylphenylamine
bencilfenol benzylphenol
bencilformamida benzylformamide
bencilhidrazina benzylhydrazine
bencilhidroquinona benzylhydroquinone
bencilidenacetofenona
    benzylideneacetophenone
bencilidenacetona benzylideneacetone
bencilideno benzylidene
bencilidietanolamina
    benzylidiethanolamine
bencilidina benzylidine
bencilisoeugenol benzyl isoeugenol
bencilisopropilamina
    benzylisopropylamine
bencilisotiocianato benzyl isothiocyanate
bencilmercaptano benzyl mercaptan
bencilmetilamina benzylmethylamine
bencilnaftaleno benzylnaphthalene
bencilo benzyl, benzil
benciloxifenol benzyloxyphenol
bencilpiridina benzylpyridine
benciltiol benzylthiol
bencimidazol benzimidazole
bencimidazolilo benzimidazolyl
bencimidoilo benzimidoyl
bencina benzin, benzine
bencina de petróleo petroleum benzin
bencindol benzindole
bencino benzyne
bencisoxasol benzisoxasole
bendiocarb bendiocarb
beneficio beneficiation
benefina benefin
benomilo benomyl
bensulida bensulide
bentazona bentazone
bentonita bentonite
benzacetina benzacetin
benzaconina benzaconine
benzacridina benzacridine
benzal benzal
benzalacetona benzalacetone
benzalazina benzalazine
benzalcloruro benzal chloride
benzalcohol benzalcohol
benzalconio benzalkonium

benzaldehído benzaldehyde
benzaldoxima benzaldoxime
benzalftalida benzalphthalide
benzamida benzamide
benzamida de plata silver benzamide
benzamidina benzamidine
benzamido benzamido
benzanilida benzanilide
benzantracina benzanthracene
benzantrona benzanthrone
benzatina benzathine
benzazida benzazide
benzazimida benzazimide
benzazina benzazine
benzfurano benzfuran
benzhidrazoína benzhydrazoin
benzhidrilo benzhydryl
benzhidrol benzhydrol
benzoato benzoate
benzoato alumínico aluminum benzoate
benzoato amónico ammonium benzoate
benzoato bárico barium benzoate
benzoato cálcico calcium benzoate
benzoato ceroso cerous benzoate
benzoato cobaltoso cobaltous benzoate
benzoato cúprico cupric benzoate
benzoato de alilo allyl benzoate
benzoato de amilo amyl benzoate
benzoato de antipireno antipyrene
    benzoate
benzoato de bencilo benzyl benzoate
benzoato dc benzoilo benzoyl benzoate
benzoato de butilo butyl benzoate
benzoato de cafeína caffeine benzoate
benzoato de cesio cesium benzoate
benzoato de cinc zinc benzoate
benzoato de cobre copper benzoate
benzoato de cocaína cocaine benzoate
benzoato de denatonio denatonium
    benzoate
benzoato de etileno ethylene benzoate
benzoato de etilo ethyl benzoate
benzoato de hidroxiquinolina
    hydroxyquinoline benzoate
benzoato de hierro iron benzoate
benzoato de isoamilo isoamyl benzoate
benzoato de isobutilo isobutyl benzoate
benzoato de litio lithium benzoate
benzoato de mercurio mercury benzoate

benzoato de metilo   methyl benzoate
benzoato de níquel   nickel benzoate
benzoato de plata   silver benzoate
benzoato de plomo   lead benzoate
benzoato de propilo   propyl benzoate
benzoato de tetrahidrofurfurilo
    tetrahydrofurfuryl benzoate
benzoato de uranilo   uranyl benzoate
benzoato fenilmercúrico   phenylmercuric
    benzoate
benzoato férrico   ferric benzoate
benzoato magnésico   magnesium
    benzoate
benzoato mercúrico   mercuric benzoate
benzoato potásico   potassium benzoate
benzoato sódico   sodium benzoate
benzocaína   benzocaine
benzodiantrona   benzodianthrone
benzodiazina   benzodiazine
benzodiazol   benzodiazole
benzodihidropirona   benzodihydropyrone
benzodioxazina   benzodioxazine
benzofenol   benzophenol
benzofenona   benzophenone
benzofluoreno   benzofluorene
benzofuranilo   benzofuranyl
benzofurano   benzofuran
benzofuranona   benzofuranone
benzoguanamina   benzoguanamine
benzohidrilo   benzohydryl
benzohidrol   benzohydrol
benzoilacetaldehído   benzoylacetaldehyde
benzoilacetato de etilo   ethyl
    benzoylacetate
benzoilacetona   benzoilacetone
benzoilación   benzoylation
benzoilamida   benzoylamide
benzoilamino   benzoylamino
benzoilanilina   benzoylaniline
benzoilazida   benzoylazide
benzoilciclopropano
    benzoylcyclopropane
benzoilferroceno   benzoylferrocene
benzoilglicina   benzoylglycin
benzoilhidrazina   benzoylhydrazine
benzoilo   benzoyl
benzoiloxi   benzoyloxy
benzoilpiridina   benzoylpyridine
benzoína   benzoin

benzol   benzol
benzomorfolina   benzomorpholine
benzonafteno   benzonaphthene
benzonaftol   benzonaphthol
benzonitrilo   benzonitrile
benzopentazol   benzopentazole
benzoperóxido   benzoperoxide
benzopinacol   benzopinacol
benzopiranilo   benzopyranyl
benzopirano   benzopyran
benzopireno   benzopyrene
benzopiridina   benzopyridine
benzopirona   benzopyrone
benzopirrol   benzopyrrole
benzopurpurina   benzopurpurine
benzoquinol   benzoquinol
benzoquinona   benzoquinone
benzoselenofurano   benzoselenofuran
benzosulfimida   benzosulfimide
benzosulfimida sódica   sodium
    benzosulfimide
benzotetrazina   benzotetrazine
benzotetrazol   benzotetrazole
benzotiacilo   benzothiazyl
benzotiazina   benzothiazine
benzotiazol   benzothiazole
benzotiofurano   benzothiofuran
benzotiopirano   benzothiopyran
benzotiopirona   benzothiopyrone
benzotoluida   benzotoluide
benzotriazepina   benzotriazepine
benzotriazina   benzotriazine
benzotriazol   benzotriazole
benzotricloruro   benzotrichloride
benzotrifluoruro   benzotrifluoride
benzotrifurano   benzotrifuran
benzoxanteno   benzoxanthene
benzoxazina   benzoxazine
benzoxazol   benzoxazole
benzoxazolilo   benzoxazolyl
benzoxazolona   benzoxozolone
benzoxtetrazina   benzoxtetrazine
benzoxtriazina   benzoxtriazine
benzozono   benzozone
benzpirazol   benzpyrazole
benztiofeno   benzthiophene
bequelita   beckelite
berberina   berberine
beresovita   beresovite

berilato beryllate
berilia beryllia
berilio beryllium
berilio-cobre beryllium-copper
beriliuro beryllide
berilo beryl
berilonita beryllonite
berkelio berkelium
bermellón vermillion
bertierina berthierine
bertolida berthollide
bertolita bertholite
bertrandita bertrandite
beta beta
betaína betaine
betametasona betamethasone
betatópico betatopic
betatrón betatron
betulina betulin
betulinol betulinol
beumlerita baeumlerite
bi bi
biaceno biacene
biacetileno biacetylene
bialilo biallyl
bianilina bianiline
bianisidina bianisidine
biantrilo bianthryl
biaxial biaxial
bibencilo bibenzyl
biborato biborate
biborato de trihexilenglicol trihexylene
    glycol biborate
bicarbonato bicarbonate
bicarbonato amónico ammonium
    bicarbonate
bicarbonato bárico barium bicarbonate
bicarbonato de cesio cesium bicarbonate
bicarbonato de colina choline
    bicarbonate
bicarbonato de estroncio strontium
    bicarbonate
bicarbonato de litio lithium bicarbonate
bicarbonato de sosa bicarbonate of soda,
    baking soda
bicarbonato potásico potassium
    bicarbonate, bicarbonate of potash
bicarbonato sódico sodium bicarbonate,
    bicarbonate of soda, baking soda

bicíclico bicyclic
biciclodecano bicyclodecane
biciclohexilo bicyclohexyl
bicloruro bichloride
bicloruro de azufre sulfur bichloride
bicloruro de bencilo benzyl bichloride
bicloruro de benzol benzol bichloride
bicloruro de mercurio mercury
    bichloride
bicloruro de paladio palladium
    bichloride
bicloruro de platino platinum bichloride
bicromato amónico ammonium
    bichromate
bicromato bárico barium bichromate
bicromato potásico potassium
    bichromate
bicromato sódico sodium bichromate
bieberita bieberite
biestabilidad bistability
bietileno biethylene
bifenilamina biphenylamine
bifenildiamina biphenyldiamine
bifenileno biphenylene
bifenililamina biphenylylamine
bifenililo biphenylyl
bifenilo biphenyl
bifenilo policlorado polychlorinated
    biphenyl
biflavonilo biflavonyl
bifluoruro amónico ammonium
    bifluoride
bifluoruro potásico potassium bifluoride
bifluoruro sódico sodium bifluoride
biformina biformin
bifosfato amónico ammonium
    biphosphate
bifosfato magnésico magnesium
    biphosphate
bifosfato sódico sodium biphosphate
biftalato potásico potassium biphthalate
bifurcación bifurcation
biguanida biguanide
bihexilo bihexyl
biimino biimino
bilateral bilateral
bilifuscina bilifuscin
bilirrubina bilirubin
bimetal bimetal

bimolecular bimolecular
binaftalenilo binaphthalenyl
binaftaleno binaphthalene
binaftilo binaphthyl
binapacrilo binapacryl
binario binary
binita binnite
binitro binitro
binoxalato potásico potassium binoxalate
binoxalato sódico sodium binoxalate
binóxido binoxide
binóxido bárico barium binoxide
binóxido de manganeso manganese binoxide
binóxido sódico sodium binoxide
binuclear binuclear
bioactivación bioactivation
bioacumulación bioaccumulation
bioanálisis bioanalysis
biocatalizador biocatalyst
biocida biocide
biocitina biocytin
biocoloide biocolloid
bioconversión bioconversion
biocromo biochrome
biodegradable biodegradable
bioelectroquímica bioelectrochemistry
bioelemento bioelement
bioensayo bioassay
bioética bioethics
biofila biophyl
bioflavonoide bioflavonoid
biofosfato cálcico calcium biophosphate
biogas biogas
biogénico biogenic
biogeoquímica biogeochemistry
bioingeniería bioengineering
biología biology
biología molecular molecular biology
biológico biological
bioluminiscencia bioluminescence
biomasa biomass
biomaterial biomaterial
biomineral biomineral
biomolécula biomolecule
biopolímero biopolymer
bioquímica biochemistry
bioquímica de radiación radiation biochemistry

bioquímico biochemical
biorización biorization
biorreactor bioreactor
biorresmetrina bioresmethrin
biortofosfato de plomo lead biorthophosphate
biosa biose
biosfera biosphere
biosíntesis biosynthesis
biostático biostat
biosterol biosterol
biotecnología biotechnology
biótico biotic
biotina biotin
biotita biotite
bioxalato amónico ammonium bioxalate
bioxalato bárico barium bioxalate
bioxilo bioxyl
biperiden biperiden
bipirámide trigonal trigonal bipyramid
bipiridilo bipyridyl
bipropenilo bipropenyl
biquinolilo biquinolyl
biquinolina biquinoline
birradical biradical
birrectificación birectification
birrefrigencia birefringence
birrefrigencia circular circular birefringence
birrefringente birefringent
birrotación birotation
bis bis
bisaboleno bisabolene
bisamida bisamide
bisdiazo bisdiazo
bisfenol bisphenol
bishidroximetilurea bishydroxymethylurea
bismanol bismanol
bismita bismite
bismutato sódico sodium bismuthate
bismutida bismuthide
bismutilo bismuthyl
bismutina bismuthine
bismutinita bismuthinite
bismutita bismuthite
bismuto bismuth
bismuto de fenol phenol bismuth
bismutoso bismuthous

bistetrazol bistetrazole
bistriazol bistriazole
bisulfato bisulfate
bisulfato amónico ammonium bisulfate
bisulfato bárico barium bisulfate
bisulfato de acetona sódico sodium
    acetone bisulfate
bisulfato potásico potassium bisulfate
bisulfato sódico sodium bisulfate
bisulfato sódico de acetona acetone
    sodium bisulfate
bisulfito bisulfite
bisulfito amónico ammonium bisulfite
bisulfito cálcico calcium bisulfite
bisulfito de formaldehído sódico sodium
    formaldehyde bisulfite
bisulfito potásico potassium bisulfite
bisulfito sódico sodium bisulfite
bisulfuro cálcico calcium bisulfide
bisulfuro de carbono carbon bisulfide
bisulfuro de estaño tin bisulfide
bisulfuro de yodo iodine bisulfide
bisulfuro sódico sodium bisulfide
bitartrato bitartrate
bitartrato amónico ammonium bitartrate
bitartrato de colina choline bitartrate
bitartrato potásico potassium bitartrate
bitartrato sódico sodium bitartrate
bitionol bithionol
bitionolato sódico sodium bithionolate
bitrex bitrex
bittern bittern
bitumen bitumen
bitumen asfáltico asphaltic bitumen
bituminoso bituminous
biurea biurea
biuret biuret
bivalencia bivalence
bivalente bivalent
bivinilo bivinyl
bizaso bisazo
blanco de cinc zinc white
blanco de China Chinese white
blanco de España Spanish white
blanco de titanio titanium white
blanco de wolframio wolfram white
blanco fijo blanc fixe
blanco perla pearl white
blanco permanente permanent white

blanqueador whitener
blanquear bleach
blanqueo bleaching
blenda blende
blenda de cinc zinc blende
bleomicina bleomycin
blindaje shielding
bloque block
bloqueo block
boehmita boehmite
boghead boghead coal
bohidruro sódico sodium borohydride
boleíta boleite
bolómetro bolometer
bolsa de filtración filter bag
bomba bomb, pump
bomba calorimétrica bomb calorimeter
bomba centrífuga centrifugal pump
bomba criogénica cryogenic pump
bomba de aire air pump
bomba de difusión diffusion pump
bomba de succión suction pump
bomba de vacío vacuum pump
bomba iónica ion pump
bomba para filtración filter pump
bombardeo bombardment
bombardeo con átomos rápidos fast-
    atom bombardment
bombardeo electrónico electron
    bombardment
bombardeo neutrónico neutron
    bombardment
bombardeo nuclear nuclear
    bombardment
bombardeo protónico proton
    bombardment
bondur bondur
boracita boracite
borano borane
boranodiilo boranediyl
borato borate
borato alumínico aluminum borate
borato amónico ammonium borate
borato bárico barium borate
borato cálcico calcium borate
borato de bismuto bismuth borate
borato de butilo butyl borate
borato de cinc zinc borate
borato de cobre copper borate

borato de cocaína cocaine borate
borato de litio lithium borate
borato de manganeso manganese borate
borato de metilo methyl borate
borato de plata silver borate
borato de plomo lead borate
borato de tributilo tributyl borate
borato de trietilo triethyl borate
borato de trimetilo trimethyl borate
borato fenilmercúrico phenylmercuric
    borate
borato magnésico magnesium borate
borato mercúrico mercuric borate
borato potásico potassium borate
borato sódico sodium borate
bórax borax
borazina borazine
borazol borazole
borde de absorción absorption edge
borina borine
bornano bornane
borneno bornene
borneol borneol
bornilo bornyl
bornita bornite
boro boron
boroetano boroethane
borofluoruro potásico potassium
    borofluoride
boroformiato alumínico aluminum
    boroformate
boroformiato sódico sodium
    boroformate
borohidruro borohydride
borohidruro alumínico aluminum
    borohydride
borohidruro de litio lithium borohydride
borohidruro potásico potassium
    borohydride
borol borol
borosalicilato borosalicylate
borosilicato borosilicate
borosilicato de plomo lead borosilicate
borotungstato bárico barium
    borotungstate
borotungstato de cadmio cadmium
    borotungstate
boroxina boroxine
bort bort

boruro boride
boruro alumínico aluminum boride
boruro bárico barium boride
boruro cobáltico cobaltic boride
boruro de cromo chromium boride
boruro de hafnio hafnium boride
boruro de manganeso manganese boride
boruro de molibdeno molybdenum
    boride
boruro de silicio silicon boride
boruro de titanio titanium boride
boruro de tungsteno tungsten boride
boruro de zirconio zirconium boride
botella bottle
botella para gotear dropping bottle
brasilina brasilin
braunita braunite
brea tar, pitch
brea de alquitrán de hulla coal-tar pitch
brea de alquitrán de pino pine-tar pitch
brea mineral mineral pitch
bretonita bretonite
breunerita breunnerite
brillante brilliant
brillo luster
britolita britholite
brocanita brochanite
bromacetato bromacetate
bromacil bromacil
bromal bromal
bromalina bromalin
bromato bromate
bromato alumínico aluminum bromate
bromato bárico barium bromate
bromato cálcico calcium bromate
bromato ceroso cerous bromate
bromato cobaltoso cobaltous bromate
bromato de alcanfor camphor bromate
bromato de cadmio cadmium bromate
bromato de cesio cesium bromate
bromato de cinc zinc bromate
bromato de estroncio strontium bromate
bromato de plata silver bromate
bromato de plomo lead bromate
bromato magnésico magnesium bromate
bromato potásico potassium bromate
bromato sódico sodium bromate
bromelaína bromelain
bromelina bromelin

bromeosina bromeosin
brometilo bromethyl
bromhidrato hydrobromide
bromhidrato de etilamina ethylamine
    hydrobromide
brominación bromination
brominado bromated
bromito sódico sodium bromite
bromo bromine
bromoacetamida bromoacetamide
bromoacetato bromoacetate
bromoacetato de etilo ethyl
    bromoacetate
bromoacetona bromoacetone
bromoalileno bromoallylene
bromoanilina bromoaniline
bromobenceno bromobenzene
bromobenzaldehído bromobenzaldehyde
bromobutano bromobutane
bromociclopentano bromocyclopentane
bromocloroetano bromochloroethane
bromoclorometano bromochloromethane
bromocloropropano
    bromochloropropane
bromododecano bromododecane
bromoestireno bromostyrene
bromoetano bromoethane
bromofenilacetonitrilo
    bromophenylacetonitrile
bromofenol bromophenol
bromofluorobenceno
    bromofluorobenzene
bromoformo bromoform
bromoformo de acetona acetone
    bromoform
bromofosgeno bromophosgene
bromohexano bromohexane
bromohidrina de etileno ethylene
    bromohydrin
bromohidrina de glicol glycol
    bromohydrin
bromol bromol
bromometano bromomethane
bromometría bromometry
bromonaftaleno bromonaphthalene
bromopentano bromopentane
bromopicrina bromopicrin
bromopiridina bromopyridine
bromopirina bromopyrine

bromopropeno bromopropene
bromosuccinimida bromosuccinimide
bromotiofenol bromothiophenol
bromotolueno bromotoluene
bromotriclorometano
    bromotrichloromethane
bromotrifluoroetileno
    bromotrifluoroethylene
bromotrifluorometano
    bromotrifluoromethane
bromuro alumínico aluminum bromide
bromuro amónico ammonium bromide
bromuro amónico de cadmio cadmium
    ammonium bromide
bromuro antimonioso antimonous
    bromide
bromuro arsenioso arsenous bromide
bromuro áurico auric bromide
bromuro bárico barium bromide
bromuro bárico mercúrico mercuric
    barium bromide
bromuro cálcico calcium bromide
bromuro ceroso cerous bromide
bromuro cobaltoso cobaltous bromide
bromuro crómico chromic bromide
bromuro cromoso chromous bromide
bromuro cúprico cupric bromide
bromuro cuproso cuprous bromide
bromuro de acetilo acetyl bromide
bromuro de alilo allyl bromide
bromuro de antimonio antimony
    bromide
bromuro de arsénico arsenic bromide
bromuro de azufre sulfur monobromide,
    sulfur bromide
bromuro de bencilo benzyl bromide
bromuro de benzhidrilo benzhydryl
    bromide
bromuro de benzoilo benzoyl bromide
bromuro de benzol benzol bromide
bromuro de berilio beryllium bromide
bromuro de bismutilo bismuthyl
    bromide
bromuro de bismuto bismuth bromide
bromuro de boro boron bromide
bromuro de butilo butyl bromide
bromuro de cadmio cadmium bromide
bromuro de carbonilo carbonyl bromide
bromuro de cesio cesium bromide

bromuro de cetilo   cetyl bromide

bromuro de cetilpiridinio
cetylpyridinium bromide

bromuro de cetiltrimetilamonio
cetyltrimethylammonium bromide

bromuro de cianógeno   cyanogen
bromide

bromuro de ciclohexilo   cyclohexyl
bromide

bromuro de cinc   zinc bromide

bromuro de cloro   chlorine bromide

bromuro de cobalto   cobalt bromide

bromuro de cobre   copper bromide

bromuro de cromo   chromium bromide

bromuro de difenilmetilo
diphenylmethyl bromide

bromuro de estaño   tin bromide

bromuro de estroncio   strontium bromide

bromuro de etileno   ethylene bromide

bromuro de etilhexilo   ethylhexyl
bromide

bromuro de etilmagnesio
ethylmagnesium bromide

bromuro de etilo   ethyl bromide

bromuro de fenilmagnesio
phenylmagnesium bromide

bromuro de fenilo   phenyl bromide

bromuro de formilo   formyl bromide

bromuro de fosfonio   phosphonium
bromide

bromuro de fosforilo   phosphoryl
bromide

bromuro de fósforo   phosphorus bromide

bromuro de gadolinio   gadolinium
bromide

bromuro de galio   gallium bromide

bromuro de hexilo   hexyl bromide

bromuro de hidrógeno   hydrogen
bromide

bromuro de hierro   iron bromide

bromuro de hierro amónico   ammonium
iron bromide

bromuro de indio   indium bromide

bromuro de iridio   iridium bromide

bromuro de isopropilo   isopropyl
bromide

bromuro de iterbio   ytterbium bromide

bromuro de itrio   yttrium bromide

bromuro de litio   lithium bromide

bromuro de manganeso   manganese
bromide

bromuro de mercurio   mercury bromide

bromuro de metileno   methylene bromide

bromuro de metilmagnesio
methylmagnesium bromide

bromuro de metilo   methyl bromide

bromuro de niobio   niobium bromide

bromuro de níquel   nickel bromide

bromuro de nitrobencilo   nitrobenzyl
bromide

bromuro de nitrosilo   nitrosyl bromide

bromuro de nonilo   nonyl bromide

bromuro de octilo   octyl bromide

bromuro de oro   gold bromide

bromuro de pentametileno
pentamethylene bromide

bromuro de pentilo   pentyl bromide

bromuro de plata   silver bromide

bromuro de plomo   lead bromide

bromuro de propargilo   propargyl
bromide

bromuro de propilmagnesio
propylmagnesium bromide

bromuro de propilo   propyl bromide

bromuro de radio   radium bromide

bromuro de rubidio   rubidium bromide

bromuro de rutenio   ruthenium bromide

bromuro de selenio   selenium bromide

bromuro de silicio   silicon bromide

bromuro de sulfinilo   sulfinyl bromide

bromuro de talio   thallium bromide

bromuro de tántalo   tantalum bromide

bromuro de telurio   tellurium bromide

bromuro de tionilo   thionyl bromide

bromuro de titanio   titanium bromide

bromuro de trimetileno   trimethylene
bromide

bromuro de uranio   uranium bromide

bromuro de vanadilo   vanadyl bromide

bromuro de vanadio   vanadium bromide

bromuro de vinilo   vinyl bromide

bromuro de xililo   xylyl bromide

bromuro de yodo   iodine bromide

bromuro de zirconilo   zirconyl bromide

bromuro de zirconio   zirconium bromide

bromuro estánnico   stannic bromide

bromuro estannoso   stannous bromide

bromuro férrico   ferric bromide

bromuro ferroso ferrous bromide
bromuro fosfórico phosphoric bromide
bromuro gálico gallic bromide
bromuro galoso gallous bromide
bromuro magnésico magnesium bromide
bromuro manganoso manganous
    bromide
bromuro mercúrico mercuric bromide
bromuro mercúrico bárico barium
    mercury bromide
bromuro mercurioso mercurous bromide
bromuro niqueloso nickelous bromide
bromuro potásico potassium bromide
bromuro samárico samaric bromide
bromuro sódico sodium bromide
bromuro talioso thallous bromide
bromuro telúrico telluric bromide
bromuro teluroso tellurous bromide
bronce bronze
bronce de aluminio aluminum bronze
bronce de campana bell metal
bronce de cañón gun metal
bronce de estaño tin bronze
bronce de manganeso manganese bronze
bronce de oro gold bronze
bronce de plata silver bronze
bronce de silicio silicon bronze
bronce de tungsteno tungsten bronze
bronce duro hard bronze
bronce fosforado phosphor bronze
broranotriilo boranetriyl
brucina brucine
brucita brucite
bruto crude
bufanina buphanine
bufotanina bufotanine
bufotenina bufotenine
bunamiodilo bunamiodyl
bunsenita bunsenite
burbuja bubble
bureta buret
bureta automática automatic buret
bureta de pesar weight buret
bureta graduada graduated buret
butadieno butadiene
butaldehído butaldehyde
butanal butanal
butano butane
butanoato cálcico calcium butanoate

butanoato de butilo butyl butanoate
butanoato de etilo ethyl butanoate
butanoato de pentilo pentyl butanoate
butanoato de propilo propyl butanoate
butanoato potásico potassium butanoate
butanodial butanedial
butanodiamida butanediamide
butanodiamina butanediamine
butanodiol butanediol
butanodiolamina butanediolamine
butanodiona butanedione
butanol butanol
butanona butanone
butanonitrilo butanenitrile
butanotiol butanethiol
butanotriol butanetriol
butenal butenal
butenilideno butenylidene
butenilo butenyl
buteno butene
butesina butesin
butetal butethal
butil-litio butyllithium
butilacrilamida butylacrylamide
butilamina butylamine
butilaminoetanol butylaminoethanol
butilanilina butylaniline
butilantraquinona butylanthraquinone
butilbenceno butylbenzene
butilbencenosulfonamida
    butylbenzenesulfonamide
butilcarbinol butyl carbinol
butildiamilamina butyldiamylamine
butilenglycol butylene glycol
butileno butylene
butiletanolamina butylethanolamine
butiletilacetaldehído
    butylethylacetaldehyde
butilfenol butylphenol
butilformamida butylformamide
butilhidroquinona butylhydroquinone
butilideno butylidene
butilisocianato butylisocyanate
butilmercaptano butylmercaptan
butilmetacrilato butylmethacrylate
butilo butyl
butilparabeno butylparaben
butilperoxipivalato butylperoxypivalate
butiltiofenol butylthiophenol

butiltolueno butyltoluene
butiltriclorosilano butyltrichlorosilane
butilurea butylurea
butino butyne
butinodiol butynediol
butiraldehído butyraldehyde
butirato bárico barium butyrate
butirato cobaltoso cobaltous butyrate
butirato de alilo allyl butyrate
butirato de amilo amyl butyrate
butirato de bencilo benzyl butyrate
butirato de butilo butyl butyrate
butirato de etilo ethyl butyrate
butirato de geranilo geranyl butyrate
butirato de isoamilo isoamyl butyrate
butirato de isopropilo isopropyl butyrate
butirato de metilo methyl butyrate
butirato de plomo lead butyrate
butirato de propilo propyl butyrate
butirato de vinilo vinyl butyrate

butirato sódico sodium butyrate
butírico butyric
butirilo butyryl
butirina butyrin
butirolactama butyrolactam
butirolactona butyrolactone
butirona butyrone
butironitrilo butyronitrile
butonato butonate
butopironoxilo butopyronoxyl
butoxi butoxy
butóxido alumínico aluminum butoxide
butóxido de piperonilo piperonyl
    butoxide
butóxido potásico potassium butoxide
butoxietanol butoxyethanol
butoxifenol butoxyphenol
butoxipropanol butoxypropanol
butoxitriglicol butoxytriglycol

C

cabrerita cabrerite
cacao cocoa
cacodilato cacodylate
cacodilato sódico sodium cacodylate
cacodilo cacodyl
cacotelina cacotheline
cactina cactine
cadaleno cadalene
cadaverina cadaverine
cadena chain
cadena abierta open chain
cadena cerrada closed chain
cadena de carbonos carbon chain
cadena de desintegración decay chain
cadena de transporte electrónico
    electron transport chain
cadena lateral side chain, lateral chain
cadena periódica periodic chain
cadena radiactiva radioactive chain
cadena ramificada branched chain
cadena recta straight chain
cadineno cadinene
cadión cadion
cadmio cadmium
cadmio anaranjado orange cadmium
cadmio de dimetilo dimethyl cadmium
cadmoso cadmous
cafeína caffeine
cainita kainite
caja con atmósfera inerte inert
    atmosphere box
caja con guantes glove box
caja de Petri petri dish
cal lime
cal apagada slaked lime
cal cáustica caustic lime
cal clorada chlorinated lime
cal hidratada hydrated lime
cal sodada soda lime
cal sulfurada sulfurated lime
cal viva quicklime, burnt lime
calabarina calabarine
calamina calamine

calandria calender
calaverita calaverite
calcantita chalcanthite
calcedonia chalcedony
calceína calcein
calciferol calciferol
calcímetro calcimeter
calcimina calcimine
calcina calcine
calcinación calcination, roasting
calcinado calcined
calcinita calcinite
calcinol calcinol
calcio calcium
calcita calcite
calcofanita chalcophanite
calcógeno chalcogen
calcona chalcone
calcopirita chalcopyrite
calcosiderita chalcosiderite
calcosina chalcocite
caldeo heating
caledonita caledonite
calentamiento heating
calentamiento adiabático adiabatic
    heating
calentamiento espontáneo spontaneous
    heating
calibración calibration
calibrado calibrated
calibrar calibrate
calicantina calycanthine
caliche caliche
californio californium
californita californite
calinita kalinite
calixareno calixarene
caliza limestone
caliza magnésica magnesium limestone
calomel calomel
calor heat
calor atómico atomic heat
calor de absorción heat of absorption

318

calor de activación heat of activation
calor de adhesión heat of adhesion
calor de adsorción heat of adsorption
calor de agregación heat of aggregation
calor de asociación heat of association
calor de atomización heat of atomization
calor de combinación heat of combination
calor de combustible heat of fuel
calor de combustión heat of combustion
calor de compresión heat of compression
calor de condensación heat of condensation
calor de cristalización heat of crystallization
calor de descomposición heat of decomposition
calor de dilución heat of dilution
calor de disociación heat of dissociation
calor de disolución heat of dissolution
calor de emisión heat of emission
calor de enfriamiento heat of cooling
calor de evaporación heat of evaporation
calor de explosión heat of explosion
calor de formación heat of formation
calor de fusión heat of fusion
calor de hidratación heat of hydration
calor de ionización heat of ionization
calor de isomerización heat of isomerization
calor de neutralización heat of neutralization
calor de oxidación heat of oxidation
calor de racemización heat of racemization
calor de reacción heat of reaction
calor de solidificación heat of solidification
calor de solución heat of solution
calor de sublimación heat of sublimation
calor de transición heat of transition
calor de vaporización heat of vaporization
calor específico specific heat
calor específico molal molal specific heat
calor específico molar molar specific heat
calor específico molecular molecular specific heat
calor latente latent heat
calor latente molar molar latent heat
calor molar molar heat
calor nuclear nuclear heat
calor total total heat
caloría calorie
calorimetría calorimetry
calorimetría de exploración diferencial differential scanning calorimetry
calorimétrico calorimetric
calorímetro calorimeter
calorímetro adiabático adiabatic calorimeter
calorímetro de Atwater Atwater calorimeter
calorización calorizing
calotropina calotropin
calutrón calutron
cámara chamber
cámara de fisión fission chamber
cámara de ionización ionization chamber
cambio adiabático adiabatic change
cambio de energía energy change
cambio de estado change of state
cambio de éster ester interchange
cambio de fase change of phase
cambio químico chemical change
campana bell glass, bell jar, hood
campana de humos fume cupboard, fume hood
campo field
campo atómico atomic field
campo molecular molecular field
canadina canadine
canalización channeling
canavanina canavanine
cancrinita cancrinite
canfano camphane
canfanol camphanol
canfanona camphanone
canfeno camphene
canfeno clorado chlorinated camphene
canforoilo camphoroyl
canforoquinona camphoroquinone
cannabidiol cannabidiol
cannabinol cannabinol
cannabis cannabis
canonita cannonite

cantaridina cantharidin
cantaxantina canthaxanthin
cantidad quantity
cantidad de factor factor quantity
cantidad de substancia amount of
    substance
cantidad física physical quantity
cantidad molal parcial partial molal
    quantity
cáñamo hemp
caolin kaolin
caolinita kaolinite
caos chaos
capa layer, shell
capa abierta open shell
capa anódica anodic layer
capa catódica cathodic layer
capa de barrera barrier layer
capa de Beilby Beilby layer
capa de difusión diffusion layer
capa de Gouy Gouy layer
capa de Helmholtz Helmholtz layer
capa de níquel nickel layer
capa de ozono ozone layer
capa de valencia valence shell
capa doble double layer
capa doble eléctrica electric double layer
capa electrónica electron shell
capa exterior outer shell
capa fina thin layer
capacidad capacity
capacidad amortiguadora buffer
    capacity
capacidad básica basic capacity
capacidad térmica heat capacity
capacidad térmica específica specific
    heat capacity
capacidad térmica molar molar heat
    capacity
capacidad térmica molecular molecular
    heat capacity
caparrosa copperas
caparrosa amarilla yellow copperas
caparrosa azul blue copperas
caparrosa blanca white copperas
capilaridad capillarity
capraldehído capraldehyde
caprato caprate
caprato de etilo ethyl caprate

caprato de metilo methyl caprate
capreomicina capreomycin
caprilamida caprylamide
caprilato caprylate
caprilato amónico ammonium caprylate
caprilato de cinc zinc caprylate
caprilato de etilo ethyl caprylate
caprilato de metilo methyl caprylate
caprilato sódico sodium caprylate
caprililo caprylyl
caprilo capryl
caproato de etilo ethyl caproate
caproato de metilo methyl caproate
caproilo caproyl
caprolactama caprolactam
caprolactona caprolactone
capronilo capronyl
cápsula capsule
cápsula de combustión combustion
    capsule
captafol captafol
captano captan
captura capture
captura electrónica electron capture
captura neutrónica neutron capture
captura neutrónica radiativa radiative
    neutron capture
captura nuclear nuclear capture
captura protónica proton capture
captura radiativa radiative capture
caracolita caracolite
característica de electrodo electrode
    characteristic
caramelano caramelan
caramelización caramelization
caramelo caramel
carano carane
carbaborano carbaborane
carbacol carbachol
carbaldehído carbaldehyde
carbamato carbamate
carbamato amónico ammonium
    carbamate
carbamato de alilo allyl carbamate
carbamato de butilo butyl carbamate
carbamato de diglicol diglycol
    carbamate
carbamato de etilo ethyl carbamate
carbamato de fenilo phenyl carbamate

carbamato de hidroxietilo hydroxyethyl carbamate
carbamato de pentilo pentyl carbamate
carbamato de propilo propyl carbamate
carbámico carbamic
carbamida carbamide
carbamidina carbamidine
carbamido carbamido
carbamilo carbamyl
carbamilurea carbamylurea
carbamita carbamite
carbamoilo carbamoyl
carbanilida carbanilide
carbanilo carbanil
carbanión carbanion
carbarilo carbaryl
carbato carbate
carbazato de etilo ethyl carbazate
carbazida carbazide
carbazoilo carbazoyl
carbazol carbazole
carbazolilo carbazolyl
carbazona carbazone
carbazotato amónico ammonium carbazotate
carbenio carbenium
carbeno carbene
carbenoide carbenoid
carbetoxiciclohexanona carbethoxycyclohexanone
carbetoxiciclopentanona carbethoxycyclopentanone
carbetoxipiperazina carbethoxypiperazine
carbilamina carbylamine
carbiloxima carbyloxime
carbino carbyne
carbinol carbinol
carbitol carbitol
carbobenzoilo carbobenzoyl
carbocatión carbocation
carbocíclico carbocyclic
carbodiazona carbodiazone
carbodihidrazida carbodihydrazide
carbodiimida carbodiimide
carboditioico carbodithioic
carboestirilo carbostyril
carbofenotión carbophenothion
carbofurano carbofuran

carbohidrasa carbohydrase
carbohidrato carbohydrate
carbohidrazida carbohydrazide
carbohidruro carbohydride
carboide carboid
carbolato sódico sodium carbolate
carbolfucsina carbolfuchsin
carbómero carbomer
carbometeno carbomethene
carbometoxi carbomethoxy
carbomicina carbomycin
carbón carbon, charcoal, coal
carbón activado activated carbon, activated charcoal
carbón activo active carbon
carbón bituminoso bituminous coal
carbón de bujía cannel coal
carbón industrial industrial carbon
carbón mineral mineral carbon
carbonación carbonation
carbonado carbonado
carbonatación carbonatation
carbonato carbonate
carbonato ácido amónico ammonium acid carbonate
carbonato ácido potásico potassium acid carbonate
carbonato ácido sódico sodium acid carbonate
carbonato alumínico aluminum carbonate
carbonato amónico ammonium carbonate
carbonato bárico barium carbonate
carbonato cálcico calcium carbonate
carbonato ceroso cerous carbonate
carbonato cobaltoso cobaltous carbonate
carbonato cromoso chromous carbonate
carbonato cúprico cupric carbonate
carbonato cuproso cuprous carbonate
carbonato de berilio beryllium carbonate
carbonato de bismutilo bismuthyl carbonate
carbonato de bismuto bismuth carbonate
carbonato de cadmio cadmium carbonate
carbonato de cerio cerium carbonate
carbonato de cesio cesium carbonate
carbonato de cinc zinc carbonate

carbonato de cobalto cobalt carbonate
carbonato de cobre copper carbonate
carbonato de dietilo diethyl carbonate
carbonato de difenilo diphenyl carbonate
carbonato de estroncio strontium carbonate
carbonato de etilbutilo ethylbutyl carbonate
carbonato de etileno ethylene carbonate
carbonato de etilo ethyl carbonate
carbonato de fenilo phenyl carbonate
carbonato de glicerina glycerin carbonate
carbonato de glicol glycol carbonate
carbonato de guanidina guanidine carbonate
carbonato de hidrógeno hydrogen carbonate
carbonato de hidrógeno amónico ammonium hydrogen carbonate
carbonato de hidrógeno de estroncio strontium hydrogen carbonate
carbonato de hidrógeno de litio lithium hydrogen carbonate
carbonato de hidrógeno magnésico magnesium hydrogen carbonate
carbonato de hidrógeno potásico potassium hydrogen carbonate
carbonato de hidrógeno sódico sodium hydrogen carbonate
carbonato de hidroximetileno hydroxymethylene carbonate
carbonato de hierro iron carbonate
carbonato de itrio yttrium carbonate
carbonato de lantano lanthanum carbonate
carbonato de litio lithium carbonate
carbonato de manganeso manganese carbonate
carbonato de metilo methyl carbonate
carbonato de metilo sódico sodium methyl carbonate
carbonato de neodimio neodymium carbonate
carbonato de níquel nickel carbonate
carbonato de plata silver carbonate
carbonato de plomo lead carbonate
carbonato de propileno propylene carbonate

carbonato de radio radium carbonate
carbonato de rubidio rubidium carbonate
carbonato de talio thallium carbonate
carbonato de zirconilo zirconyl carbonate
carbonato de zirconilo amónico ammonium zirconyl carbonate
carbonato de zirconio zirconium carbonate
carbonato ferroso ferrous carbonate
carbonato magnésico magnesium carbonate
carbonato manganoso manganous carbonate
carbonato mercúrico mercuric carbonate
carbonato niqueloso nickelous carbonate
carbonato potásico potassium carbonate
carbonato sódico sodium carbonate
carbonato sódico potásico potassium sodium carbonate
carbonato talioso thallous carbonate
carbonatomineral carbonate mineral
carbónico carbonic
carbonilación carbonylation
carbonilato carbonylate
carbonildiimidazol carbonyldiimidazole
carbonilo carbonyl
carbonilo de cobalto cobalt carbonyl
carbonilo de cromo chromium carbonyl
carbonilo de hierro iron carbonyl
carbonilo de manganeso manganese carbonyl
carbonilo de níquel nickel carbonyl
carbonilo de rutenio ruthenium carbonyl
carbonilo de tungsteno tungsten carbonyl
carbonimidoilo carbonimidoyl
carbonio carbonium
carbonita carbonite
carbonitrilo carbonitrile
carbonización carbonization
carbonizado carbonized
carbonizar carbonize
carbonldioxi carbonyldioxy
carbono carbon
carbono asimétrico asymmetric carbon
carbono combinado combined carbon
carbono cuaternario quaternary carbon

carbono divalente divalent carbon
carbono fijo fixed carbon
carbono secundario secondary carbon
carbono terciario tertiary carbon
carbono tetraédrico tetrahedral carbon
carbonohidrazida carbonohydrazide
carbonohidrazido carbonohydrazido
carbonómetro carbometer
carborano carborane
carbotioico carbothioic
carbotiónico carbothionic
carboxamida carboxamide
carboxamidina carboxamidine
carboxi carboxy
carboxibenceno carboxybenzene
carboxilación carboxylation
carboxilación de Koch-Haaf Koch-Haaf
    carboxylation
carboxilasa carboxylase
carboxílico carboxylic
carboxilo carboxyl
carboximetilcelulosa
    carboxymethylcellulose
carboximetilcelulosa sódica sodium
    carboxymethylcellulose
carboxipeptidasa carboxypeptidase
carburita carburite
carburización carburization
carburo carbide
carburo alumínico aluminum carbide
carburo bárico barium carbide
carburo cálcico calcium carbide
carburo crómico chromic carbide
carburo de berilio beryllium carbide
carburo de boro boron carbide
carburo de cromo chromium carbide
carburo de ditungsteno ditungsten
    carbide
carburo de hafnio hafnium carbide
carburo de hierro iron carbide
carburo de litio lithium carbide
carburo de manganeso manganese
    carbide
carburo de niobio niobium carbide
carburo de silicio silicon carbide
carburo de tántalo tantalum carbide
carburo de titanio titanium carbide
carburo de torio thorium carbide
carburo de tungsteno tungsten carbide

carburo de uranio uranium carbide
carburo de vanadio vanadium carbide
carburo de zirconio zirconium carbide
carburo sódico sodium carbide
carcinógeno carcinogen
cardenillo verdigris
cardenillo azul blue verdigris
cardenolida cardenolide
careno carene
carga charge
carga atómica atomic charge
carga de electrón charge of electron
carga efectiva effective charge
carga electrónica electron charge
carga elemental elementary charge
carga equivalente equivalent charge
carga específica specific charge
carga específica de electrón electron
    specific charge
carga formal formal charge
carga iónica ionic charge
carga libre free charge
carga nuclear nuclear charge
carga química chemical load
cargado charged
cariofileno caryophyllene
carmín carmine
carminato sódico sodium carminate
carminita carminite
carnalita carnallite
carnegina carnegine
carnina carnine
carnitina carnitine
carnosina carnosine
carnotita carnotite
carobina carobine
carona carone
caroteno carotene
carotenoide carotenoid
carpaína carpain
carraginina carrageenan
carreno carreen
cartamina carthamin
carvacrol carvacrol
carveol carveol
carvestreno carvestrene
carvona carvone
cascada cascade
caseína casein

caseinasa caseinase
caseinato cálcico calcium caseinate
caseinato sódico sodium caseinate
casiterita cassiterite, tinstone
catabiótico catabiotic
catabólico catabolic
catabolismo catabolism
cataforesis cataphoresis
catalasa catalase
catálisis catalysis
catálisis ácido-base acid-base catalysis
catálisis de radiación radiation catalysis
catálisis fotoquímica photochemical
    catalysis
catálisis heterogénea heterogeneous
    catalysis
catálisis homogénea homogeneous
    catalysis
catálisis negativa negative catalysis
catálisis protolítica protolytic catalysis
catalítico catalytic
catalizador catalyst
catalizador alfin alfin catalyst
catalizador de Adams Adams catalyst
catalizador de Raney Raney catalyst
catalizador de Wilkinson Wilkinson
    catalyst
catalizador de zeolita zeolite catalyst
catalizador de Ziegler Ziegler catalyst
catalizador estereoespecífico
    stereospecific catalyst
catalizador redox redox catalyst
catalizador selectivo de tamaño shape-
    selective catalyst
catarómetro catharometer
catartina cathartin
catecol catechol
catecolborano catecholborane
catena catena
catenación catenation
catenano catenane
catenario catenary
catenilo catenyl
catequina catechin
catetómetro cathetometer
catina cathine
catión cation
catiónico cationic
cationotropía cationotropy

cationtrofía cationtrophy
catlinita catlinite
catódico cathodic
cátodo cathode
cátodo electrolítico electrolytic cathode
católito catholyte
caucho rubber
caucho artificial artificial rubber
caucho butilo butyl rubber
caucho clorado chlorinated rubber
caucho de clorohidrina chlorohydrin
    rubber
caucho de estireno-butadieno styrene-
    butadiene rubber
caucho de poliuretano polyurethane
    rubber
caucho de silicio silicon rubber
caucho de silicona silicone rubber
caucho frío cold rubber
caucho nitrilo nitrile rubber
caucho poliéster polyester rubber
caucho sintético synthetic rubber
causticidad causticity
cáustico caustic
cavitación cavitation
cebadilla sabadilla
cedazo sieve, screen
cedazo molecular molecular sieve
cedrina cedrin
cedrol cedrol
cefalina cephalin
cefalosporina cephalosporin
cefamicina cephamycin
celasa cellase
celda de plomo lead cell
celda electrolítica electrolytic cell
celestina celestine
celestita celestite
celobiosa cellobiose
celodextrina cellodextrin
celofán cellophane
celoidina celloidin
celosía lattice
célula cell
célula de absorción absorption cell
célula de Bénard Bénard cell
célula de redox redox cell
célula electroquímica electrochemical
    cell

célula fotoeléctrica photoelectric cell
célula fotovoltaica photovoltaic cell
célula osmótica osmotic cell
célula solar solar cell
celulasa cellulase
celulosa cellulose
celulosa alfa alpha cellulose
celulosa de madera wood cellulose
celulosato sódico sodium cellulosate
celulósico cellulosic
cementación cementation
cementita cementite
cemento cement
cemento portland portland cement
ceniza ash
ceniza caustificada causticized ash
ceniza de estaño tin ash
ceniza de huesos bone ash
ceniza de humero fly ash
ceniza de madera wood ash
ceniza de sosa soda ash
cenizas de perla pearl ash
centelleador scintillator
centelleante scintillating
centelleo scintillation
centi centi
centígrado centigrade
centinormal centinormal
centrado en el cubo cubic-centered
centrífuga centrifuge
centrifugación centrifugation
centrífugo centrifugal
centrípeto centripetal
centro activo active center
centro de simetría center of symmetry
control quiral chiral center
centrón centron
cera wax
cera de abejas beeswax
cera de Brasil Brazil wax
cera de candelilla candelilla wax
cera de caña de azúcar sugarcane wax
cera de carnauba carnauba wax
cera de ceresina ceresin wax
cera de China Chinese wax
cera de parafina paraffin wax
cera de petrolato petrolatum wax
cera de petróleo petroleum wax
cera de tierra earth wax

cera microcristalina microcrystalline wax
cera mineral mineral wax
cera montana montan wax
cerámica ceramic
ceramida ceramide
cerato cerate
cerebrosa cerebrose
cerebrósidos cerebrosides
ceresina ceresin
ceria ceria
cérico ceric
cerílico cerylic
cerilo ceryl
cerina cerin
cerio cerium
cerita cerite
cermet cermet
cero absoluto absolute zero
ceroato de metilo methyl ceroate
cerosina cerosin
cerosinilo cerosinyl
ceroso cerous
ceroteno cerotene
cerusita cerussite
cervantita cervantite
cerveza beer
cesio cesium
cespitina cespitine
cetal ketal
cetano cetane
cetena ketene
ceteno cetene
cetilato cetylate
cetilbetamina cetylbetamine
cetileno cetylene
cetilmercaptano cetyl mercaptan
cetilo cetyl
cetimida ketimide
cetimina ketimine
cetina ketine, cetin
ceto keto
cetoácido keto acid
cetoamina ketoamine
cetobenzotriazina ketobenzotriazine
cetohexosa ketohexose
cetol ketol, cetol
cetona ketone
cetona aromática aromatic ketone

cetona de aldehído aldehyde ketone

cetona etílica bencílica benzyl ethyl
    ketone

cetopentosa ketopentose

cetosa ketose

cetosis ketosis

cetoxima ketoxime

cetrimida cetrimide

cianamida cyanamide

cianamida cálcica calcium cyanamide

cianamida de dialilo diallyl cyanamide

cianamida de dimetilo dimethyl
    cyanamide

cianamida sódica sodium cyanamide

cianato cyanate

cianato bárico barium cyanate

cianato de clorobecilo chlorobenzyl
    cyanate

cianato de etilo ethyl cyanate

cianato de metilo methyl cyanate

cianato de plata silver cyanate

cianato de plomo lead cyanate

cianato mercúrico mercuric cyanate

cianato potásico potassium cyanate

cianato sódico sodium cyanate

cianaurita cyanaurite

cianazina cyanazine

cianhidrina cyanohydrin

cianhidrina de acetaldehído
    acetaldehyde cyanohydrin

cianhidrina de acetona acetone
    cyanohydrin

cianhidrina de benzaldehído
    benzaldehyde cyanohydrin

cianhidrina de bromoacetona
    bromoacetone cyanohydrin

cianhidrina de etileno ethylene
    cyanohydrin

cianidina cyanidine

cianina cyanine

cianina de alizarina alizarin cyanine

cianita cyanite

ciano cyano

cianoacetamida cyanoacetamide

cianoacetato de etilhexilo ethylhexyl
    cyanoacetate

cianoacetato de etilo ethyl cyanoacetate

cianoacetato de metilo methyl
    cyanoacetate

cianoacrilato cyanoacrylate

cianoacrilato de isobutilo isobutyl
    cyanoacrylate

cianoaldehído cyanoaldehyde

cianoanilina cyanoaniline

cianoaurita sódica sodium cyanoaurite

cianobenceno cyanobenzene

cianobencilo cyanobenzyl

cianobenzamida cyanobenzamide

cianoborohidruro sódico sodium
    cyanoborohydride

cianocarburo cyanocarbon

cianocobalamina cyanocobalamin

cianocuprato sódico sodium
    cyanocuprate

cianoetanoato de etilo ethyl
    cyanoethanoate

cianoetanosulfonato sódico sodium
    cyanoethanesulfonate

cianoetilación cyanoethylation

cianoferrato cyanoferrate

cianoformiato cyanoformate

cianoformiato de metilo methyl
    cyanoformate

cianógeno cyanogen

cianopiridina cyanopyridine

cianoplatinato cyanoplatinate

cianoplatinita bárica barium
    cyanoplatinite

cianurato de trialilo triallyl cyanurate

cianúrico cyanuric

cianuro cyanide

cianuro amónico ammonium cyanide

cianuro áurico auric cyanide

cianuro bárico barium cyanide

cianuro cálcico calcium cyanide

cianuro cobaltoso cobaltous cyanide

cianuro cúprico cupric cyanide

cianuro cuproso cuprous cyanide

cianuro de alilo allyl cyanide

cianuro de bencilo benzyl cyanide

cianuro de benzoilo benzoyl cyanide

cianuro de bromo bromine cyanide

cianuro de bromobencilo bromobenzyl
    cyanide

cianuro de butilo butyl cyanide

cianuro de cadmio cadmium cyanide

cianuro de carbonilo carbonyl cyanide

cianuro de cesio cesium cyanide

cianuro de cinc   zinc cyanide
cianuro de cloro   chlorine cyanide
cianuro de clorometilo   chloromethyl cyanide
cianuro de cobalto   cobalt cyanide
cianuro de cobre   copper cyanide
cianuro de cobre sódico   sodium copper cyanide
cianuro de etileno   ethylene cyanide
cianuro de etilo   ethyl cyanide
cianuro de fenilo   phenyl cyanide
cianuro de flúor   fluorine cyanide
cianuro de hidrógeno   hydrogen cyanide
cianuro de isopropilo   isopropyl cyanide
cianuro de mercurio   mercury cyanide
cianuro de metilmercurio   methylmercury cyanide
cianuro de metilo   methyl cyanide
cianuro de níquel   nickel cyanide
cianuro de nitrobencilo   nitrobenzyl cyanide
cianuro de oro   gold cyanide
cianuro de oro sódico   sodium gold cyanide
cianuro de pentilo   pentyl cyanide
cianuro de plata   silver cyanide
cianuro de plata sódico   sodium silver cyanide
cianuro de plomo   lead cyanide
cianuro de propilo   propyl cyanide
cianuro de vinilo   vinyl cyanide
cianuro de yodo   iodine cyanide
cianuro ferripotásico   ferripotassium cyanide
cianuro mercúrico   mercuric cyanide
cianuro negro   black cyanide
cianuro palmítico   palmitic cyanide
cianuro potásico   potassium cyanide
cianuro potásico cuproso   cuprous potassium cyanide
cianuro potásico de cadmio   cadmium potassium cyanide
cianuro potásico de plata   silver potassium cyanide
cianuro potásico mercúrico   mercuric potassium cyanide
cianuro sódico   sodium cyanide
cianuro sódico de plata   silver sodium cyanide

cibotáctico   cybotactic
ciclamato   cyclamate
ciclamato cálcico   calcium cyclamate
ciclamato potásico   potassium cyclamate
ciclamato sódico   sodium cyclamate
cicleno   cyclene
cíclico   cyclic
ciclitol   cyclitol
ciclización   cyclization
ciclización de Volhard-Erdmann   Volhard-Erdmann cyclization
ciclo   cycle
ciclo de ácido cítrico   citric acid cycle
ciclo de ácido tricarboxílico   tricarboxylic acid cycle
ciclo de azufre   sulfur cycle
ciclo de Born-Haber   Born-Haber cycle
ciclo de carbono   carbon cycle
ciclo de carbono-nitrógeno   carbon-nitrogen cycle
ciclo de Carnot   Carnot cycle
ciclo de fósforo   phosphorus cycle
ciclo de glioxilato   glyoxylate cycle
ciclo de Krebs   Krebs cycle
ciclo de nitrógeno   nitrogen cycle
ciclo de oxidación-reducción   oxidation-reduction cycle
ciclo de urea   urea cycle
ciclo neutrónico   neutron cycle
ciclo protónico   proton cycle
ciclo térmico   heat cycle
cicloadición   cycloaddition
cicloalcano   cycloalkane
cicloalifático   cycloaliphatic
cicloalqueno   cycloalkene
cicloalquilo   cycloalkyl
ciclobarbital   cyclobarbital
ciclobarbitona   cyclobarbitone
ciclobutadieno   cyclobutadiene
ciclobutano   cyclobutane
ciclobutanol   cyclobutanol
ciclobuteno   cyclobutene
ciclobutilo   cyclobutyl
ciclofano   cyclophane
ciclofosfamida   cyclophosphamide
cicloheptadeceno   cycloheptadecene
cicloheptano   cycloheptane
cicloheptanol   cycloheptanol
cicloheptanona   cycloheptanone

ciclohepteno cycloheptene
cicloheptilo cycloheptyl
ciclohexano cyclohexane
ciclohexanodiol cyclohexanediol
ciclohexanol cyclohexanol
ciclohexanona cyclohexanone
ciclohexeniltriclorosilano
    cyclohexenyltrichlorosilane
ciclohexeno cyclohexene
ciclohexenol cyclohexenol
ciclohexilamina cyclohexylamine
ciclohexilbenceno cyclohexylbenzene
ciclohexilfenol cyclohexylphenol
ciclohexilideno cyclohexylidene
ciclohexilpiperdina cyclohexylpiperdine
cicloheximida cycloheximide
ciclón cyclone
ciclonita cyclonite
ciclononano cyclononane
ciclooctadieno cyclooctadiene
ciclooctano cyclooctane
ciclooctatetraeno cyclooctatetraene
cicloocteno cyclooctene
cicloolefina cycloolefin
cicloparafina cycloparaffin
ciclopentadecanona cyclopentadecanone
ciclopentadienida sódica sodium
    cyclopentadienide
ciclopentadieno cyclopentadiene
ciclopentano cyclopentane
ciclopentanol cyclopentanol
ciclopentanona cyclopentanone
ciclopentenilacetona
    cyclopentenylacetone
ciclopentenilo cyclopentenyl
ciclopenteno cyclopentene
ciclopentilacetona cyclopentylacetone
ciclopentileno cyclopentylene
ciclopentilo cyclopentyl
ciclopropano cyclopropane
ciclopropilo cyclopropyl
ciclosilano cyclosilane
ciclotetraeno cyclotetraene
ciclotrón cyclotron
cicuta hemlock
cicutoxina cicutoxine
cierre de anillo ring closure
ciguatoxina ciguatoxin
cilindrita cylindrite

cilindro cylinder
cilindro de filtración filter cylinder
cilíndro de gas gas cylinder
cilindro graduado graduated cylinder
cimeno cymene
cimetidina cimetidine
cimilo cymyl
cinabrio cinnabar
cinamaldehído cinnamaldehyde
cinamaldehído de hexilo hexyl
    cinnamaldehyde
cinamamida cinnamamide
cinamato cinnamate
cinamato cálcico calcium cinnamate
cinamato de alilo allyl cinnamate
cinamato de bencilo benzyl cinnamate
cinamato de cadmio cadmium cinnamate
cinamato de cinamilo cinnamyl
    cinnamate
cinamato de etilo ethyl cinnamate
cinamato de isobutilo isobutyl cinnamate
cinamato de metilo methyl cinnamate
cinamato potásico potassium cinnamate
cinamato sódico sodium cinnamate
cinameína cinnamein
cinamenilo cinnamenyl
cinameno cinnamene
cinamida cinnamide
cinamilideno cinnamylidene
cinamoilo cinnamoyl
cinc zinc
cincato zincate
cincato cobalto cobalt zincate
cincita zincite
cincocaína cinchocain
cinconamina cinchonamine
cinconidina cinchonidin
cinconina cinchonine
cincoso zincous
cineno cinene
cineol cineol
cinerina cinerin
cinerolona cinerolone
cinética kinetics
cinética de enzima enzyme kinetics
cinética de reacciones reaction kinetics
cinética química chemical kinetics
cinético kinetic
cinolina cinnoline

ciprita cyprite
cis cis
cisco breeze
cisoide cisoid
cistamina cystamine
cisteína cysteine
cistina cystine
citidina cytidine
citiolona citiolone
citocromo cytochrome
citoplasma cytoplasm
citoquímica cytochemistry
citosina cytosine
citotóxico cytotoxic
citotoxina cytotoxin
citraconoilo citraconoyl
citral citral
citramida citramide
citrato citrate
citrato ácido magnésico acid magnesium citrate
citrato amónico ammonium citrate
citrato amónico férrico ferric ammonium citrate
citrato bárico barium citrate
citrato cálcico calcium citrate
citrato ceroso cerous citrate
citrato cobaltoso cobaltous citrate
citrato cúprico cupric citrate
citrato de bismuto bismuth citrate
citrato de butilo butyl citrate
citrato de cafeína caffeine citrate
citrato de cinc zinc citrate
citrato de codeína codeine citrate
citrato de etilo ethyl citrate
citrato de fentanilo fentanyl citrate
citrato de hierro iron citrate
citrato de hierro amónico ammonium iron citrate
citrato de litio lithium citrate
citrato de manganeso manganese citrate
citrato de plata silver citrate
citrato de plomo lead citrate
citrato de tributilo tributyl citrate
citrato de trietilo triethyl citrate
citrato disódico disodium citrate
citrato estannoso stannous citrate
citrato férrico ferric citrate
citrato magnésico magnesium citrate

citrato manganoso manganous citrate
citrato potásico potassium citrate
citrato sódico sodium citrate
citrato tricálcico tricalcium citrate
citrato trisódico trisodium citrate
citreno citrene
citresia citresia
citrinina citrinin
citronelal citronellal
citronelol citronellol
citronilo citronyl
citrulina citrulline
civetal civettal
civeto civet
clarificación clarification
clarificador clarifier
clasificación classification
clasificación de Ehrenfest Ehrenfest classification
clasificación periódica periodic classification
clatrato clathrate
clebrio clebrium
clinohedrita clinohedrite
clinoptilolita clinoptilolite
clioquinol clioquinol
clomercuriferroceno chloromercuriferrocene
clomifeno clomiphene
clonazepam clonazepam
clopidol clopidol
clorado chlorinated
cloral chloral
clorálcali chloralkali
cloralosa chloralose
cloramina chloramine
cloranfenicol chloramphenicol
cloranil chloranil
cloranilato chloranilate
clorapatita chlorapatite
clorato chlorate
clorato alumínico aluminum chlorate
clorato amónico ammonium chlorate
clorato bárico barium chlorate
clorato cálcico calcium chlorate
clorato cobaltoso cobaltous chlorate
clorato de cadmio cadmium chlorate
clorato de cinc zinc chlorate
clorato de estroncio strontium chlorate

clorato de estroncio-potasio strontium-potassium chlorate

clorato de litio lithium chlorate

clorato de plata silver chlorate

clorato de plomo lead chlorate

clorato magnésico magnesium chlorate

clorato mercurioso mercurous chlorate

clorato potásico potassium chlorate

clorato potásico bárico barium potassium chlorate

clorato sódico sodium chlorate

cloraurita chloraurite

clorazina chlorazine

clorbensida chlorbenside

clorbutanol chlorbutanol

clordano chlordane

clordimeformo chlordimeform

clorendato de dialilo diallyl chlorendate

clorfenvinfos chlorfenvinphos

clorhexidina chlorhexidine

clorhidrato hydrochloride

clorhidrato de acetamida acetamide hydrochloride

clorhidrato de acetamidina acetamidine hydrochloride

clorhidrato de aminoazobenceno aminoazobenzene hydrochloride

clorhidrato de anilina aniline hydrochloride

clorhidrato de betaína betaine hydrochloride

clorhidrato de buclizina buclizine hydrochloride

clorhidrato de butetamina butethamine hydrochloride

clorhidrato de cafeína caffeine hydrochloride

clorhidrato de caucho rubber hydrochloride

clorhidrato de ciclizina cyclizine hydrochloride

clorhidrato de clordiazepóxido chlordiazepoxide hydrochloride

clorhidrato de cocaína cocaine hydrochloride

clorhidrato de codeína codeine hydrochloride

clorhidrato de diaminofenol diaminophenol hydrochloride

clorhidrato de dipenteno dipentene hydrochloride

clorhidrato de efedrina ephedrine hydrochloride

clorhidrato de etilmorfina ethylmorphine hydrochloride

clorhidrato de fenacaína phenacaine hydrochloride

clorhidrato de fenilciclideno phenylcyclidene hydrochloride

clorhidrato de galactosamina galactosamine hydrochloride

clorhidrato de guanidina guanidine hydrochloride

clorhidrato de hidralazina hydralazine hydrochloride

clorhidrato de hidroxilamina hydroxylamine hydrochloride

clorhidrato de meclizina meclizine hydrochloride

clorhidrato de meperidina meperidine hydrochloride

clorhidrato de metadona methadone hydrochloride

clorhidrato de metopón metopon hydrochloride

clorhidrato de naftilamina naphthylamine hydrochloride

clorhidrato de nicotina nicotine hydrochloride

clorhidrato de nilidrina nylidrin hydrochloride

clorhidrato de papaverina papaverine hydrochloride

clorhidrato de pineno pinene hydrochloride

clorhidrato de piridoxal pyridoxal hydrochloride

clorhidrato de procaína procaine hydrochloride

clorhidrato de promazina promazine hydrochloride

clorhidrato de semicarbazida semicarbazide hydrochloride

clorhidrato de tetracaína tetracaine hydrochloride

clorhidrato de tiamina thiamine hydrochloride

clorhidrato de vancomicina vancomycin

hydrochloride
clorhídrico hydrochloric
cloridión chloridion
clorilo chloryl
clorimida chlorimide
clorinación chlorination
clorinidad chlorinity
clorinólisis chlorinolysis
cloritión chlorition
clorito chlorite
clorito cálcico calcium chlorite
clorito de plata silver chlorite
clorito sódico sodium chlorite
clormetazanona chlormethazanone
cloro chlorine
cloroacetal de etilo ethyl chloroacetal
cloroacetaldehído chloroacetaldehyde
cloroacetamida chloroacetamide
cloroacetato chloroacetate
cloroacetato de cadmio cadmium
    chloroacetate
cloroacetato de etilo ethyl chloroacetate
cloroacetato de metilo methyl
    chloroacetate
cloroacetato sódico sodium chloroacetate
cloroacetilo chloroacetyl
cloroacetiluretano chloroacetylurethane
cloroacetoacetanilida
    chloroacetoacetanilide
cloroacetoacetato de etilo ethyl
    chloroacetoacetate
cloroacetofenona chloroacetophenone
cloroacetona chloroacetone
cloroacetonitrilo chloroacetonitrile
cloroacrilato chloroacrylate
cloroacrilonitrilo chloroacrylonitrile
cloroacroleína chloroacrolein
cloroaluminato sódico sodium
    chloroaluminate
cloroanilato de lantano lanthanum
    chloroanilate
cloroanilina chloroaniline
cloroantraquinona chloroanthraquinone
cloroargentato chloroargentate
cloroaurato sódico sodium chloroaurate
clorobecilpiridina chlorobenzylpyridine
clorobenceno chlorobenzene
clorobencenosulfonamida
    chlorobenzenesulfonamide

clorobencenotiol chlorobenzenethiol
clorobenzal chlorobenzal
clorobenzaldehído chlorobenzaldehyde
clorobenzantrona chlorobenzanthrone
clorobenzhidrol chlorobenzhydrol
clorobenzoato chlorobenzoate
clorobenzofenona chlorobenzophenone
clorobenzohidrol chlorobenzohydrol
clorobenzoilo chlorobenzoyl
clorobenzotricloruro
    chlorobenzotrichloride
clorobenzotrifluoruro
    chlorobenzotrifluoride
clorobromo chlorobromo
clorobromuro de etileno ethylene
    chlorobromide
clorobutadieno chlorobutadiene
clorobutano chlorobutane
clorobutanol chlorobutanol
clorobutirilo chlorobutyryl
clorocarbonato chlorocarbonate
clorocarbonato de bencilo benzyl
    chlorocarbonate
clorocarbonato de etilo ethyl
    chlorocarbonate
clorocarbonato de pentilo pentyl
    chlorocarbonate
clorocarbónico chlorocarbonic
clorocarburo chlorocarbon
clorocianógeno chlorocyanogen
clorocianuro chlorocyanide
clorocosano chlorocosane
clorocromato chlorochromate
clorocromato potásico potassium
    chlorochromate
clorocumarina chlorocoumarin
clorodecona chlorodecone
clorodifenilo chlorodiphenyl
clorodifluoroetano chlorodifluoroethane
clorodifluorometano
    chlorodifluoromethane
clorodinitrobenceno
    chlorodinitrobenzene
cloroestireno chlorostyrene
cloroetanamida chloroethanamide
cloroetano chloroethane
cloroetanol chloroethanol
cloroeteno chloroethene
cloroetilcloroformiato

chloroethylchloroformate

cloroetileno chloroethylene

clorofenetol chlorofenethol

clorofenicita chlorophoenicite

clorofeniltriclorosilano
chlorophenyltrichlorosilane

clorofenol chlorophenol

clorofila chlorophyll

clorofilina chlorophyllin

clorofluorocarburo chlorofluorocarbon

clorofluoruro chlorofluoride

cloroformiato chloroformate

cloroformiato de bencilo benzyl
chloroformate

cloroformiato de etilo ethyl
chloroformate

cloroformiato de isobutilo isobutyl
chloroformate

cloroformiato de metilo methyl
chloroformate

cloroformiato de triclorometilo
trichloromethyl chloroformate

cloroformo chloroform

cloroformo de acetona acetone
chloroform

cloroformoxima chloroformoxime

clorofosfato de dibutilo dibutyl
chlorophosphate

clorofosfato de dioctilo dioctyl
chlorophosphate

cloroftalato sódico sodium
chlorophthalate

clorohidrato alumínico aluminum
chlorohydrate

clorohidrina chlorohydrin

clorohidrina de etileno ethylene
chlorohydrin

clorohidrina de propileno propylene
chlorohydrin

clorohidrina de trimetileno trimethylene
chlorohydrin

clorohidroquinona chlorohydroquinone

clorohidroxibenceno
chlorohydroxybenzene

clorohidroxitolueno
chlorohydroxytoluene

clorometano chloromethane

clorometilación chloromethylation

clorometilanilina chloromethylaniline

clorometilbenceno chloromethylbenzene

clorometilcloroformiato
chloromethylchloroformate

clorometilclorosulfonato
chloromethylchlorosulfonate

clorometilnaftaleno
chloromethylnaphthalene

clorometilo chloromethyl

cloronaftaleno chloronaphthalene

cloronitroanilina chloronitroaniline

cloronitrobenceno chloronitrobenzene

cloronitrotolueno chloronitrotoluene

cloropentafluoroacetona
chloropentafluoroacetone

cloropentano chloropentane

cloropicrina chloropicrin

cloropiridina chloropyridine

cloroplatinato chloroplatinate

cloroplatinato bárico barium
chloroplatinate

cloroplatinato potásico potassium
chloroplatinate

cloroplatinato sódico sodium
chloroplatinate

cloroplatinita sódica sodium
chloroplatinite

cloropreno chloroprene

cloropropano chloropropane

cloropropeno chloropropene

cloropropileno chloropropylene

cloropropionato de etilo ethyl
chloropropionate

cloropropionitrilo chloropropionitrile

cloroquina chloroquine

cloroquinaldina chloroquinaldine

clorosalicilanilida chlorosalicylanilide

clorosfosfato de dietilo diethyl
chlorophosphate

clorosilano de metilo methyl
chlorosilane

clorosilo chlorosyl

cloroso chlorous

clorosuccinimida chlorosuccinimide

clorosulfonato de etilo ethyl
chlorosulfonate

clorosulfonato de metilo methyl
chlorosulfonate

clorosulfonato de propilo propyl
chlorosulfonate

clorosulfonato sódico sodium chlorosulfonate
clorotiazida chlorothiazide
clorotimol chlorothymol
clorotiofenol chlorothiophenol
clorotolueno chlorotoluene
clorotrifluoroetileno chlorotrifluoroethylene
clorotrifluorometano chlorotrifluoromethane
clorovinildicloroarsina chlorovinyldichloroarsine
cloroxileno chloroxylene
cloroxina chloroxine
cloroyoduro chloroiodide
cloroyoduro de cinc zinc chloroiodide
clorpromazina chlorpromazine
clorquinaldol chlorquinaldol
clortetraciclina chlortetracycline
cloruro chloride
cloruro alumínico aluminum chloride
cloruro amónico ammonium chloride
cloruro amónico de níquel nickel ammonium chloride
cloruro amónico de platino platinum ammonium chloride
cloruro amónico platínico platinic ammonium chloride
cloruro antimónico antimonic chloride
cloruro antimonioso antimonous chloride
cloruro arsenioso arsenous chloride
cloruro áurico auric chloride
cloruro bárico barium chloride
cloruro cálcico calcium chloride
cloruro ceroso cerous chloride
cloruro cianofórmico cyanoformic chloride
cloruro cianúrico cyanuric chloride
cloruro cobáltico cobaltic chloride
cloruro cobaltoso cobaltous chloride
cloruro crómico chromic chloride
cloruro cromoso chromous chloride
cloruro cúprico cupric chloride
cloruro cuproso cuprous chloride
cloruro de acetamida acetamide chloride
cloruro de acetilo acetyl chloride
cloruro de ácido acid chloride
cloruro de acriloilo acryloyl chloride

cloruro de alilo allyl chloride
cloruro de amida amide chloride
cloruro de amilo amyl chloride
cloruro de anilina aniline chloride
cloruro de anisoilo anisoyl chloride
cloruro de antimonio antimony chloride
cloruro de arsénico arsenic chloride
cloruro de azufre sulfur chloride
cloruro de bencenocarbonilo benzenecarbonyl chloride
cloruro de bencenodiazonio benzenediazonium chloride
cloruro de bencetonio benzethonium chloride
cloruro de bencilo benzyl chloride
cloruro de benciltrimetilamonio benzyltrimethylammonium chloride
cloruro de benzalconio benzalkonium chloride
cloruro de benzhidrilo benzhydryl chloride
cloruro de benzoilo benzoyl chloride
cloruro de benzol benzol chloride
cloruro de berilio beryllium chloride
cloruro de bismutilo bismuthyl chloride
cloruro de bismuto bismuth chloride
cloruro de boro boron chloride
cloruro de bromo bromine chloride
cloruro de butanoilo butanoyl chloride
cloruro de butilmagnesio butylmagnesium chloride
cloruro de butilo butyl chloride
cloruro de butirilo butyryl chloride
cloruro de butiroilo butyroyl chloride
cloruro de cacodilo cacodyl chloride
cloruro de cadmio cadmium chloride
cloruro de carbamida carbamide chloride
cloruro de carbonilo carbonyl chloride
cloruro de cerio cerium chloride
cloruro de cesio cesium chloride
cloruro de cetilpiridinio cetylpyridinium chloride
cloruro de cetiltrimetilamonio cetyltrimethylammonium chloride
cloruro de cianógeno cyanogen chloride
cloruro de ciclohexilo cyclohexyl chloride
cloruro de cinamoilo cinnamoyl chloride

cloruro de cinc zinc chloride
cloruro de cloroacetilo chloroacetyl chloride
cloruro de clorobecilo chlorobenzyl chloride
cloruro de clorobenzoilo chlorobenzoyl chloride
cloruro de cloroformilo chloroformyl chloride
cloruro de cobalto cobalt chloride
cloruro de cobre copper chloride
cloruro de cobre sódico sodium copper chloride
cloruro de cocaína cocaine chloride
cloruro de colina choline chloride
cloruro de cromilo chromyl chloride
cloruro de cromo chromium chloride
cloruro de decanoilo decanoyl chloride
cloruro de diazobenceno diazobenzene chloride
cloruro de dicloroacetilo dichloroacetyl chloride
cloruro de diclorobencilo dichlorobenzyl chloride
cloruro de diclorobenzoilo dichlorobenzoyl chloride
cloruro de didimio didymium chloride
cloruro de dimetilbencilo dimethylbenzyl chloride
cloruro de dinitrobenzoilo dinitrobenzoyl chloride
cloruro de escandio scandium chloride
cloruro de estaño tin chloride
cloruro de estearoilo stearoyl chloride
cloruro de estroncio strontium chloride
cloruro de etanoilo ethanoyl chloride
cloruro de etilbencilo ethylbenzyl chloride
cloruro de etileno ethylene chloride
cloruro de etilhexilo ethylhexyl chloride
cloruro de etilideno ethylidene chloride
cloruro de etilmagnesio ethylmagnesium chloride
cloruro de etilo ethyl chloride
cloruro de europio europium chloride
cloruro de fenacilo phenacyl chloride
cloruro de fenarsazina phenarsazine chloride
cloruro de fenilacetilo phenylacetyl

chloride
cloruro de fenilcarbilamina phenylcarbylamine chloride
cloruro de fenilmagnesio phenylmagnesium chloride
cloruro de fenilo phenyl chloride
cloruro de fenilpropilo phenylpropyl chloride
cloruro de ferrocenoilo ferrocenoyl chloride
cloruro de formilo formyl chloride
cloruro de fosfonio phosphonium chloride
cloruro de fosforilo phosphoryl chloride
cloruro de fósforo phosphorus chloride
cloruro de ftaloilo phthaloyl chloride
cloruro de fumarilo fumaryl chloride
cloruro de fumaroilo fumaroyl chloride
cloruro de furoilo furoyl chloride
cloruro de gadolinio gadolinium chloride
cloruro de glicerilo glyceryl chloride
cloruro de heptanoilo heptanoyl chloride
cloruro de hexadecanoilo hexadecanoyl chloride
cloruro de hexametonio hexamethonium chloride
cloruro de hexanoilo hexanoyl chloride
cloruro de hexilo hexyl chloride
cloruro de hidrazina hydrazine chloride
cloruro de hidrógeno hydrogen chloride
cloruro de hidrógeno anhidro anhydrous hydrogen chloride
cloruro de hidroxilamonio hydroxylammonium chloride
cloruro de hierro iron chloride
cloruro de hierro amónico ammonium iron chloride
cloruro de holmio holmium chloride
cloruro de indio indium chloride
cloruro de iridio iridium chloride
cloruro de isoamilo isoamyl chloride
cloruro de isobornilo isobornyl chloride
cloruro de isobutiroilo isobutyroyl chloride
cloruro de isodecilo isodecyl chloride
cloruro de isoftaloilo isophthaloyl chloride
cloruro de isopropenilo isopropenyl chloride

cloruro de isopropilo isopropyl chloride
cloruro de iterbio ytterbium chloride
cloruro de itrio yttrium chloride
cloruro de lantano lanthanum chloride
cloruro de laurilpiridinio
    laurylpyridinium chloride
cloruro de lauroilo lauroyl chloride
cloruro de litio lithium chloride
cloruro de lutecio lutetium chloride
cloruro de manganeso manganese
    chloride
cloruro de mercurio mercury chloride
cloruro de mesilo mesyl chloride
cloruro de mesitilensulfonilo
    mesitylenesulfonyl chloride
cloruro de metacriloilo methacryloyl
    chloride
cloruro de metanosulfonilo
    methanesulfonyl chloride
cloruro de metilalilo methylallyl
    chloride
cloruro de metileno methylene chloride
cloruro de metilmagnesio
    methylmagnesium chloride
cloruro de metilo methyl chloride
cloruro de miristilo myristyl chloride
cloruro de molibdeno molybdenum
    chloride
cloruro de neodimio neodymium
    chloride
cloruro de niobio niobium chloride
cloruro de níquel nickel chloride
cloruro de níquel amónico ammonium
    nickel chloride
cloruro de nitrilo nitryl chloride
cloruro de nitrobenzoilo nitrobenzoyl
    chloride
cloruro de nitrógeno nitrogen chloride
cloruro de nitrosilo nitrosyl chloride
cloruro de nonanoilo nonanoyl chloride
cloruro de nonilo nonyl chloride
cloruro de octanoilo octanoyl chloride
cloruro de octilmagnesio
    octylmagnesium chloride
cloruro de octilo octyl chloride
cloruro de oleoilo oleoyl chloride
cloruro de oro gold chloride
cloruro de oro sódico sodium gold
    chloride

cloruro de osmio osmium chloride
cloruro de oxalilo oxalyl chloride
cloruro de paladio palladium chloride
cloruro de paladio sódico sodium
    palladium chloride
cloruro de palmitoilo palmitoyl chloride
cloruro de pelargonilo pelargonyl
    chloride
cloruro de pentilo pentyl chloride
cloruro de picrilo picryl chloride
cloruro de pirosulfurilo pyrosulfuryl
    chloride
cloruro de plata silver chloride
cloruro de plata sódico sodium silver
    chloride
cloruro de platino platinum chloride
cloruro de plomo lead chloride
cloruro de polietilenglicol polyethylene
    glycol chloride
cloruro de polivinilideno polyvinylidene
    chloride
cloruro de polivinilo polyvinyl chloride
cloruro de praseodimio praseodymium
    chloride
cloruro de propanoilo propanoyl
    chloride
cloruro de propargilo propargyl chloride
cloruro de propileno propylene chloride
cloruro de propilo propyl chloride
cloruro de propionilo propionyl chloride
cloruro de radio radium chloride
cloruro de renio rhenium chloride
cloruro de rodio rhodium chloride
cloruro de rodiocarbonilo rhodium
    carbonyl chloride
cloruro de rubidio rubidium chloride
cloruro de rutenio ruthenium chloride
cloruro de samario samarium chloride
cloruro de santalilo santalyl chloride
cloruro de sebacoilo sebacoyl chloride
cloruro de selenio selenium chloride
cloruro de silicio silicon chloride
cloruro de succinilo succinyl chloride
cloruro de sulfinilo sulfinyl chloride
cloruro de sulfonilo sulfonyl chloride
cloruro de sulfurilo sulfuryl chloride
cloruro de suxametonio suxamethonium
    chloride
cloruro de talio thallium chloride

cloruro de tántalo tantalum chloride

cloruro de telurio tellurium chloride

cloruro de terbio terbium chloride

cloruro de tereftaloilo terephthaloyl chloride

cloruro de tetrabutilamonio tetrabutylammonium chloride

cloruro de tetradecilo tetradecyl chloride

cloruro de tetrametilamonio tetramethylammonium chloride

cloruro de tionilo thionyl chloride

cloruro de titanio titanium chloride

cloruro de tolilo tolyl chloride

cloruro de toluensulfonilo toluenesulfonyl chloride

cloruro de torio thorium chloride

cloruro de tributilestaño tributyltin chloride

cloruro de triclorometilsulfenilo trichloromethylsulfenyl chloride

cloruro de trifenilestaño triphenyltin chloride

cloruro de tulio thulium chloride

cloruro de tungsteno tungsten chloride

cloruro de uranilo uranyl chloride

cloruro de uranio uranium chloride

cloruro de vanadilo vanadyl chloride

cloruro de vanadio vanadium chloride

cloruro de vinilideno vinylidene chloride

cloruro de vinilmagnesio vinylmagnesium chloride

cloruro de vinilo vinyl chloride

cloruro de xililo xylyl chloride

cloruro de yodo iodine chloride

cloruro de zirconilo zirconyl chloride

cloruro de zirconio zirconium chloride

cloruro de zirconio potásico potassium zirconium chloride

cloruro dietilalumínico diethylaluminum chloride

cloruro estánnico stannic chloride

cloruro estannoso stannous chloride

cloruro fenilmercúrico phenylmercuric chloride

cloruro férrico ferric chloride

cloruro ferroso ferrous chloride

cloruro fosfórico phosphoric chloride

cloruro gálico gallic chloride

cloruro germánico germanic chloride

cloruro hidroxifenilmercúrico hydroxyphenylmercuric chloride

cloruro irídico iridic chloride

cloruro magnésico magnesium chloride

cloruro magnésico amónico ammonium magnesium chloride

cloruro manganoso manganous chloride

cloruro mercúrico mercuric chloride

cloruro mercúrico amoniacal ammoniated mercury chloride

cloruro mercúrico amónico ammonium mercuric chloride

cloruro mercurioso mercurous chloride

cloruro niqueloso nickelous chloride

cloruro osmoso osmous chloride

cloruro paladioso palladous chloride

cloruro platínico platinic chloride

cloruro platinoso platinous chloride

cloruro potásico potassium chloride

cloruro potásico alumínico aluminum potassium chloride

cloruro potásico de iridio iridium potassium chloride

cloruro potásico de osmio osmium potassium chloride

cloruro potásico de paladio palladium potassium chloride

cloruro potásico de platino platinum potassium chloride

cloruro potásico paladioso palladous potassium chloride

cloruro ruténico ruthenic chloride

cloruro samárico samaric chloride

cloruro samaroso samarous chloride

cloruro sódico sodium chloride

cloruro sódico alumínico aluminum sodium chloride

cloruro sódico de paladio palladium sodium chloride

cloruro sódico de plata silver sodium chloride

cloruro sódico de platino platinum sodium chloride

cloruro sódico paladioso palladous sodium chloride

cloruro sódico platínico platinic sodium chloride

cloruro sulfúrico sulfuric chloride

cloruro talioso thallous chloride

cloruro tantálico tantalic chloride
cloruro tantaloso tantalous chloride
cloruro teluroso tellurous chloride
cloruro titánico titanic chloride
cloruro titanoso titanous chloride
cloruro uránico uranic chloride
cloruro uranoso uranous chloride
cloruro vanadílico vanadylic chloride
cloruro vanadioso vanadous chloride
closo closo
coacervación coacervation
coagulación coagulation
coagulante coagulant
coalescencia coalescence
coalescente coalescent
cobalamina cobalamin
cobaltamina cobaltammine
cobáltico cobaltic
cobaltita cobaltite
cobalto cobalt
cobaltoceno cobaltocene
cobre copper
cobre azul blue copper
cobustible de cohete rocket fuel
coca coca
cocaína cocaine
cocarboxilasa cocarboxylase
cocer roast
cocido cooking, roasting
coco coccus
cocromo cochrome
coculina cocculin
cochinilina cochinilin
cochinilla cochineal
codeína codeine
coeficiente coefficient
coeficiente crítico critical coefficient
coeficiente de absorción absorption coefficient
coeficiente de absorción atómica atomic absorption coefficient
coeficiente de acidez acidity coefficient
coeficiente de actividad activity coefficient
coeficiente de adsorción adsorption coefficient
coeficiente de atenuación attenuation coefficient
coeficiente de conversión conversion coefficient
coeficiente de distribución distribution coefficient
coeficiente de Einstein Einstein coefficient
coeficiente de expansión térmica thermal expansion coefficient
coeficiente de extinción extinction coefficient
coeficiente de extracción extraction coefficient
coeficiente de fenol phenol coefficient
coeficiente de ionización ionization coefficient
coeficiente de Joule-Thomas Joule-Thomas coefficient
coeficiente de partición partition coeffcient
coeficiente de sedimentación sedimentation coefficient
coeficiente de selectividad selectivity coefficient
coeficiente de solubilidad solubility coefficient
coeficiente de temperatura temperature coefficient
coeficiente de transferencia transfer coefficient
coeficiente de transporte transport coefficient
coeficiente estequiométrico stoichiometric coefficient
coeficiente osmótico osmotic coefficient
coenzima coenzyme
coercitividad coercivity
cofactor cofactor
coherencia coherence
coherente coherent
cohesión cohesion
cohesión dieléctrica dielectric cohesion
cohobación cohobation
cola glue, cola
colagenasa collagenase
colágeno collagen
colamina colamine
colano cholane
colantreno cholanthrene
colapez isinglass
colargol collargol

colas tailings
colecalciferol cholecalciferol
colector collector
colemanita colemanite
colestérico cholesteric
colesterol cholesterol
colestipol colestipol
colestiramina cholestyramine
colidina collidine
coligar colligate
coligativo colligative
colina choline
colinesterasa cholinesterase
colisión collision
colisión de intercambio exchange
    collision
colisión desactivante deactivating
    collision
colisión elástica elastic collision
colisión electrónica electron collision
colisión inelástica inelastic collision
colisión iónica molecular molecular ion
    collision
colisión nuclear nuclear collision
colisión térmica thermal collision
colistina colistin
colmena beehive
colodión collodion
colofonia colophony, rosin
colofonia de madera wood rosin
colofonita colophonite
coloidal colloidal
coloide colloid
coloide hidrofílico hydrophilic colloid
coloide hidrofóbico hydrophobic colloid
coloide irreversible irreversible colloid
coloide molecular molecular colloid
colometría visual visual colorimetry
color certificado certified color
color de alimento food color
coloración coloration
colorante colorant, dye, stain
colorante ácido acid dye
colorante al azufre sulfur dye
colorante al sulfuro sulfide dye
colorante azoico azo dye
colorante azoico primario primary azo
    dye
colorante de alizarina alizarin dye

colorante de alquitrán de hulla coal-tar
    dye
colorante de anilina aniline dye
colorante de antraquinona
    anthraquinone dye
colorante de azina azine dye
colorante de bencidina benzidine dye
colorante de cianina cyanine dye
colorante de cromo chrome dye
colorante de estilbeno stilbene dye
colorante de quinolina quinoline dye
colorante de tiazol thiazole dye
colorante de tina vat dye
colorante de triarilmetano
    triarylmethane dye
colorante de trifenilmetano
    triphenylmethane dye
colorante de xanteno xanthene dye
colorante de Ziehl Ziehl stain
colorante desarrollado developed dye
colorante diazoico diazo dye
colorante directo direct dye
colorante fluorescente fluorescent dye
colorante incoloro colorless dye
colorante indirecto indirect dye
colorante nitro nitro dye
colorante nitroso nitroso dye
colorante oxiazoico oxyazo dye
colorante reactivo reactive dye
colorante substantivo substantive dye
colorante vegetal vegetable dye
colorimetría colorimetry
colorimétrico colorimetric
colorímetro colorimeter
colorímetro fotoeléctrico photoelectric
    colorimeter
colorización colorization
colquicina colchicine
columbato columbate
columbio columbium
columbita columbite
columna column
columna de Clusius Clusius column
columna de destilación distillation
    column
columna de gradiente de densidad
    density-gradient column
columna de intercambio iónico ion-
    exchange column

columna de placas plate column

columna empaquetada packed column

columna fraccionante fractionating column

columna rectificante rectifying column

columna térmica thermal column

colza colza

combinación combination

combinación molecular molecular combination

combinación química chemical combination

combinado combined

combustibilidad combustibility

combustible fuel

combustible de boro boron fuel

combustible fósil fossil fuel

combustible nuclear nuclear fuel

combustible sin humo smokeless fuel

combustible sintético synthetic fuel

combustión combustion

combustión espontánea spontaneous combustion

combustión fraccionaria fractional combustion

combustión homogénea homogeneous combustion

combustión lenta slow combustion

combustión orgánica organic combustion

combustión superficial surface combustion

comicelización comicellization

comirina comirin

comparador comparator

comparador de colores color comparator

comparador de densidad density comparator

compatibilidad compatibility

compatible compatible

compensación interna internal compensation

compensación intramolecular intramolecular compensation

complejación complexing

complejo complex

complejo activado activated complex

complejo coordinado coordinated complex

complejo de acetileno acetylene complex

complejo de adición addition complex

complejo de inclusión inclusion complex

complejo de Meisenheimer Meisenheimer complex

complejo de Werner Werner complex

complejo enzima-substrato enzyme-substrate complex

complejo excitado exciplex

complejo inerte inert complex

complejo lábil labile complex

complejo pi pi complex

complejo vitamínico vitamin complex

complexona complexone

componente component

componente activo active component

composición composition

composición constante constant composition

composición isotópica isotopic composition

compresión adiabática adiabatic compression

compresor de aire air compressor

comprimido compressed

comprobación de calidad quality assurance

compuesto compound

compuesto acíclico acyclic compound

compuesto alicíclico alicyclic compound

compuesto alifático aliphatic compound

compuesto antimónico antimonic compound

compuesto antimonioso antimonous compound

compuesto argéntico argentic compound

compuesto aromático aromatic compound

compuesto asimétrico asymmetric compound

compuesto áurico auric compound

compuesto bicíclico bicyclic compound

compuesto binario binary compound

compuesto cíclico cyclic compound

compuesto complejo complex compound

compuesto de acabado finishing compound

compuesto de carbono carbon compound

compuesto de coordinación coordination

compound

compuesto de encerramiento enclosure compound

compuesto de Herz Herz compound

compuesto de inclusión inclusion compound

compuesto de intercalación intercalation compound

compuesto de Vaska Vaska compound

compuesto electrovalente electrovalent compound

compuesto en anillo ring compound

compuesto en jaula cage compound

compuesto estequiométrico stoichiometric compound

compuesto exotérmico exothermic compound

compuesto fotocrómico photochromic compound

compuesto heterocíclico heterocyclic compound

compuesto impermeabilizante waterproofing compound

compuesto insaturado unsaturated compound

compuesto intermetálico intermetallic compound

compuesto marcado labeled compound

compuesto monístico monistic compound

compuesto monofuncional monofunctional compound

compuesto nativo native compound

compuesto neutro neutral compound

compuesto no polar nonpolar compound

compuesto orgánico organic compound

compuesto orgánico volátil volatile organic compound

compuesto organoalumínico organoaluminum compound

compuesto organocíncico organozinc compound

compuesto organofosforado organophosphorus compound

compuesto organolítico organolithium compound

compuesto organomagnesiano organomagnesium compound

compuesto organomercúrico organomercury compound

compuesto organometálico organometallic compound

compuesto polar polar compound

compuesto químico chemical compound

compuesto saturado saturated compound

compuesto simétrico symmetrical compound

compuesto superficial surface compound

compuesto ternario ternary compound

compuesto tetraédrico tetrahedral compound

compuesto trazador tracer compound

compuesto volátil volatile compound

cóncavo concave

concentración concentration

concentración de ion hidrógeno hydrogen-ion concentration

concentración de substancia substance concentration

concentración de volmen de pigmento pigment volume concentration

concentración electrónica electron concentration

concentración molal molal concentration

concentración molecular molecular concentration

concentrado concentrated

concentrador concentrator

concentrar concentrate

concoidal conchoidal

concreto concrete

condensación condensation

condensación aciloínica acyloin condensation

condensación aldólica aldol condensation

condensación capilar capillary condensation

condensación de aldehído aldehyde condensation

condensación de benzoína benzoin condensation

condensación de Claisen Claisen condensation

condensación de Claisen-Schmidt Claisen-Schmidt condensation

condensación de Darzens Darzens condensation

condensación de Friedel-Crafts Friedel-Crafts condensation

condensación de Guareschi-Thorpe Guareschi-Thorpe condensation
condensación de Michael Michael condensation
condensación de pinacol pinacol condensation
condensación de Stobbe Stobbe condensation
condensación fraccionaria fractional condensation
condensación intramolecular intramolecular condensation
condensación retrógrada retrograde condensation
condensado condensed
condensador condenser
condensador de Friedrichs Friedrichs condenser
condensador de Liebig Liebig condenser
condensador de reflujo reflux condenser
condensador en espiral coil condenser
condensador parcial partial condenser
condestilación condistillation
condiciones ambientales environmental conditions
condiciones críticas critical conditions
condiciones de laboratorio laboratory conditions
condiciones normales normal conditions, standard conditions
condiciones óptimas optimum conditions
conducción conduction
conducción electrolítica electrolytic conduction
conducción electrónica electron conduction
conducción metálica metallic conduction
conductancia conductance
conductancia de electrodo electrode conductance
conductancia electrolítica electrolytic conductance
conductancia equivalente equivalent conductance
conductancia específica specific conductance
conductancia iónica ionic conductance
conductancia térmica thermal conductance

conductividad conductivity
conductividad de solución solution conductivity
conductividad eléctrica electric conductivity
conductividad electrolítica electrolytic conductivity
conductividad equivalente equivalent conductivity
conductividad específica specific conductivity
conductividad molal molal conductivity
conductividad molar molar conductivity
conductividad molecular molecular conductivity
conductividad térmica thermal conductivity, heat conductivity
conductométrico conductometric
conductor conductor
conductor electrónico electron conductor
conesina conessine
configuración configuration
configuración absoluta absolute configuration
configuración electrónica electron configuration
conformación conformation
conformación eclipsada eclipsed conformation
conformación en bote boat conformation
conformación en cadena chain conformation
conformación en silla chair conformation
congelación freezing, congelation
congelar freeze
conglomerado conglomerate
coniceína conicein
coniferilo coniferyl
coniferina coniferin
coniferol coniferol
conina conine
conjugación conjugation
conjugado conjugated
cono cone
cono de filtración filter cone
cono de Seger Seger cone
cono pirométrico pyrometric cone
conproporción conproportionation

conquiolina conchiolin
conservación de energía conservation of energy
conservación de materia conservation of matter
consistencia consistency
consistómetro consistometer
consoluto consolute
constantán constantan
constante constant
constante atómica atomic constant
constante crioscópica cryoscopic constant
constante crítica critical constant
constante de absorción absorption constant
constante de acidez acidity constant
constante de acoplamiento coupling constant
constante de apantallamiento screening constant
constante de Avogadro Avogadro constant
constante de basicidad basicity constant
constante de Boltzmann Boltzmann constant
constante de célula cell constant
constante de desintegración decay constant
constante de desintegración radiactiva radioactive decay constant
constante de difusión diffusion constant
constante de disociación dissociation constant
constante de dissociación básica base dissociation constant
constante de equilibrio equilibrium constant
constante de estabilidad stability constant
constante de Faraday Faraday constant
constante de Fermi Fermi constant
constante de formación formation constant
constante de fuerza force constant
constante de gas gas constant
constante de ionización ionization constant
constante de Madelung Madelung

constant
constante de masa atómica unificada unified atomic mass constant
constante de Michaelis Michaelis constant
constante de Planck Planck constant
constante de producto de solubilidad solubility-product constant
constante de reacción reaction constant
constante de Rydberg Rydberg constant
constante de sedimentación sedimentation constant
constante de van der Waals van der Waals constant
constante de velocidad velocity constant
constante dieléctrica dielectric constant
constante ebulloscópica ebullioscopic constant
constante elástica elastic constant
constante física physical constant
constante fundamental fundamental constant
constante química chemical constant
constante radiactiva radioactive constant
constante rotacional rotational constant
constante térmica thermal constant
constantes de agua water constants
constitución química chemical constitution
constitucional constitutional
constituyente constituent
contacto contact
contador con halógeno halogen counter
contador de burbujas bubble counter
contador de centelleos scintillation counter
contador de Geiger Geiger counter
contador de Geiger-Müller Geiger-Müller counter
contador de radiación radiation counter
contaminación contamination, pollution
contaminación ambiental environmental contamination
contaminación atmosférica atmospheric pollution
contaminación de agua water pollution
contaminación de aire air pollution
contaminación química chemical pollution

contaminación radiactiva radioactive contamination
contaminación térmica thermal pollution
contaminante contaminant
contenedor container
contenido ácido acid content
contenido de uranio uranium content
contenido térmico heat content
conteo de partículas particle counting
continuidad de estado continuity of state
continuo continuous
continuo de Lyman Lyman continuum
contracción contraction
contracción de actinoides actinoid contraction
contracción lantánida lanthanide contraction
contracolorante counterstain
contracorriente countercurrent
contraion counterion
contraste contrast
control control
control automático automatic control
control de calidad quality control
control de desperdicios waste control
control de flujo de gas gas-flow control
control de llama flame control
control de reacción reaction control
control de temperatura temperature control
controlado por cuarzo quartz-controlled
convección convection
conversión conversion
conversión de bencidina benzidine conversion
conversión de energía energy conversion
conversión geométrica geometric conversion
conversión interna internal conversion
convertidor converter
convertidor catalítico catalytic converter
convertidor de energía energy converter
convertidor de neutrones neutron converter
convexo convex
coordinación coordination
coordinación bipiramidal trigonal trigonal bipyramidal coordination
coordinación tetraédrica tetrahedral

coordination
coordinado coordinate
copal copal
copelación cupellation
copigmento copigment
copirina copyrine
copolimerización copolymerization
copolímero copolymer
copolímero aleatorio random copolymer
copolímero con injerto graft copolymer
copolímero de acrilamida acrylamide copolymer
copolímero de acrilonitrilo acrylonitrile copolymer
copolímero de bloques block copolymer
copolímero olefínico olefin copolymer
coprastanol coprastanol
coprecipitación coprecipitation
coprosterol coprosterol
coque coke
coque de petróleo petroleum coke
coquización coking
coral coral
corcho cork
cordita cordite
coriandrol coriandrol
corindón corundum
coritoide choritoid
cornezuelo ergot
corona corona
coroneno coronene
corrección de Rydberg Rydberg correction
corrección radiativa radiative correction
correlación correlation
corriente constante constant current
corriente de difusión diffusion current
corriente de electrodo electrode current
corriente de gas gas current
corriente de ionización ionization current
corriente de migración migration current
corriente de polarización polarization current
corriente eléctrica electric current
corriente electrónica electron current, electron stream
corriente galvánica galvanic current

corriente neutrónica neutron current
corrosión corrosion
corrosión electrolítica electrolytic corrosion
corrosión electroquímica electrochemical corrosion
corrosión galvánica galvanic corrosion
corrosión química chemical corrosion
corrosivo corrosive
corsita corsite
corticoide corticoid
corticosterona corticosterone
corticotropina corticotropin
cortisol cortisol
cortisona cortisone
cosanol cosanol
cosilo cosyl
cosmético cosmetic
costra scale
coulombio coulomb
covalencia covalence
covalencia coordinada coordinate covalence
covalente covalent
covelina coveline
covelita covellite
covolumen de van der Waals van der Waals covolume
craquear crack
craqueo cracking
craqueo catalítico catalytic cracking
craqueo térmico thermal cracking
creatina creatine
creatinasa creatinase
creatinina creatinine
creatininasa creatininase
crecimiento de cristal crystal growth
crémor tártaro cream of tartar
creosol creosol
creosota creosote
cresilato cresylate
cresilo cresyl
cresol cresol
cresolftaleína cresolphthalein
cresorcina cresorcin
creta chalk
criba screen
cribado screening
criogenia cryogenics

criogénica cryogenics
criogénico cryogenic
criogenina cryogenin
criohidrato cryohydrate
criolita cryolite
criómetro cryometer
crioquímica cryochemistry
crioscopia cryoscopy
crioscopio de Hortvet Hortvet cryoscope
crióstato cryostat
criptando cryptand
criptato cryptate
criptidina cryptidine
criptocianina cryptocyanine
criptohalita cryptohalite
criptón krypton
criptoxantina cryptoxanthin
crisanilina chrysaniline
criseno chrysene
crisol crucible
crisol de filtración filter crucible
crisol de Gooch Gooch crucible
crisólito chrysolite
crisoquinona chrysoquinone
crisotilo chrysotile
cristal crystal
cristal covalente covalent crystal
cristal cúbico cubic crystal
cristal cultivado grown crystal
cristal de cuarzo quartz crystal
cristal de roca rock crystal
cristal homopolar homopolar crystal
cristal ideal ideal crystal
cristal iónico ionic crystal
cristal líquido liquid crystal
cristal macromolecular macromolecular crystal
cristal metálico metallic crystal
cristal mixto mixed crystal
cristal nemático nematic crystal
cristal no polar nonpolar crystal
cristal óptico optical crystal
cristal oscilante oscillating crystal
cristal positivo positive crystal
cristalino crystalline
cristalito crystallite
cristalización crystallization
cristalización fraccionaria fractional crystallization

cristalizador al vacío vacuum crystallizer

cristalizar crystallize

cristalografía crystallography

cristalografía electrónica electron crystallography

cristalografía por rayos X X-ray crystallography

cristalograma crystallogram

cristalograma de Bragg Bragg crystallogram

cristaloide crystalloid

cristaloluminiscencia crystalloluminescence

cristobalita cristobalite

crítico critical

croceína brillante brilliant crocein

crocetina crocetin

crocidolita crocidolite

croma chroma

cromano chroman

cromaticidad chromaticity

cromático chromatic

cromatina chromatin

cromato chromate

cromato ácido sódico sodium acid chromate

cromato amónico ammonium chromate

cromato bárico barium chromate

cromato cálcico calcium chromate

cromato cobaltoso cobaltous chromate

cromato cúprico cupric chromate

cromato de bismuto bismuth chromate

cromato de butilo butyl chromate

cromato de cesio cesium chromate

cromato de cinc zinc chromate

cromato de cobalto cobalt chromate

cromato de cobre copper chromate

cromato de estaño tin chromate

cromato de estroncio strontium chromate

cromato de hierro iron chromate

cromato de hierro amónico ammonium iron chromate

cromato de litio lithium chromate

cromato de manganeso manganese chromate

cromato de mercurio mercury chromate

cromato de plata silver chromate

cromato de plomo lead chromate

cromato de rubidio rubidium chromate

cromato de talio thallium chromate

cromato estánnico stannic chromate

cromato estannoso stannous chromate

cromato férrico ferric chromate

cromato magnésico magnesium chromate

cromato manganoso manganous chromate

cromato mercúrico mercuric chromate

cromato mercurioso mercurous chromate

cromato potásico potassium chromate

cromato potásico bárico barium potassium chromate

cromato potásico rojo red potassium chromate

cromato sódico sodium chromate

cromatografía chromatography

cromatografía bidimensional two-dimensional chromatography

cromatografía circular circular chromatography

cromatografía de capa fina thin-layer chromatography

cromatografía de exclusión iónica ion-exclusion chromatography

cromatografía de gases gas chromatography

cromatografía de gel gel chromatography

cromatografía de intercambio iónico ion-exchange chromatography

cromatografía de partición partition chromatography

cromatografía de película fina thin-film chromatography

cromatografía de permeación en gel gel-permeation chromatography

cromatografía descendente descending chromatography

cromatografía en fase vapor vapor-phase chromatography

cromatografía gas-líquido gas-liquid chromatography

cromatografía horizontal horizontal chromatography

cromatografía líquida liquid

chromatography

cromatografía líquida de alto rendimiento high-performance liquid chromatography

cromatografía por adsorción adsorption chromatography

cromatografía por afinidad affinity chromatography

cromatografía por desplazamiento displacement chromatography

cromatografía radial radial chromatography

cromatografía sobre papel paper chromatography

cromatograma chromatogram

crómico chromic

cromilo chromyl

crominancia chrominance

cromita chromite

cromita sódica sodium chromite

cromitita chromitite

cromo chromium, chrome

cromo-cobre chromium-copper

cromofílico chromophilic

cromofórico chromophoric

cromóforo chromophore

cromogénico chromogenic

cromógeno chromogen

cromoisómero chromoisomer

cromona chromone

cromosana chromosan

cromoso chromous

cromosoma chromosome

cromotropía chromotropy

cronopotenciometría chronopotentiometry

crotonal crotonal

crotonaldehído crotonaldehyde

crotonato de butilo butyl crotonate

crotonato de etilo ethyl crotonate

crotonileno crotonylene

crotonilo crotonyl

crotonoide crotonoid

crotonoilo crotonoyl

crotoxina crotoxin

crudo crude

cuadri quadri

cuadrimolecular quadrimolecular

cuadripolo quadrupole

cuadrivalencia quadrivalence

cuadrivalente quadrivalent

cuajo rennet

cualitativo qualitative

cuantificación quantization

cuantificado quantized

cuantificar quantize

cuantitativo quantitative

cuantivalencia quantivalence

cuanto quantum

cuanto de energía quantum of energy

cuanto de luz quantum of light

cuanto de radiación quantum of radiation

cuanto virtual virtual quantum

cuarcita quartzite, quartz rock

cuarteto quartet

cuarzo quartz

cuarzo fundido fused quartz

cuasia quassia

cuasicristal quasicrystal

cúbico cubical

cúbico centrado en el cuerpo body-centered cubic

cúbico centrado en las caras face-centered cubic

cubierto coated

cubo cube

cuello de vidrio esmerilado ground-glass neck

cuenta count

cuero leather

cuerpo body

cuerpo cetónico ketone body

cumal cumal

cumaldehído cumaldehyde

cumarina coumarin

cumarona coumarone

cumenilo cumenyl

cumeno cumene

cumidina cumidine

cumilo cumyl

cuminol cuminol

cumol cumol

cumuleno cumulene

cupferrón cupferron

cuprato cuprate

cuprato amónico ammonium cuprate

cupreína cupreine

cupreno cuprene

cúprico cupric

cuprita cuprite

cuprocianuro cuprocyanide

cuprocianuro potásico potassium cuprocyanide

cuprocúprico cuprocupric

cuproína cuproine

cuproníquel cupronickel

cuproso cuprous

cuprotungsteno cuprotungsten

cúpula dome

curado curing

curare curare

curativo curative

curcumina curcumin

curie curie

curio curium

curtido tanning

curva de actividad activity curve

curva de afinidad affinity curve

curva de Bragg Bragg curve

curva de energía electrónica electronic energy curve

curva de excitación excitation curve

curva de ionización de Bragg Bragg ionization curve

curva de Sargent Sargent curve

curva de solubilidad solubility curve

curva espinodal spinodal curve

cusparina cusparine

CH

chabasita chabazite
chapado cladding
chavicol chavicol
chi chi

chicle chicle
chispa spark
chisporroteo sparking, sputter,
    sputtering

D

dacita dacite
dactilina dactylin
dactinomicina dactinomycin
dafnetina daphnetin
dafnita daphnite
dahllita dahllite
dalapón dalapon
dalina dahlin
dalton dalton
damascinina damascinine
dambonita dambonite
damenita dahmenite
damourita damourite
danaíta danaite
danalita danalite
dansilo dansyl
dapsona dapsone
darapskita darapskite
datación dating
datación por carbono carbon dating
datación por radiocarbono radiocarbon
    dating
datación química chemical dating
datación radiactiva radioactive dating
datación radiométrica radiometric
    dating
datolita datolite
datos químicos chemical data
daubreelita daubreelite
daunomicina daunomycin
daviesita daviesite
dawsonita dawsonite
deanol deanol
débil weak
debye debye
decaborano decaborane
decahidrato decahydrate
decahidrato de carbonato sódico sodium
    carbonate decahydrate
decahidrato de pentaborato sódico
    sodium pentaborate decahydrate
decahidrato de sulfato sódico sodium
    sulfate decahydrate

decahidronaftaleno
    decahydronaphthalene
decaimiento decay
decalina decalin
decametiltetrasiloxano
    decamethyltetrasiloxane
decametrina decamethrin
decanal decanal
decanamida decanamide
decano decane
decanoato decanoate
decanoato de etilo ethyl decanoate
decanodiol decanediol
decanoilo decanoyl
decanol decanol
decanonitrilo decanenitrile
decantación decantation
decantar decant
decapado pickling
decatilo decatyl
deceno decene
decicaína decicaine
decilamida decylamide
decilamina decylamine
decileno decylene
deciletileno decylethylene
decilmercaptano decyl mercaptan
decilo decyl
deciltriclorosilano decyltrichlorosilane
decimolar decimolar
decino decyne
decinormal decinormal
decocción decoction
decoloración decoloration
decolorante decolorant
decolorización decolorization
decrepitación decrepitation
decrepitar decrepitate
defecto defect
defecto de Frenkel Frenkel defect
defecto de masa mass defect
defecto de Schottky Schottky defect
defervescencia defervescence

deficiencia electrónica electron deficiency
deflagración deflagration
deflegmación dephlegmation
deflexión magnética magnetic deflection
deformabilidad deformability
deformación deformation, strain
degeneración degeneration
degenerado degenerate
degradación degradation
degradación de Barbier-Wieland Barbier-Wieland degradation
degradación de Bergmann Bergmann degradation
degradación de Edman Edman degradation
degradación de Emde Emde degradation
degradación de Grignard Grignard degradation
degradación de Hofmann Hofmann degradation
degradación de Krafft Krafft degradation
degradación de Miescher Miescher degradation
degradación de Ruff-Fenton Ruff-Fenton degradation
degradación de Strecker Strecker degradation
degradación de Weerman Weerman degradation
degradación de Wohl Wohl degradation
degradación térmica thermal degradation
degradar degrade
delanio delanium
delcosina delcosine
delessita delessite
delfinina delphinin
delicuescencia deliquescence
delicuescente deliquescent
delta delta
delvauxita delvauxite
demal demal
demanda química de oxígeno chemical oxygen demand
demeclociclina demeclocycline
demetileno demethylene
demetona demeton

demolización demolization
demulsibilidad demulsibility
dendrita dendrite
dendrobina dendrobine
densidad density
densidad absoluta absolute density
densidad aparente apparent density
densidad crítica critical density
densidad de carga charge density
densidad de carga electrónica electron-charge density
densidad de colisiones collision density
densidad de corriente current density
densidad de energía energy density
densidad de gas gas density
densidad de neutrones neutron density
densidad de vapor vapor density
densidad electrónica electron density
densidad limitadora limiting density
densidad molecular molecular density
densidad nuclear nuclear density
densidad óptica optical density
densidad relativa relative density
densitometría ultravioleta ultraviolet densitometry
densitómetro densitometer
deoxi deoxy
deposición deposition
deposición al vacío vacuum deposition
deposición electrolítica electrolytic deposition
deposición electrónica sputter, sputtering
deposición electroquímica electrochemical deposition
deposición húmeda wet deposition
deposición metálica metal deposition
deposición química chemical deposition
deposición rítmica rhythmic deposition
deposición seca dry deposition
depositado plated
depósito deposit, plating
depósito radiactivo radioactive deposit
depresión depression
depresión de punto de congelación freezing-point depression
depresión de punto de fusión melting-point depression
depresión molecular molecular depression

depresor depressor
depsida depside
depurador depurator, scavenger, scrubber
depurador de gas getter
derivación derivation
derivado derivative
derretir melt
derritol derritol
des des
desacetilación deacetylation
desacoplar decouple
desactivación deactivation
desactivador deactivator
desaglomeración disagglomeration
desalinación desalination
desamidación deamidation
desaminación deamination
desarrollador developer
desarrollo development
descalcificación decalcification
descarbonizar decarbonize
descarboxilación decarboxylation
descarboxilar decarboxylate
descarboxilasa decarboxylase
descarburización decarburization
descarburizar decarburize
descarga discharge
descarga gaseosa gaseous discharge
desclizita desclizite
desclorinación dechlorination
descomponer decompose
descomposición decomposition
descomposición doble double decomposition
descomposición electrolítica electrolytic decomposition
descomposición subatómica subatomic decomposition
descomposición térmica thermal decomposition
descontaminación decontamination
desecado desiccated
desecador desiccator
desecador de Fresenius Fresenius desiccator
desecante desiccant
deseleniuro diselenide
desemulsificación demulsification

desespumante defoamer
desfloculación deflocculation
desfloculante deflocculant
desgaseamiento degassing, outgassing
desgasificación degasification
deshidratación dehydration
deshidratado dehydrated
deshidratador dehydrator
deshidratante dehydrant
deshidro dehydro
deshidroacetato cálcico calcium dehydroacetate
deshidroacetato sódico sodium dehydroacetate
deshidrociclización dehydrocyclization
deshidrocolesterol dehydrocholesterol
deshidrogenación dehydrogenation
deshidrogenasa dehydrogenase
deshidroisoandrosterona dehydroisoandrosterone
deshidrólisis dehydrolysis
deshumidificación dehumidification
desilo desyl
desinfección disinfection
desinfectante disinfectant
desintegración decay, disintegration
desintegración alfa alpha decay
desintegración atómica atomic disintegration
desintegración beta beta decay
desintegración en cadena chain decay
desintegración espontánea spontaneous decay
desintegración exponencial exponential decay
desintegración gamma gamma decay
desintegración múltiple multiple decay
desintegración nuclear nuclear decay
desintegración radiactiva radioactive decay
desionización deionization
desionizado deionized
deslanosida deslanoside
deslizamiento shift, shifting
deslizamiento batocromático bathochromatic shift
deslizamiento de Stokes Stokes shift
deslizamiento espectral spectral shift
deslizamiento isomérico isomeric shift

deslizamiento isotópico isotope shift
deslizamiento químico chemical shift
deslocalización delocalization
desmagnetización demagnetization
desmetilación demethylation
desmina desmin
desmineralización demineralization
desmotropismo desmotropism
desmótropo desmotrope
desmuriación demuriation
desnaturalización denaturation
desnaturalizado denatured
desnaturalizante denaturant
desnaturalizar denature
desnitración denitration
desnitrificación denitrification
desnitrogenación denitrogenation
desodorante deodorant
desorción desorption
desoxi deoxy, desoxy
desoxianisoína deoxyanisoin
desoxibenzoína deoxybenzoin
desoxidación deoxidation
desoxidante deoxidizer
desoxidar deoxidize
desoxigenación deoxygenation
desoxirribonucleasa deoxyribonuclease
desoxirribosa deoxyribose
desperdicios waste
desperdicios agrícolas agricultural waste
desperdicios industriales industrial
    waste
desperdicios nucleares nuclear waste
desperdicios peligrosos hazardous waste
desperdicios químicos chemical waste
desperdicios radiactivos radioactive
    waste
desplazamiento displacement, shift,
    shifting
desplazamiento de Zeeman Zeeman
    displacement
desplazamiento doble double
    displacement
despolarización depolarization
despolarizador depolarizer
despolmerización depolymerization
desproporción disproportionation
destello flash
destello de luz light flash

destilación distillation
destilación a alto vacío high-vacuum
    distillation
destilación al vacío vacuum distillation
destilación al vapor steam distillation
destilación analítica analytical
    distillation
destilación azeotrópica azeotropic
    distillation
destilación continua continuous
    distillation
destilación de agua water distillation
destilación destructiva destructive
    distillation
destilación extractiva extractive
    distillation
destilación fraccionaria fractional
    distillation
destilación instantánea flash distillation
destilación molecular molecular
    distillation
destilación seca dry distillation
destilación simple simple distillation
destilado distilled
destilado de alquitrán de hulla coal-tar
    distillate
destilador distiller, still
destilador de agua water still
destilador molecular molecular still
destoxificación detoxification
desulfuración desulfuration
desvitrificación devitrification
detección de agua water detection
detector de fuego fire detector
detector de gases gas detector, gas
    sniffer
detector de humedad humidity detector
detector de infrarrojo infrared detector
detector de ionización por llama flame-
    ionization detector
detector de llama flame detector
detector de radiación radiation detector
detector de umbral threshold detector
detector iónico ion detector
detector químico chemical detector
detector térmico heat detector
detergente detergent
detergente iónico ionic detergent
detergente sintético synthetic detergent

deterioro electroquímico
electrochemical deterioration
determinación determination
determinación de ion hidrógeno
hydrogen-ion determination
determinación de Van Slyke Van Slyke
determination
determinación de Zeisel Zeisel
determination
determinación de Zerewitinoff
Zerewitinoff determination
detonación detonation
detonador detonator
deuteración deuteration
deuterado deuterated
deuterar deuterate
deuterio deuterium
deutero deutero
deuterón deuteron
deuteroxilo deuteroxyl
deuteruro deuteride
deuteruro alumínico de litio lithium
aluminum deuteride
deuteruro de litio lithium deuteride
deweylita deweylite
dexanfetamina dexamphetamine
dextranasa dextranase
dextrano dextran
dextrina dextrin
dextrinosa dextrinose
dextro dextro
dextrógiro dextrorotatory, dextrogyric
dextrosa dextrose
di di
diabasa diabase
diacetamida diacetamide
diacetanilida diacetanilide
diacetato diacetate
diacetato alumínico aluminum diacetate
diacetato cálcico calcium diacetate
diacetato de celulosa cellulose diacetate
diacetato de dibutilestaño dibutyltin
diacetate
diacetato de etilenglicol ethylene glycol
diacetate
diacetato de etileno ethylene diacetate
diacetato de etilideno ethylidene
diacetate
diacetato de glicerilo glyceryl diacetate

diacetato de glicerina glycerol diacetate
diacetato de glicerol glycerol diacetate
diacetato de metalidina methallydine
diacetate
diacetato de paladio palladium diacetate
diacetato sódico sodium diacetate
diacetilamida diacetylamide
diacetilanilida diacetylanilide
diacetileno diacetylene
diacetiletano diacetylethane
diacetilmorfina diacetylmorphine
diacetilo diacetyl
diacetilperóxido diacetylperoxide
diacetilurea diacetylurea
diacetina diacetin
diacetona diacetone
diacetonaacrilamida diacetone
acrylamide
diacetonaalcohol diacetone alcohol
diacetonamina diacetonamine
diácido diacid
diacolación diacolation
diadoquita diadochite
diaestereoisómero diastereoisomer
diaestereotópico diastereotopic
diaforita diaphorite
diafragma diaphragm
diagénesis diagenesis
diagrama centroide centroid diagram
diagrama de correlación correlation
diagram
diagrama de Cox Cox chart
diagrama de Ellingham Ellingham
diagram
diagrama de equilibrio equilibrium
diagram
diagrama de fases phase diagram
diagrama de flujo flow diagram
diagrama de Grotrian Grotrian diagram
diagrama de Jablonski Jablonski
diagram
diagrama de Moseley Moseley diagram
diagrama de niveles de energía energy-
level diagram
diagrama de Orgel Orgel diagram
diagrama de puntos de congelación
freezing-point diagram
diagrama de puntos de ebullición
boiling-point diagram

diagrama de Tanabe-Sugano Tanabe-Sugano diagram

diagrama molecular molecular diagram

diagrama triangular triangular diagram

dialdehído dialdehyde

dialdehído de almidón starch dialdehyde

dialilamina diallylamine

dialilmelanina diallylmelamine

dialilo diallyl

dialilurea diallylurea

diálisis dialysis

diálisis de equilibrio equilibrium dialysis

dializado dialyzed

dializador dialyzer

dialozita dialozite

dialqueno dialkene

dialquileno dialkylene

dialquilo dialkyl

diamagnetismo diamagnetism

diamagnetismo atómico atomic diamagnetism

diamagnetismo molecular molecular diamagnetism

diamante diamond

diamante industrial industrial diamond

diámetro atómico atomic diameter

diámetro molecular molecular diameter

diámetro molecular efectivo effective molecular diameter

diamida diamide

diamidina diamidine

diamido diamido

diamilamina diamylamine

diamilanilina diamylaniline

diamileno diamylene

diamilfenol diamyl phenol

diamilo diamyl

diamina diamine

diamino diamino

diaminoacridina diaminoacridine

diaminoazobenceno diaminoazobenzene

diaminoazoxitolueno diaminoazoxytoluene

diaminobenceno diaminobenzene

diaminobencidina diaminobenzidine

diaminobenzofenona diaminobenzophenone

diaminobutano diaminobutane

diaminodifenilamina diaminodiphenylamine

diaminodifenilmetano diaminodiphenylmethane

diaminodifenilo diaminodiphenyl

diaminoetano diaminoethane

diaminofenazina diaminophenazine

diaminofenol diaminophenol

diaminohexano diaminohexane

diaminopentano diaminopentane

diaminopropano diaminopropane

diaminotolueno diaminotoluene

diamonio diammonium

diamorfina diamorphine

dianhídrido piromelítico pyromellitic dianhydride

dianilina dianiline

dianilinoetano dianilinoethane

dianisidina dianisidine

diarabinosa diarabinose

diarsano diarsane

diarsenito diarsenite

diarsilo diarsyl

diasólisis diasolysis

diásporo diaspore

diastasa diastase

diatermia diathermy

diatómico diatomic

diatomita diatomite

diatrizoato sódico sodium diatrizoate

diazaciclo diazacyclo

diazepam diazepam

diazina diazine

diazinón diazinon

diazo diazo

diazoacetato diazoacetate

diazoacetato de etilo ethyl diazoacetate

diazoamina diazoamine

diazoamino diazoamino

diazoaminobenceno diazoaminobenzene

diazoaminonaftaleno diazoaminonaphthalene

diazoato diazoate

diazobenceno diazobenzene

diazoetano diazoethane

diazoetanoato de etilo ethyl diazoethanoate

diazofenol diazophenol

diazohidrato diazohydrate

diazohidróxido diazohydroxide

diazoimida diazoimide
diazoimido diazoimido
diazol diazole
diazometano diazomethane
diazonio diazonium
diazonitrofenol diazonitrophenol
diazosulfonato diazosulfonate
diazotización diazotization
diazoxi diazoxy
diazóxido diazoxide
dibásico dibasic
dibenalacetona dibenalacetone
dibencenilo dibenzenyl
dibenceno dibenzene
dibencenocromo dibenzenechromium
dibencilamina dibenzylamine
dibencilideno dibenzylidene
dibencilidino dibenzylidyne
dibencilmetilamina dibenzylmethylamine
dibencilo dibenzyl
dibenzamida dibenzamide
dibenzantraceno dibenzanthracene
dibenzantrona dibenzanthrone
dibenzantronilo dibenzanthronyl
dibenzo dibenzo
dibenzoantraceno dibenzoanthracene
dibenzociclohepatadienona
    dibenzocyclohepatadienone
dibenzoiletano dibenzoylethane
dibenzoiletileno dibenzoylethylene
dibenzoilo dibenzoyl
dibenzopirona dibenzopyrone
dibenzopirrol dibenzopyrrole
dibenzotiacina dibenzothiazine
diborano diborane
diboruro de titanio titanium diboride
diboruro de zirconio zirconium diboride
dibromhidrato dihydrobromide
dibromo dibromo
dibromoacetato de etilo ethyl
    dibromoacetate
dibromoacetileno dibromoacetylene
dibromoantraceno dibromoanthracene
dibromobenceno dibromobenzene
dibromoclorometano
    dibromochloromethane
dibromocloropropano
    dibromochloropropane
dibromodifluorometano

dibromodifluoromethane
dibromoetano dibromoethane
dibromofluorobenceno
    dibromofluorobenzene
dibromometano dibromomethane
dibromopentano dibromopentane
dibromopropano dibromopropane
dibromopropanol dibromopropanol
dibromuro dibromide
dibromuro de acetileno acetylene
    dibromide
dibromuro de benceno benzene
    dibromide
dibromuro de bencilideno benzylidene
    dibromide
dibromuro de etileno ethylene dibromide
dibromuro de etilideno ethylidene
    dibromide
dibromuro de galio gallium dibromide
dibromuro de germanio germanium
    dibromide
dibromuro de paladio palladium
    dibromide
dibromuro de pentametileno
    pentamethylene dibromide
dibromuro de platino platinum
    dibromide
dibromuro de selenio selenium
    dibromide
dibromuro de telurio tellurium
    dibromide
dibucaína dibucaine
dibultilamina dibutylamine
dibutilditiocarbamato sódico sodium
    dibutyldithiocarbamate
dibutilestaño dibutyltin
dibutilo dibutyl
dibutiltiourea dibutylthiourea
dibutirato de etilenglicol ethylene glycol
    dibutyrate
dibutoxianilina dibutoxyaniline
dicaprilo dicapryl
dicarbilamina dicarbylamine
dicarbocianina dicarbocyanine
dicarbonato dicarbonate
dicarboxilo dicarboxyl
dicarburo dicarbide
dicarburo cálcico calcium dicarbide
dicarburo de uranio uranium dicarbide

dicetena diketene
dicetilo dicetyl
dicetona diketone
dicetopiperazina diketopiperazine
dicetopirrolidina diketopyrrolidine
diciandiamida dicyandiamide
diciano dicyano
dicianoargenato potásico potassium
    dicyanoargenate
dicianoaurato potásico potassium
    dicyanoaurate
dicianoaurato sódico sodium
    dicyanoaurate
dicianógeno dicyanogen
dicianuro dicyanide
dicianuro de etileno ethylene dicyanide
dicianuro de paladio palladium
    dicyanide
dicíclico dicyclic
diciclohexilamina dicyclohexylamine
diciclohexilcarbodiimida
    dicyclohexylcarbodiimide
diciclohexilo dicyclohexyl
diciclopentadienilo dicyclopentadienyl
diciclopentadieno dicyclopentadiene
diclona dichlone
dicloramina dichloramine
diclorhidrato dihydrochloride
diclorhidrato de hidrazina hydrazine
    dihydrochloride
diclorhidrato de piperazina piperazine
    dihydrochloride
dicloro dichlorine
dicloroacetal dichloroacetal
dicloroacetato de etilo ethyl
    dichloroacetate
dicloroacetato de metilo methyl
    dichloroacetate
dicloroacetileno dichloroacetylene
dicloroanilina dichloroaniline
dicloroantraceno dichloroanthracene
diclorobenceno dichlorobenzene
diclorobencidina dichlorobenzidine
diclorobenzaldehído
    dichlorobenzaldehyde
diclorobutano dichlorobutane
diclorobuteno dichlorobutene
diclorocarbeno dichlorocarbene
diclorodifluorurometano

dichlorodifluoromethane
dicloroestearato de metilo methyl
    dichlorostearate
dicloroetano dichloroethane
dicloroeteno dichloroethene
dicloroetileno dichloroethylene
diclorofeno dichlorophene
diclorofenol dichlorophenol
diclorohidrina dichlorohydrin
diclorohidrina de glicerol glycerol
    dichlorohydrin
dicloroisocianurato potásico potassium
    dichloroisocyanurate
dicloroisocianurato sódico sodium
    dichloroisocyanurate
diclorometano dichloromethane
diclorometilsilano dichloromethylsilane
dicloronaftaleno dichloronaphthalene
dicloronitrobenceno
    dichloronitrobenzene
dicloropentano dichloropentane
dicloropropano dichloropropane
dicloropropeno dichloropropene
dicloropropionato sódico sodium
    dichloropropionate
diclororiofeno dichlorothiophene
diclorotolueno dichlorotoluene
diclorotrifluoruro de antimonio
    antimony dichlorotrifluoride
dicloruro dichloride
dicloruro benceno benzene dichloride
dicloruro clorometilfosfónico
    chloromethylphosphonic dichloride
dicloruro de acetileno acetylene
    dichloride
dicloruro de amileno amylene dichloride
dicloruro de azufre sulfur dichloride
dicloruro de bencilideno benzylidene
    dichloride
dicloruro de bencilo benzyl dichloride
dicloruro de bismuto bismuth dichloride
dicloruro de carbono carbon dichloride
dicloruro de dibutilestaño dibutyltin
    dichloride
dicloruro de dimetilestaño dimethyltin
    dichloride
dicloruro de estaño tin dichloride
dicloruro de etilaluminio ethylaluminum
    dichloride

dicloruro de etileno ethylene dichloride
dicloruro de ferrocenoilo ferrocenoyl
dichloride
dicloruro de galio gallium dichloride
dicloruro de germanio germanium
dichloride
dicloruro de glicol glycol dichloride
dicloruro de hierro iron dichloride
dicloruro de indio indium dichloride
dicloruro de metileno methylene
dichloride
dicloruro de molibdeno molybdenum
dichloride
dicloruro de oro gold dichloride
dicloruro de osmio osmium dichloride
dicloruro de oxalilo oxalyl dichloride
dicloruro de platino platinum dichloride
dicloruro de polivinilo polyvinyl
dichloride
dicloruro de propileno propylene
dichloride
dicloruro de rutenio ruthenium
dichloride
dicloruro de selenio selenium dichloride
dicloruro de telurio tellurium dichloride
dicloruro de tetraglicol tetraglycol
dichloride
dicloruro de titanio titanium dichloride
dicloruro de titanoceno titanocene
dichloride
dicloruro de trietilenglicol triethylene
glycol dichloride
dicloruro de triglicol triglycol dichloride
dicloruro de tungsteno tungsten
dichloride
dicloruro de vanadio vanadium
dichloride
dicloruro de xililo xylyl dichloride
dicloruro de zirconoceno zirconocene
dichloride
dicrilo dicryl
dicroico dichroic
dicroismo dichroism
dicroismo circular circular dichroism
dicromático dichromatic
dicromato dichromate
dicromato alumínico aluminum
dichromate
dicromato amónico ammonium

dichromate
dicromato bárico barium dichromate
dicromato cálcico calcium dichromate
dicromato de cesio cesium dichromate
dicromato de cinc zinc dichromate
dicromato de litio lithium dichromate
dicromato de mercurio mercury
dichromate
dicromato de plata silver dichromate
dicromato de plomo lead dichromate
dicromato de rubidio rubidium
dichromate
dicromato férrico ferric dichromate
dicromato magnésico magnesium
dichromate
dicromato mercúrico mercuric
dichromate
dicromato potásico potassium
dichromate
dicromato sódico sodium dichromate
dicrómico dichromic
didecanoato de trietilenglicol triethylene
glycol didecanoate
didimio didymium
didimolita didymolite
diectilo diectyl
dieldrina dieldrin
dieléctrico dielectric
dienestrol dienestrol
dieno diene
dienófilo dienophile
diestearato alumínico aluminum
distearate
diestearato de poliglicol polyglycol
distearate
diestearato de propilenglicol propylene
glycol distearate
diéster diester
diesterato de glicerol glycerol distearate
diestireno distyrene
dietanolamina diethanolamine
dietanolamina de butilo butyl
diethanolamine
diéter diether
dietilacetal diethylacetal
dietilamida de ácido lisérgico lysergic
acid diethylamide
dietilamina diethylamine
dietilamino diethylamino

dietilaminobenzaldehído
  diethylaminobenzaldehyde
dietilaminoetanol diethylaminoethanol
dietilaminopropilamina
  diethylaminopropylamine
dietilanilina diethylaniline
dietilbenceno diethylbenzene
dietilcadmio diethylcadmium
dietilcarbamazina diethylcarbamazine
dietilcarbinol diethylcarbinol
dietilcetona diethyl ketone
dietilciclohexilamina
  diethylcyclohexylamine
dietilcinc diethylzinc
dietilditiocarbamato de cadmio
  cadmium diethyldithiocarbamate
dietilditiocarbamato de selenio selenium
  diethyldithiocarbamate
dietilendiamina diethylenediamine
dietilenglicol diethylene glycol
dietileno diethylene
dietilenotriamina diethylenetriamine
dietilestilbestrol diethylstilbestrol
dietiletileno diethylethylene
dietilideno diethylidene
dietilmalonato de dietilo diethyl
  diethylmalonate
dietilmetilmetano diethylmethylmethane
dietilo diethyl
dietiloxamida diethyloxamide
dietilurea diethylurea
dietoxianilina diethoxyaniline
dietoxietano diethoxyethane
dietzeita dietzeite
difenilamina diphenylamine
difenilamino diphenylamino
difenilaminosulfonato bárico barium
  diphenylamine sulfonate
difenilbenceno diphenylbenzene
difenilcarbazida diphenylcarbazide
difenilcetona diphenyl ketone
difenilcloroarsina diphenylchloroarsine
difenildiclorosilano
  diphenyldichlorosilane
difenildiimida diphenyldiimide
difenileno diphenylene
difenilestaño diphenyltin
difeniletano diphenylethane
difeniletileno diphenylethylene

difenilglioxal diphenylglyoxal
difenilguanidina diphenylguanidine
difenilhidantoína sódica sodium
  diphenylhydantoin
difenilmetano diphenylmethane
difenilmetilo diphenylmethyl
difenilo diphenyl
difenilo clorado chlorinated diphenyl
difenilsulfona diphenylsulfone
difenilurea diphenylurea
difenol diphenol
difenolato diphenolate
difluoro difluoro
difluoroamino difluoroamino
difluorobenceno difluorobenzene
difluorodiazina difluorodiazine
difluoroetano difluoroethane
difluorometano difluoromethane
difluoruro difluoride
difluoruro de cobalto cobalt difluoride
difluoruro de estaño tin difluoride
difluoruro de hidrógeno amónico
  ammonium hydrogen difluoride
difluoruro de hidrógeno sódico sodium
  hydrogen difluoride
difluoruro de vanadio vanadium
  difluoride
diformiato alumínico aluminum
  diformate
diformiato de etilenglicol ethylene
  glycol diformate
diformiato de glicol glycol diformate
difosfano diphosphane
difosfato diphosphate
difosfato de adenosina adenosine
  diphosphate
difosfato de fructosa fructose
  diphosphate
difosfato disódico disodium diphosphate
difosfato potásico potassium diphosphate
difosfato tetrasódico tetrasodium
  diphosphate
difosfina diphosphine
difosgeno diphosgene
difracción diffraction
difracción de rayos X X-ray diffraction
difracción electrónica electron
  diffraction
difracción electrónica de baja energía

low-energy electron diffraction

difracción neutrónica neutron diffraction

difracción por transmisión transmission
diffraction

difusión diffusion

difusión electroquímica electrochemical
diffusion

difusión gaseosa gaseous diffusion

difusión térmica thermal diffusion

difuso diffuse

digestión digestion

digitalina digitalin

digitalis digitalis

digitalosa digitalose

digitonina digitonin

digitoxigenina digitoxigenin

digitoxina digitoxin

diglicérido diglyceride

diglicerol diglycerol

diglicol diglycol

diglicólido diglycolide

diglime diglyme

diglime de butilo butyl diglyme

digol digol

digoxina digoxin

dihedral dihedral

dihedrón dihedron

dihexilamina dihexylamine

dihexilo dihexyl

dihidoabietato de trietilenglicol
triethylene glycol dihydroabietate

dihidrato dihydrate

dihidrazona dihydrazone

dihidro dihydro

dihidroazirina dihydroazirine

dihidrobenceno dihydrobenzene

dihidrocalcona dihydrochalcone

dihidrocolesterol dihydrocholesterol

dihidrogenofosfato sódico sodium
dihydrogenphosphate

dihidrogenoortoperyodato sódico
sodium dihydrogenorthoperiodate

dihidrogenosulfato magnésico
magnesium dihydrogensulfate

dihidronaftaleno dihydronaphthalene

dihidropirano dihydropyran

dihidropropeno dihydropropene

dihidroxi dihydroxy

dihidroxiacetona dihydroxyacetone

dihidroxiamina dihydroxyamine

dihidroxiantraquinona
dihydroxyanthraquinone

dihidroxibenceno dihydroxybenzene

dihidroxidifenilsulfona
dihydroxydiphenylsulfone

dihidroxietano dihydroxyethane

dihidroxietilglicina sódica sodium
dihydroxyethylglycine

dihidroxinaftaleno dihydroxynaphthalene

dihidroxipropano dihydroxypropane

dihidruro dihydride

diimida diimide

diimina diimine

diimino diimino

diisoamilo diisoamyl

diisobutilamina diisobutylamine

diisobutilcetona diisobutyl ketone

diisobutileno diisobutylene

diisobutilo diisobutyl

diisobutilofenol diisobutylphenol

diisocianato diisocyanate

diisocianato de hexametileno
hexamethylene diisocyanate

diisopentilo diisopentyl

diisopropilbenceno diisopropylbenzene

diisopropilo diisopropyl

diisopropóxido de cloroaluminio
chloroaluminum diisopropoxide

dilactona dilactone

dilatación dilatation

dilatómetro dilatometer

dilaurato de dibutilestaño dibutyltin
dilaurate

dilaurilamina dilaurylamine

dilaurilfosfito dilauryl phosphite

dilaurilo dilauryl

dilaurilperóxido dilauryl peroxide

dilaurilsulfuro dilauryl sulfide

dilución dilution

dilución infinita infinite dilution

diluido dilute

diluyente diluent

dimedona dimedone

dimercaptoacetato de glicol glycol
dimercaptoacetate

dimérico dimeric

dimerización dimerization

dímero dimer

dímero de acrilonitrilo acrylonitrile dimer

dímero de acroleína acrolein dimer

dimetacrilato de tetraetilenglicol tetraethylene glycol dimethacrylate

dimetano dimethano

dimetilacetal dimethylacetal

dimetilacetal de fenilacetaldehído phenylacetaldehyde dimethylacetal

dimetilamina dimethylamine

dimetilamino dimethylamino

dimetilaminoetanol dimethylaminoethanol

dimetilaminopropanol dimethylaminopropanol

dimetilán dimetilan

dimetilanilina dimethylaniline

dimetilantraceno dimethylanthracene

dimetilarseniato sódico sodium dimethylarsenate

dimetilarsina dimethylarsine

dimetilbenceno dimethylbenzene

dimetilberilio dimethylberyllium

dimetilbutadieno dimethylbutadiene

dimetilbutano dimethylbutane

dimetilbuteno dimethylbutene

dimetilcarbinol dimethylcarbinol

dimetilciclohexano dimethylcyclohexane

dimetilciclopentano dimethylcyclopentane

dimetilcloroacetal dimethylchloroacetal

dimetildiclorosilano dimethyldichlorosilane

dimetilditiocarbamato sódico sodium dimethyldithiocarbamate

dimetileno dimethylene

dimetiletileno dimethylethylene

dimetilfenol dimethylphenol

dimetilformamida dimethylformamide

dimetilfosfina dimethylphosphine

dimetilglioxima dimethylglyoxime

dimetilglioxima de níquel nickel dimethylglyoxime

dimetilhepteno dimethylheptene

dimetilhexadieno dimethylhexadiene

dimetilhexanodiol dimethylhexanediol

dimetilhexinodiol dimethylhexynediol

dimetilhidantoína dimethylhydantoin

dimetilhidrazina dimethylhydrazine

dimetilhidroxibenceno dimethylhydroxybenzene

dimetilisopropanolamina dimethylisopropanolamine

dimetilmercurio dimethylmercury

dimetilmetano dimethylmethane

dimetilnitrosamina dimethylnitrosamine

dimetilo dimethyl

dimetiloctadieno dimethyloctadiene

dimetilpentaldehído dimethylpentaldehyde

dimetilpentano dimethylpentane

dimetilpropano dimethylpropane

dimetilurea dimethylurea

dimetilxantina dimethylxanthine

dimetoato dimethoate

dimetoxi dimethoxy

dimetoxianilina dimethoxyaniline

dimetoxibenceno dimethoxybenzene

dimetoxietano dimethoxyethane

dimetoximetano dimethoxymethane

dimetoxitetraglicol dimethoxytetraglycol

dimetracrilato de butileno butylene dimethacrylate

dimétrico dimetric

dimórfico dimorphic

dimorfismo dimorphism

dinada dynad

dinaftileno dinaphthylene

dinaftilo dinaphthyl

dinafto dinaphtho

dinaftol dinaphthol

dinamita dynamite

dinenciléter dibenzylether

dinérico dineric

dineutrón dineutron

dinitrato dinitrate

dinitrato de dietilenglicol diethylene glycol dinitrate

dinitrato de etilenglicol ethylene glycol dinitrate

dinitrato de etileno ethylene dinitrate

dinitrato de glicol glycol dinitrate

dinitrato de paladio palladium dinitrate

dinitrato de propilenglicol propylene glycol dinitrate

dinitrilo malónico malonic dinitrile

dinitrito dinitrite

dinitro dinitro

dinitroanilina dinitroaniline
dinitroantraquinona
    dinitroanthraquinone
dinitrobenceno dinitrobenzene
dinitrocresol dinitrocresol
dinitrocresolato sódico sodium
    dinitrocresolate
dinitrofenilhidrazina
    dinitrophenylhydrazine
dinitrofenol dinitrophenol
dinitrofluorobenceno
    dinitrofluorobenzene
dinitrógeno dinitrogen
dinitronaftaleno dinitronaphthalene
dinitronaftol dinitronaphthol
dinitroso dinitroso
dinitrotolueno dinitrotoluene
dinucleótido dinucleotide
dioctilamina dioctylamine
dioctilmetano dioctylmethane
dioctilo dioctyl
dioctoato de trietilenglicol triethylene
    glycol dioctoate
diol diol
diolefina diolefin
diona dione
diópsido diopside
dioptasa dioptase
diorita diorite
dioxadieno dioxadiene
dioxalano dioxalane
dioxana dioxan
dioxano dioxane
dioxazol dioxazole
dioxdiazol dioxdiazole
dioxi dioxy
dióxido dioxide
dióxido bárico barium dioxide
dióxido cálcico calcium dioxide
dióxido de azufre sulfur dioxide
dióxido de carbono carbon dioxide
dióxido de cerio cerium dioxide
dióxido de cesio cesium dioxide
dióxido de cinc zinc dioxide
dióxido de cloro chlorine dioxide
dióxido de cromo chromium dioxide
dióxido de dietileno diethylene dioxide
dióxido de dipenteno dipentene dioxide
dióxido de estaño tin dioxide

dióxido de estroncio strontium dioxide
dióxido de germanio germanium dioxide
dióxido de hidrógeno hydrogen dioxide
dióxido de manganeso manganese
    dioxide
dióxido de molibdeno molybdenum
    dioxide
dióxido de neptunio neptunium dioxide
dióxido de nitrógeno nitrogen dioxide
dióxido de osmio osmium dioxide
dióxido de paladio palladium dioxide
dióxido de platino platinum dioxide
dióxido de plomo lead dioxide
dióxido de selenio selenium dioxide
dióxido de silicio silicon dioxide
dióxido de telurio tellurium dioxide
dióxido de titanio titanium dioxide
dióxido de torio thorium dioxide
dióxido de tricarbono tricarbon dioxide
dióxido de uranio uranium dioxide
dióxido de vanadio vanadium dioxide
dióxido de vinilciclohexeno
    vinylcyclohexene dioxide
dióxido de yodo iodine dioxide
dióxido de zirconio zirconium dioxide
dióxido líquido liquid dioxide
dióxido magnésico magnesium dioxide
dióxido sódico sodium dioxide
dioxigenilo dioxygenyl
dioxima dioxime
dioxina dioxin
dioxo dioxo
dioxolano dioxolane
dipalmitato dipalmitate
dipalmitilamina dipalmitylamine
dipelargonato de trietilenglicol
    triethylene glycol dipelargonate
dipentaeritritol dipentaerythritol
dipenteno dipentene
dipentenoglicol dipentene glycol
dipentilo dipentyl
dipéptido dipeptide
dipiridilo dipyridyl
dipiridina dipyridine
diplomo dilead
dipnona dypnone
dipolar dipolar
dipolo dipole
dipolo molecular molecular dipole

dipropilberilio dipropylberyllium
dipropilenglicol dipropylene glycol
dipropilo dipropyl
dipropionato de trietilenglicol
    triethylene glycol dipropionate
diprotón diproton
dirradical diradical
dirresorcina diresorcinol
disacárido disaccharide
disacrilo discaryl
disco de estallido bursting disk
discrasita dyscrasite
discriminación de filtro filter
    discrimination
diselano diselane
diseleniuro de niobio niobium diselenide
diseleniuro de tungsteno tungsten
    diselenide
disilano disilane
disilicato disilicate
disiliciuro disilicide
disiliciuro de molibdeno molybdenum
    disilicide
disiliciuro de titanio titanium disilicide
disiliciuro de zirconio zirconium
    disilicide
disililo disilyl
disiloxano disiloxane
disimetría dissymmetry
disimétrico dissymmetrical
disinalilo disilanyl
disipación dissipation
disipación de electrodo electrode
    dissipation
disipación térmica heat dissipation
dislfuro de dietilo diethyl disulfide
dislocación dislocation
dismutación dismutation
dismutasa superóxida superoxide
    dismutase
disociación dissociation
disociación de ácido acid dissociation
disociación electrolítica electrolytic
    dissociation
disociación fotoquímica photochemical
    dissociation
disociación hidrolítica hydrolytic
    dissociation
disociación molecular molecular

    dissociation
disociado dissociated
disociar dissociate
disolución solution, dissolution
disolvente dissolvent, solvent
disolver dissolve
dispersador dispersator
dispersante dispersing
dispersar disperse
dispersión dispersion, scattering
dispersión anómala anomalous
    dispersion
dispersión coloidal colloidal dispersion
dispersión de Delbrück Delbrück
    scattering
dispersión de luz light scattering
dispersión de partículas particle
    scattering
dispersión de Rutherford Rutherford
    scattering
dispersión molecular molecular
    dispersion
dispersión nuclear nuclear scattering
dispersión rotatoria rotatory dispersion
dispersión rotatoria óptica optical
    rotatory dispersion
dispersión sódica sodium dispersion
disperso dispersed, scattered
dispersoide dispersoid
disponibilidad availability
disponibilidad biológica biological
    availability
disponible available
dispositivo de laboratorio laboratory
    device
dispositivo de seguridad safety device
disprosio dysprosium
distancia distance
distancia atómica atomic distance
distancia de enlace bond distance
distancia de enlazado bonding distance
distancia interatómica interatomic
    distance
distribución de Maxwell-Boltzmann
    Maxwell-Boltzmann distribution
distribución de partículas particle
    distribution
distribución electrónica electron
    distribution

disuelto dissolved

disulfato disulfate

disulfato de hierro iron disulfate

disulfato potásico potassium disulfate

disulfato sódico sodium disulfate

disulfiram disulfiram

disulfito disulfite

disulfito potásico potassium disulfite

disulfito sódico sodium disulfite

disulfo disulfo

disulfonato disulfonate

disulfuro disulfide

disulfuro de arsénico arsenic disulfide

disulfuro de bencilo benzyl disulfide

disulfuro de benzoilo benzoyl disulfide

disulfuro de benzotiacilo benzothiazyl disulfide

disulfuro de carbono carbon disulfide

disulfuro de dibencilo dibenzyl disulfide

disulfuro de dimetilo dimethyl disulfide

disulfuro de germanio germanium disulfide

disulfuro de hafnio hafnium disulfide

disulfuro de hidrógeno hydrogen disulfide

disulfuro de hierro iron disulfide

disulfuro de iridio iridium disulfide

disulfuro de molibdeno molybdenum disulfide

disulfuro de osmio osmium disulfide

disulfuro de paladio palladium disulfide

disulfuro de silicio silicon disulfide

disulfuro de tántalo tantalum disulfide

disulfuro de telurio tellurium disulfide

disulfuro de titanio titanium disulfide

disulfuro de torio thorium disulfide

disulfuro de tungsteno tungsten disulfide

disulfuro de vanadio vanadium disulfide

disulfuro de yodo iodine disulfide

disulfuro de zirconio zirconium disulfide

ditanato de bismuto bismuth ditannate

diterpeno diterpene

diterpenoide diterpenoid

ditiano dithiane

ditiazina dithiazine

ditienilo dithienyl

ditieno dithiene

ditio dithio

ditiocarbamato dithiocarbamate

ditiocarbamato amónico ammonium dithiocarbamate

ditiocarbonato dithiocarbonate

ditiocarbonato de mercurio mercury dithiocarbonate

ditiocarboxi dithiocarboxy

ditiol dithiol

ditioleno dithiolene

ditionato dithionate

ditionato amónico ammonium dithionate

ditionato bárico barium dithionate

ditionato de cinc zinc dithionate

ditionato de estroncio strontium dithionate

ditionato de manganeso manganese dithionate

ditionato de plata silver dithionate

ditionato de plomo lead dithionate

ditionato potásico potassium dithionate

ditionato sódico sodium dithionate

ditionita dithionite

ditionito sódico sodium dithionite

ditiooxamida dithiooxamide

ditiosalicilato sódico sodium dithiosalicylate

ditizona dithizone

ditolilo ditolyl

diuranato sódico sodium diuranate

diuriniato bárico barium diurinate

divalencia divalence

divalente divalent

divanadato sódico sodium divanadate

divinilacetileno divinylacetylene

divinilbenceno divinylbenzene

divinilo divinyl

división splitting, cleavage

división de grasas fat splitting

dixililetano dixylylethane

dixililo dixylyl

diyodo diiodo

diyodoacetileno diiodoacetylene

diyodometano diiodomethane

diyoduro diiodide

diyoduro de arsénico arsenic diiodide

diyoduro de etileno ethylene diiodide

diyoduro de etilideno ethylidene diiodide

diyoduro de galio gallium diiodide

diyoduro de germanio germanium diiodide

diyoduro de metileno methylene diiodide
diyoduro de paladio palladium diiodide
diyoduro de platino platinum diiodide
doblete doublet
docosano docosane
docosanol docosanol
dodecaedral dodecahedral
dodecaedro dodecahedron
dodecahedro rómbico rhombic
 dodecahedron
dodecahidrato dodecahydrate
dodecanal dodecanal
dodecano dodecane
dodecanoato de butilo butyl dodecanoate
dodecanoilo dodecanoyl
dodecanol dodecanol
dodecenal dodecenal
dodeceno dodecene
dodecilamina dodecylamine
dodecilbenceno dodecylbenzene
dodecilbencenosulfonato sódico sodium
 dodecylbenzenesulfonate
dodecileno dodecylene
dodecilfenol dodecylphenol
dodecilo dodecyl
dodeciltriclorosilano
 dodecyltrichlorosilane
dolerita dolerite
dolicol dolichol
dolomita dolomite
dolomol dolomol
domesticina domesticine
dominguita domingite
donador donor

donador electrónico electron donor
donador protónico proton donor
donarita donarite
dopado doping
dopado con oro gold doping
dopante dope
dorado gold-plating
dosis de radiación radiation dose
dosis letal lethal dose
dotriacontano dotriacontane
draconilo draconyl
drimina drimin
droga drug
droga psicotrópica psychotropic drug
droga sulfa sulfa drug
dúctil ductile
ductilidad ductility
dufrenita dufrenite
dufrenoisita dufrenoysite
dulcitol dulcitol
dumortierita dumortierite
dunita dunnite
duplete duplet
duraluminio duralumin
dureno durene
dureza hardness
dureza de agua hardness of water
dureza de Brinell Brinell hardness
dureza de Knoop Knoop hardness
dureza permanente permanent hardness
durileno durylene
durilo duryl
duro hard
durol durol

E

easina easin
ebonita ebonite
ebullición ebullition, boiling
ebullidor ebullator
ebulliómetro ebulliometer
ebullometría ebulliometry
ebullómetro diferencial differential ebulliometer
ebulloscopia ebullioscopy
ebulloscopio ebullioscope
ecdisona ecdysone
ecgonina ecgonine
eclipsado eclipsed
eclogita eclogite
ecología ecology
economizador economizer
ecosistema ecosystem
ecuación equation
ecuación de adsorción de Gibbs Gibbs adsorption equation
ecuación de Arrhenius Arrhenius equation
ecuación de Beattie-Bridgman Beattie-Bridgman equation
ecuación de Berthelot Berthelot equation
ecuación de Boltzmann Boltzmann equation
ecuación de Bragg Bragg equation
ecuación de Butler-Volmer Butler-Volmer equation
ecuación de Clapeyron Clapeyron equation
ecuación de Clapeyron-Clausius Clapeyron-Clausius equation
ecuación de Clausius-Mosotti Clausius-Mosotti equation
ecuación de de Broglie de Broglie equation
ecuación de Dieterici Dieterici equation
ecuación de Duhem Duhem equation
ecuación de estado equation of state
ecuación de Eyring Eyring equation
ecuación de Flood Flood equation

ecuación de Fokker-Planck Fokker-Planck equation
ecuación de gases gas equation
ecuación de Gibbs-Duhem Gibbs-Duhem equation
ecuación de Gibbs-Helmholtz Gibbs-Helmholtz equation
ecuación de Gibbs-Poynting Gibbs-Poynting equation
ecuación de Haggenmacher Haggenmacher equation
ecuación de Hammett Hammett equation
ecuación de Helmholtz Helmholtz equation
ecuación de Henderson-Hasselbach Henderson-Hasselbach equation
ecuación de Ilkovic Ilkovic equation
ecuación de Karplus Karplus equation
ecuación de Kohlrausch Kohlrausch equation
ecuación de Mark-Houwink Mark-Houwink equation
ecuación de Morse Morse equation
ecuación de Nernst Nernst equation
ecuación de Nernst-Einstein Nernst-Einstein equation
ecuación de Onsager Onsager equation
ecuación de Perrin Perrin equation
ecuación de Pitzer Pitzer equation
ecuación de Ramsay-Shields Ramsay-Shields equation
ecuación de Ramsay-Young Ramsay-Young equation
ecuación de reacción reaction equation
ecuación de Sackur-Tetrode Sackur-Tetrode equation
ecuación de Schomaker-Stevenson Schomaker-Stevenson equation
ecuación de Schrödinger Schrödinger equation
ecuación de van der Waals van der Waals equation
ecuación de van't Hoff van't Hoff

equation

ecuación de Watson Watson equation

ecuación maestra master equation

ecuación molecular molecular equation

ecuación nuclear nuclear equation

ecuación química chemical equation

ecuatorial equatorial

edeato edeate

edenita edenite

edestina edestin

edetato edetate

edetato disódico edetate disodium

edetato disódico cálcico calcium disodium edetate

edetato sódico sodium edetate

edingtonita edingtonite

efectividad de filtro filter effectiveness

efecto effect

efecto alostérico allosteric effect

efecto anódico anodic effect

efecto celostérico cellosteric effect

efecto cinético kinetic effect

efecto de apantallamiento screening effect

efecto de Back-Goudsmit Back-Goudsmit effect

efecto de Baker-Nathan Baker-Nathan effect

efecto de Baker-Venkataraman Baker-Venkataraman effect

efecto de Becquerel Becquerel effect

efecto de campo eléctrico electric field effect

efecto de Cerenkov Cerenkov effect

efecto de Compton Compton effect

efecto de Cotton Cotton effect

efecto de Chadwick-Goldhaber Chadwick-Goldhaber effect

efecto de Draper Draper effect

efecto de empaquetamiento packing effect

efecto de Faraday Faraday effect

efecto de Hall Hall effect

efecto de invernadero greenhouse effect

efecto de ion común common-ion effect

efecto de ionización ionization effect

efecto de Jahn-Teller Jahn-Teller effect

efecto de Joule Joule effect

efecto de Joule-Kelvin Joule-Kelvin

effect

efecto de Joule-Thomson Joule-Thomson effect

efecto de Kelvin Kelvin effect

efecto de Kundt Kundt effect

efecto de Mössbauer Mössbauer effect

efecto de orientación orientation effect

efecto de Overhauser Overhauser effect

efecto de Paschen-Back Paschen-Back effect

efecto de Pasteur Pasteur effect

efecto de Pfeiffer Pfeiffer effect

efecto de presión pressure effect

efecto de quelato chelate effect

efecto de radiación radiation effect

efecto de Raman Raman effect

efecto de Ramsauer Ramsauer effect

efecto de relajación relaxation effect

efecto de saturación saturation effect

efecto de Soret Soret effect

efecto de Stark Stark effect

efecto de Stark-Luneland Stark-Luneland effect

efecto de Szilard-Chalmers Szilard-Chalmers effect

efecto de transición transition effect

efecto de Tyndall Tyndall effect

efecto de Weigert Weigert effect

efecto de Weissenberg Weissenberg effect

efecto de Wien Wien effect

efecto de xenón xenon effect

efecto de Zeeman Zeeman effect

efecto electroforético electrophoretic effect

efecto electroquímico electrochemical effect

efecto estérico steric effect

efecto fotoquímico photochemical effect

efecto inductivo inductive effect

efecto isotópico isotope effect

efecto piezoeléctrico piezoelectric effect

efecto químico chemical effect

efecto térmico heat effect, thermal effect

efecto termiónico thermionic effect

efecto termomagnético thermomagnetic effect

efecto trans trans effect

efecto túnel tunnel effect

efedrina ephedrine
efervescencia effervescence
efervescente effervescent
eficiencia cuántica quantum efficiency
eficiencia de fluorescencia fluorescence efficiency
eficiencia de placas plate efficiency
eficiencia química chemical efficiency
eflorescencia efflorescence
efluente effluent
efusiómetro effusiometer
efusión effusion
eicosano eicosane
eicosanoide eicosanoid
eicosanol eicosanol
einstein einstein
einstenio einsteinium
eje axis
eje de simetría axis of symmetry
eje óptico optical axis
eje principal principal axis
eka eka
ekahafnio ekahafnium
elaboración elaboration
elaidato de metilo methyl elaidate
elaidinización elaidinization
elastasa elastase
elasticidad elasticity
elástico elastic
elastina elastin
elastómero elastomer
elaterita elaterite
electodializador electrodialyzer
electreto electret
electricidad atómica atomic electricity
electricidad galvánica galvanic electricity
eléctrico electric
electrificación electrification
electroanálisis electroanalysis
electrocinética electrokinetics
electrocrático electrocratic
electrocromatografía electrochromatography
electrocromatografía sobre papel paper electrochromatography
electrodeposición electrodeposition, plating
electrodepositado plated

electrodepósito electrodeposit
electrodiálisis electrodialysis
electrodisolución electrodissolution
electrodispersión electrodispersion
electrodo electrode
electrodo acelerador accelerating electrode
electrodo con gas gas electrode
electrodo de cadmio cadmium electrode
electrodo de calomel calomel electrode
electrodo de carbono carbon electrode
electrodo de Clark Clark electrode
electrodo de hidrógeno hydrogen electrode
electrodo de oxígeno oxygen electrode
electrodo de quinhidrona quinhydrone electrode
electrodo de referencia reference electrode
electrodo de vidrio glass electrode
electrodo metálico metal electrode
electrodo polarizado polarized electrode
electroenfoque electrofocusing
electroestricción electrostriction
electrofílico electrophilic
electrófilo electrophile
electroflotación electroflotation
electroforesis electrophoresis
electroforesis de gel gel electrophoresis
electroforesis por zonas zone electrophoresis
electroforesis sobre papel paper electrophoresis
electroforético electrophoretic
electroforetograma electrophoretogram
electroformación electroforming
electroforograma electrophorogram
electrografito electrographite
electrogravimetría electrogravimetry
electrólisis electrolysis
electrolítico electrolytic
electrólito electrolyte
electrólito anfótero amphoteric electrolyte
electrólito coloidal colloidal electrolyte
electrólito de batería battery electrolyte
electrólito débil weak electrolyte
electrólito fuerte strong electrolyte
electrólito fundido fused electrolyte

electrolizador electrolyzer
electrolizar electrolyze
electroluminiscencia
    electroluminescence
electromagnético electromagnetic
electromerismo electromerism
electrometalurgia electrometallurgy
electrómetro electrometer
electromigración electromigration
electrón electron
electrón de Auger Auger electron
electrón de conducción conduction
    electron
electrón de conversión conversion
    electron
electrón de enlazado bonding electron
electrón disponible available electron
electrón exterior outer electron
electrón giratorio spinning electron
electrón libre free electron
electrón ligado bound electron
electrón metaestable metastable electron
electrón metastásico metastasic electron
electrón orbital orbital electron
electrón pesado heavy electron
electrón pi pi electron
electrón positivo positive electron
electrón primario primary electron
electrón sigma sigma electron
electrón-volt electron-volt
electronegatividad electronegativity
electronegativo electronegative
electrones de valencia valence electrons
electrones dispersos scattered electrons
electrones emparejados paired electrons
electrones equivalentes equivalent
    electrons
electrones no emparejados unpaired
    electrons
electrones ópticos optical electrons
electrones planetarios planetary
    electrons
electroósmosis electroosmosis
electropositivo electropositive
electropotencial electropotential
electropulido electropolishing
electroquímica electrochemistry
electroquímico electrochemical
electrorrecubrimiento electroplating

electrorrevestimiento electrocoating
electroscopia electroscopy
electroscopio electroscope
electrosíntesis electrosynthesis
electrosol electrosol
electrovalencia electrovalence
electrovalente electrovalent
electroviscocidad electroviscosity
electrum electrum
electruro electride
elemento element
elemento activo active element
elemento artificial artificial element
elemento combustible fuel element
elemento de simetría element of
    symmetry
elemento de tierras raras rare-earth
    element
elemento de transición transition
    element
elemento electronegativo electronegative
    element
elemento electropositivo electropositive
    element
elemento enmascarado masked element
elemento inestable unstable element
elemento magnético magnetic element
elemento nativo native element
elemento negativo negative element
elemento neutro neutral element
elemento pesado heavy element
elemento positivo positive element
elemento químico chemical element
elemento radiactivo radioactive element
elemento radiactivo artificial artificial
    radioactive element
elemento siderófilo siderophile element
elemento superpesado superheavy
    element
elemento transactínido transactinide
    element
elemento transuránico transuranic
    element
elemento trazador tracer element
eleonorita eleonorite
elevación de punto de congelación
    freezing-point elevation
elevación de punto de ebullición
    boiling-point elevation

elevación molecular   molecular elevation
elíxir   elixir
elpidita   elpidite
elución   elution
eluido   eluate
elutriación   elutriation
eluyente   eluant
emanación   emanation
emanación radiactiva   radioactive
     emanation
embolita   embolite
embudo   funnel
embudo de Buchner   Buchner funnel
embudo de Bunsen   Bunsen funnel
embudo de filtración   filter funnel
embudo de separación   separatory funnel
emetamina   emetamine
emetina   emetine
emisión   emission
emisión alfa   alpha emission
emisión de fotones   photon emission
emisión de luz   light emission
emisión de partículas   particle emission
emisión de positrón   positron emission
emisión electrónica   electron emission
emisión espontánea   spontaneous
     emission
emisión gamma   gamma emission
emisión inducida   induced emission
emisión iónica de campo   field-ion
     emission
emisión primaria   primary emission
emisión radiactiva   radioactive emission
emisión secundaria   secondary emission
emisor alfa   alpha emitter
emisor beta   beta emitter
emisor gamma   gamma emitter
emodina   emodin
empaque   packaging
empaquetado   packaging
empaquetado ampollado   blister
     packaging
empaquetamiento   packing
empírico   empirical
emplectita   emplectite
emulagador   emulgator
emulsión   emulsion
emulsionador   emulsifier
emulsionamiento   emulsification

emulsionar   emulsify
en polvo   powdered
enamina   enamine
enantilo   enanthyl
enantiomérico   enantiomeric
enantiómero   enantiomer
enantiomórfico   enantiomorphic
enantiomorfismo   enantiomorphism
enantiomorfo   enantiomorph
enantiotópico   enantiotopic
enantiotropía   enantiotropy
enantiotrópico   enantiotropic
enantiótropo   enantiotropic
enantol   enanthol
enargita   enargite
encadenamiento cruzado   cross-linking
encapsulación   encapsulation
encefalina   enkephalin
endo   endo
endógeno   endogenous
endomicina   endomycin
endopeptidasa   endopeptidase
endorfina   endorphin
endósmosis   endosmosis
endotérmico   endothermic
endrina   endrin
endulcorante   sweetener
endulcorante no nutritivo   nonnutritive
     sweetner
endulzador   sweetener
endulzar   sweeten
endurecedor   hardener
endurecimiento   hardening
endurecimiento de grasa   fat hardening
endurecimiento electroquímico
     electrochemical hardening
endurecimiento por llama   flame
     hardening
energía   energy
energía al cero absoluto   zero-point
     energy
energía atómica   atomic energy
energía cinética   kinetic energy
energía de activación   activation energy
energía de deformación   deformation
     energy
energía de disociación   dissociation
     energy
energía de enlace   bond energy

energía de estabilización stabilization energy

energía de fisión fission energy

energía de fusión fusion energy

energía de ionización ionization energy

energía de partícula particle energy

energía de reacción reaction energy

energía de resonancia resonance energy

energía de separación separation energy

energía de separación de neutrón neutron binding energy

energía de unión nuclear nuclear binding energy

energía de Zeeman Zeeman energy

energía electrónica electron energy

energía equipotencial equipotential energy

energía específica specific energy

energía interna internal energy

energía latente latent energy

energía libre free energy

energía libre de Gibbs Gibbs free energy

energía libre de Helmholtz Helmholtz free energy

energía neutrónica neutron energy

energía nuclear nuclear energy

energía potencial nuclear nuclear potential energy

energía química chemical energy

energía reticular lattice energy

energía rotacional rotational energy

energía solar solar energy

energía superficial surface energy

energía superficial interfacial interfacial surface energy

energía térmica thermal energy, heat energy

energía termonuclear thermonuclear energy

energía umbral threshold energy

energía vibracional vibrational energy

enflurano enflurane

enfriamiento cooling

enfriamiento adiabático adiabatic cooling

enilo enyl

enlace bond, linkage

enlace coordinado coordinate bond

enlace covalente covalent bond

enlace covalente dativo dative covalent bond

enlace covalente polar polar covalent bond

enlace dativo dative bond

enlace de alta energía high-energy bond

enlace de carbono carbon bond

enlace de hidrógeno hydrogen bond

enlace de par de electrones electron-pair bond

enlace de péptido peptide bond, peptide linkage

enlace de valencia valence bond

enlace delta delta bond

enlace doble double bond

enlace doble aislado isolated double bond

enlace electrostático electrostatic bond

enlace electrovalente electrovalent bond

enlace glicosídico glycosidic bond

enlace heteropolar heteropolar bond

enlace homopolar homopolar bond

enlace intermedio intermediate bond

enlace iónico ionic bond

enlace metal-metal metal-metal bond

enlace metálico metallic bond

enlace múltiple multiple bond

enlace pi pi bond

enlace químico chemical bond

enlace reactivo reactive bond

enlace sencillo single bond

enlace sigma sigma bond

enlace triple triple bond

enlaces dobles acumulativos cumulalive double bonds

enlaces dobles conjugados conjugated double bonds

enlazado bonding

enlazado delta delta bonding

enlazado hidrofóbico hydrophobic bonding

enlazado iónico ionic bonding

enlazado metálico metallic bonding

enlazado múltiple multiple bonding

enlazado pi pi bonding

enmascaramiento masking

eno ene

enodiol enediol

enol enol

enolasa enolase

enriquecimiento enrichment

enriquecimiento de uranio uranium enrichment

ensayo assay, test

enstatita enstatite

entalpía enthalpy

entalpía de evaporación enthalpy of evaporation

entalpía de fusión enthalpy of fusion

entalpía de reacción enthalpy of reaction

entidad química chemical entity

entrada de gas gas inlet

entrada de hidrógeno hydrogen inlet

entrada de oxígeno oxygen inlet

entropía entropy

entropía de activación entropy of activation

entropía de mezcla entropy of mixing

entropía de transición entropy of transition

envejecimiento aging

enzima enzyme

enzima alostérica allosteric enzyme

enzima proteolítica proteolytic enzyme

eosina eosin

epi epi

epialcanfor epicamphor

epicianhidrina epicyanohydrin

epiclorita epichlorite

epiclorohidrina epichlorohydrin

epididimita epididymite

epidioxi epidioxy

epidióxido epidioxide

epiditio epidithio

epiestilbita epistilbite

epigenita epigenite

epihidrina epihydrin

epimerasa epimerase

epimérico epimeric

epimerida epimeride

epimerismo epimerism

epimerización epimerization

epímero epimer

epinefrina epinephrine

epitaxia epitaxy

epitio epithio

epitioximino epithioximino

epoxi epoxy

epoxibutano epoxybutane

epoxidación epoxidation

epóxido epoxide

epóxido de hexafluoropropileno hexafluoropropylene epoxide

epóxido de tetrafluoroetileno tetrafluoroethylene epoxide

epoxiestearato de butilo butyl epoxystearate

epoxietano epoxyethane

epoxiimino epoxyimino

epoxinitrilo epoxynitrilo

epoxipropano epoxypropane

epoxipropanol epoxypropanol

epoxitio epoxythio

epsilón epsilon

epsomita epsomite

equilenina equilenin

equilibrio equilibrium

equilibrio ácido-base acid-base equilibrium

equilibrio de Donnan Donnan equilibrium

equilibrio de fases phase equilibrium

equilibrio de Gibbs-Donnan Gibbs-Donnan equilibrium

equilibrio de oxígeno oxygen balance

equilibrio de redox redox equilibrium

equilibrio de sedimentación sedimentation equilibrium

equilibrio dinámico dynamic equilibrium

equilibrio energético energy balance

equilibrio estable stable equilibrium

equilibrio incompleto incomplete equilibrium

equilibrio iónico ionic equilibrium

equilibrio líquido-sólido liquid-solid equilibrium

equilibrio líquido-vapor liquid-vapor equilibrium

equilibrio químico chemical equilibrium

equilibrio radiactivo radioactive equilibrium

equilibrio sólido-líquido solid-liquid equilibrium

equilibrio térmico thermal equilibrium

equilibrio vapor-líquido vapor-liquid equilibrium

equinomicina echinomycin
equinopsina echinopsine
equipartición equipartition
equipartición de energía equipartition of energy
equipo de laboratorio laboratory equipment
equipo de manejo remoto remote-handling equipment
equivalencia equivalence
equivalencia cuántica quantum equivalence
equivalente equivalent
equivalente coloidal colloid equivalent
equivalente de hidrógeno hydrogen equivalent
equivalente de neutralización neutralization equivalent
equivalente de nitrógeno nitrogen equivalent
equivalente de saponificación saponification equivalent
equivalente electroquímico electrochemical equivalent
equivalente fotoquímico photochemical equivalent
equivalente-gramo gram-equivalent
equivalente osmótico osmotic equivalent
equivalente químico chemical equivalent
equivalente térmico heat equivalent
erbia erbia
erbio erbium
erbón erbon
ergina ergine
ergocalciferol ergocalciferol
ergódico ergodic
ergometrina ergometrine
ergonovina ergonovine
ergosinina ergosinine
ergosterol ergosterol
ergotamina ergotamine
ergotinina ergotinine
ergotoxina ergotoxine
erilita erilite
erinita erinite
eriodictiol eriodictyol
eritorbato sódico sodium erythorbate
eritrina erythrin
eritrita erythrite

eritritilo erythrityl
eritritol erythritol
eritro erythro
eritroidina erythroidine
eritrol erythrol
eritromicina erythromycin
eritrosa erythrose
eritrosiderita erythrosiderite
eritrosina erythrosin
erosión erosion
erosión química chemical erosion
error absoluto absolute error
error de calibración calibration error
error de medida measurement error
error de paralaje parallax error
error determinado determinate error
error relativo relative error
erubescita erubescite
escala scale
escala atómica atomic scale
escala Celsius Celsius scale
escala centígrada centigrade scale
escala cromática chromatic scale
escala de Knoop Knoop scale
escala de Mohs Mohs scale
escala de pH pH scale
escala de temperatura temperature scale
escala de temperatura absoluta absolute temperature scale
escala Fahrenheit Fahrenheit scale
escala Kelvin Kelvin scale
escala Rankine Rankine scale
escala termométrica thermometer scale
escandio scandium
escape de gas gas escape
escapolita scapolite
escatol skatole
escisión scission
escleroscopio scleroscope
escolecita scolecite
escopolamina scopolamine
escopolina scopoline
escoria slag
escoria de fosfato phosphate slag
escorodita scorodite
escualano squalane
escualeno squalene
esencia de perla pearl essence
esencial essential

eserina eserine
esfalerita sphalerite
esfingomielina sphingomyelin
esfingosina sphingosine
esmalte enamel
esmalte de porcelana porcelain enamel
esmalte de vidrio glass enamel
esmaltín smalt
esméctico smectic
esmeralda emerald
esmeril emery
esmerilado ground
espacio space
espacio de configuración configuration space
espacio de Hilbert Hilbert space
espacio elemental elementary space
espacio interatómico interatomic space
espacio intraatómico intraatomic space
espalación spallation
espandex spandex
espato spar
espato de Islandia Iceland spar
espato flúor fluorspar
espato pesado heavy spar
espato satinado satin spar
especialidad química chemical specialty
especie species
especie atómica atomic species
especie nuclear nuclear species
específico specific
espectral spectral
espectro spectrum
espectro atómico atomic spectrum
espectro continuo continuous spectrum
espectro de absorción absorption spectrum
espectro de absorción de infrarrojo infrared absorption spectrum
espectro de acción action spectrum
espectro de arco arc spectrum
espectro de Aston Aston spectrum
espectro de bandas band spectrum
espectro de difracción diffraction spectrum
espectro de emisión emission spectrum
espectro de excitación excitation spectrum
espectro de explosión explosion spectrum
espectro de fisión fission spectrum
espectro de Fraunhofer Fraunhofer spectrum
espectro de hidrógeno hydrogen spectrum
espectro de infrarrojo infrared spectrum
espectro de líneas line spectrum
espectro de líneas brillantes bright-line spectrum
espectro de luz light spectrum
espectro de llama flame spectrum
espectro de masa mass spectrum
espectro de microondas microwave spectrum
espectro de primer orden first-order spectrum
espectro de Raman Raman spectrum
espectro de rayos gamma gamma-ray spectrum
espectro de rayos X X-ray spectrum
espectro de resonancia resonance spectrum
espectro de Rydberg Rydberg spectrum
espectro discreto discrete spectrum
espectro electromagnético electromagnetic spectrum
espectro electrónico electron spectrum
espectro electroquímico electrochemical spectrum
espectro magnético magnetic spectrum
espectro molecular molecular spectrum
espectro neutrónico neutron spectrum
espectro nuclear nuclear spectrum
espectro óptico optical spectrum
espectro paramagnético paramagnetic spectrum
espectro químico chemical spectrum
espectro rotacional rotational spectrum
espectro ultravioleta ultraviolet spectrum
espectro vibracional vibrational spectrum
espectro visible visible spectrum
espectroanálisis spectroanalysis
espectroanalítco spectroanalytical
espectrofluorometría spectrofluorometry
espectrofluorométrico spectrofluorometric

espectrofluorómetro spectrofluorometer
espectrofotometría spectrophotometry
espectrofotometría de absorción
ultravioleta ultraviolet-absorption
spectrophotometry
espectrofotometría de absorción visible
visible absorption spectrophotometry
espectrofotometría de llama flame
spectrophotometry
espectrofotometría de matriz matrix
spectrophotometry
espectrofotometría de Raman Raman
spectrophotometry
espectrofotometría diferencial
differential spectrophotometry
espectrofotometría ultravioleta
ultraviolet spectrophotometry
espectrofotometría visible visible
spectrophotometry
espectrofotométrico spectrophotometric
espectrofotómetro spectrophotometer
espectrofotómetro de doble haz double-
beam spectrophotometer
espectrofotómetro de filtro filter
spectrophotometer
espectrofotómetro de llama flame
spectrophotometer
espectrografía spectrography
espectrografía de imagen de rayos X X-
ray image spectrography
espectrografía de rayos X X-ray
spectrography
espectrográficamente spectrographically
espectrográfico spectrographic
espectrógrafo spectrograph
espectrógrafo de masa mass
spectrograph
espectrógrafo de rayos X X-ray
spectrograph
espectrógrafo óptico optical
spectrograph
espectrograma spectrogram
espectrometría spectrometry
espectrometría de llama flame
spectrometry
espectrometría de masa mass
spectrometry
espectrometría de masa de iones
secundarios secondary-ion mass

spectrometry
espectrometría de neutrones lentos
slow-neutron spectrometry
espectrometría de neutrones rápidos
fast-neutron spectrometry
espectrometría de rayos X X-ray
spectrometry
espectrometría neutrónica neutron
spectrometry
espectrométricamente spectrometrically
espectrométrico spectrometric
espectrómetro spectrometer
espectrómetro de Bragg Bragg
spectrometer
espectrómetro de centelleos scintillation
spectrometer
espectrómetro de helio helium
spectrometer
espectrómetro de infrarrojo infrared
spectrometer
espectrómetro de llama flame
spectrometer
espectrómetro de masa mass
spectrometer
espectrómetro de masa de helio helium
mass spectrometer
espectrómetro de microondas
microwave spectrometer
espectrómetro de radiofrecuencia
radiofrequency spectrometer
espectrómetro de rayos beta beta-ray
spectrometer
espectrómetro de rayos gamma gamma-
ray spectrometer
espectrómetro de rayos X X-ray
spectrometer
espectrómetro de rejilla grid
spectrometer
espectrómetro neutrónico neutron
spectrometer
espectrómetro óptico optical
spectrometer
espectrómetro ultravioleta ultraviolet
spectrometer
espectromicroscopia de fotoelectrones
photoelectron spectromicroscopy
espectromicroscópico
spectromicroscopic
espectromicroscopio spectromicroscope

espectroquímica spectrochemistry
espectroquímico spectrochemical
espectrorradiómetro spectroradiometer
espectroscopia spectroscopy
espectroscopia atómica atomic
  spectroscopy
espectroscopia de absorción absorption
  spectroscopy
espectroscopia de absorción atómica
  atomic absorption spectroscopy
espectroscopia de correlación
  correlation spectroscopy
espectroscopia de destello flash
  spectroscopy
espectroscopia de emisión emission
  spectroscopy
espectroscopia de emisión atómica
  atomic emission spectroscopy
espectroscopia de emisión de llama
  flame-emission spectroscopy
espectroscopia de fluorescencia
  fluorescence spectroscopy
espectroscopia de fotoelectrones
  photoelectron spectroscopy
espectroscopia de infrarrojo infrared
  spectroscopy
**espectroscopia de ionización en
  resonancia** resonance ionization
  spectroscopy
espectroscopia de láser laser
  spectroscopy
espectroscopia de masa mass
  spectroscopy
espectroscopia de microondas
  microwave spectroscopy
espectroscopia de modulación
  modulation spectroscopy
espectroscopia de Mössbauer
  Mössbauer spectroscopy
espectroscopia de radiofrecuencia
  radiofrequency spectroscopy
espectroscopia de Raman Raman
  spectroscopy
espectroscopia de rayos gamma gamma-
  ray spectroscopy
espectroscopia de saturación saturation
  spectroscopy
espectroscopia electrónica electron
  spectroscopy

espectroscopia fotoacústica
  photoacoustic spectroscopy
espectroscopia gamma gamma
  spectroscopy
espectroscopia molecular molecular
  spectroscopy
espectroscopia neutrónica neutron
  spectroscopy
espectroscopia nuclear nuclear
  spectroscopy
espectroscopia óptica optical
  spectroscopy
espectroscopia quiroóptica chirooptical
  spectroscopy
espectroscopia ultravioleta ultraviolet
  spectroscopy
espectroscopia ultravioleta-visible
  ultraviolet-visible spectroscopy
espectroscópicamente spectroscopically
espectroscópico spectroscopic
espectroscopio spectroscope
espectroscopio de masa mass
  spectroscope
espectroscopio de microondas
  microwave spectroscope
espectroscopio óptico optical
  spectroscope
especular specular
esperrilita sperrylite
espesador thickener
espesamiento thickening
espetropolarimétrico spectropolarimetric
espetropolarímetro spectropolarimeter
espín spin
espín antiparalelo antiparallel spin
espín de electrón electron spin
espín isobárico isobaric spin
espín isotópico isotopic spin
espín nuclear nuclear spin
espinela spinel
espines paralelos parallel spins
espirano spiran
espíritu spirit
espironolactona spironolactone
espiropentano spiropentane
espiroqueta spirochete
espita unidireccional one-way tap
espodumena spodumene
esponja sponge

esponja de caucho rubber sponge
esponja de hierro iron sponge
esponja de platino platinum sponge
esponja de titanio titanium sponge
espontáneo spontaneous
espuma foam, froth
espuma de plástico plastic foam
espuma poliéter polyether foam
estabilidad stability
estabilidad ambiental environmental stability
estabilidad limitada limited stability
estabilidad nuclear nuclear stability
estabilidad térmica thermal stability
estabilidad termodinámica thermodynamic stability
estabilización stabilization
estabilizador stabilizer
estabilizador de emulsión emulsion stabilizer
estabilizador ultravioleta ultraviolet stabilizer
estable stable
estadística cuántica quantum statistics
estadística de Bose-Einstein Bose-Einstein statistics
estadística de Fermi-Dirac Fermi-Dirac statistics
estado state
estado activado activated state
estado amorfo amorphous state
estado coloidal colloidal state
estado correspondiente corresponding state
estado cristalino crystalline state
estado crítico critical state
estado cuántico quantum state
estado de oxidación oxidation state
estado de transición transition state
estado disociado dissociated state
estado estacionario steady state, stationary state
estado excitado excited state
estado fundamental ground state
estado fundamental nuclear nuclear ground state
estado gaseoso gaseous state
estado ligado bound state
estado líquido liquid state

estado mesomórfico mesomorphic state
estado metaestable metastable state
estado normal normal state
estado orbital orbital state
estado reducido reduced state
estado sólido solid state
estado virtual virtual state
estalagmometría stalagmometry
estalagmómetro stalagmometer
estándar standard
estandarización standardization
estandarizado standardized
estandarizar standardize
estannano stannane
estannato stannate
estannato bárico barium stannate
estannato cálcico calcium stannate
estannato cobaltoso cobaltous stannate
estannato de bismuto bismuth stannate
estannato de níquel nickel stannate
estannato de plomo lead stannate
estannato magnésico magnesium stannate
estannato potásico potassium stannate
estannato sódico sodium stannate
estánnico stannic
estannilo stannyl
estannita stannite
estannita sódica sodium stannite
estanniuro magnésico magnesium stannide
estannoso stannous
estañado tin-plating, tinning
estañar tin-plate
estaño tin, stannum
estático static
estearaldehído stearaldehyde
estearamida stearamide
estearamida de butilo butyl stearamide
estearato stearate
estearato alumínico aluminum stearate
estearato amónico ammonium stearate
estearato bárico barium stearate
estearato cálcico calcium stearate
estearato de atropina atropine stearate
estearato de butilo butyl stearate
estearato de butoxietilo butoxyethyl stearate
estearato de cadmio cadmium stearate

estearato de cerio cerium stearate
estearato de ciclohexilo cyclohexyl stearate
estearato de cinc zinc stearate
estearato de cobre copper stearate
estearato de dietilenglicol diethylene glycol stearate
estearato de diglicol diglycol stearate
estearato de glicol glycol stearate
estearato de isobutilo isobutyl stearate
estearato de isocetilo isocetyl stearate
estearato de litio lithium stearate
estearato de metilo methyl stearate
estearato de plata silver stearate
estearato de plomo lead stearate
estearato de trihidroxietilamina trihydroxyethylamine stearate
estearato de vinilo vinyl stearate
estearato férrico ferric steareate
estearato magnésico magnesium stearate
estearato mercúrico mercuric stearate
estearato potásico potassium stearate
estearato sódico sodium stearate
estearilmercaptano stearyl mercaptan
estearilo stearyl
estearina stearin
estearoilaminofenol stearoylaminophenol
estearoilo stearoyl
estearoilolactilato sódico sodium stearoyl lactylate
estearona stearone
estearonitrilo stearonitrile
esteatita steatite, soapstone
estequiometría stoichiometry
estequiométrico stoichiometric
éster ester
éster acético acetic ester
éster acetoacético acetoacetic ester
éster acrílico acrylic ester
éster amilacético amylacetic ester
éster carbónico carbonic ester
éster cianoacético cyanoacetic ester
éster cianúrico cyanuric ester
éster de ácido bórico boric acid ester
éster de brosilato brosylate ester
éster de poliglicerol polyglycerol ester
éster de titanio titanium ester
éster diazoacético diazoacetic ester
éster ftálico phthalic ester

éster glicólico glycol ester
éster graso fatty ester
éster malónico malonic ester
éster malónico sódico sodium malonic ester
éster metílico methyl ester
éster ortofórmico orthoformic ester
esterasa esterase
estereoespecífico stereospecific
estereoisomerismo stereoisomerism
estereoisómero stereoisomer
estereoquímica stereochemistry
estereoquímica absoluta absolute stereochemistry
estereoquímica relativa relative stereochemistry
estereorregular stereoregular
estereoscopio stereoscope
estereoselectividad stereoselectivity
estérico steric
esterificación esterification
esteril sterile
esterilización sterilization
esterilización de agua water sterilization
esteriode steroid
esterol sterol
estersil estersil
estibina stibine
estibino stibino
estibinoso stibinoso
estibnita stibnite
estibo stibo
estibonio stibonium
estibono stibono
estibonoso stibonoso
estiboso stiboso
estigmasterol stigmasterol
estilbeno stilbene
estilbestrol stilbestrol
estilbita stilbite
estimulación stimulation
estiracina styracin
estirenglicol styrene glycol
estireno styrene
estirenosulfonato sódico sodium styrenesulfonate
estirilo styryl
estocástico stochastic
estradiol estradiol

estragol estragole
estrano estrane
estreptolina streptolin
estreptomicina streptomycin
estrés stress
estricnidina strychnidine
estricnina strychnine
estriol estriol
estrógeno estrogen
estrona estrone
estroncia strontia
estroncianita strontianite
estroncio strontium
estructura structure
estructura antifluorita antifluorite structure
estructura asimétrica asymmetric structure
estructura atómica atomic structure
estructura centrada en el cuerpo body-centered structure
estructura centrada en las caras face-centered structure
estructura con defectos defect structure
estructura cristalina crystal structure
estructura cúbica centrada en el cuerpo body-centered cubic structure
estructura cúbica centrada en las caras face-centered cubic structure
estructura de anillo ring structure
estructura de benceno benzene structure
estructura de capas shell structure
estructura de Dewar Dewar structure
estructura de Kekulé Kekulé structure
estructura de Lewis Lewis structure
estructura de van der Waals van der Waals structure
estructura fina fine structure
estructura hiperfina hyperfine structure
estructura molecular molecular structure
estructura nuclear nuclear structure
estructura por rayos X X-ray structure
estructural structural
eta eta
etal ethal
etaldehído ethaldehyde
etamina ethamine
etanal ethanal

etanaldehído ethanaldehyde
etanamida ethanamide
etanamidina ethanamidine
etanilo ethanyl
etanita ethanite
etano ethane
etanoato ethanoate
etanoato alumínico aluminum ethanoate
etanoato cálcico calcium ethanoate
etanoato de amilo amyl ethanoate
etanoato de cinc zinc ethanoate
etanoato de etilo ethyl ethanoate
etanoato de hierro iron ethanoate
etanoato de plomo lead ethanoate
etanoato fenilmercúrico phenylmercuric ethanoate
etanoato potásico potassium ethanoate
etanoato sódico sodium ethanoate
etanodial ethanedial
etanodiamida ethanediamide
etanodiamina ethanediamine
etanodinitrilo ethanedinitrile
etanoditiol ethanedithiol
etanoilación ethanoylation
etanoilo ethanoyl
etanol ethanol
etanolamina ethanolamine
etanolato ethanolate
etanolato de talio thallium ethanolate
etanolato sódico sodium ethanolate
etanolurea ethanolurea
etanonitrilo ethanenitrile
etanotial ethanethial
etanotiol ethanethiol
etenilo ethenyl
etenio ethenium
eteno ethene
etenol cthcnol
etenona ethenone
éter ether
éter acético acetic ether
éter alílico allyl ether
éter amílico amyl ether
éter bencílico benzyl ether
éter bencílico de hidroquinona hydroquinone benzyl ether
éter bencílico de isoamilo isoamyl benzyl ether
éter butilfenílico butylphenyl ether

éter butílico butyl ether

éter carbónico carbonic ether

éter celulósico cellulose ether

éter cetílico cetyl ether

éter clorhídrico hydrochloric ether

éter de petróleo petroleum ether

éter de vino wine ether

éter diacético diacetic ether

éter dicloroetílico dichloroethyl ether

éter didecílico didecyl ether

éter diectílico diectyl ether

éter dietílico diethyl ether

éter dietílico de hidroquinona
    hydroquinone diethyl ether

éter difenílico diphenyl ether

éter diisopropílico diisopropyl ether

éter dimetílico dimethyl ether

éter dimetílico de glicol glycol dimethyl
    ether

éter dimetílico de hidroquinona
    hydroquinone dimethyl ether

éter dimetílico de resorcina resorcinol
    dimethyl ether

éter dimetílico de trietilenglicol
    triethylene glycol dimethyl ether

éter dioctílico dioctyl ether

éter dipropílico dipropyl ether

éter divinílico divinyl ether

éter en corona crown ether

éter etílico ethyl ether

éter etílico bencílico benzyl ethyl ether

éter etílico de dietilenglicol diethylene
    glycol ethyl ether

éter etílico polivinílico polyvinyl ethyl
    ether

éter etilmetílico ethyl methyl ether

éter fenílico phenyl ether

éter fenílico de propilenglicol propylene
    glycol phenyl ether

éter fenólico phenol ether

éter fórmico formic ether

éter glicidílico glycidyl ether

éter glicólico glycol ether

éter hexaclorometílico hexachloromethyl
    ether

éter hexílico hexyl ether

éter isoamílico isoamyl ether

éter isobutílico isobutyl ether

éter isobutílico polivinílico polyvinyl

isobutyl ether

éter isopropílico isopropyl ether

éter metilbencílico methylbenzyl ether

éter metilfenílico methylphenyl ether

éter metílico methyl ether

éter metílico polivinílico polyvinyl
    methyl ether

éter monobutílico de etilenglicol
    ethylene glycol monobutyl ether

éter monoetílico de etilenglicol ethylene
    glycol monoethyl ether

éter monoetílico de hidroquinona
    hydroquinone monoethyl ether

éter monometílico de etilenglicol
    ethylene glycol monomethyl ether

éter nítrico nitric ether

éter pentilacético pentylacetic ether

éter pentílico pentyl ether

éter polivinílico polyvinyl ether

éter propiónico propionic ether

etéreo ethereal

eterificación etherification

etida ethide

etidina ethidine

etil-litio ethyllithium

etilacetamida ethylacetamide

etilacetanilida ethylacetanilide

etilacetileno ethylacetylene

etilacetoacetato de cobre copper
    ethylacetoacetate

etilacetona ethylacetone

etilación ethylation

etilamilcetona ethyl amyl ketone

etilamina ethylamine

etilamino ethylamino

etilaminoacetato ethylaminoacetate

etilanilina ethylaniline

etilantraquinona ethylanthraquinone

etilato ethylate

etilato alumínico aluminum ethylate

etilato de vanadio vanadium ethylate

etilato sódico sodium ethylate

etilaziridina ethylaziridine

etilbenceno ethylbenzene

etilbencilanilina ethylbenzylaniline

etilbutanol ethylbutanol

etilbuteno ethylbutene

etilbutilamina ethylbutylamine

etilbutilcetona ethylbutyl ketone

etilbutilo ethylbutyl
etilbutiraldehído ethylbutyraldehyde
etilcarbazol ethylcarbazole
etilcarbinol ethylcarbinol
etilcelulosa ethylcellulose
etilciclohexano ethylcyclohexane
etilciclohexilamina ethylcyclohexylamine
etilciclopentano ethylcyclopentane
etilciclopentanona ethylcyclopentanone
etildiclorosilano ethyldichlorosilane
etildietanolamina ethyldiethanolamine
etilenamina ethyleneamine
etilendiamina ethylenediamine
etilendiaminotetraacetonitrilo ethylenediaminetetraacetonitrile
etilenditiol ethylenedithiol
etilenglicol ethylene glycol
etilenimina ethyleneimine
etileno ethylene
etileno-propileno ethylene-propylene
etilenonaftaleno ethylenenaphthalene
etiletanolamina ethylethanolamine
etiletilenimina ethylethyleneimine
etiletileno ethylethylene
etilfenilcetona ethyl phenyl ketone
etilfenol ethylphenol
etilfluorosulfonato ethylfluorosulfonate
etilfurano ethylfuran
etilheptano ethylheptane
etilhexaldehído ethylhexaldehyde
etilhexanal ethylhexanal
etilhexanodiol ethylhexanediol
etilhexanol ethylhexanol
etilhexenal ethylhexenal
etilhexilamina ethylhexylamine
etilhexilanilina ethylhexylaniline
etilhexilo ethylhexyl
etilhexoato cálcico calcium cthylhcxoatc
etilhexoato de cinc zinc ethylhexoate
etilhexoato de cobalto cobalt ethylhexoate
etilhexoato estannoso stannous ethylhexoate
etilhidrazina ethylhydrazine
etilideno ethylidene
etilidino ethylidyne
etilina ethylin
etilisobutilmetano ethylisobutylmethane
etilisohexanol ethylisohexanol

etilmalonato de dietilo diethyl ethylmalonate
etilmercaptano ethyl mercaptan
etilmetilacetileno ethylmethylacetylene
etilmetilcetona ethyl methyl ketone
etilmorfolina ethylmorpholine
etilnaftilamina ethylnaphthylamine
etilo ethyl
etiloico ethyloic
etilol ethylol
etilparabeno ethylparaben
etilpentano ethylpentane
etilsalicilato de bencilo benzyl ethylsalicylate
etilsulfato bárico barium ethylsulfate
etilsulfato de cinc zinc ethylsulfate
etiltriclorosilano ethyltrichlorosilane
etiluretano ethylurethane
etilviniléter ethyl vinyl ether
etina ethine
etinilación ethynylation
etinilestradiol ethynylestradiol
etinilo ethynyl
etino ethyne
etión ethion
etionamida ethionamide
etiqueta label
etisterona ethisterone
etohexadiol ethohexadiol
etolida etholide
etoxalilo ethoxalyl
etoxi ethoxy
etoxiacetanilida ethoxyacetanilide
etoxiacetona ethoxyacetone
etoxianilina ethoxyaniline
etoxicarbonilo ethoxycarbonyl
etóxido ethoxide
etóxido alumínico aluminum ethoxide
etoxido sódico sodium ethoxide
etoxietanol ethoxyethanol
etoxifenol ethoxyphenol
etoxilo ethoxyl
etoximetilenomalononitrilo ethoxymethylenemalononitrile
etoxitriglicol ethoxytriglycol
ettringita ettringite
eucairita eucairite
eucaliptol eucalyptol
eucarvona eucarvone

eucriptita eucryptite
eucroíta euchroite
eudaleno eudalene
eudialita eudialite
eudiómetro eudiometer
eudiómetro de Bunsen Bunsen
    eudiometer
eugenol eugenol
eulitina eulytin
eupirina eupyrine
europio europium
eutéctico eutectic
eutectoide eutectoid
eutroficación eutrophication
eutropía eutropy
euxenita euxenite
evansita evansite
evaporación evaporation
evaporación catódica cathodic
    evaporation, cathodic sputtering,
    sputter, sputtering
evaporación retrógrada retrograde
    evaporation
evaporador evaporator
evaporar evaporate
evaporímetro evaporimeter
evaporita evaporite
evolución evolution
exactitud accuracy
examen microscópico microscopic
    examination
excéntrico eccentric
exceso de neutrones neutron excess
excímero excimer
excitación excitation
excitación atómica atomic excitation
excitación de Coulomb Coulomb
    excitation
excitación fotoquímica photochemical
    excitation
excitación nuclear nuclear excitation
excitación por colisión collision
    excitation
excitación térmica thermal excitation
excitado excited
excitador exciter
excitar excite
excitón exciton
exclusión exclusion

exclusión iónica ion exclusion
exilxantato sódico sodium ethylxanthate
exo exo
exocarburo exocarbon
exocondensación exocondensation
exógeno exogenous
exopeptidasa exopeptidase
exósmosis exosmosis
exotérmico exothermic
expandirse expand
expansión expansion
expansión adiabática adiabatic
    expansion
expansión térmica thermal expansion
expansor expander
experimentación experimentation
experimental experimental
experimento experiment
experimento de laboratorio laboratory
    experiment
exploración scan, scanning
explosión explosion
explosión de Coulomb Coulomb
    explosion
explosivo explosive
explosivo aceptable acceptable explosive
explosivo de Nobel Nobel explosive
explosivo iniciador initiating explosive
explosivo permisible permissible
    explosive
explotar explode
exposición exposure
expresión expression
expresión fraccionaria fractional
    expression
extensor extender
extinción extinction
extinguir extinguish, quench
extintor de fuego fire extinguisher
extracción extraction
extracción analítica analytical extraction
extracción electrolítica electrowinning
extracción líquido-líquido liquid-liquid
    extraction
extracción por solvente solvent
    extraction
extractante extractant
extracto extract
extracto de Göulard Göulard extract

extracto de malta malt extract
extractor extractor
extractor de Soxhlet Soxhlet extractor
extranuclear extranuclear
extrapolación extrapolation

extrapolación de Birge-Sponer Birge-Sponer extrapolation
extrusión extrusion
exudar exude

F

fabianol fabianol
fabricación de cerveza brewing
fabulita fabulite
fac fac
factor factor
factor de atenuación attenuation factor
factor de Cabannes Cabannes factor
factor de conversión conversion factor
factor de dispersión atómica atomic scattering factor
factor de van't Hoff van't Hoff factor
factor limitador limiting factor
factor volumétrico volumetric factor
factores ambientales environmental factors
fagarina fagarine
familia de desintegración decay family
familia radiactiva radioactive family
fantasma de Rowland Rowland ghost
farad farad
faraday faraday
faradiol faradiol
farinosa farinose
farmacéutico pharmaceutical
farmacocinética pharmacokinetics
farmacolita pharmacolite
farmacología pharmacology
farneseno farnesene
farnesol farnesol
fase phase
fase continua continuous phase
fase discontinua discontinuous phase
fase dispersa dispersed phase
fase esméctica smectic phase
fase estacionaria stationary phase
fase interna internal phase
fase metaestable metastable phase
fase nemática nematic phase
fasotropía phasotropy
faujasita faujasite
fayalita fayalite
felandreno phellandrene
feldespato feldspar

feldespato potásico potassium feldspar
feldespatoide feldspathoid
felita felite
felvación felvation
fémico femic
femtoquímica femtochemistry
fen phen, fen
fenacetina phenacetin
fenacetol phenacetol
fenacilideno phenacylidene
fenacilo phenacyl
fenantraquinona phenanthraquinone
fenantrenilo phenanthrenyl
fenantreno phenanthrene
fenantrenoquinona phenanthrenequinone
fenantridina phenanthridine
fenantrileno phenanthrylene
fenantrilo phenanthryl
fenantrofenacina phenanthrophenazine
fenantrol phenanthrol
fenantrolina phenanthroline
fenantrona phenanthrone
fenato phenate
fenato de cinc zinc phenate
fenato sódico sodium phenate
fenazina phenazine
fenazona phenazone
fenazonio phenazonium
fencano fenchane
fencanol fenchanol
fencanona fenchanone
fencol fenchol
fencona fenchone
fencoxima fenchoxime
fenenilo phenenyl
fenetidina phenetidine
fenetidino phenetidino
fenetilamina phenethylamine
fenetileno phenethylene
fenetilo phenethyl
fenetol phenetole
fenidona phenidone
fenil-litio phenyllithium

fenilacetaldehído phenylacetaldehyde
fenilacetamida phenylacetamide
fenilacetato de bencilo benzyl
    phenylacetate
fenilacetato de etilo ethyl phenylacetate
fenilacetato de isobutilo isobutyl
    phenylacetate
fenilacetato sódico sodium phenylacetate
fenilacetilo phenylacetyl
fenilacetonitrilo phenylacetonitrile
fenilación phenylation
fenilalanina phenylalanine
fenilamina phenylamine
fenilanilina phenylaniline
fenilato cálcico calcium phenylate
fenilazo phenylazo
fenilbarbital phenylbarbital
fenilbenceno phenylbenzene
fenilbencilo phenylbenzyl
fenilbenzamida phenylbenzamide
fenilbenzoato phenylbenzoate
fenilbenzoilo phenylbenzoyl
fenilbutano phenylbutane
fenilbutazona phenylbutazone
fenilbuteno phenylbutene
fenilbutinol phenylbutynol
fenilcarbamoilo phenylcarbamoyl
fenilcarbetoxipirazolona
    phenylcarbethoxypyrazolone
fenilcarbimida phenylcarbimide
fenilcarbinol phenylcarbinol
fenilcetona phenyl ketone
fenilciclohexano phenylcyclohexane
fenilciclohexanol phenylcyclohexanol
fenilcloroformo phenylchloroform
fenildicloroarsina phenyldichloroarsine
fenildietanolamina phenyldiethanolamine
fenildimetilurea phenyldimethylurea
fenilendiamina phenylenediamine
fenileno phenylene
feniletano phenylethane
feniletilamina phenylethylamine
feniletilenglicol phenylethylene glycol
feniletileno phenylethylene
feniletiletanolamina
    phenylethylethanolamine
feniletiléter phenylethylether
feniletilmercaptano phenylethyl
    mercaptan

feniletilo phenylethyl
fenilfenato sódico sodium phenylphenate
fenilfenol phenylphenol
fenilfenolato sódico sodium
    phenylphenolate
fenilformamida phenylformamide
fenilfosfina phenylphosphine
fenilfosfinato sódico sodium
    phenylphosphinate
fenilglicina phenylglycine
fenilglicol phenylglycol
fenilhidrazina phenylhydrazine
fenilhidrazina de pentilo pentyl
    phenylhydrazine
fenilhidrazona phenylhydrazone
fenilhidrazona de benzaldehído
    benzaldehyde phenylhydrazone
fenilhidroxilamina phenylhydroxylamine
fenílico phenylic
fenilideno phenylidene
fenilmagnesio phenylmagnesium
fenilmercaptano phenyl mercaptan
fenilmercúrico phenylmercuric
fenilmetano phenylmethane
fenilmetanol phenylmethanol
fenilmetilcarbinol phenylmethylcarbinol
fenilmetilcetona phenyl methyl ketone
fenilmetileno phenylmethylene
fenilmetiletanolamina
    phenylmethylethanolamine
fenilmetilo phenylmethyl
fenilmorfolina phenylmorpholine
fenilnaftaleno phenylnaphthalene
fenilnaftilamina phenylnaphthylamine
fenilnitroamina phenylnitroamine
fenilnonano phenylnonane
fenilo phenyl
fenilpentano phenylpentane
fenilpiperazina phenylpiperazine
fenilpirazolida phenylpyrazolide
fenilpirazolidona phenylpyrazolidone
fenilpiridina phenylpyridine
fenilpropano phenylpropane
fenilpropanol phenylpropanol
fenilpropanona phenylpropanone
fenilpropenal phenylpropenal
fenilpropenol phenylpropenol
fenilpropilcetona phenylpropyl ketone
fenilpropilo phenylpropyl

fenilpropilpiridina phenylpropylpyridine
fenilsulfonilo phenylsulfonyl
feniltiourea phenylthiourea
feniltolueno phenyltoluene
feniltriclorosilano phenyltrichlorosilane
feniltridecano phenyltridecane
feniluretano phenylurethane
fenilvinilcetona phenyl vinyl ketone
fenindiona phenindione
fenitoína phenytoin
fenitrotión fenitrothion
fenixina phenixin
fenmetilo phenmethyl
fenoacetato sódico sodium phenoacetate
fenobarbital phenobarbital
fenobarbital sódico sodium
    phenobarbital
fenobarbitona phenobarbitone
fenodiazina phenodiazine
fenol phenol
fenol polihídrico polyhydric phenol
fenolato phenolate
fenolato alumínico aluminum phenolate
fenolato cálcico calcium phenolate
fenolato de hierro iron phenolate
fenolato de resorcina resorcinol
    phenolate
fenolato férrico ferric phenolate
fenolato sódico sodium phenolate
fenolftaleína phenolphthalein
fenolftalida phenolphthalide
fenólico phenolic
fenolsafranina phenolsafranin
fenolsulfato sódico mercúrico mercuric
    sodium phenolsulfate
fenolsulfonaftaleína
    phenolsulfonephthalein
fenolsulfonato phenolsulfonate
fenolsulfonato amónico ammonium
    phenolsulfonate
fenolsulfonato de cinc zinc
    phenolsulfonate
fenolsulfonato de cobre copper
    phenolsulfonate
fenolsulfonato sódico sodium
    phenolsulfonate
fenómeno físico physical phenomenon
fenómeno lambda lambda phenomenon
fenómeno sigma sigma phenomenon

fenona phenone
fenotiazina phenothiazine
fenotol phenotole
fenotrina phenothrin
fenoxazina phenoxazine
fenoxi phenoxy
fenoxibenceno phenoxybenzene
fenoxidihidroxipropano
    phenoxydihydroxypropane
fenóxido phenoxide
fenoxietanol phenoxyethanol
fenoxietilpenicilina
    phenoxyethylpenicillin
fenoxipropanodiol phenoxypropanediol
fenqueno fenchene
fenquilo fenchyl
fensulfotiona fensulfothion
fentión fenthion
ferberita ferberite
fergusonita fergusonite
fermentable fermentable
fermentación fermentation
fermentación alcohólica alcoholic
    fermentation
fermentar ferment
fermio fermium
feromona pheromone
ferrato ferrate
ferrato bárico barium ferrate
ferrato potásico potassium ferrate
ferredoxina ferredoxin
ferri ferri
ferricianuro ferricyanide
ferricianuro de plata silver ferricyanide
ferricianuro férrico ferric ferricyanide
ferricianuro ferroso ferrous ferricyanide
ferricianuro potásico potassium
    ferricyanide
ferricianuro sódico sodium ferricyanide
férrico ferric
ferricromo ferrichrome
ferriferroso ferriferrous
ferrimagnetismo ferrimagnetism
ferrimanganeso ferrimanganese
ferrimangánico ferrimanganic
ferripirina ferripyrine
ferripotasio ferripotassium
ferrisodio ferrisodium
ferrita ferrite

ferrita bárica barium ferrite
ferrita cobaltosa cobaltous ferrite
ferrita ferrosa ferrous ferrite
ferrita potásica potassium ferrite
ferrítico ferritic
ferro ferro
ferroaleación ferroalloy
ferroaluminio ferroaluminum
ferroamonio ferroammonium
ferroboro ferroboron
ferrocenilborano ferrocenylborane
ferroceno ferrocene
ferroceno de clorocarbonilo
    chlorocarbonyl ferrocene
ferroceno de titanio titanium ferrocene
ferrocenoilo ferrocenoyl
ferrocerio ferrocerium
ferrocianato de etilo ethyl ferrocyanate
ferrocianuro ferrocyanide
ferrocianuro amónico ammonium
    ferrocyanide
ferrocianuro bárico barium ferrocyanide
ferrocianuro cálcico calcium
    ferrocyanide
ferrocianuro de cinc zinc ferrocyanide
ferrocianuro de cobre copper
    ferrocyanide
ferrocianuro de hierro iron ferrocyanide
ferrocianuro férrico ferric ferrocyanide
ferrocianuro ferroso ferrous
    ferrocyanide
ferrocianuro potásico potassium
    ferrocyanide
ferrocianuro sódico sodium ferrocyanide
ferrocromo ferrochromium
ferrodinámico ferrodynamic
ferrodolomita ferrodolomite
ferroelectricidad ferroelectricity
ferroeléctrico ferroelectric
ferroferricianuro ferroferricyanide
ferroférrico ferroferric
ferroferrocianuro ferroferrocyanide
ferrofósforo ferrophosphorus
ferroína ferroin
ferromagnesio ferromagnesium
ferromagnesita ferromagnesite
ferromagnético ferromagnetic
ferromagnetismo ferromagnetism
ferromanganeso ferromanganese

ferromangánico ferromanganic
ferromanganoso ferromanganous
ferromolibdeno ferromolybdenum
ferrón ferron
ferroniobio ferroniobium
ferroníquel ferronickel
ferropirina ferropyrine
ferropotasio ferropotassium
ferrosilicio ferrosilicon
ferrosilita ferrosilite
ferroso ferrous
ferrosoférrico ferrosoferric
ferrotitanio ferrotitanium
ferrotungsteno ferrotungsten
ferrovanadio ferrovanadium
ferroverdina ferroverdin
ferroxilo ferroxyl
ferrozirconio ferrozirconium
ferruginoso ferruginous
ferrum ferrum
fertilizante fertilizer
ferulaldehído ferulaldehyde
feruleno ferulene
fervanita fervanite
fibra fiber
fibra de carbono carbon fiber
fibra de celulosa cellulose fiber
fibra de fluorocarburo fluorocarbon
    fiber
fibra de grafito graphite fiber
fibra de vidrio glass fiber, fiberglass
fibra olefínica olefin fiber
fibra óptica optical fiber
fibra poliacrílica polyacrylic fiber
fibra poliéster polyester fiber
fibra sintética synthetic fiber
fibrido fibrid
fibrilla fibril
fibrina fibrin
fibrinógeno fibrinogen
fibroína fibroin
fibrolita fibrolite
fibroóptico fiber-optic
ficina ficin
ficocoloide phycocolloid
fictelita fichtelite
fíctil fictile
fijación fixation
fijación de nitrógeno nitrogen fixation

fijar fix, bind
fijativo fixative
filamento filament
filipita phillipite
filtración filtration
filtración al vacío vacuum filtration
filtración bajo presión filtration under
    pressure
filtración de gel gel filtration
filtración fraccionaria fractional
    filtration
filtración por bordes edge filtration,
    streamline filtration
filtrado filtrate
filtro filter
filtro acelerador accelofilter
filtro de aire air filter
filtro de Bechhold Bechhold filter
filtro de Berkefeld Berkefeld filter
filtro de gas gas filter
filtro de gelatina gelatin filter
filtro de Pasteur Pasteur filter
filtro de presión pressure filter
filtros de Zsigmondy Zsigmondy filters
fisetina fisetin
física electrónica electron physics
física nuclear nuclear physics
fisicoquímica physical chemistry
fisicoquímico physicochemical
físil fissile
fisión fission
fisión atómica atomic fission
fisión de uranio uranium fission
fisión espontánea spontaneous fission
fisión homolítica homolytic fission
fisión inducida induced fission
fisión nuclear nuclear fission
fisión por neutrones rápidos fast-
    neutron fission
fisión radiactiva radioactive fission
fisión térmica thermal fission
fisionable fissionable
fisioquímica fissiochemistry
fisostigmina physostigmine
fitano phytane
fitato cálcico calcium phytate
fitato hexacálcico hexacalcium phytate
fitato sódico sodium phytate
fitilo phytyl

fitina phytin
fitol phytol
fitoquímica phytochemistry
fitotóxico phytotoxic
flagstafita flagstaffite
flavanilina flavaniline
flavanol flavanol
flavanona flavanone
flavazina flavazine
flavilio flavylium
flavina flavin
flavofenina flavophenine
flavoglaucina flavoglaucin
flavol flavol
flavona flavone
flavonoide flavonoid
flavonol flavonol
flavopanina flavopannin
flavoproteína flavoprotein
flavopurpurina flavopurpurin
flavoxantina flavoxanthin
floculación flocculation
floculado flocculate
floculante flocculant
floculento flocculent
flor flower
floroglucinol phloroglucinol
floruro de alilo allyl fluoride
flotación flotation
flox flox
flucloxacilina flucloxacillin
fluctuación fluctuation
fluellita fluellite
fluidez fluidity
fluidización fluidization
fluido fluid
fluido de Bouin Bouin fluid
fluido de Carnoy Carnoy fluid
fluido de Condy Condy fluid
fluido hidráulico hydraulic fluid
fluido newtoniano Newtonian fluid
fluido supercrítico supercritical fluid
fluido volátil volatile fluid
flujo flow, flux
flujo de gas gas flow
flujo de Knudsen Knudsen flow
flujo electrónico electron flow
flujo en frío cold flow
flujo molecular molecular flow

flujo neutrónico neutron flux
flujo plástico plastic flow
flujo térmico heat flow
fluo fluo
fluocerita fluocerite
fluoflavina fluoflavine
flúor fluorine
fluoracetamida fluoracetamide
fluoración fluoration
fluorán fluoran
fluorandiol fluorandiol
fluoranteno fluoranthene
fluorantraquinona fluoranthraquinone
fluorapatita fluorapatite
fluorenilamina fluorenylamine
fluorenilideno fluorenylidene
fluorenilo fluorenyl
fluoreno fluorene
fluorenol fluorenol
fluorenona fluorenone
fluoresceína fluorescein
fluoresceína sódica sodium fluorescein
fluorescencia fluorescence
fluorescencia de rayos X X-ray
    fluorescence
fluorescencia de resonancia resonance
    fluorescence
fluorescente fluorescent
fluorescina fluorescin
fluorgermanato fluorgermanate
fluorilidina fluorylidine
fluorilo fluoryl
fluorimetría fluorimetry
fluorímetro fluorimeter
fluorinión fluorinion
fluorión fluorion
fluorita fluorite
fluorización fluoridation
fluoro fluoro
fluoroacetamida fluoroacetamide
fluoroacetanilida fluoroacetanilide
fluoroacetato fluoroacetate
fluoroacetato sódico sodium
    fluoroacetate
fluoroacetofenona fluoroacetophenone
fluoroalcano fluoroalkane
fluoroanilina fluoroaniline
fluoroapatita fluoroapatite
fluorobenceno fluorobenzene

fluoroborato fluoroborate
fluoroborato amónico ammonium
    fluoroborate
fluoroborato de cinc zinc fluoroborate
fluoroborato de plata silver fluoroborate
fluoroborato de plomo lead fluoroborate
fluoroborato férrico ferric fluoroborate
fluoroborato potásico potassium
    fluoroborate
fluoroborato sódico sodium fluoroborate
fluorocarburo fluorocarbon
fluorocromato fluorochromate
fluorodiclorometano
    fluorodichloromethane
fluoroelastómero fluoroelastomer
fluorofenol fluorophenol
fluoroformo fluoroform
fluoróforo fluorophore
fluorofosfato cálcico calcium
    fluorophosphate
fluorofosfato de litio lithium
    fluorophosphate
fluorofosfato sódico sodium
    fluorophosphate
fluorofosfonato fluorophosphonate
fluorofotómetro fluorophotometer
fluorógeno fluorogen
fluorohidrocarburo fluorohydrocarbon
fluorometano fluoromethane
fluorometolona fluorometholone
fluorometría fluorometry
fluorómetro fluorometer
fluorona fluorone
fluoronio fluoronium
fluoropolímero fluoropolymer
fluoroquímico fluorochemical
fluoroscopio fluoroscope
fluorosilicato fluorosilicate
fluorosilicato cálcico calcium
    fluorosilicate
fluorosilicato de cinc zinc fluorosilicate
fluorosilicato de cobalto cobalt
    fluorosilicate
fluorosilicato de plomo lead
    fluorosilicate
fluorosilicato férrico ferric fluorosilicate
fluorosilicato ferroso ferrous
    fluorosilicate
fluorosilicato sódico sodium

fluorosilicate

fluorosulfonato de metilo methyl
fluorosulfonate

fluoroteno fluorothene

fluorotriclorometano
fluorotrichloromethane

fluoroyodato fluoroiodate

fluoruro fluoride

fluoruro ácido amónico ammonium acid
fluoride

fluoruro ácido potásico potassium acid
fluoride

fluoruro ácido sódico sodium acid
fluoride

fluoruro alumínico aluminum fluoride

fluoruro alumínico potásico potassium
aluminum fluoride

fluoruro amónico ammonium fluoride

fluoruro antimónico antimonic fluoride

fluoruro antimonioso antimonous
fluoride

fluoruro arsenioso arsenous fluoride

fluoruro bárico barium fluoride

fluoruro cálcico calcium fluoride

fluoruro cérico ceric fluoride

fluoruro ceroso cerous fluoride

fluoruro cobáltico cobaltic fluoride

fluoruro cobaltoso cobaltous fluoride

fluoruro crómico chromic fluoride

fluoruro cromoso chromous fluoride

fluoruro cúprico cupric fluoride

fluoruro de acetilo acetyl fluoride

fluoruro de anilina aniline fluoride

fluoruro de antimonio antimony fluoride

fluoruro de arsénico arsenic fluoride

fluoruro de azufre sulfur fluoride

fluoruro de bencilo benzyl fluoride

fluoruro de benzoilo benzoyl fluoride

fluoruro de berilio beryllium fluoride

fluoruro de berilio sódico sodium
beryllium fluoride

fluoruro de bismutilo bismuthyl fluoride

fluoruro de boro boron fluoride

fluoruro de bromo bromine fluoride

fluoruro de cadmio cadmium fluoride

fluoruro de carbonilo carbonyl fluoride

fluoruro de carbono carbon fluoride

fluoruro de cesio cesium fluoride

fluoruro de cianógeno cyanogen fluoride

fluoruro de cinc zinc fluoride

fluoruro de cobre copper fluoride

fluoruro de cromilo chromyl fluoride

fluoruro de cromo chromium fluoride

fluoruro de escandio scandium fluoride

fluoruro de estaño tin fluoride

fluoruro de estroncio strontium fluoride

fluoruro de europio europium fluoride

fluoruro de fenilo phenyl fluoride

fluoruro de formilo formyl fluoride

fluoruro de fosforilo phosphoryl fluoride

fluoruro de gadolinio gadolinium
fluoride

fluoruro de hidrógeno hydrogen fluoride

fluoruro de hidrógeno amónico
ammonium hydrogen fluoride

fluoruro de hidrógeno potásico
potassium hydrogen fluoride

fluoruro de hidrógeno sódico sodium
hydrogen fluoride

fluoruro de hierro iron fluoride

fluoruro de holmio holmium fluoride

fluoruro de iterbio ytterbium fluoride

fluoruro de lantano lanthanum fluoride

fluoruro de litio lithium fluoride

fluoruro de lutecio lutetium fluoride

fluoruro de manganeso manganese
fluoride

fluoruro de metilo methyl fluoride

fluoruro de neodimio neodymium
fluoride

fluoruro de niobio niobium fluoride

fluoruro de níquel nickel fluoride

fluoruro de nitrilo nitryl fluoride

fluoruro de nitrógeno nitrogen fluoride

fluoruro de nitrosilo nitrosyl fluoride

fluoruro de oxígeno oxygen fluoride

fluoruro de perclorilo perchloryl
fluoride

fluoruro de plata silver fluoride

fluoruro de plomo lead fluoride

fluoruro de polivinilideno
polyvinylidene fluoride

fluoruro de polivinilo polyvinyl fluoride

fluoruro de renio rhenium fluoride

fluoruro de rubidio rubidium fluoride

fluoruro de rutenio ruthenium fluoride

fluoruro de silicio silicon fluoride

fluoruro de sulfinilo sulfinyl fluoride

fluoruro de sulfonilo sulfonyl fluoride
fluoruro de sulfurilo sulfuryl fluoride
fluoruro de talio thallium fluoride
fluoruro de tántalo tantalum fluoride
fluoruro de telurio tellurium fluoride
fluoruro de titanio titanium fluoride
fluoruro de titanio potásico potassium
 titanium fluoride
fluoruro de torio thorium fluoride
fluoruro de tungsteno tungsten fluoride
fluoruro de uranilo uranyl fluoride
fluoruro de uranio uranium fluoride
fluoruro de vanadio vanadium fluoride
fluoruro de vinilideno vinylidene
 fluoride
fluoruro de vinilo vinyl fluoride
fluoruro de yodo iodine fluoride
fluoruro de zirconio zirconium fluoride
fluoruro estánnico stannic fluoride
fluoruro estannoso stannous fluoride
fluoruro férrico ferric fluoride
fluoruro ferroso ferrous fluoride
fluoruro magnésico magnesium fluoride
fluoruro mangánico manganic fluoride
fluoruro manganoso manganous fluoride
fluoruro mercúrico mercuric fluoride
fluoruro monocálcico monocalcium
 fluoride
fluoruro potásico potassium fluoride
fluoruro potásico berílico beryllium
 potassium fluoride
fluoruro potásico de germanio
 germanium potassium fluoride
fluoruro potásico de tántalo tantalum
 potassium fluoride
fluoruro potásico de titanio titanium
 potassium fluoride
fluoruro potásico de zirconio zirconium
 potassium fluoride
fluoruro sódico sodium fluoride
fluoruro sódico alumínico aluminum
 sodium fluoride
fluoruro sódico berílico beryllium
 sodium fluoride
fluoryodo fluoriodine
fluosilicato fluosilicate
fluosilicato amónico ammonium
 fluosilicate
fluosilicato bárico barium fluosilicate

fluosilicato de plata silver fluosilicate
fluosilicato magnésico magnesium
 fluosilicate
fluosilicato potásico potassium
 fluosilicate
fluoximesterona fluoxymesterone
fluozirconato potásico potassium
 fluozirconate
folacina folacin
folato sódico sodium folate
fomómetro fomometer
fómula de Briet-Wigner Briet-Wigner
 formula
fongisterol fongisterol
fonoquímica phonochemistry
forato phorate
forbol phorbol
forilo phoryl
forma canónica canonical form
forma cristalina crystal form
formacilo formazyl
formación formation
formación al vacío vacuum forming
formación de anillo ring formation
formal formal
formaldehído formaldehyde
formaldoxima formaldoxime
formalidad formality
formalina formalin
formamida formamide
formamida de etanol ethanol formamide
formamida de etilo ethyl formamide
formamidina formamidine
formamido formamido
formamilo formamyl
formamina formamine
formanilida formanilide
formanilina formaniline
formazán formazan
formiato formate, formiate
formiato alumínico aluminum formate
formiato amónico ammonium formate
formiato bárico barium formate
formiato cálcico calcium formate
formiato cobaltoso cobaltous formate
formiato cromoso chromous formate
formiato de amilo amyl formate
formiato de bencilo benzyl formate
formiato de bornilo bornyl formate

formiato de butilo butyl formate
formiato de cadmio cadmium formate
formiato de cinc zinc formate
formiato de estroncio strontium formate
formiato de etilo ethyl formate
formiato de geranilo geranyl formate
formiato de heptilo heptyl formate
formiato de hexilo hexyl formate
formiato de hidrazina hydrazine formate
formiato de hierro iron formate
formiato de isoamilo isoamyl formate
formiato de linalilo linalyl formate
formiato de metilo methyl formate
formiato de níquel nickel formate
formiato de octilo octyl formate
formiato de pentilo pentyl formate
formiato de plomo lead formate
formiato de propilo propyl formate
formiato de talio thallium formate
formiato de uranilo uranyl formate
formiato férrico ferric formate
formiato magnésico magnesium formate
formiato manganoso manganous formate
formiato mercurioso mercurous formate
formiato niqueloso nickelous formate
formiato potásico potassium formate
formiato sódico sodium formate
formilación formylation
formilamida formylamide
formilamina formylamine
formilamino formylamino
formileno formylene
formilhidrazina formylhydrazine
formilo formyl
formiloxi formyloxy
formilpiperidina formylpiperidine
formimidoilo formimidoyl
formina formine
formohidrazida formohydrazide
formoilo formoyl
formol formol
formolita formolite
formonitrilo formonitrile
formotión formothion
formoxi formoxy
formoxilo formoxyl
formoxima formoxime
fórmula formula
fórmula constitucional constitutional

formula
fórmula de Balmer Balmer formula
fórmula de Bamberger Bamberger
formula
fórmula de Cook Cook formula
fórmula de London London formula
fórmula de Meyer Meyer formula
fórmula de reacción reaction formula
fórmula de Rydberg Rydberg formula
fórmula de van't Hoff van't Hoff
formula
fórmula electrónica electron formula
fórmula empírica empirical formula
fórmula estructural structural formula
fórmula general general formula
fórmula genérica generic formula
fórmula gráfica graphic formula
fórmula molecular molecular formula
fórmula química chemical formula
fórmula racional rational formula
formulación formulation
forona phorone
forsterita forsterite
fosalona phosalone
fosfa phospha
fosfaceno phosphazene
fosfamida phosphamide
fosfamidón phosphamidon
fosfano phosphane
fosfatasa phosphatase
fosfatasa ácida acid phosphatase
fosfátido phosphatide
fosfato phosphate
fosfato ácido acid phosphate
fosfato ácido amónico ammonium acid
phosphate
fosfato ácido de amilo amyl acid
phosphate
fosfato ácido potásico potassium acid
phosphate
fosfato ácido sódico sodium acid
phosphate
fosfato alumínico aluminum phosphate
fosfato alumínico sódico sodium
aluminum phosphate
fosfato amónico ammonium phosphate
fosfato amónico cobaltoso cobaltous
ammonium phosphate
fosfato amónico magnésico magnesium

ammonium phosphate
fosfato amónico sódico sodium
ammonium phosphate
fosfato bárico barium phosphate
fosfato bicálcico bicalcium phosphate
fosfato cálcico calcium phosphate
fosfato cálcico monobásico monobasic
calcium phosphate
fosfato cálcico tribásico tribasic calcium
phosphate
fosfato ceroso cerous phosphate
fosfato cobaltoso cobaltous phosphate
fosfato crómico chromic phosphate
fosfato cúprico cupric phosphate
fosfato de adenosina adenosine
phosphate
fosfato de almidón starch phosphate
fosfato de betaína betaine phosphate
fosfato de bismuto bismuth phosphate
fosfato de boro boron phosphate
fosfato de cadmio cadmium phosphate
fosfato de cinc zinc phosphate
fosfato de cobalto cobalt phosphate
fosfato de cobre copper phosphate
fosfato de codeína codeine phosphate
fosfato de cresilo cresyl phosphate
fosfato de cromo chromium phosphate
fosfato de dibutilo dibutyl phosphate
fosfato de dimetilo dimethyl phosphate
fosfato de estroncio strontium phosphate
fosfato de etilo ethyl phosphate
fosfato de galio gallium phosphate
fosfato de hidrógeno potásico potassium
hydrogen phosphate
fosfato de hidrógeno sódico sodium
hydrogen phosphate
fosfato de hierro iron phosphate
fosfato de huesos bone phosphate
fosfato de litio lithium phosphate
fosfato de manganeso manganese
phosphate
fosfato de manganeso amónico
ammonium manganese phosphate
fosfato de mercurio mercury phosphate
fosfato de neodimio neodymium
phosphate
fosfato de níquel nickel phosphate
fosfato de octilo octyl phosphate
fosfato de piridoxal pyridoxal phosphate

fosfato de plata silver phosphate
fosfato de plomo lead phosphate
fosfato de riboflavina riboflavin
phosphate
fosfato de tetraetilo tetraethyl phosphate
fosfato de trialilo triallyl phosphate
fosfato de tributilo tributyl phosphate
fosfato de tricresilo tricresyl phosphate
fosfato de trietilo triethyl phosphate
fosfato de trifenilo triphenyl phosphate
fosfato de trimagnesio trimagnesium
phosphate
fosfato de trimetilo trimethyl phosphate
fosfato de trioctilo trioctyl phosphate
fosfato de trisetilhexilo trisethylhexyl
phosphate
fosfato de tritolilo tritolyl phosphate
fosfato de urea-amonio urea-ammonium
phosphate
fosfato de uridina uridine phosphate
fosfato de zirconilo zirconyl phosphate
fosfato de zirconio zirconium phosphate
fosfato dicálcico dicalcium phosphate
fosfato disódico disodium phosphate
fosfato férrico ferric phosphate
fosfato ferroso ferrous phosphate
fosfato magnésico magnesium phosphate
fosfato magnésico amónico ammonium
magnesium phosphate
fosfato manganoso manganous phosphate
fosfato mercúrico mercuric phosphate
fosfato mercurioso mercurous phosphate
fosfato monocálcico monocalcium
phosphate
fosfato niqueloso nickelous phosphate
fosfato potásico potassium phosphate
fosfato sódico sodium phosphate
fosfato sódico amónico ammonium
sodium phosphate
fosfato tricálcico tricalcium phosphate
fosfato tripotásico tripotassium
phosphate
fosfato trisódico trisodium phosphate
fosfato trisódico clorado chlorinated
trisodium phosphate
fosfenilo phosphenyl
fosfina phosphine
fosfina de dietilo diethyl phosphine
fosfina de etilo ethyl phosphine

fosfina de tributilo tributyl phosphine
fosfinato phosphinate
fosfinato alumínico aluminum phosphinate
fosfinato bárico barium phosphinate
fosfinato cálcico calcium phosphinate
fosfinato de manganeso manganese phosphinate
fosfinato de plomo lead phosphinate
fosfinato magnésico magnesium phosphinate
fosfinato potásico potassium phosphinate
fosfinato sódico sodium phosphinate
fosfinico phosphinico
fosfinita phosphinite
fosfino phosphino
fosfinoilo phosphinoyl
fosfinoso phosphinoso
fosfito phosphite
fosfito amónico ammonium phosphite
fosfito cálcico calcium phosphite
fosfito de dialilo diallyl phosphite
fosfito de dibutilo dibutyl phosphite
fosfito de dietilo diethyl phosphite
fosfito de difenilo diphenyl phosphite
fosfito de dioctilo dioctyl phosphite
fosfito de fenildidecilo phenyldidecyl phosphite
fosfito de fenilneopentilo phenylneopentyl phosphite
fosfito de plomo lead phosphite
fosfito de tributilo tributyl phosphite
fosfito de tricresilo tricresyl phosphite
fosfito de trietilo triethyl phosphite
fosfito de trifenilo triphenyl phosphite
fosfito de triisooctilo triisooctyl phosphite
fosfito de triisopropilo triisopropyl phosphite
fosfito de trimetilo trimethyl phosphite
fosfito de trioctadecilo trioctadecyl phosphite
fosfito de triscloroetilo trischloroethyl phosphite
fosfito de trisetilhexilo trisethylhexyl phosphite
fosfito potásico potassium phosphite
fosfito sódico sodium phosphite
fosfo phospho

fosfoaluminato sódico sodium phosphoaluminate
fosfobenceno phosphobenzene
fosfocreatina phosphocreatine
fosfodiéster phosphodiester
fosfodiesterasa phosphodiesterase
fosfoglicérido phosphoglyceride
fosfolipasa phospholipase
fosfolípido phospholipid
fosfomicina fosfomycin
fosfomolibdato phosphomolybdate
fosfomolibdato amónico ammonium phosphomolybdate
fosfomolibdato sódico sodium phosphomolybdate
fosfonato phosphonate
fosfonato de cinc zinc phosphonate
fosfonato de dibutilbutilo dibutylbutyl phosphonate
fosfonato magnésico magnesium phosphonate
fosfonato potásico potassium phosphonate
fosfonato sódico sodium phosphonate
fosfonio phosphonium
fosfonita phosphonite
fosfonitrilo phosphonitrile
fosfono phosphono
fosfonoilo phosphonoyl
fosfonoso phosphonoso
fosfoproteína phosphoprotein
fosforado phosphorated
fosforano phosphorane
fosforato phosphorate
fosforescencia phosphorescence
fosforescente phosphorescent
fosfórico phosphoric
fosforilación phosphorylation
fosforilasa phosphorylase
fosforilo phosphoryl
fosforimetría phosphorimetry
fosforita phosphorite
fósforo phosphorus, phosphor
fósforo amarillo yellow phosphorus
fósforo blanco white phosphorus
fósforo de Baldwin Baldwin phosphorus
fósforo rojo red phosphorus
fosforoso phosphoroso
fosforotioato de trietilo triethyl

phosphorothioate

fosforotiolato de tributilo tributyl
  phosphorothiolate

fosfosilicato bárico barium
  phosphosilicate

fosfotungstato phosphotungstate

fosfotungstato amónico ammonium
  phosphotungstate

fosfotungstato sódico sodium
  phosphotungstate

fosfovanadato sódico sodium
  phosphovanadate

fosfuro phosphide

fosfuro alumínico aluminum phosphide

fosfuro arsenioso arsenous phosphide

fosfuro bárico barium phosphide

fosfuro cálcico calcium phosphide

fosfuro cuproso cuprous phosphide

fosfuro de boro boron phosphide

fosfuro de cesio cesium phosphide

fosfuro de cinc zinc phosphide

fosfuro de cobalto cobalt phosphide

fosfuro de cobre copper phosphide

fosfuro de cromo chromium phosphide

fosfuro de galio gallium phosphide

fosfuro de hidrógeno hydrogen
  phosphide

fosfuro de hierro iron phosphide

fosfuro de indio indium phosphide

fosfuro de itrio yttrium phosphide

fosfuro de lantano lanthanum phosphide

fosfuro de plata silver phosphide

fosfuro estánnico stannic phosphide

fosfuro ferroso ferrous phosphide

fosfuro magnésico magnesium phosphide

fosfuro potásico potassium phosphide

fosfuro sódico sodium phosphide

fosgeno phosgene

foshagita foshagite

fósil fossil

fosmet phosmet

fostato diamónico diammonium
  phosphate

fostión fostion

fotocatálisis photocatalysis

fotocatalizador photocatalyst

fotocátodo photocathode

fotoconducción photoconduction

fotocrómico photochromic

fotocromismo photochromism

fotodegradación photodegradation

fotodeposición photodeposition

fotodescomposición photodecomposition

fotodesintegración photodisintegration

fotodimerización photodimerization

fotodisociación photodissociation

fotoeléctrico photoelectric

fotoelectrón photoelectron

fotoemisión photoemission

fotoemisivo photoemissive

fotoemisor photoemitter

fotoestabilización photostabilization

fotográfico photographic

fotoionización photoionization

fotólisis photolysis

fotólisis instantánea flash photolysis

fotolítico photolytic

fotólito photolyte

fotoluminiscencia photoluminescence

fotometría con filtro filter photometry

fotometría de llama flame photometry

fotométrico photometric

fotómetro photometer

fotómetro de llama flame photometer

fotón photon

fotoneutrón photoneutron

fotonuclear photonuclear

fotooxidación photooxidation

fotopolimerización photopolymerization

fotopolímero photopolymer

fotoprotón photoproton

fotoquímica photochemistry

fotoquímico photochemical

fotorreacción photoreaction

fotorreducción photoreduction

fotosensibilidad photosensitivity

fotosensibilización photosensitization

fotosensibilizador photosensitizer

fotosensible photosensitive

fotosíntesis photosynthesis

fototropia phototropy

fototropismo phototropism

fotovoltaico photovoltaic

fowlerita fowlerite

foyaíta foyaite

fracción fraction

fracción de empaquetamiento packing
  fraction

fracción de masa mass fraction
fracción molar mole fraction
fraccionación fractionation
fraccionamiento cracking
fraccionante fractionating
fraccionar crack
fraccionario fractional
fractura fracture
fracturación hidráulica hydraulic fracturing
fragancia fragrance
fragilidad fragility
fragilización cáustica caustic cracking
fragmentación térmica thermal fragmentation
fragmento atómico atomic fragment
fragmento nuclear nuclear fragment
francio francium
francolita francolite
frangulina frangulin
franklinita franklinite
franqueíta franckeite
frasco lavador wash bottle
fraunhofer fraunhofer
fraxetina fraxetin
fraxina fraxin
fraxinol fraxinol
frecuencia frequency
frecuencia atómica atomic frequency
frecuencia molecular molecular frequency
freibergita freibergite
freierslebenita freierslebenite
freyalita freyalite
fricción friction
frigorífico frigorific
frita frit
fritado fritting
fructofuranosidasa fructofuranosidase
fructosa fructose
fructosida fructoside
ftalamida phthalamide
ftalato phthalate
ftalato de bistridecilo bistridecyl phthalate
ftalato de butilbencilo butylbenzyl phthalate
ftalato de butilo butyl phthalate
ftalato de dialilo diallyl phthalate

ftalato de dibutilo dibutyl phthalate
ftalato de dietilo diethyl phthalate
ftalato de difenilo diphenyl phthalate
ftalato de dimetilo dimethyl phthalate
ftalato de dinonilo dinonyl phthalate
ftalato de dioctilo dioctyl phthalate
ftalato de etilo ethyl phthalate
ftalato de fenilo phenyl phthalate
ftalato de glicerilo glyceryl phthalate
ftalato de hidrógeno potásico potassium hydrogen phthalate
ftalato de isoamilo isoamyl phthalate
ftalato de pentilo pentyl phthalate
ftalato de plomo lead phthalate
ftalato sódico sodium phthalate
ftalazina phthalazine
ftaldiamida phthaldiamide
ftaleína phthalein
ftalida phthalide
ftalidilideno phthalidylidene
ftalidilo phthalidyl
ftalilo phthalyl
ftalimida phthalimide
ftalimido phthalimido
ftalocianina phthalocyanine
ftalodinitrilo phthalodinitrile
ftaloilo phthaloyl
ftalonitrilo phthalonitrile
fuberidazol fuberidazole
fucitol fucitol
fucosa fucose
fucosita fucosite
fucsina fuchsin
fucsina ácida acid fuchsin
fucusina fucusine
fucusol fucusol
fuchsita fuchsite
fuente source
fuente de electrones electron source
fuente de energía energy source
fuente de infrarrojo infrared source
fuente de neutrones neutron source
fuente de radiación radiation source
fuente de rayos gamma gamma-ray source
fuente iónica ion source
fuente radiactiva radioactive source
fuerte strong
fuerza force, strength

fuerza catalítica catalytic force
fuerza centrífuga centrifugal force
fuerza de campo field strength
fuerza de Majorana Majorana force
fuerza de Yukawa Yukawa force
fuerza dieléctrica dielectric strength
fuerza electromotriz electromotive force
fuerza intermolecular intermolecular force
fuerza iónica ionic strength
fuerza nuclear nuclear force
fuerzas de dispersión dispersion forces
fuerzas de van der Waals van der Waals forces
fuerzas ligantes metálicas metallic binding forces
fuga leakage, leak
fuga de gas gas leak
fugacidad fugacity
fulereno fullerene, buckyballs
fulerita fullerite
fulguración fulguration
fulgurador fulgurator
fulgurita fulgurite
fulminato fulminate
fulminato de mercurio mercury fulminate
fulminato de plata silver fulminate
fulvaleno fulvalene
fulveno fulvene
fumagilina fumagillin
fumante fuming
fumaramida fumaramide
fumarato fumarate
fumarato cálcico calcium fumarate
fumarato de bencilo benzyl fumarate
fumarato de cadmio cadmium fumarate
fumarato de dibutilo dibutyl fumarate
fumarato de plomo lead fumarate
fumarato ferroso ferrous fumarate
fumarhidrazida fumarhydrazide
fumarilo fumaryl
fumaroilo fumaroyl
fumigante fumigant
función de correlación correlation function
función de excitación excitation function
función de Gibbs Gibbs function
función de onda wave function

función de partición partition function
función de Patterson Patterson function
función sigma sigma function
función trabajo work function
fundamental fundamental, ground
fundente magnésico magnesium flux
fundido fused, molten
fundir fuse, melt, smelt
fungicida fungicide
fural fural
furaldehído furaldehyde
furamida furamide
furancarbinol furancarbinol
furandiona furandione
furanilmetilo furanylmethyl
furanmetanotiol furanmethanethiol
furanmetilamina furanmethylamine
furano furan
furanosa furanose
furanosida furanoside
furazán furazan
furazanilo furazanyl
furazolidona furazolidone
furfural furfural
furfuraldehído furfuraldehyde
furfuramida furfuramide
furfurano furfuran
furfurilamida furfurylamide
furfurilamina furfurylamine
furfurilideno furfurylidene
furfurilo furfuryl
furfurol furfurol
furil furil
furilcarbinol furylcarbinol
furildioxima furildioxime
furilidina furylidine
furilmetilo furylmethyl
furilo furyl
furoamida furoamide
furoato furoate
furoato de butilo butyl furoate
furoato de etilo ethyl furoate
furoato de isoamilo isoamyl furoate
furoato de isobutilo isobutyl furoate
furoato de pentilo pentyl furoate
furoato de propilo propyl furoate
furoilo furoyl
furoína furoin
furol furol

fusibilidad fusibility
fusible fusible
fusidato sódico sodium fusidate
fusión fusion
fusión atómica atomic fusion

fusión nuclear nuclear fusion
fusión termonuclear thermonuclear
  fusion
fusionable fusionable

G

gadinina gadinine
gadolinio gadolinium
gadolinita gadolinite
gafas de protección goggles
gafas protectoras safety goggles
gahnita gahnite
gailusita gaylussite
galacetofenona gallacetophenone
galactano galactan
galactosa galactose
galactosidasa galactosidase
galamina gallamine
galato gallate
galato amónico ammonium gallate
galato de bismuto bismuth gallate
galato de metilo methyl gallate
galato de octilo octyl gallate
galato de propilo propyl gallate
galeína gallein
galena galena
galenita galenite
gálico gallic
galina gallin
galio gallium
galocianina gallocyanine
galógeno gallogen
galoilo galloyl
galoso gallous
galotanino gallotannin
galvánico galvanic
galvanismo galvanism
galvanización galvanizing
galvanizado galvanized
galvanizar galvanize
galvanomagnético galvanomagnetic
galvanometría galvanometry
galvanométrico galvanometric
galvanómetro galvanometer
galvanoplasteado plated
galvanoplastia galvanoplasty,
    electroplating, plating
gamexano gammexane
gamilo gammil

gamma gamma
gammaglobulina gamma-globulin
ganga gangue
ganofilita ganophyllite
ganomalita ganomalite
ganomatita ganomatite
garnierita garnierite
gas gas
gas asfixiante asphyxiant gas
gas atmosférico atmospheric gas
gas comprimido compressed gas
gas de agua water gas
gas de coque coke gas
gas de hidrocarburos hydrocarbon gas
gas de hulla coal gas
gas de pantano marsh gas
gas de petróleo petroleum gas, oil gas
gas de petróleo licuado liquefied
    petroleum gas
gas de refinería refinery gas
gas de síntesis synthesis gas
gas detonante detonating gas
gas electrolítico electrolytic gas
gas electrónico electron gas
gas hilarante laughing gas
gas ideal ideal gas
gas inerte inert gas
gas ionizado ionized gas
gas licuado liquefied gas
gas líquido liquid gas
gas molecular molecular gas
gas monoatómico monoatomic gas
gas mostaza mustard gas
gas natural natural gas
gas natural licuado liquefied natural gas
gas natural sintético synthetic natural
    gas
gas neurotóxico nerve gas
gas noble noble gas
gas nocivo noxious gas
gas perfecto perfect gas
gas permanente permanent gas
gas pobre producer gas

gas poliatómico polyatomic gas
gas portador carrier gas
gas raro rare gas
gas real real gas
gas rico rich gas
gas sintético synthetic gas
gas venenoso poison gas
gasa de alambre wire gauze
gaseamiento gassing
gaseoso gaseous
gasificación gasification
gasohol gasohol
gasol gasol
gasóleo gas oil
gasolina gasoline
gasolina blanca white gasoline
gasométrico gasometric
gastrina gastrin
gaulterina gaultherin
gehlenita gehlenite
geisina geissine
gel gel
gel de alúmina alumina gel
gel de sílice silica gel
gel incendiario incendiary gel
gel iónico ionic gel
gelatina gelatin
gelatina explosiva blasting gelatin
gelatinización gelatinization
gelatinizar gelatinize
gelatinoso gelatinous
gelificación gelling
gelignita gelignite
gema gem, precious stone
geminal geminal
gen gene
genalcaloide genalkaloid
generación generation
generador de gas gas generator
generar generate
genérico generic
genético genetic
genistina genistin
gentamicina gentamicin
gentiobiosa gentiobiose
geocronita geocronite
geometría estérica steric geometry
geoquímica geochemistry
geranial geranial

geranialdehído geranialdehyde
geranilo geranyl
geraniol geraniol
germanato germanate
germanato de litio lithium germanate
germanato sódico sodium germanate
germánico germanic
germanio germanium
germanita germanite
germaniuro germanide
germano germane
germanoso germanous
germicida germicidal
germinación germination
gibbsita gibbsite
giberelina gibberellin
gibrel gibrel
gigatolita gigantolite
gilpinita gilpinite
gilsonita gilsonite
ginocardina gynocardine
girolita gyrolite
gismondita gismondite
gitonina gitonin
gitoxina gitoxin
glacial glacial
glaserita glaserite
glauberita glauberite
glaucocroíta glaucochroite
glaucodot glaucodot
glaucofana glaucophane
glauconita glauconite
glaucopicrina glaucopicrine
gliadina gliadin
glical glycal
gliceral glyceral
gliceraldehído glyceraldehyde
glicerato glycerate
glicerato de etilo ethyl glycerate
glicérico glyceric
glicérido glyceride
glicerilo glyceryl
glicerina glycerin, glycerol
glicerinato glycerinate
glicero glycero
glicerofosfato glycerophosphate
glicerofosfato cálcico calcium
    glycerophosphate
glicerofosfato de manganeso manganese

glycerophosphate
glicerofosfato férrico ferric
  glycerophosphate
glicerofosfato potásico potassium
  glycerophosphate
gliceroilo glyceroyl
glicerol glycerol, glycerin
glicerosa glycerose
glicida glycide
glicidilo glycidyl
glicidol glycidol
glicilglicina glycylglycine
glicilo glycyl
glicina glycine
glicinato de cobre copper glycinate
glicinio glycinium
glico glyco
glicogenasa glycogenase
glicogénesis glycogenesis
glicógeno glycogen
glicol glycol
glicolal glycolal
glicolaldehído glycolaldehyde
glicolato cálcico calcium glycolate
glicolato de etilo ethyl glycolate
glicolato de zirconio zirconium glycolate
glicolato de zirconio sódico sodium
  zirconium glycolate
glicolato sódico sodium glycolate
glicólido glycolide
glicolilo glycolyl
glicolípido glycolipid
glicólisis glycolysis
glicoloilo glycoloyl
glicolonitrilo glycolonitrile
gliconitrilo glyconitrile
glicoproteína glycoprotein
glicosaminoglicano glycosaminoglycan
glicosidasa glycosidase
glicosídico glycosidic
glicósido glycoside
glicosilación glycosylation
gliftal glyphtal
glime glyme
glioxal glyoxal
glioxalasa glyoxalase
glioxalina glyoxaline
glioxilasa glyoxylase
glioxima glyoxime

globulina globulin
glucagón glucagon
glucano glucan
glucasa glucase
glucinio glucinium
glucógeno glycogen
glucoheptonato sódico sodium
  glucoheptonate
gluconato gluconate
gluconato alumínico aluminum gluconate
gluconato amónico ammonium gluconate
gluconato cálcico calcium gluconate
gluconato de cinc zinc gluconate
gluconato de cobre copper gluconate
gluconato de manganeso manganese
  gluconate
gluconato ferroso ferrous gluconate
gluconato magnésico magnesium
  gluconate
gluconato potásico potassium gluconate
gluconato sódico sodium gluconate
glucopiranosa glucopyranose
glucosa glucose
glucosamina glucosamine
glucosazona glucosazone
glucosidasa glucosidase
glucósido glucoside, glycoside
glucosina glucosin
glucuronida glucuronide
glucuronidasa glucuronidase
glucuronolactona glucuronolactone
glusida gluside
glutamato amónico ammonium
  glutamate
glutamato cálcico calcium glutamate
glutamato monosódico monosodium
  glutamate
glutamato potásico potassium glutamate
glutamato sódico sodium glutamate
glutamilcisteinilglicina
  glutamylcysteinylglycine
glutamilo glutamyl
glutamina glutamine
glutamoilo glutamoyl
glutaraldehído glutaraldehyde
glutárico glutaric
glutaronitrilo glutaronitrile
glutationa glutathione
glutelina glutelin

gluten gluten
glutenina glutenin
glutetimida glutethimidc
glutosa glutose
gmelinita gmelinite
gneis gneiss
gnoscopina gnoscopine
goetita goethite
golpeteo knock
goma arábiga arabic gum
goma de acacia acacia gum
goma de celulosa cellulose gum
goma de éster ester gum
goma de tragacanto tragacanth gum
goma de xantano xanthan gum
goma de yaca yacca gum
goma guar guar gum
goma karaya karaya gum
goma laca shellac
goma soluble en agua water-soluble gum
goma vegetal vegetable gum
gonadotropina coriónica chorionic
　　gonadotropin
gosipol gossypol
goslarita goslarite
gradación grading
gradiente gradient
gradiente de concentración
　　concentration gradient
gradiente osmótico osmotic gradient
grado degree, grade
grado de Brix Brix degree
grado de dilución degree of dilution
grado de disociación degree of
　　dissociation
grado de dureza degree of hardness
grado de hidrólisis degree of hydrolysis
grado de ionización ionization degree
grado de polimerización degree of
　　polymerization
grado de substitución degree of
　　substitution
grado de temperatura degree of
　　temperature
grado de vacío degree of vacuum
grados de libertad degrees of freedom
graduación graduation
graduado graduated
gradual gradual

grafitización graphitization
grafito graphite
gramicidina gramicidin
gramina gramine
gramnegativo gram-negative
gramo gram
grampositivo gram-positive
granate garnet
granito granite
granulación granulation
granular granular
gránulo granule
granulosa granulose
granulosidad granularity
grasa fat, grease
grasa de lana wool fat
grasa mineral mineral fat
grasa poliinsaturada polyunsaturated fat
grasa saturada saturated fat
gravedad específica specific gravity
gravilla de ebullición boiling chips
gravimetría gravimetry
gravimétrico gravimetric
gray gray
greenockita greenockite
grifita griphite
griolita gryolite
grosularita grossularite
grunerita grunerite
grupo group
grupo acídico acidic group
grupo ácido acid group
grupo acilo acyl group
grupo activo active group
grupo alifático aliphatic group
grupo alilo allyl group
grupo amida amide group
grupo amino amino group
grupo atómico atomic group
grupo básico basic group
grupo carbonilo carbonyl group
grupo carboxilo carboxyl group
grupo cuántico quantum group
grupo de elementos group of elements
grupo funcional functional group
grupo negativo negative group
grupo positivo positive group
grupo prostético prosthetic group
grupo protector protecting group

grupo substituyente substituent
guaiol guaiol
guanacilo guanazyl
guanamina guanamine
guanasa guanase
guanidina guanidine
guanidinio guanidinium
guanidino guanidino
guanido guanido

guanilato sódico sodium guanylate
guanina guanine
guano guano
guanosina guanosine
guantes protectores protective gloves
guayacol guaiacol
guerra química chemical warfare
gusto taste
gutapercha gutta-percha

H

hábito habit
hachís hashish
hafnio hafnium
halazona halazone
halita halite
halo halo
haloalcano haloalkane
haloamina haloamine
halocarburo halocarbon
halocromismo halochromism
haloformo haloform
halogenación halogenation
halogenado halogenated
halogenar halogenate
halogeniuro halogenide
halógeno halogen
halohidrina halohydrin
haloide haloid
halonio halonium
haloperidol haloperidol
halotano halothane
haluro halide
haluro ácido acid halide
haluro de acilo acyl halide
haluro de alquilo alkyl halide
haluro de cadmio cadmium halide
haluro de carbonilo carbonyl halide
haluro de cloro chlorine halide
haluro de cobalto cobalt halide
haluro de erbio erbium halide
haluro de flúor fluorine halide
haluro de galio gallium halide
haluro de germanio germanium halide
haluro de litio lithium halide
haluro de manganeso manganese halide
haluro de oro gold halide
haluro de plomo lead halide
haluro magnésico magnesium halide
haluro sódico sodium halide
halloysita halloysite
hamartita hamartite
hanksita hanksite
hapto hapto

harina de aceite de linaza linseed oil
  meal
harina de aceite de maní peanut oil meal
harina de madera wood flour, wood
  meal
harina de pizarra slate flour
harina de semilla de algodón cottonseed
  meal
harmina harmine
harmotoma harmotome
hartita hartite
hausmannita hausmannite
hauyna hauyne
haz electrónico electron beam
haz iónico molecular molecular ion
  beam
haz molecular molecular beam
haz neutrónico neutron beam
hectorita hectorite
hedenbergita hedenbergite
hedonal hedonal
helcosol helcosol
helenina helenine
helenita helenite
hélice helix
hélice alfa alpha helix
helicoidal helical
helio helium
helvita helvite
hem heme
hemateína hematein
hematina hematin
hematita hematite
hematoxilina hematoxylin
hemel hemel
hemi hemi
hemiacetal hemiacetal
hemicelulosa hemicellulose
hemihedral hemihedral
hemimeliteno hemimellitene
hemina hemin
hemiquinonoide hemiquinonoid
hemoglobina hemoglobin

403

hempa hempa
hendecano hendecane
hendidura cleavage
heneicosano heneicosane
heneicosanoato de metilo methyl heneicosanoate
hentriacontano hentriacontane
heparina heparin
hepta hepta
heptabarbital heptabarbital
heptaclorepóxido heptachlorepoxide
heptacloro heptachlor
heptacosano heptacosane
heptadecano heptadecane
heptadecanoato de metilo methyl heptadecanoate
heptadecanol heptadecanol
heptadecanona heptadecanone
heptadecilglioxalidina heptadecylglyoxalidine
heptadieno heptadiene
heptahidrato heptahydrate
heptaldehído heptaldehyde
heptaleno heptalene
heptametildisilazano heptamethyldisilazane
heptametileno heptamethylene
heptametilnonano heptamethylnonane
heptanal heptanal
heptano heptane
heptanoato heptanoate
heptanoato de etilo ethyl heptanoate
heptanodioilo heptanedioyl
heptanoilo heptanoyl
heptanol heptanol
heptanona heptanone
heptasulfuro de fósforo phosphorus heptasulfide
heptasulfuro de renio rhenium heptasulfide
heptavalente heptavalent
heptenilo heptenyl
hepteno heptene
heptilamina heptylamine
heptileno heptylene
heptilo heptyl
heptino heptyne
heptoato de heptilo heptyl heptoate
heptosa heptose

heptóxido heptoxide
heptóxido de cloro chlorine heptoxide
heptóxido de manganeso manganese heptoxide
heptóxido de renio rhenium heptoxide
herapatita herapathite
herbicida herbicide
hercinita hercynite
herderita herderite
hermético hermetic
hermético al agua watertight
heroína heroin
herrerita herrerite
herrumbre rust
hervir boil
hesperidina hesperidin
hessita hessite
hessonita hessonite
hetero hetero
heteroaromático heteroaromatic
heteroatómico heteroatomic
heteroauxina heteroauxin
heteroazeótropo heteroazeotrope
heterobárico heterobaric
heterocíclico heterocyclic
heterogéneo heterogeneous
heteroléptico heteroleptic
heterólisis heterolysis
heterolítico heterolytic
heterometría heterometry
heteromolibdato heteromolybdate
heteronuclear heteronuclear
heteropolar heteropolar
heteropoli heteropoly
heteropolímero heteropolymer
heterotipo heterotype
heterotópico heterotopic
heterótopo heterotope
heulandita heulandite
hexa hexa
hexaamina hexaamine
hexaborano hexaborane
hexabromo hexabromo
hexabromoetano hexabromoethane
hexabromoplatinato amónico ammonium hexabromoplatinate
hexabromuro de benceno benzene hexabromide
hexacarbonilo de cromo chromium

hexacarbonyl

hexacarbonilo de molibdeno
molybdenum hexacarbonyl

hexacarbonilo de tungsteno tungsten
hexacarbonyl

hexacarbonilo de vanadio vanadium
hexacarbonyl

hexacarbonilo potásico potassium
hexacarbonyl

hexaciano hexacyano

hexacianoferrato amónico ammonium
hexacyanoferrate

hexacianoferrato bárico barium
hexacyanoferrate

hexacianoferrato de cinc zinc
hexacyanoferrate

hexacianoferrato de cobre copper
hexacyanoferrate

hexacianoferrato de hierro iron
hexacyanoferrate

hexacianoferrato de manganeso
manganese hexacyanoferrate

hexacianoferrato de mercurio mercury
hexacyanoferrate

hexacianoferrato de plata silver
hexacyanoferrate

hexacianoferrato de plomo lead
hexacyanoferrate

hexacianoferrato de uranilo uranyl
hexacyanoferrate

hexacianoferrato potásico potassium
hexacyanoferrate

hexacianoferrato sódico sodium
hexacyanoferrate

hexacianógeno hexacyanogen

hexacloro hexachloro

hexacloroacetona hexachloroacetone

hexaclorobenceno hexachlorobenzene

hexaclorobutadieno hexachlorobutadiene

hexaclorociclohexano
hexachlorocyclohexane

hexaclorociclopentadieno
hexachlorocyclopentadiene

hexacloroestannato amónico ammonium
hexachlorostannate

hexacloroetano hexachloroethane

hexaclorofeno hexachlorophene

hexaclorometilcarbonato
hexachloromethylcarbonate

hexacloronaftaleno
hexachloronaphthalene

hexacloroosmiato sódico sodium
hexachloroosmate

hexacloroplatinato hexachloroplatinate

hexacloroplatinato amónico ammonium
hexachloroplatinate

hexacloroplatinato bárico barium
hexachloroplatinate

hexacloroplatinato de cesio cesium
hexachloroplatinate

hexacloroplatinato de litio lithium
hexachloroplatinate

hexacloroplatinato de rubidio rubidium
hexachloroplatinate

hexacloroplatinato potásico potassium
hexachloroplatinate

hexacloroplatinato sódico sodium
hexachloroplatinate

hexacloropropano hexachloropropane

hexacloropropileno hexachloropropylene

hexacloruro de benceno benzene
hexachloride

hexacloruro de carbono carbon
hexachloride

hexacloruro de tungsteno tungsten
hexachloride

hexacontano hexacontane

hexacosano hexacosane

hexacosanol hexacosanol

hexacosilo hexacosyl

hexadecanal hexadecanal

hexadecano hexadecane

hexadecanoato hexadecanoate

hexadecanoilo hexadecanoyl

hexadecanol hexadecanol

hexadeceno hexadecene

hexadecenolida hexadecenolide

hexadecilmercaptano hexadecyl
mercaptan

hexadecilo hexadecyl

hexadeciltriclorosilano
hexadecyltrichlorosilane

hexadecino hexadecyne

hexadieno hexadiene

hexadiíno hexadiyne

hexafenildisilano hexaphenyldisilane

hexafeniletano hexaphenylethane

hexafluoroacetona hexafluoroacetone

hexafluoroaluminato amónico
ammonium hexafluoroaluminate
hexafluoroaluminato sódico sodium
hexafluoroaluminate
hexafluorobenceno hexafluorobenzene
hexafluorodisilano hexafluorodisilane
hexafluoroetano hexafluoroethane
hexafluorofosfato amónico ammonium
hexafluorophosphate
hexafluorofosfato potásico potassium
hexafluorophosphate
hexafluorogermantano bárico barium
hexafluorogermantane
hexafluoropropeno hexafluoropropene
hexafluoropropileno
hexafluoropropylene
hexafluorosilicato hexafluorosilicate
hexafluorosilicato amónico ammonium
hexafluorosilicate
hexafluorosilicato cálcico calcium
hexafluorosilicate
hexafluorosilicato cobaltoso cobaltous
hexafluorosilicate
hexafluorosilicato cúprico cupric
hexafluorosilicate
hexafluorosilicato cuproso cuprous
hexafluorosilicate
hexafluorosilicato de cesio cesium
hexafluorosilicate
hexafluorosilicato de cinc zinc
hexafluorosilicate
hexafluorosilicato de estaño tin
hexafluorosilicate
hexafluorosilicato de estroncio
strontium hexafluorosilicate
hexafluorosilicato de hidrógeno
hydrogen hexafluorosilicate
hexafluorosilicato de hierro iron
hexafluorosilicate
hexafluorosilicato de litio lithium
hexafluorosilicate
hexafluorosilicato de manganeso
manganese hexafluorosilicate
hexafluorosilicato de mercurio mercury
hexafluorosilicate
hexafluorosilicato de níquel nickel
hexafluorosilicate
hexafluorosilicato de plata silver
hexafluorosilicate

hexafluorosilicato de plomo lead
hexafluorosilicate
hexafluorosilicato de rubidio rubidium
hexafluorosilicate
hexafluorosilicato de talio thallium
hexafluorosilicate
hexafluorosilicato magnésico
magnesium hexafluorosilicate
hexafluorosilicato sódico sodium
hexafluorosilicate
hexafluoruro de azufre sulfur
hexafluoride
hexafluoruro de telurio tellurium
hexafluoride
hexafluoruro de uranio uranium
hexafluoride
hexafluorurosilicato de cadmio
cadmium hexafluorosilicate
hexaglicerol hexaglycerol
hexagonita hexagonite
hexágono hexagon
hexahelicina hexahelicine
hexahidrato hexahydrate
hexahidrato de piperazina piperazine
hexahydrate
hexahídrico hexahydric
hexahidro hexahydro
hexahidroanilina hexahydroaniline
hexahidroantraceno
hexahydroanthracene
hexahidrobenceno hexahydrobenzene
hexahidrocresol hexahydrocresol
hexahidrofenol hexahydrophenol
hexahidrometilo hexahydromethyl
hexahidropiridina hexahydropyridine
hexahidrotolueno hexahydrotoluene
hexahidroxi hexahydroxy
hexahidroxiciclohexano
hexahydroxycyclohexane
hexahidroxileno hexahydroxylene
hexaldehído hexaldehyde
hexalina hexalin
hexametafosfato hexametaphosphate
hexametafosfato sódico sodium
hexametaphosphate
hexametilbenceno hexamethylbenzene
hexametildisilano hexamethyldisilane
hexametildisilazano
hexamethyldisilazane

hexametilendiamina hexamethylenediamine
hexametilenglicol hexamethylene glycol
hexametilenimina hexamethyleneimine
hexametileno hexamethylene
hexametilentetramina hexamethylenetetramine
hexametilmelamina hexamethylmelamine
hexamina hexamine
hexanafteno hexanaphthene
hexanal hexanal
hexanitrato de manitol mannitol hexanitrate
hexanitro hexanitro
hexanitrocobaltato potásico potassium hexanitrocobaltate
hexanitrocobaltato sódico sodium hexanitrocobaltate
hexanitruro bárico barium hexanitride
hexano hexane
hexanoato hexanoate
hexanoato de etilo ethyl hexanoate
hexanoato de propenilo propenyl hexanoate
hexanodiamida hexanediamide
hexanodinitrilo hexanedinitrile
hexanodioato hexanedioate
hexanodiol hexanediol
hexanodiona hexanedione
hexanoilo hexanoyl
hexanol hexanol
hexanolactama hexanelactam
hexanona hexanone
hexanotriol hexanetriol
hexaprismo hexaprismo
hexatriacontano hexatriacontane
hexayodo hexaiodo
hexayoduro de azufre sulfur hexaiodide
hexenal hexenal
hexénico hexenic
hexenilo hexenyl
hexeno hexene
hexenol hexenol
hexestrol hexestrol
hexetidina hexetidine
hexilamina hexylamine
hexilenglicol hexylene glycol
hexileno hexylene

hexilfenol hexylphenol
hexilmercaptano hexyl mercaptan
hexilo hexyl
hexilresorcina hexylresorcinol
hexiltriclorosilano hexyltrichlorosilane
hexino hexyne
hexinol hexynol
hexobarbital hexobarbital
hexobarbitona hexobarbitone
hexoico hexoic
hexona hexone
hexoquinasa hexokinase
hexosa hexose
hexotriosa hexotriose
hialofano hyalophane
hialosiderita hyalosiderite
hialuronidasa hyaluronidase
hibridización hybridization
hibridización de enlace bond hybridization
hidantoína hydantoin
hiddenita hiddenite
hidracilo hydrazyl
hidramina hydramine
hidrargillita hydrargillite
hidrasa hydrase
hidrastina hydrastine
hidratación hydration
hidratado hydrated
hidrato hydrate
hidrato alumínico aluminum hydrate
hidrato amónico ammonium hydrate
hidrato bárico barium hydrate
hidrato cálcico calcium hydrate
hidrato de amileno amylene hydrate
hidrato de amilo amyl hydrate
hidrato de berilio beryllium hydrate
hidrato de bismuto bismuth hydrate
hidrato de carbono carbohydrate
hidrato de cerio cerium hydrate
hidrato de cesio cesium hydrate
hidrato de cloral chloral hydrate
hidrato de cloro chlorine hydrate
hidrato de cobalto cobalt hydrate
hidrato de cromo chromium hydrate
hidrato de diamida diamide hydrate
hidrato de etano ethane hydrate
hidrato de etileno ethylene hydrate
hidrato de fenilo phenyl hydrate

hidrato de gas gas hydrate
hidrato de hidrazina hydrazine hydrate
hidrato de manganeso manganese
    hydrate
hidrato de pentilo pentyl hydrate
hidrato de propano propane hydrate
hidrato de terpina terpin hydrate
hidrato férrico ferric hydrate
hidrato potásico potassium hydrate
hidrato sódico sodium hydrate
hidrazi hydrazi
hidrazida hydrazide
hidrazida maleica maleic hydrazide
hidrazidina hydrazidine
hidrazido hydrazido
hidrazina hydrazine
hidrazina de etanol ethanol hydrazine
hidrazina de metilo methyl hydrazine
hidrazinio hydrazinium
hidrazino hydrazino
hidrazinobenceno hydrazinobenzene
hidrazo hydrazo
hidrazobenceno hydrazobenzene
hidrazona hydrazone
hidrazono hydrazono
hidrazotolueno hydrazotoluene
hidrdato de butilcloral butyl chloral
    hydrate
hidrido hydrido
hidrina hydrin
hidro hydro
hidrobenzoína hydrobenzoin
hidrobiotita hydrobiotite
hidroboración hydroboration
hidrocarbonilo de cobalto cobalt
    hydrocarbonyl
hidrocarburo hydrocarbon
hidrocarburo alifático aliphatic
    hydrocarbon
hidrocarburo aromático aromatic
    hydrocarbon
hidrocarburo cíclico cyclic hydrocarbon
hidrocarburo clorado chlorinated
    hydrocarbon
hidrocarburo de acetileno acetylene
    hydrocarbon
hidrocarburo de cadena recta straight-
    chain hydrocarbon
hidrocarburo insaturado unsaturated

hydrocarbon
hidrocarburo ligero light hydrocarbon
hidrocarburo normal normal
    hydrocarbon
hidrocarburo policíclico polycyclic
    hydrocarbon
hidrocarburo policíclico alifático
    aliphatic polycyclic hydrocarbon
hidrocarburo polinuclear polynuclear
    hydrocarbon
hidrocarburo saturado saturated
    hydrocarbon
hidrocelulosa hydrocellulose
hidrocianuro hydrocyanide
hidrocincita hydrozincite
hidrocoloide hydrocolloid
hidrocortisona hydrocortisone
hidrocraqueo hydrocracking
hidrodesalquilación hydrodealkylation
hidrodestilación hydrodistillation
hidroextractor hydroextractor
hidrofílico hydrophilic
hidrofilita hydrophilite
hidrófilo hydrophile
hidroflumetiazida hydroflumethiazide
hidrofóbico hydrophobic
hidroformación hydroforming
hidroformilación hydroformylation
hidrofuramida hydrofuramide
hidrogasificación hydrogasification
hidrogel hydrogel
hidrogenación hydrogenation
hidrogenado hydrogenated
hidrogenar hydrogenate
hidrógeno hydrogen
hidrógeno acídico acidic hydrogen
hidrógeno atómico atomic hydrogen
hidrógeno ligero light hydrogen
hidrógeno nasciente nascent hydrogen
hidrógeno pesado heavy hydrogen
hidrógenofosfato disódico disodium
    hydrogenphosphate
hidrogenólisis hydrogenolysis
hidrogenosufuro de indio indium
    hydrogensulfide
hidrogenosulfato de cesio cesium
    hydrogensulfate
hidrohematita hydrohematite
hidrol hydrol

hidrolasa hydrolase
hidrolicuefacción hydroliquefaction
hidrólisis hydrolysis
hidrólisis de Donnan Donnan hydrolysis
hidrólisis por membrana membrane hydrolysis
hidrolítico hydrolytic
hidrólito hydrolyte
hidrolizado hydrolized
hidrolizar hydrolyze
hidromagnesita hydromagnesite
hidrómetro hydrometer
hidrona hydrone
hidroperóxido hydroperoxide
hidroperóxido de cumeno cumene hydroperoxide
hidropolisulfuro hydropolysulfide
hidropónica hydroponics
hidroquinol hydroquinol
hidroquinona hydroquinone
hidrorrefinación hydrofining
hidroseleno hydroseleno
hidrosol hydrosol
hidrosoluble water soluble
hidrosolvatación hydrosolvation
hidrostático hydrostatic
hidrosulfito hydrosulfite
hidrosulfito de cinc zinc hydrosulfite
hidrosulfito-formaldehído hydrosulfite-formaldehyde
hidrosulfito sódico sodium hydrosulfite
hidrosulfuro hydrosulfide
hidrosulfuro amónico ammonium hydrosulfide
hidrosulfuro bárico barium hydrosulfide
hidrosulfuro cálcico calcium hydrosulfide
hidrosulfuro de etilo ethyl hydrosulfide
hidrosulfuro potásico potassium hydrosulfide
hidrosulfuro sódico sodium hydrosulfide
hidrotetrazona hydrotetrazone
hidrótropo hydrotrope
hidroxi hydroxy
hidroxiacetal hydroxyacetal
hidroxiacetanilida hydroxyacetanilide
hidroxiacetofenona hydroxyacetophenone
hidroxiacetona hydroxyacetone

hidroxiadipaldehído hydroxyadipaldehyde
hidroxialanina hydroxyalanine
hidroxiamino hydroxyamino
hidroxianilina hydroxyaniline
hidroxianisol hydroxyanisole
hidroxianisol butilado butylated hydroxyanisole
hidroxiantraceno hydroxyanthracene
hidroxiapatita hydroxyapatite
hidroxibenceno hydroxybenzene
hidroxibenzaldehído hydroxybenzaldehyde
hidroxibenzamida hydroxybenzamide
hidroxibenzofenona hydroxybenzophenone
hidroxibutanal hydroxybutanal
hidroxibutiraldehído hydroxybutyraldehyde
hidroxibutiranilida hydroxybutyranilide
hidroxicanfano hydroxycamphane
hidroxicaroteno hydroxycarotene
hidroxicerusita hydroxycerussite
hidroxicitronelal hydroxycitronellal
hidroxicloruro de zirconilo zirconyl hyroxychloride
hidroxicobalamina hydroxycobalamin
hidroxicolestano hydroxycholestane
hidroxicolina hydroxycholine
hidroxicorticosterona hydroxycorticosterone
hidroxidifenilamina hydroxydiphenylamine
hidroxidifenilo hydroxydiphenyl
hidroxidimetilbenceno hydroxydimethylbenzene
hidróxido hydroxide
hidróxido alumínico aluminum hydroxide
hidróxido amónico ammonium hydroxide
hidróxido anfótero amphoteric hydroxide
hidróxido áurico auric hydroxide
hidróxido bárico barium hydroxide
hidróxido cálcico calcium hydroxide
hidróxido cérico ceric hydroxide
hidróxido ceroso cerous hydroxide
hidróxido cobáltico cobaltic hydroxide

hidróxido cobaltoso cobaltous hydroxide
hidróxido crómico chromic hydroxide
hidróxido cromoso chromous hydroxide
hidróxido cúprico cupric hydroxide
hidróxido cuproso cuprous hydroxide
hidróxido de bencenodiazonio
  benzenediazonium hydroxide
hidróxido de berilio beryllium hydroxide
hidróxido de bismutilo bismuthyl
  hydroxide
hidróxido de bismuto bismuth hydroxide
hidróxido de boro boron hydroxide
hidróxido de cadmio cadmium
  hydroxide
hidróxido de cesio cesium hydroxide
hidróxido de cinc zinc hydroxide
hidróxido de cobalto cobalt hydroxide
hidróxido de cobre copper hydroxide
hidróxido de cromo chromium
  hydroxide
hidróxido de diazobenceno
  diazobenzene hydroxide
hidróxido de escandio scandium
  hydroxide
hidróxido de estaño tin hydroxide
hidróxido de estroncio strontium
  hydroxide
hidróxido de fenilo phenyl hydroxide
hidróxido de fosfonio phosphonium
  hydroxide
hidróxido de gadolinio gadolinium
  hydroxide
hidróxido de galio gallium hydroxide
hidróxido de germanio germanium
  hydroxide
hidróxido de hafnio hafnium hydroxide
hidróxido de hierro iron hydroxide
hidróxido de indio indium hydroxide
hidróxido de itrio yttrium hydroxide
hidróxido de litio lithium hydroxide
hidróxido de manganeso manganese
  hydroxide
hidróxido de molibdeno molybdenum
  hydroxide
hidróxido de neodimio neodymium
  hydroxide
hidróxido de oro gold hydroxide
hidróxido de paladio palladium
  hydroxide

hidróxido de plomo lead hydroxide
hidróxido de rodio rhodium hydroxide
hidróxido de rubidio rubidium
  hydroxide
hidróxido de rutenio ruthenium
  hydroxide
hidróxido de talio thallium hydroxide
hidróxido de telurio tellurium hydroxide
hidróxido de terbio terbium hydroxide
hidróxido de tetraetanolamonio
  tetraethanolammonium hydroxide
hidróxido de titanio titanium hydroxide
hidróxido de torio thorium hydroxide
hidróxido de trifenilestaño triphenyltin
  hydroxide
hidróxido de uranilo uranyl hydroxide
hidróxido de uranio uranium hydroxide
hidróxido de vanadio vanadium
  hydroxide
hidróxido de zirconilo zirconyl
  hydroxide
hidróxido de zirconio zirconium
  hydroxide
hidróxido estánnico stannic hydroxide
hidróxido estannoso stannous hydroxide
hidroxido fenilmercúrico
  phenylmercuric hydroxide
hidróxido férrico ferric hydroxide
hidróxido ferroso ferrous hydroxide
hidróxido gálico gallic hydroxide
hidróxido magnésico magnesium
  hydroxide
hidróxido mangánico manganic
  hydroxide
hidróxido manganoso manganous
  hydroxide
hidróxido mercúrico mercuric hydroxide
hidróxido niquélico nickelic hydroxide
hidróxido niqueloso nickelous hydroxide
hidróxido potásico potassium hydroxide
hidróxido samárico samaric hydroxide
hidróxido sódico sodium hydroxide
hidróxido talioso thallous hydroxide
hidróxido titánico titanic hydroxide
hidróxido vanadioso vanadous hydroxide
hidroxietilacetamida
  hydroxyethylacetamide
hidroxietilamina hydroxyethylamine
hidroxietilcelulosa hydroxyethylcellulose

hidroxietilcelulosa de etilo ethyl hydroxyethylcellulose

hidroxietiletilendiamina hydroxyethylethylenediamine

hidroxietilhidrazina hydroxyethylhydrazine

hidroxifencano hydroxyfenchane

hidroxifenilalanina hydroxyphenylalanine

hidroxifenilo hydroxyphenyl

hidroxifenol hydroxyphenol

hidroxihidrazida hydroxyhydrazide

hidroxiimino hydroxyimino

hidroxilación hydroxylation

hidroxilación de Woodward Woodward hydroxylation

hidroxilamina hydroxylamine

hidroxilo hydroxyl

hidroximercuricresol hydroxymercuricresol

hidroximesitileno hydroxymesitylene

hidroximetano hydroxymethane

hidroximetileno hydroxymethylene

hidroximetilfuraldehído hydroxymethylfuraldehyde

hidroximetilo hydroxymethyl

hidroxinaftaleno hydroxynaphthalene

hidroxinaftoquinona hydroxynaphthoquinone

hidroxipiperidina hydroxypiperidine

hidroxiprogesterona hydroxyprogesterone

hidroxiprolina hydroxyproline

hidroxipropanona hydroxypropanone

hidroxipropilamina hydroxypropylamine

hidroxipropilcelulosa hydroxypropyl cellulose

hidroxipropilmetilcelulosa hydroxypropyl methylcellulose

hidroxipropionitrilo hydroxypropionitrile

hidroxiquinolina hydroxyquinoline

hidroxitolueno hydroxytoluene

hidroxitolueno butilado butylated hydroxytoluene

hidroxitriacontano hydroxytriacontane

hidroxitriptamina hydroxytryptamine

hidroxo hydroxo

hidroxonio hydroxonium

hidruro hydride

hidruro alumínico aluminum hydride

hidruro alumínico cálcico calcium aluminum hydride

hidruro alumínico de litio lithium aluminum hydride

hidruro alumínico sódico sodium aluminum hydride

hidruro antimonioso antimonous hydride

hidruro arsenioso arsenous hydride

hidruro bárico barium hydride

hidruro cálcico calcium hydride

hidruro cálcico alumínico aluminum calcium hydride

hidruro de amilo amyl hydride

hidruro de antimonio antimony hydride

hidruro de arsénico arsenic hydride

hidruro de boro boron hydride

hidruro de bromo bromine hydride

hidruro de carbonilo carbonyl hydride

hidruro de cinamilo cinnamyl hydride

hidruro de decilo decyl hydride

hidruro de estaño tin hydride

hidruro de etileno ethylene hydride

hidruro de flúor fluorine hydride

hidruro de formacilo formazyl hydride

hidruro de formoxilo formoxyl hydride

hidruro de fósforo phosphorus hydride

hidruro de galio gallium hydride

hidruro de germanio germanium hydride

hidruro de litio lithium hydride

hidruro de litio alumínico aluminum lithium hydride

hidruro de metilo methyl hydride

hidruro de nitrógeno nitrogen hydride

hidruro de nonilo nonyl hydride

hidruro de paladio palladium hydride

hidruro de pentilo pentyl hydride

hidruro de rubidio rubidium hydride

hidruro de silicio silicon hydride

hidruro de titanio titanium hydride

hidruro de uranio uranium hydride

hidruro de zirconio zirconium hydride

hidruro magnésico magnesium hydride

hidruro potásico potassium hydride

hidruro sódico sodium hydride

hielmita hielmite

hielo ice

hielo seco dry ice
hierro iron
hierro amoniacal ammoniated iron
hierro beta beta iron
hierro de lingote ingot iron
hierro delta delta iron
hierro electrolítico electrolytic iron
hierro en lingotes pig iron
hierro fundido cast iron
hierro galvanizado galvanized iron
hierro gamma gamma iron
hierro pasivo passive iron
higrométrico hygrometric
higrómetro hygrometer
higromicina hygromycin
higroscópico hygroscopic
hiosciamina hyoscyamine
hioscina hyoscine
hiper hyper
hiperconjugación hyperconjugation
hiperconjugación isovalente isovalent
    hyperconjugation
hipercrómico hyperchromic
hiperón hyperon
hiperóxido hyperoxide
hipersorción hypersorption
hipertónico hypertonic
hipo hypo
hipoalergénico hypoallergenic
hipobromato hypobromate
hipobromito hypobromite
hipobromito sódico sodium hypobromite
hipoclorito hypochlorite
hipoclorito cálcico calcium hypochlorite
hipoclorito de litio lithium hypochlorite
hipoclorito de plata silver hypochlorite
hipoclorito potásico potassium
    hypochlorite
hipoclorito sódico sodium hypochlorite
hipocrómico hypochromic
hipofamina hypophamine
hipofluorita hypofluorite
hipofosfato hypophosphate
hipofosfato amónico ammonium
    hypophosphate
hipofosfato bárico barium
    hypophosphate
hipofosfato de plata silver
    hypophosphate

hipofosfito hypophosphite
hipofosfito bárico barium hypophosphite
hipofosfito cálcico calcium
    hypophosphite
hipofosfito de cinc zinc hypophosphite
hipofosfito de manganeso manganese
    hypophosphite
hipofosfito férrico ferric hypophosphite
hipofosfito ferroso ferrous
    hypophosphite
hipofosfito potásico potassium
    hypophosphite
hipofosfito sódico sodium hypophosphite
hiponitrilo hyponitrile
hiponitrito sódico sodium hyponitrite
hiposulfato hyposulfate
hiposulfato bárico barium hyposulfate
hiposulfato ferroso ferrous hyposulfate
hiposulfato potásico potassium
    hyposulfate
hiposulfato sódico sodium hyposulfate
hiposulfito hyposulfite
hiposulfito bárico barium hyposulfite
hiposulfito de estroncio strontium
    hyposulfite
hiposulfito de plomo lead hyposulfite
hiposulfito de plomo sódico sodium lead
    hyposulfite
hiposulfito magnésico magnesium
    hyposulfite
hiposulfito potásico potassium
    hyposulfite
hiposulfito sódico sodium hyposulfite
hipótesis hypothesis
hipótesis de Avogadro Avogadro
    hypothesis
hipotónico hypotonic
hipoxantina hypoxanthine
hipoyodato hypoiodate
hipurato hippurate
hipurato potásico potassium hippurate
hipuroilo hippuroyl
histamina histamine
histaminasa histaminase
histéresis hysteresis
histidina histidine
histona histone
histoquímica histochemistry
hogbomita hogbomite

hoja leaf, foil
hoja de oro gold foil, gold leaf
hoja metálica metal foil
hojalata tin, tinplate
hojuela foil
holmio holmium
holocelulosa holocellulose
holografía holography
holohedral holohedral
hollín soot
homatropina homatropine
homilita homilite
homo homo
homocéntrico homocentric
homocíclico homocyclic
homocrómico homochromic
homogéneo homogeneous
homogenización homogenization
homoisohídrico homoisohydric
homoléptico homoleptic
homólisis homolysis
homomorfo homomorph
homonuclear homonuclear
homopolar homopolar
homopolímero homopolymer
homosalato homosalate
homótopo homotope
hongo fungus
hopcalita hopcalite
hormona hormone

hormona corticoide corticoid hormone
hormona tirotrópica thyrotropic hormone
hornablenda hornblende
horno oven, furnace, kiln
horno de combustión combustion furnace
horno de coque coke oven
horno de Fletcher Fletcher furnace
horno eléctrico electric furnace
horno solar solar furnace
horsfordita horsfordite
howlita howlite
hubnerita hübnerite
hueco hole
hulla coal
humeante fuming
humectante humectant
humedad humidity, moisture
humedad absoluta absolute humidity
humedad atmosférica atmospheric humidity
humedad crítica critical humidity
humedad relativa relative humidity
humero flue
humidificación humidification
humidímetro humidity meter
humo smoke, fume
humo químico chemical smoke
humus humus

I

ibogaína ibogaine
ibuprofeno ibuprofen
ico ic
icosa icosa
icosaedro icosahedron
icosano icosane
icosanol icosanol
icosilo icosyl
ictamol ichthammol
ictiocola ichthyocolla
ictiolato ichthyolate
ideno idene
idino idyne
idiocromático idiochromatic
ido ide
idosa idose
ignición ignition
ignición espontánea spontaneous ignition
ignífugo flame-resistant
ilideno ylidene
ilidino ylidyne
ilinio illinium
ilmenita ilmenite
ilo yl
iluro ylide
ilvaíta ilvaite
imagen de espejo mirror-image
imagen electrónica electron image
imazalil imazalil
imbibición imbibition
imida imide
imida de alcanfor camphor imide
imidazol imidazole
imidazolidona imidazolidone
imidazolilo imidazolyl
imidazolio imidazolium
imidazoltriona imidazoletrione
imidina imidine
imido imido
imina imine
iminio iminium
imino imino
iminoácido imino acid

iminobispropilamina
    iminobispropylamine
iminodiacetonitrilo iminodiacetonitrile
iminourea iminourea
imodoilo imidoyl
impacto impact
impacto electrónico electron impact
impalpable impalpable
impedimento estérico steric hindrance
impermeable waterproof
implosión implosion
impregnación impregnation, permeation
impregnar impregnate
impulso pulse
impulsor impeller
impureza impurity
in vitro in vitro
in vivo in vivo
ina ine
inactivo inactive
incandescencia incandescence
incandescente incandescent
incendiario incendiary
incidencia incidence
incineración incineration, ashing
incombustible flameproof
incompatibilidad química chemical
    incompatibility
incompatible incompatible
incompresible incompressible
incrustación de caldera boiler
    incrustation
incubación incubation
indamina indamine
indandiona indandione
indanilo indanyl
indano indan
indantreno indanthrene
indantrona indanthrone
indazol indazole
indenilo indenyl
indeno indene
indenona indenone

414

independencia de carga charge independence
indeterminado indeterminate
indicador indicator
indicador ácido-base acid-base indicator
indicador acromático achromatic indicator
indicador de absorción absorption indicator
indicador de adsorción adsorption indicator
indicador de color color indicator
indicador de Degener Degener indicator
indicador de fenantrolina phenanthroline indicator
indicador de fluorescencia fluorescence indicator
indicador de Formanek Formanek indicator
indicador de humedad humidity indicator
indicador de ion hidrógeno hydrogen-ion indicator
indicador de oxidación-reducción oxidation-reduction indicator
indicador de pH pH indicator
indicador de precipitación precipitation indicator
indicador de reacción reaction indicator
indicador de Sörensen Sörensen indicator
indicador isotópico isotopic indicator
indicador metalocrómico metallochromic indicator
indicador mixto mixed indicator
indicador químico chemical indicator
indicador radiactivo radioactive indicator, tracer
indicador redox redox indicator
indicador universal universal indicator
indicador visual display
indicadores de Clark-Lubs Clark-Lubs indicators
indicán indican
índice index
índice amortiguador buffer index
índice de actividad óptica optical-activity index
índice de Bellier Bellier index

índice de excitación excitation index
índice de octano octane rating, octane number
índice de refracción refractive index
índice de retención retention index
índice de viscosidad viscosity index
índice térmico heat index
índigo indigo
indigotina indigotin
indilo indyl
indio indium
indofenol indophenol
indógeno indogen
indogenuro indogenide
indol indole
indolilo indolyl
indolizina indolizine
indolol indolol
indometacina indomethacin
indoxilo indoxyl
inducción induction
inducción asimétrica asymmetric induction
inducción fotoquímica photochemical induction
inducido por neutrones neutron-induced
inducido por radiación radiation-induced
indurita indurite
industrial industrial
inelástico inelastic
inerte inert
inestable unstable
inflamabilidad inflammability, flammability
inflamable inflammable, flammable
inflamador inflamer
inflamar inflame
infrarrojo infrared
infrarrojo con transformación de Fourier Fourier transform infrared
infrarrojo de cromatografía de gases gas chromatography infrared
infusión infusion
ingeniería de seguridad safety engineering
ingeniería química chemical engineering
inhibición inhibition
inhibición no competitiva

noncompetitive inhibition
inhibidor inhibitor
inhibidor de colinesterasa cholinesterase
  inhibitor
inhibidor de reacción reaction inhibitor
inhibidor químico chemical inhibitor
iniciador initiator
inicio de cadena chain initiation
injerto grafting
inmersión immersion
inmiscible immiscible
inmunoglobulina immunoglobulin
inmunoquímica immunochemistry
ino yne, ino
inorgánico inorganic
inosamina inosamine
inosina inosine
inosinato sódico sodium inosinate
inositol inositol
insaturación unsaturation
insaturado unsaturated
insecticida insecticide
insolubilidad insolubility
insoluble insoluble
instrumentación instrumentation
instrumento instrument
instrumento espectrométrico
  spectrometric instrument
insulina insulin
intensidad intensity, strength
intensidad de campo field intensity
intensidad de olor odor intensity
intensidad de radiación radiation
  intensity
intensidad térmica thermal intensity
interacción de Gamow-Teller Gamow-
  Teller interaction
interacción espín-espín spin-spin
  interaction
interacción espín-orbital spin-orbit
  interaction
interacción nuclear nuclear interaction
interatómico interatomic
intercambiador térmico heat exchanger
intercambio exchange
intercambio de cationes cation exchange
intercambio electrónico electron
  exchange
intercambio iónico ion exchange

intercambio químico chemical exchange
intercambio térmico heat exchange
interdifusión interdiffusion
interelectródico interelectrode
interfacial interfacial
interfase interface
interferencia interference
interferón interferon
interhalógeno interhalogen
interiónico interionic
intermetálico intermetallic
intermolecular intermolecular
interno internal
intersticial interstitial
intersticio interstice
intervalo amortiguador buffer range
intervalo de indicador indicator range
intervalo de puntos de congelación
  freezing-point range
intervalo de puntos de ebullición
  boiling-point range
intervalo de transición transition
  interval
intoxicación intoxication
intraanular intraannular
intraatómico intraatomic
intramolecular intramolecular
intranuclear intranuclear
intrínseco intrinsic
introfacción introfaction
intrón intron
inulina inulin
inulinasa inulinase
inundación química chemical flooding
invención invention
inversión inversion, reversal
inversión de fase phase reversal
inversión de Walden Walden inversion
invertasa invertase
investigación research
investigación aplicada applied research
investigación fundamental fundamental
  research
investigación química chemical research
investigación y desarrollo research and
  development
iolita iolite
ion ion
ion amónico ammonium ion

ion carbenio carbenium ion
ion carbonio carbonium ion
ion cloronio chloronium ion
ion complejo complex ion
ion dipolar dipolar ion
ion excitado excited ion
ion-gramo gram-ion
ion hidrógeno hydrogen ion
ion hidronio hydronium ion
ion hidroxilo hydroxyl ion
ion libre free ion
ion metaestable metastable ion
ion negativo negative ion
ion oxonio oxonium ion
ion positivo positive ion
ion primario primary ion
iónico ionic
ionización ionization
ionización de campo field ionization
ionización de gas gas ionization
ionización de Penning Penning
    ionization
ionización electrolítica electrolytic
    ionization
ionización por colisión collision
    ionization
ionización por radiación radiation
    ionization
ionización primaria primary ionization
ionización proporcional proportional
    ionization
ionización térmica thermal ionization
ionizado ionized
ionizante ionizing
ionizar ionize
ionogénico ionogenic
ionógeno ionogen
ionografía ionography
ionol ionol
ionómero ionomer
ionona ionone
iotalamato sódico sodium iothalamate
iperita yperite
iprodiona iprodione
iridio iridium
irido irido
iridosmina iridosmine
irona irone
irradiación irradiation

irradiante irradiating
irreversible irreversible
isatina isatin
isazol isazol
iso iso
isoamildicloroarsina
    isoamyldichloroarsine
isoamileno isoamylene
isoamilmercaptano isoamyl mercaptan
isoamilo isoamyl
isoascorbato sódico sodium isoascorbate
isóbaro isobar
isóbaro estable stable isobar
isoborneol isoborneol
isobornilo isobornyl
isobutano isobutane
isobutanol isobutanol
isobutanolamina isobutanolamine
isobuteno isobutene
isobutilamina isobutylamine
isobutilaminobenzoato
    isobutylaminobenzoate
isobutilbenceno isobutylbenzene
isobutilcarbinol isobutyl carbinol
isobutileno isobutylene
isobutilmercaptano isobutyl mercaptan
isobutilo isobutyl
isobutilxantato sódico sodium
    isobutylxanthate
isobutiraldehído isobutyraldehyde
isobutirato cálcico calcium isobutyrate
isobutirato de etilo ethyl isobutyrate
isobutirato de fenetilo phenethyl
    isobutyrate
isobutirato de isobutilo isobutyl
    isobutyrate
isobutirato de linalilo linalyl isobutyrate
isobutirato de tolilo tolyl isobutyrate
isobutírico isobutyric
isobutirilo isobutyryl
isobutiroilo isobutyroyl
isobutironitrilo isobutyronitrile
isobutoxi isobutoxy
isocianato isocyanate
isocianato de alilo allyl isocyanate
isocianato de ciclohexilo cyclohexyl
    isocyanate
isocianato de clorofenilo chlorophenyl
    isocyanate

isocianato de etilo ethyl isocyanate
isocianato de fenilo phenyl isocyanate
isocianato de metilo methyl isocyanate
isocianato de octadecilo octadecyl isocyanate
isociano isocyano
isocianurato isocyanurate
isocianuro isocyanide
isocianuro de etilo ethyl isocyanide
isocianuro de fenilo phenyl isocyanide
isocianuro de pentilo pentyl isocyanide
isocíclico isocyclic
isócora isochore
isócora de van't Hoff van't Hoff isochore
isodecaldehído isodecaldehyde
isodecano isodecane
isodecanol isodecanol
isodecilo isodecyl
isodinámico isodynamic
isodipersión isodispersion
isodrina isodrin
isodureno isodurene
isoeléctrico isoelectric
isoelectrónico isoelectronic
isoenzima isoenzyme
isoestructural isostructural
isoeugenol isoeugenol
isoforona isophorone
isoftalato de dialilo diallyl isophthalate
isoftalato de etilo ethyl isophthalate
isoheptano isoheptane
isohexano isohexane
isoléptico isoleptic
isoleucina isoleucine
isólogo isolog
isomerasa isomerase
isomerasa de glucosa glucose isomerase
isomérico isomeric
isomerismo isomerism
isomerismo cis-trans cis-trans isomerism
isomerismo de anillo ring isomerism
isomerismo de enlace doble double-bond isomerism
isomerismo de saturación saturation isomerism
isomerismo dinámico dynamic isomerism
isomerismo estructural structural isomerism

isomerismo geométrico geometrical isomerism
isomerismo nuclear nuclear isomerism
isomerismo por coordinación coordination isomerism
isomerización isomerization
isomerización de cadena chain isomerization
isomerización de valencia valence isomerization
isómero isomer
isómero cis cis-isomer
isómero geométrico geometric isomer
isómero nuclear nuclear isomer
isómero óptico optical isomer
isómero trans trans-isomer
isométrico isometric
isomolécula isomolecule
isomórfico isomorphic
isomorfismo isomorphism
isomorfo isomorph
isonicotinato de metilo methyl isonicotinate
isonitrilo isonitrile
isonitro isonitro
isonitroso isonitroso
isonitrosocetona isonitrosoketone
isooctano isooctane
isoocteno isooctene
isoparafina isoparaffin
isopentaldehído isopentaldehyde
isopentano isopentane
isopoliácido isopolyacid
isopoliéster isopolyester
isopolimolibdato isopolymolybdate
isopolimorfismo isopolymorphism
isopolitungstato isopolytungstate
isopreno isoprene
isoprenoide isoprenoid
isopropanol isopropanol
isopropanolamina isopropanolamine
isopropilacetona isopropylacetone
isopropilamina isopropylamine
isopropilaminoetanol isopropylaminoethanol
isopropilanilina isopropylaniline
isopropilato alumínico aluminum isopropylate

isopropilato de titanio titanium
isopropylate
isopropilbenceno isopropylbcnzene
isopropilbencilo isopropylbenzyl
isopropilcarbinol isopropylcarbinol
isopropilcresol isopropylcresol
isopropiletileno isopropylethylene
isopropilfenol isopropylphenol
isopropilideno isopropylidene
isopropilmercaptano isopropyl
mercaptan
isopropilmetilbenceno
isopropylmethylbenzene
isopropilnaftaleno isopropylnaphthalene
isopropilo isopropyl
isopropiltolueno isopropyltoluene
isopropóxido alumínico aluminum
isopropoxide
isopropoxietanol isopropoxyethanol
isopulegol isopulegol
isoquinolina isoquinoline
isosafrol isosafrole
isostérico isosteric
isosterismo isosterism
isosulfocianato de alilo allyl
isosulfocyanate
isotáctico isotactic
isoterma isotherm
isoterma de adsorción de Gibbs Gibbs
adsorption isotherm
isoterma de adsorción de Langmuir
Langmuir adsorption isotherm
isoterma de Freundlich Freundlich
isotherm
isoterma de van't Hoff van't Hoff
isotherm

isotiocianato isothiocyanate
isotiocianato de etoxicarbonilo
ethoxycarbonyl isothiocyanate
isotiocianato de fenilo phenyl
isothiocyanate
isotiocianato de metilo methyl
isothiocyanate
isotiocianato de propenilo propenyl
isothiocyanate
isotiocinato de alilo allyl isothiocyanate
isotónico isotonic
isótono isotone
isotópico isotopic
isótopo isotope
isótopo estable stable isotope
isótopo inestable unstable isotope
isótopo pesado heavy isotope
isótopo radiactivo radioactive isotope
isotrón isotron
isotrópico isotropic
isovalente isovalent
isovaleraldehído isovaleraldehyde
isovalerato de bornilo bornyl isovalerate
isovalerato de etilo ethyl isovalerate
isovalerato de isoamilo isoamyl
isovalerate
isovalerato sódico sodium isovalerate
isoxazolilo isoxazolyl
iterbia ytterbia
iterbio ytterbium
itria yttria
itrialita yttrialite
itrio yttrium
itrocerita yttrocerite
itrotantalita yttrotantalite
iuro ide

J

jabón soap
jabón de alquitrán tar soap
jabón metálico metallic soap
jaborina jaborine
jade jade
jadeíta jadeite
jalea jelly
jalea mineral mineral jelly
jarabe syrup

jarabe de almidón starch syrup
jarabe de cacao cacao syrup
jarabe de maíz corn syrup
jardín químico chemical garden
jaspe jasper
jazmona jasmone
johimbina johimbine
joseíta joseite
jugo gástrico gastric juice

K

kanamicina kanamycin
kappa kappa
kernita kernite
kerógeno kerogen
kieselguhr kieselguhr
kieserita kieserite
kilogramo kilogram

kimberlita kimberlite
kleinita kleinite
koppita koppite
kovar kovar
kremersita kremersite
krugita krügite
kunzita kunzite

L

lábil labile
laboratorio laboratory
laboratorio de investigación research
  laboratory
labradorita labradorite
laca lacquer, lake
laca amarilla yellow lake
laca de tungsteno tungsten lake
lacasa laccase
lacatato de plata silver lactate
lacmoide lacmoid
lacrimógeno lachrymatory
lactalbúmina lactalbumin
lactama lactam
lactamida lactamide
lactasa lactase
lactato lactate
lactato alumínico aluminum lactate
lactato amónico ammonium lactate
lactato bárico barium lactate
lactato cálcico calcium lactate
lactato ceroso cerous lactate
lactato de antimonio antimony lactate
lactato de butilo butyl lactate
lactato de cadmio cadmium lactate
lactato de cinc zinc lactate
lactato de cobre copper lactate
lactato de estroncio strontium lactate
lactato de etilo ethyl lactate
lactato de litio lithium lactate
lactato de manganeso manganese lactate
lactato de metilo methyl lactate
lactato de plomo lead lactate
lactato de zirconio zirconium lactate
lactato de zirconio sódico sodium
  zirconium lactate
lactato fenilmercúrico phenylmercuric
  lactate
lactato ferroso ferrous lactate
lactato magnésico magnesium lactate
lactato mercúrico mercuric lactate
lactato sódico sodium lactate
lactida lactide

lactilo lactyl
lactima lactim
lactobiosa lactobiose
lactogénico lactogenic
lactoglobulina lactoglobulin
lactoilo lactoyl
lactol lactol
lactolida lactolide
lactona lactone
lactona beta beta lactone
lactona de angélica angelica lactone
lactona de nonilo nonyl lactone
lactonitrilo lactonitrile
lactopreno lactoprene
lactosa lactose
lagoriolita lagoriolite
lambda lambda
lambert lambert
lámina beta beta sheet
laminado laminated
laminarina laminarin
lámpara de halógeno halogen lamp
lámpara de neón neon lamp
lámpara de sodio sodium lamp
lana wool
lana de vidrio glass wool
lanarkita lanarkite
landsbergita landsbergite
langbeinita langbeinite
lanolina lanolin
lanosterol lanosterol
lansfordita lansfordite
lantana lanthana
lantánido lanthanide
lantano lanthanum
lantionina lanthionine
lapislázuli lapis lazuli
larvicida larvicide
láser laser
laterita laterite
látex latex
latón brass
latón amarillo yellow brass

latón de aluminio aluminum brass
latón duro hard brass
latón rojo red brass
laudanidina laudanidine
laudanina laudanine
láudano laudanum
laudanosina laudanosine
laumonita laumonite
lauraldehído lauraldehyde
laurato amónico ammonium laurate
laurato de butoxietilo butoxyethyl laurate
laurato de cinc zinc laurate
laurato de diglicol diglycol laurate
laurato de isocetilo isocetyl laurate
laurato de metilo methyl laurate
laurato de plata silver laurate
laurato de plomo lead laurate
laurato de tetrahidrofurfurilo tetrahydrofurfuryl laurate
laurato magnésico magnesium laurate
laurato potásico potassium laurate
laurencio lawrencium
laurilacetato lauryl acetate
laurilbromuro lauryl bromide
laurilcloruro lauryl chloride
laurilmercaptano lauryl mercaptan
laurilmetacrilato lauryl methacrylate
laurilo lauryl
laurilsulfato lauryl sulfate
laurilsulfato de trietanolamina triethanolamine lauryl sulfate
laurilsulfato magnésico magnesium lauryl sulfate
laurilsulfato sódico sodium lauryl sulfate
laurita laurite
lauroilo lauroyl
lauroilsarcosina lauroylsarcosine
lauroleato de metilo methyl lauroleate
laurona laurone
lautal lautal
lavado washing
lavado cáustico caustic wash
lavador de ojos eye-wash
lazulita lazulite
lazurita lazurite
lecitina lecithin
lectina lectin
lechada de cal whitewash

leche de magnesia milk of magnesia
lejía lye
lenacil lenacil
leno lene
leonardita leonardite
leonita leonite
lepidina lepidine
lepidolita lepidolite
lepidona lepidone
leptón lepton
letal lethal
leucilo leucyl
leucina leucine
leucita leucite
leuco leuco
leucoanilina leucoaniline
leucolina leucoline
leuconita leukonite
leucovorina leucovorin
levadura yeast
levadura de cerveza brewer's yeast
levadura de tórula torula yeast
levo levo
levorrotatorio levorotatory
levulosa levulose
lewisita lewisite
ley cero de termodinámica zeroth law of thermodynamics
ley de acción de masas mass action law
ley de alternación alternation law
ley de Avogadro Avogadro law
ley de Babo Babo law
ley de Beer Beer law
ley de Beer-Lambert Beer-Lambert law
ley de Blagden Blagden law
ley de Boltzmann Boltzmann law
ley de Bouguer-Lambert Bouguer-Lambert law
ley de Boyle Boyle law
ley de Bragg Bragg law
ley de Bunsen-Kirchhoff Bunsen-Kirchhoff law
ley de Clausius Clausius law
ley de Clausius-Mosotti Clausius-Mosotti law
ley de Coehn Coehn law
ley de composición definida definite composition law
ley de conservación conservation law

ley de conservación de masa mass conservation law

ley de Coppet Coppet law

ley de Curie Curie law

ley de Curie-Weiss Curie-Weiss law

ley de Charles Charles law

ley de Dalton Dalton law

ley de desplazamiento displacement law

ley de desplazamiento de Soddy Soddy displacement law

ley de desplazamiento radiactivo radioactive displacement law

ley de Despretz Despretz law

ley de difusión diffusion law

ley de difusión de Fick Fick diffusion law

ley de dilución dilution law

ley de dilución de Ostwald Ostwald dilution law

ley de distribución distribution law

ley de Draper Draper law

ley de Einstein Einstein law

ley de equilibrio equilibrium law

ley de equivalente fotoquímico photochemical equivalent law

ley de esterificación esterification law

ley de Fajans-Soddy Fajans-Soddy law

ley de Faraday Faraday law

ley de Gay-Lussac Gay-Lussac law

ley de Goldschmidt Goldschmidt law

ley de Graham Graham law

ley de Grottius-Draper Grottius-Draper law

ley de Henry Henry law

ley de Hess Hess law

ley de Joule Joule law

ley de Kopp Kopp law

ley de Lambert Lambert law

ley de Lambert-Beer Lambert-Beer law

ley de Mendeleev Mendeleev law

ley de Meyer Meyer law

ley de Moseley Moseley law

ley de Nernst Nernst law

ley de Oudeman Oudeman law

ley de partición partition law

ley de Pascal Pascal law

ley de proporciones definidas definite proportions law

ley de proporciones múltiples multiple proportions law

ley de proporciones recíprocas reciprocal proportions law

ley de Ramsay-Young Ramsay-Young law

ley de Raoult Raoult law

ley de reacción reaction law

ley de Retger Retger law

ley de Richter Richter law

ley de Sievert Sievert law

ley de Stark-Einstein Stark-Einstein law

ley de Stokes Stokes law

ley de termodinámica thermodynamics law

ley de van't Hoff van't Hoff law

ley de Wullner Wullner law

ley periódica periodic law

leyes de gases gas laws

leyes de termodinámica laws of thermodynamics

leyes químicas chemical laws

liberación liberation, release

licopeno lycopene

licopodio lycopodium

licor liquor

licuación liquation

licuado liquefied

licuefacción liquefaction

lidocaína lidocaine

ligador binder

ligadura binding

ligando ligand

ligando cuadridentado quadridentate ligand

ligasa ligase

lignina lignin

lignita lignite

lignocaína lignocaine

lignocerato de metilo methyl lignocerate

lignosulfonato lignosulfonate

ligroína ligroin

límite de absorción absorption limit

límite de concentración concentration limit

límite de convergencia convergence limit

límite de identificación identification limit

límite de inflamabilidad flammability

limit
limoneno limonene
limonita limonite
limpieza de gas gas cleanup
limpieza electrolítica electrolytic
cleaning
linalilo linalyl
linalol linalool
linamarina linamarin
linarita linarite
lincomicina lincomycin
lindano lindane
lindgrenita lindgrenite
línea de absorción absorption line
línea de Becke Becke line
línea de hidrógeno hydrogen line
línea de isoactividad isoactivity line
línea de Rayleigh Rayleigh line
línea de Stokes Stokes line
línea espectral spectral line, spectrum
line
línea prohibida forbidden line
línea realzada enhanced line
lineal linear
líneas de Balmer Balmer lines
líneas de emisión emission lines
líneas de Fraunhofer Fraunhofer lines
líneas de Raman Raman lines
líneas de Schmidt Schmidt lines
linnaeíta linnaeite
linoleato linoleate
linoleato amónico ammonium linoleate
linoleato cobaltoso cobaltous linoleate
linoleato de cinc zinc linoleate
linoleato de cobalto cobalt linoleate
linoleato de manganeso manganese
linoleate
linoleato de metilo methyl linoleate
linoleato de plomo lead linoleate
linoleato potásico potassium linoleate
linolenato de metilo methyl linolenate
liofílico lyophilic
liofilización lyophilization, freeze drying
liofóbico lyophobic
liogel lyogel
lionita lionite
liotópico lyotopic
liótropo lyotrope
lipasa lipase

lípido lipid
lipofílico lipophilic
lipólisis lipolysis
lipoproteína lipoprotein
lipoxidasa lipoxidase
lipoxigenasa lipoxygenase
liquen islándico Iceland moss
líquido liquid
líquido polar polar liquid
líquido saturado saturated liquid
líquido subenfriado supercooled liquid
líquidos asociados associated liquids
liroconita liroconite
lisidina lysidine
lisina lysine
lisis lysis
lisozima lysozyme
litarge litharge
litidionita litidionite
litina lithia
litio lithium
litro liter, litre
livingstonita livingstoneite
lixiviación lixiviation, leaching
lixosa lyxose
lobelina lobeline
localización localization
loción blanca white lotion
lodo anódico anodic mud, anodic slime
longitud de onda de de Broglie de
Broglie wavelength
lorandita lorandite
lorazepam lorazepam
lote matriz master batch
lubricación lubrication
lubricante lubricant
luciferina luciferin
ludwigita ludwigite
lugar activo active site
lugar de unión binding site
luminancia luminance
luminífero luminiferous
luminiscencia luminescence, glow
luminiscencia de resonancia resonance
luminescence
luminiscencia persistente afterglow
luminiscencia residual afterglow
luminiscente luminescent
luminol luminol

luminometría luminometry
luminosidad luminosity
luminoso luminous
lupinidina lupinidine
lúpulo hops
lutecia lutetia
lutecio lutetium
luteína lutein
luteotropina luteotropin

lutidina lutidine
luz coherente coherent light
luz fría cold light
luz no polarizada unpolarized light
luz polarizada polarized light
luz polarizada en un plano plane-
   polarized light
luz química chemical light
luz ultravioleta ultraviolet light

LL

llama flame
llama de Bunsen Bunsen flame
llama oxiacetilénica oxyacetylene flame
llama oxidante oxidizing flame

llama reductora reducing flame
lluvia ácida acid rain
lluvia amarilla yellow rain

M

maclado twinning
macro macro
macroanálisis macroanalysis
macrocíclico macrocyclic
macrólido macrolide
macromolécula macromolecule
macromolecular macromolecular
macroquímica macrochemistry
macrosa macrose
macroscópico macroscopic
maduración ripening
maduración de Ostwald Ostwald
    ripening
magenta magenta
magnalio magnalium
magnalita magnalite
magnesia magnesia
magnesio magnesium
magnesioferrita magnesioferrite
magnesita magnesite
magnético magnetic
magnetismo magnetism
magnetita magnetite
magnetización magnetization
magnetohidrodinámica
    magnetohydrodynamics
magnetón magneton
magnetón de Bohr Bohr magneton
magnetón nuclear nuclear magneton
magnetoquímica magnetochemistry
magnolio magnolium
magnolita magnolite
malamida malamide
malaquita malachite
malatión malathion
malato malate
malato bárico barium malate
malato cálcico calcium malate
malato de cinc zinc malate
malato de dietilo diethyl malate
malato de etilo ethyl malate
malato de plomo lead malate
maldonita maldonite

maleabilidad malleability
maleable malleable
maleato maleate
maleato de dialilo diallyl maleate
maleato de dibutilo dibutyl maleate
maleato de dietilo diethyl maleate
maleato de feniramina pheniramine
    maleate
maleato de plomo lead maleate
malenoide malenoid
maleoilo maleoyl
maloilo maloyl
malonamida malonamide
malonato malonate
malonato bárico barium malonate
malonato de dimetilo dimethyl malonate
malonato de etilo ethyl malonate
malonilo malonyl
malonilurea malonyl urea
malononitrilo malononitrile
malquita malchite
malta malt
maltasa maltase
malteno malthene
maltol maltol
maltosa maltose
mallardita mallardite
maná manna
mandelonitrilo mandelonitrile
maneb maneb
manejo de materiales materials handling
manejo remoto remote handling
manganato manganate
manganato bárico barium manganate
manganato potásico potassium
    manganate
manganato sódico sodium manganate
manganeso manganese
manganeso-boro manganese-boron
manganeso rojo red manganese
manganeso-titanio manganese-titanium
mangánico manganic
manganina manganin

manganita manganite
manganoestilbita manganostilbite
manganosita manganosite
manganoso manganous
manita mannite
manitol mannitol
manitosa mannitose
mano de mortero pestle
manómetro manometer
manosa mannose
manteca lard
manteca de cacao cocoa butter, cacao butter
mantequilla butter
maquinaria de laboratorio laboratory machinery
marcado tagged, labeled
marcador tag, label
marcasita marcasite
margarina margarine
margarita margarite
marialita marialite
mariguana marijuana
mármol marble
martensita martensite
martita martite
martonita martonite
masa mass
masa activa active mass
masa atómica atomic mass
masa atómica relativa relative atomic mass
masa crítica critical mass
masa de electrón electron mass
masa de partícula particle mass
masa en reposo de electrón electron rest mass
masa en reposo de neutrón neutron rest mass
masa isotópica isotopic mass
masa mesónica meson mass
masa molecular molecular mass
masa molecular relativa relative molecular mass
masa nuclear nuclear mass
masa reducida reduced mass
mascagnita mascagnite
máser maser
masicote massicot

masilla putty
masticación mastication
mata matte
materia matter
materia colorante dyestuff
materia prima raw material
material material
material a prueba de fuego fireproof material
material ablativo ablative material
material antideflagrante flameproof material
material combustible combustible material
material corrosivo corrosive material
material de acabado finishing material
material de blanqueo bleaching material
material fértil fertile material
material incombustible flameproof material
material inflamable flammable material
material no combustible noncombustible material
material oxidante oxidizing material
material peligroso hazardous material
material pirofórico pyrophoric material
material radiactivo radioactive material
material reductor reducing material
materialización materialization
matildita matildite
matlockita matlockite
matraz flask
matraz de base plana flat-bottom flask
matraz de Claisen Claisen flask
matraz de cuello largo long-necked flask
matraz de destilación distillation flask
matraz de Erlenmeyer Erlenmeyer flask
matraz de filtración filter flask
matraz de Giles Giles flask
matraz de Kjeldahl Kjeldahl flask
matraz de medida measuring flask
matraz de tres cuellos three-necked flask
matraz de Wurtz Wurtz flask
matraz volumétrico volumetric flask
matriz matrix
maxivalencia maxivalence
meaformaldehído metaformaldehyde
mecánica cuántica quantum mechanics
mecánica estadística statistical

mechanics
mecanismo mechanism
mecanismo de reacción reaction mechanism
mecanismo de reacción orgánica organic reaction mechanism
mecanismo en cadena chain mechanism
mecanoquímica mechanochemistry
meconina meconin
mechero burner
mechero de Boyce Boyce burner
mechero de Bunsen Bunsen burner
mechero de Teclu Teclu burner
media reacción half-reaction
media vida half-life
media vida radiactiva radioactive half-life
medida measurement
medida cuantitativa quantitative measurement
medida electroquímica electrochemical measurement
medidor meter, gage
medidor de ion hidrógeno hydrogen-ion meter
medidor de pH pH meter
medio medium
medio de dispersión dispersion medium
medio para filtración filter medium
megnesilo magnesyl
meimacita meymacite
melamina melamine
melanilina melaniline
melanina melanin
melaza molasses
melfalano melphalan
melibiosa melibiose
melitato mellitate
melonita melonite
membrana membrane
membrana parcialmente permeable partially permeable membrane
membrana permeable permeable membrane
membrana semipermeable semipermeable membrane
mena ore
menacanita menachanite
menadiona menadione

mendelevio mendelevium
mendocita mendozite
meneghinita meneghinite
menisco meniscus
mensuración mensuration
menta verde spearmint
mentadieno menthadiene
mentalcanfor menthacamphor
mentano menthane
mentanodiamina menthanediamine
mentanol menthanol
mentanona menthanone
mentenilo menthenyl
menteno menthene
mentenol menthenol
mentilo menthyl
mentol menthol
mentona menthone
mentonafteno menthonaphthene
meprobamato meprobamate
mercaptal mercaptal
mercaptamina mercaptamine
mercaptano mercaptan
mercapto mercapto
mercaptoacetato sódico sodium mercaptoacetate
mercaptobenzotiazol mercaptobenzothiazole
mercaptobenzotiazol sódico sodium mercaptobenzothiazole
mercaptoetanol mercaptoethanol
mercaptofenilo mercaptophenyl
mercaptol mercaptol
mercaptotiazolina mercaptothiazoline
mercuración mercuration
mercurial mercurial
mercúrico mercuric
mercurificación mercurification
mercurio mercury
mercurio amoniacal ammoniated mercury
mercurioso mercurous
mercurización mercurization
mercururo mercuride
meridional meridional
merwinita merwinite
mescalina mescaline
mesilato de benztropina benztropine mesylate

mesilo mesyl
mesitileno mesitylene
mesitilo mesityl
mesitita mesitite
mesitol mesitol
meso meso
mesocoloide mesocolloid
mesoiónico mesoionic
mesomería mesomerism
mesomérico mesomeric
mesómero mesomer
mesomórfico mesomorphic
mesón meson
mesónico mesonic
mesoscópico mesoscopic
mesotomía mesotomy
mesotorio mesothorium
mesoxalilo mesoxalyl
mesoxalo mesoxalo
meta meta
metaarsenito de cinc zinc metaarsenite
metaarsenito potásico potassium
     metaarsenite
metabisulfito potásico potassium
     metabisulfite
metabisulfito sódico sodium
     metabisulfite
metabólico metabolic
metabolismo metabolism
metabolito metabolite
metaborato metaborate
metaborato de cobre copper metaborate
metaborato de litio lithium metaborate
metaborato sódico sodium metaborate
metacetina methacetin
metacrilamida methacrylamide
metacrilato methacrylate
metacrilato de estearilo stearyl
     methacrylate
metacrilato de etilo ethyl methacrylate
metacrilato de hexilo hexyl methacrylate
metacrilato de hidroxietilo hydroxyethyl
     methacrylate
metacrilato de isobutilo isobutyl
     methacrylate
metacrilato de metilo methyl
     methacrylate
metacrilato sódico sodium methacrylate
metacrilonitrilo methacrylonitrile

metacroleína methacrolein
metacromasia metachromasia
metacromatismo metachromatism
metadona methadone
metaestable metastable
metaestructura metastructure
metafenileno metaphenylene
metafosfato metaphosphate
metafosfato alumínico aluminum
     metaphosphate
metafosfato amónico ammonium
     metaphosphate
metafosfato bárico barium
     metaphosphate
metafosfato cálcico calcium
     metaphosphate
metafosfato de berilio beryllium
     metaphosphate
metafosfato potásico potassium
     metaphosphate
metafosfato sódico sodium
     metaphosphate
metal metal
metal alcalino alkali metal
metal alcalinotérreo alkaline-earth metal
metal blanco white metal
metal coloidal colloidal metal
metal de Aich Aich metal
metal de Babbitt Babbitt metal
metal de base base metal
metal de Borcher Borcher metal
metal de Devarda Devarda metal
metal de Frary Frary metal
metal de Lewis Lewis metal
metal de tierras raras rare-earth metal
metal ligero light metal
metal Muntz Muntz metal
metal nativo native metal
metal noble noble metal
metal pasivo passive metal
metal pesado heavy metal
metal precioso precious metal
metal radiactivo radioactive metal
metal raro rare metal
metal rojo red metal
metalación metalation
metaldehído metaldehyde
metálico metallic
metalificar metallify

metalilo methallyl
metalización metallization, metallizing, plating
metalización al vacío vacuum metalizing
metalización electrolítica electrolytic metallization
metalización galvánica galvanic metallization
metalizado metallized, metal-coated, plated
metalizar metallize
metaloceno metallocene
metalocrómico metallochromic
metalografía metallography
metaloide metalloid
metalorgánico metalorganic
metalurgia metallurgy
metamerismo metamerism
metámero metamer
metanación methanation
metanal methanal
metanamida methanamide
metanilato sódico sodium metanilate
metanloato sódico sodium methanolate
metano methane
metanoato methanoate
metanoato de etilo ethyl methanoate
metanoato sódico sodium methanoate
metanoilo methanoyl
metanol methanol
metanolato methanolate
metanosulfonato methanesulfonate
metanotiol methanethiol
metaperyodato sódico sodium metaperiodate
metasilicato bárico barium metasilicate
metasilicato cálcico calcium metasilicate
metasilicato de litio lithium metasilicate
metasilicato de plomo lead metasilicate
metasilicato sódico sodium metasilicate
metasomatosis metasomatosis
metástasis metastasis
metavanadato amónico ammonium metavanadate
metavanadato sódico sodium metavanadate
metenamina methenamine
metenilo methenyl
meteno methene

meteórico meteoric
metepa metepa
meticilina methicillin
metida methide
metilacetileno methylacetylene
metilacetofenona methylacetophenone
metilacetona methyl acetone
metilación de Haworth Haworth methylation
metilación de Purdie Purdie methylation
metilacrilamida methylacrylamide
metilado methylated
metilal methylal
metilalanina methylalanine
metilaluminio methylaluminum
metilamilcetona methylamyl ketone
metilamilo methylamyl
metilamina methylamine
metilamino methylamino
metilaminofenol methylaminophenol
metilanilina methylaniline
metilanisol methylanisole
metilantraceno methylanthracene
metilantranilato methylanthranilate
metilantraquinona methylanthraquinone
metilar methylate
metilato de litio lithium methylate
metilato magnésico magnesium methylate
metilato sódico sodium methylate
metilbenceno methylbenzene
metilbencilamina methylbenzylamine
metilbencilo methylbenzyl
metilbutano methylbutane
metilbutanol methylbutanol
metilbuteno methylbutene
metilbutilamina methylbutylamine
metilbutilcetona methylbutyl ketone
metilbutinol methylbutynol
metilcelulosa methylcellulose
metilciclohexanilo methylcyclohexanyl
metilciclohexano methylcyclohexane
metilciclohexanol methylcyclohexanol
metilciclohexanona methylcyclohexanone
metilciclohexilamina methylcyclohexylamine
metilciclopentano methylcyclopentane
metilcolantreno methylcholanthrene

metildicloroarsina methyldichloroarsine
metildiclorosilano methyldichlorosilane
metildietanolamina
    methyldiethanolamine
metildifenilamina methyldiphenylamine
metildioxolano methyldioxolane
metildipropilmetano
    methyldipropylmethane
metileno methylene
metilenociclopentadieno
    methylenecyclopentadiene
metilestireno methylstyrene
metiletilcelulosa methylethylcellulose
metiletilcetona methylethyl ketone
metilfenilcetona methylphenyl ketone
metilfenildiclorosilano
    methylphenyldichlorosilane
metilformanilida methylformanilide
metilfurano methylfuran
metilfurfurilamina methylfurfurylamine
metilglicol methyl glycol
metilheptano methylheptane
metilheptenona methylheptenone
metilheptilamina methylheptylamine
metilhexano methylhexane
metilhexilcetona methylhexyl ketone
metilhidrazina methylhydrazine
metilhidrazona methylhydrazone
metilhidroxibutanona
    methylhydroxybutanone
metílico methylic
metilidino methylidyne
metilindol methylindole
metilisoamilcetona methylisoamyl
    ketone
metilisobutilcarbinol methylisobutyl
    carbinol
metilisobutilcetona methylisobutyl
    ketone
metilmetano methylmethane
metilnaftaleno methylnaphthalene
metilnitrobenceno methylnitrobenzene
metilo methyl
metilol methylol
metilolurea methylolurea
metilparabeno methylparaben
metilpentadieno methylpentadiene
metilpentaldehído methylpentaldehyde
metilpentano methylpentane

metilpentanol methylpentanol
metilpenteno methylpentene
metilpentosa methylpentose
metilpiperazina methylpiperazine
metilpirrol methylpyrrole
metilpirrolidina methylpyrrolidine
metilpropano methylpropane
metilpropeno methylpropene
metilpropilcetona methylpropyl ketone
metilquinolina methylquinoline
metilsulfato bárico barium methylsulfate
metilsulfonilo methylsulfonyl
metiltaurina methyltaurine
metiltriclorosilano methyltrichlorosilane
metiltrinitrobenceno
    methyltrinitrobenzene
metilundecanoato methylundecanoate
metina methine
metino methyne
metiodal sódico sodium methiodal
metionilo methionyl
metionina methionine
método científico scientific method
método conductométrico conductometric
    method
método de acetilación de Barnett
    Barnett acetylation method
método de Boeseken Boeseken method
método de Born-Oppenheimer Born-
    Oppenheimer method
método de Bouveault-Blanc Bouveault-
    Blanc method
método de Bragg Bragg method
método de Carius Carius method
método de Craig Craig method
método de Debye-Scherrer Debye-
    Scherrer method
método de Dumas Dumas method
método de Frankland Frankland method
método de Freund Freund method
método de Helferich Helferich method
método de Kjeldahl Kjeldahl method
método de Kolbe Kolbe method
método de Körner Körner method
método de Mohr Mohr method
método de Rast Rast method
método de Regnault Regnault method
método de Sonn-Muller Sonn-Muller
    method

método de Van Slyke Van Slyke method
método de Volhard Volhard method
método de Ziegler Ziegler method
método gasométrico gasometric method
metomilo methomyl
metosa methose
metotrexato methotrexate
metoxalilo methoxalyl
metoxi methoxy
metoxiacetaldehído methoxyacetaldehyde
metoxiacetanilida methoxyacetanilide
metoxiacetofenona methoxyacetophenone
metoxiamina methoxyamine
metoxibenceno methoxybenzene
metoxibenzaldehído methoxybenzaldehyde
metoxibutanol methoxybutanol
metóxido methoxide
metóxido de benciltrimetilamonio benzyltrimethylammonium methoxide
metóxido de litio lithium methoxide
metóxido magnésico magnesium methoxide
metóxido potásico potassium methoxide
metóxido sódico sodium methoxide
metoxietilo methoxyethyl
metoxifenilo methoxyphenyl
metoxifenol methoxyphenol
metoxilo methoxyl
metoxinaftaleno methoxynaphthalene
metoxipropanol methoxypropanol
metoxipropilamina methoxypropylamine
metro meter, metre
mezcla mixture, blend
mezcla azeotrópica azeotropic mixture
mezcla binaria binary mixture
mezcla de ebullición constante constant-boiling mixture
mezcla de Eschka Eschka mixture
mezcla distéctica dystectic mixture
mezcla efervescente effervescent mixture
mezcla estequiométrica stoichiometric mixture
mezcla eutéctica eutectic mixture
mezcla frigorífica frigorific mixture
mezcla gaseosa gaseous mixture
mezcla racémica racemic mixture

mezcla rica rich mixture
mezclador de Banbury Banbury mixer
miargirita miargyrite
miazina miazine
mica mica, isinglass
micela micelle
micotoxina mycotoxin
micrilo micril
micro micro
microanálisis microanalysis
microanálisis con rayos X X-ray microanalysis
microanálisis cuantitativo quantitative microanalysis
microanalizador microanalyzer
microanalizador electrónico electron microanalyzer
microbalanza microbalance
microbicida microbicide
microbio microbe
microcápsula microcapsule
microcristal microcrystal
microcristalino microcrystalline
microcuerpo microbody
microcurie microcurie
microdensitómetro microdensitometer
microdifusión microdiffusion
microelectrodo microelectrode
microelectroforesis microelectrophoresis
microelectrólisis microelectrolysis
microelemento trace element
microencapsulación microencapsulation
microespectrofotómetro microspectrophotometer
microespectroscopio microspectroscope
microfotómetro microphotometer
microgalvanómetro microgalvanometer
microgamilo microgammil
micrografía micrography
microgramo microgram
microincineración microincineration
microlina microline
micrómetro micrometer
micrón micron
micronutriente micronutrient
microonda microwave
microorganismo microorganism
micropipeta micropipet
microporoso microporous

microquímica microchemistry
microquímico microchemical
microrreacción microreaction
microscopia microscopy
microscopia de fluorescencia
    fluorescence microscopy
microscopia electrónica electron
    microscopy
microscopia electrónica de barrido
    scanning electron microscopy
microscopia química chemical
    microscopy
microscópico microscopic
microscopio microscope
microscopio de emisión de campo field-
    emission microscope
microscopio de fuerza atómica atomic
    force microscope
microscopio de infrarrojo infrared
    microscope
microscopio electrónico electron
    microscope
microscopio electrónico de barrido
    scanning electron microscope
microscopio iónico de campo field-ion
    microscope
microscopio óptico optical microscope
microscopio protónico proton
    microscope
microsegundo microsecond
microsonda microprobe
microsonda electrónica electron
    microprobe
miel honey
miemita miemite
migración migration
migración atómica atomic migration
migración de plata silver migration
migración electroquímica
    electrochemical migration
migración iónica ion migration
mili milli
milibarra millibar
milicurie millicurie
miliequivalente milliequivalent
miligramo milligram
mililitro milliliter
milímetro millimeter
milímetro de mercurio millimeter of

mercury
milimol millimole
milinormal millinormal
mimetita mimetite
mineral mineral, ore
mineral anaranjado orange mineral
mineral de carbonato carbonate mineral
mineral de hierro iron ore
mineral de plata silver ore
mineral de sulfato sulfate mineral
mineral de titanio titanium ore
mineral de uranio uranium ore
mineral radiactivo radioactive mineral
mineralización mineralization
mineralogía mineralogy
minio minium
minivalencia minivalence
minulita minulite
mioglobina myoglobin
mioquinasa myokinase
miosina myosin
mirabilita mirabilite
airceno myrcene
miricilo myricyl
miristato de butilo butyl myristate
miristato de etilo ethyl myristate
miristato de isocetilo isocetyl myristate
miristato de litio lithium myristate
miristato de metilo methyl myristate
miristato de plata silver myristate
miristato magnésico magnesium
    myristate
miristoilo myristoyl
mirra myrrh
miscibilidad miscibility
miscible miscible
mischmetal mischmetal
mispiquel mispickel
mitomicina mitomycin
mixina myxin
mixita mixite
mixo mixo
moción electrónica electron motion
modelo model
modelo atómico atomic model
modelo de Frenkel-Kontorowa Frenkel-
    Kontorowa model
modelo de Gouy-Chapman Gouy-
    Chapman model

modelo de Helmholtz Helmholtz model
modelo de Ising Ising model
modelo de Thomas-Fermi Thomas-Fermi model
modelo vectorial de estructura atómica vector model of atomic structure
moderador moderator
modificación modification
modificación de Eschweiler-Clarke Eschweiler-Clarke modification
modificación química chemical modification
modo mode
modo de desintegración decay mode
modo de desintegración nuclear nuclear-decay mode
módulo de elasticidad modulus of elasticity
moho mildew, mold, rust
mol mole, mol
molal molal
molalidad molality
molar molar
molaridad molarity
molécula molecule
molécula activada activated molecule
molécula compuesta compound molecule
molécula de van der Waals van der Waals molecule
molécula diatómica diatomic molecule
molécula elemental elementary molecule
molécula-gramo gram-molecule
molécula heteronuclear heteronuclear molecule
molécula homonuclear homonuclear molecule
molécula libre free molecule
molécula lineal linear molecule
molécula marcada tagged molecule
molécula monoatómica monoatomic molecule
molécula neutra neutral molecule
molécula no lineal nonlinear molecule
molécula no polar nonpolar molecule
molécula polar polar molecule
molécula poliatómica polyatomic molecule
molécula triatómica triatomic molecule
molecular molecular

molecularidad molecularity
moler grind
molibdato molybdate
molibdato amónico ammonium molybdate
molibdato bárico barium molybdate
molibdato cálcico calcium molybdate
molibdato de cinc zinc molybdate
molibdato de cobalto cobalt molybdate
molibdato de cobre copper molybdate
molibdato de estroncio strontium molybdate
molibdato de litio lithium molybdate
molibdato de plomo lead molybdate
molibdato magnésico magnesium molybdate
molibdato potásico potassium molybdate
molibdato sódico sodium molybdate
molibdenilo molybdenyl
molibdenita molybdenite
molibdeno molybdenum
molíbdico molybdic
molibdita molybdite
molibdofosfato sódico sodium molybdophosphate
molibdosilicato sódico sodium molybdosilicate
molido grinding
molienda química chemical milling
molino mill
molino coloidal colloid mill
molino de bolas ball mill
molisita molysite
molybdato de cadmio cadmium molybdate
momento angular nuclear nuclear angular momentum
momento angular orbital orbital angular momentum
momento cuadripolar nuclear nuclear quadripole moment
momento dipolar dipole moment
momento dipolar electrónico electron dipole moment
momento magnético magnetic moment
momento magnético atómico atomic magnetic moment
momento magnético de espín spin magnetic moment

momento magnético electrónico electron magnetic moment

momento magnético neutrónico neutron magnetic moment

momento magnético nuclear nuclear magnetic moment

momento nuclear nuclear moment

momento orbital orbital moment

monacita monazite

mónada monad

monetita monetite

monístico monistic

mono mono

monoacetato monoacetate

monoacetato de celulosa cellulose monoacetate

monoacetato de etilenglicol ethylene glycol monoacetate

monoacetato de glicerilo glyceryl monoacetate

monoacetato de glicerol glycerol monoacetate

monoacetato de glicol glycol monoacetate

monoacetato de resorcina resorcinol monoacetate

monoacetina monoacetin

monoácido monoacid

monoamina monoamine

monoaminoácido monoamino acid

monoamonio monoammonium

monoatómico monoatomic

monobásico monobasic

monobenzoato de resorcina resorcinol monobenzoate

monobromado monobromated

monobromo monobromo

monobromuro de antipireno antipyrene monobromide

monobromuro de hidrazina hydrazine monobromide

monobromuro de oro gold monobromide

monobromuro de selenio selenium monobromide

monobromuro de yodo iodine monobromide

monobromuro sulfúrico sulfuric monobromide

monocapa monolayer

monocarburo de uranio uranium monocarbide

monocloro monochloro

monocloroacetona monochloroacetone

monoclorobenceno monochlorobenzene

monocloroetano monochloroethane

monoclorofenol monochlorophenol

monoclorohidrina de glicerol glycerol monochlorohydrin

monoclorometano monochloromethane

monocloruro de azufre sulfur monochloride

monocloruro de hidrazina hydrazine monochloride

monocloruro de oro gold monochloride

monocloruro de selenio selenium monochloride

monocloruro de yodo iodine monochloride

monocloruro sulfúrico sulfuric monochloride

monocromático monochromatic

monodisperso monodisperse

monoestearato alumínico aluminum monostearate

monoestearato de glicerol glycerol monostearate

monoestearato de polioxietileno polyoxyethylene monostearate

monoestearato de propilenglicol propylene glycol monostearate

monoestearato de sucrosa sucrose monostearate

monoestearato de tetraetilenglicol tetraethylene glycol monostearate

monoéster monoester

monoetanolamina monoethanolamine

monoetilamina monoethylamine

monofluoruro de azufre sulfur monofluoride

monofluoruro de cloro chlorine monofluoride

monofosfato monophosphate

monofosfato bárico barium monophosphate

monofosfato de adenosina adenosine monophosphate

monofosfato de adenosina cíclico cyclic

adenosine monophosphate

monofosfato de guanosina guanosine monophosphate

monofosfato de uridina uridine monophosphate

monofosfato potásico potassium monophosphate

monofuncional monofunctional

monoglicérido monoglyceride

monoglime monoglyme

monohidrato monohydrate

monohidrato bárico barium monohydrate

monohidrato de carbonato sódico sodium carbonate monohydrate

monohidrato de perborato sódico sodium perborate monohydrate

monohidrato de sulfato de cinc zinc sulfate monohydrate

monohidrato de sulfato sódico sodium sulfate monohydrate

monohídrico monohydric

monolaurato de glicerilo glyceryl monolaurate

monolaurato de glicerol glycerol monolaurate

monómero monomer

monómero de estireno styrene monomer

monometilamina monomethylamine

monométrico monometric

monomolecular monomolecular

monomorfo monomorphous

mononitrilo malónico malonic mononitrile

mononitrorresorcinato de plomo lead mononitroresorcinate

mononuclear mononuclear

monooleato de glicerol glycerol monooleate

monorrefringente monorefringent

monorricinoleato de propilenglicol propylene glycol monoricinoleate

monosacárido monosaccharide

monosulfito de paladio palladium monosulfite

monosulfuro monosulfide

monosulfuro bárico barium monosulfide

monosulfuro de carbono carbon monosulfide

monosulfuro de estaño tin monosulfide

monosulfuro de estroncio strontium monosulfide

monosulfuro de platino platinum monosulfide

monosulfuro de plomo lead monosulfide

monosulfuro potásico potassium monosulfide

monosulfuro sódico sodium monosulfide

monotartrato de etilo ethyl monotartrate

monoterpeno monoterpene

monotropía monotropy

monotrópico monotropic

monovalente monovalent

monoxicloro monoxychlor

monóxido monoxide

monóxido bárico barium monoxide

monóxido de azufre sulfur monoxide

monóxido de carbono carbon monoxide

monóxido de cloro chlorine monoxide

monóxido de cobalto cobalt monoxide

monóxido de cobre copper monoxide

monóxido de dipenteno dipentene monoxide

monóxido de estroncio strontium monoxide

monóxido de galio gallium monoxide

monóxido de germanio germanium monoxide

monóxido de hierro iron monoxide

monóxido de manganeso manganese monoxide

monóxido de nitrógeno nitrogen monoxide

monóxido de osmio osmium monoxide

monóxido de paladio palladium monoxide

monóxido de plomo lead monoxide

monóxido de silicio silicon monoxide

monóxido de talio thallium monoxide

monóxido de telurio tellurium monoxide

monóxido de titanio titanium monoxide

monóxido de vanadio vanadium monoxide

monóxido de vinilciclohexeno vinylcyclohexene monoxide

monóxido sódico sodium monoxide

monóxido talioso thallous monoxide

monoxima de benzoquinona
   benzoquinone monoxime
monoyoduro de oro gold monoiodide
montanita montanite
monticelita monticellite
montmorillonita montmorillonite
montroidita montroydite
mordiente mordant
morenosita morenosite
morfina morphine
morfolina morpholine
morfología morphology
morfosán morphosan
morina morin
mortero mortar
mostasa de nitrógeno nitrogen mustard
movilidad mobility
movimiento browniano Brownian
   motion
mu mu
mucopéptido mucopeptide
mucopolisacárido mucopolysaccharide
muestra sample, sampling

muestreo sampling
multi multi
multiplete multiplet
multiplicación electrónica electron
   multiplication
multiplicación neutrónica neutron
   multiplication
multiplicador electrónico electron
   multiplier
multiplicidad electrónica electron
   multiplicity
multivalente multivalent
mullita mullite
muon muon
murexida murexide
muscarina muscarine
muscovita muscovite
musgo de Irlanda Irish moss
mutamerismo mutamerism
mutámero mutamer
mutarrotación mutarotation
mutasa mutase
mutualidad de fases mutuality of phases

N

nabam nabam
nácar nacre
nacrita nacrite
nadorita nadorite
nafta naphtha
nafta de petróleo petroleum naphtha
naftaceno naphthacene
naftacridina naphthacridine
naftal naphthal
naftaldehído naphthaldehyde
naftalendiamina naphthalenediamine
naftalendiol naphthalenediol
naftalenilo naphthalenyl
naftaleno naphthalene
naftaleno clorado chlorinated
   naphthalene
naftalenosulfonato de plomo lead
   naphthalenesulfonate
naftalenosulfonato sódico sodium
   naphthalenesulfonate
naftalentiol naphthalenethiol
naftalida naphthalide
naftalimido naphthalimido
naftalina naphthalin
naftamida naphthamide
naftamina naphthamine
naftano naphthane
naftecetol naphthacetol
naftenato cálcico calcium naphthenate
naftenato cobaltoso cobaltous
   naphthenate
naftenato de cinc zinc naphthenate
naftenato de cobalto cobalt naphthenate
naftenato de cobre copper naphthenate
naftenato de cromo chromium
   naphthenate
naftenato de manganeso manganese
   naphthenate
naftenato de plomo lead naphthenate
naftenato de zirconio zirconium
   naphthenate
naftenato fenilmercúrico
   phenylmercuric naphthenate

naftenato ferroso ferrous naphthenate
naftenato potásico potassium
   naphthenate
naftenato sódico sodium naphthenate
naftenilo naphthenyl
nafteno naphthene
naftieno naphthieno
naftilamina naphthylamine
naftilbenceno naphthylbenzene
naftilbenzoato naphthylbenzoate
naftilendiamina naphthylenediamine
naftileno naphthylene
naftiletiléter naphthylethyl ether
naftilmercúrico naphthylmercuric
naftilmetanol naphthylmethanol
naftilmetileno naphthylmethylene
naftilmetilo naphthylmethyl
naftilnaftilo naphthylnaphthyl
naftilo naphthyl
naftiloxi naphthyloxy
naftiltiourea naphthylthiourea
naftindeno naphthindene
naftionato sódico sodium naphthionate
naftiridina naphthyridine
nafto naphtho
naftodiantreno naphthodianthrene
naftoilo naphthoyl
naftol naphthol
naftolato naphtholate
naftolén naftolen
naftonitrilo naphthonitrile
naftoquinona naphthoquinone
naftorresorcina naphthoresorcinol
naftotiazol naphthothiazole
naftoxi naphthoxy
naled naled
nalorfina nalorphine
naloxona naloxone
nanómetro nanometer
nanotecnología nanotechnology
nantoquita nantokite
napalina napalin
napalm napalm

naproxén naproxen
naptalam naptalam
narceína narceine
narcótico narcotic
naringina naringin
nasciente nascent
natrolita natrolite
natrón natron
natural natural
nectar nectar
nefelita nephelite
nefelometría nephelometry
nefelómtero nephelometer
nefrita nephrite
negativo negative
negro de acetileno acetylene black
negro de alizarina alizarin black
negro de anilina aniline black
negro de antimonio antimony black
negro de carbón carbon black
negro de cobalto cobalt black
negro de gas gas black
negro de hierro iron black
negro de horno furnace black
negro de huesos bone black
negro de humo lampblack
negro de manganeso manganese black
negro de paladio palladium black
negro de petróleo oil black
negro de platino platinum black
negro de renio rhenium black
negro de rodio rhodium black
nemático nematic
neo neo
neocianita neocianite
neodimio neodymium
neohexano neohexane
neolita neolite
neomicina neomycin
neón neon
neopentano neopentane
neopentilglicol neopentyl glycol
neopentilo neopentyl
neopreno neoprene
neptunilo neptunyl
neptunio neptunium
nerilo neryl
nerol nerol
nerolidol nerolidol

nervona nervone
nesquehonita nesquehonite
neurina neurine
neutralidad neutrality
neutralización neutralization
neutralizado neutralized
neutralizante neutralizing
neutralizar neutralize
neutrino neutrino
neutro neutral
neutrón neutron
neutrón de alta energía high-energy
    neutron
neutrón de baja energía low-energy
    neutron
neutrón de fisión fission neutron
neutrón de resonancia resonance
    neutron
neutrón lento slow neutron
neutrón nuclear nuclear neutron
neutrón rápido fast neutron
neutrón retardado delayed neutron
neutrón térmico thermal neutron
neutron termonuclear thermonuclear
    neutron
neutrónica neutronics
neutrónico neutronic
niacina niacin
niacinamida niacinamide
nialamida nialamide
niccolita niccolite
nicotina nicotine
nicotinamida nicotinamide
nicotinoilo nicotinoyl
nicromo nichrome
nido nido
niebla fog
niebla contaminante smog
nieve de dióxido de carbono carbon
    dioxide snow
nilón nylon
ninhidrina ninhydrin
niobato niobate
niobato de litio lithium niobate
niobato sódico sodium niobate
nióbico niobic
niobilo niobyl
niobio niobium
niobio-estaño niobium-tin

niobio-titanio niobium-titanium
niobio-uranio niobium-uranium
niobita niobite
niobo niobus
nioboxi nioboxy
nionel nionel
níquel nickel
níquel-cadmio nickel-cadmium
níquel de Raney Raney nickel
níquel-plata nickel-silver
níquel-rodio nickel-rhodium
niquélico nickelic
niquelina nickeline
niqueloceno nickelocene
niqueloso nickelous
niranio niranium
nisina nisin
nitobifenilo nitrobiphenyl
nitracidio nitracidium
nitración nitration
nitración de Zincke Zincke nitration
nitralina nitralin
nitramida nitramide
nitramina nitramine
nitranilato sódico sodium nitranilate
nitranilida nitranilide
nitranilina nitraniline
nitrato nitrate
nitrato alumínico aluminum nitrate
nitrato alumínico cálcico calcium
    aluminum nitrate
nitrato amónico ammonium nitrate
nitrato amónico cérico ceric ammonium
    nitrate
nitrato amónico de lantano lanthanum
    ammonium nitrate
nitrato amónico de neodimio
    neodymium ammonium nitrate
nitrato bárico barium nitrate
nitrato cálcico calcium nitrate
nitrato cérico ceric nitrate
nitrato ceroso cerous nitrate
nitrato cobaltoso cobaltous nitrate
nitrato crómico chromic nitrate
nitrato cúprico cupric nitrate
nitrato de acetamida acetamide nitrate
nitrato de acetilo acetyl nitrate
nitrato de amilo amyl nitrate
nitrato de antimonilo sódico sodium

antimonyl nitrate
nitrato de atropina atropine nitrate
nitrato de berilio beryllium nitrate
nitrato de bismutilo bismuthyl nitrate
nitrato de bismuto bismuth nitrate
nitrato de brucina brucine nitrate
nitrato de butilo butyl nitrate
nitrato de cadmio cadmium nitrate
nitrato de celulosa cellulose nitrate
nitrato de cerio cerium nitrate
nitrato de cesio cesium nitrate
nitrato de cinc zinc nitrate
nitrato de cinconina cinchonine nitrate
nitrato de cobalto cobalt nitrate
nitrato de cobre copper nitrate
nitrato de cocaína cocaine nitrate
nitrato de cromo chromium nitrate
nitrato de diazobenceno diazobenzene
    nitrate
nitrato de didimio didymium nitrate
nitrato de disprosio dysprosium nitrate
nitrato de erbio erbium nitrate
nitrato de estroncio strontium nitrate
nitrato de etileno ethylene nitrate
nitrato de etilo ethyl nitrate
nitrato de europio europium nitrate
nitrato de flúor fluorine nitrate
nitrato de gadolinio gadolinium nitrate
nitrato de galio gallium nitrate
nitrato de guanidina guanidine nitrate
nitrato de hidrazina hydrazine nitrate
nitrato de hierro iron nitrate
nitrato de indio indium nitrate
nitrato de isopropilo isopropyl nitrate
nitrato de itrio yttrium nitrate
nitrato de lantano lanthanum nitrate
nitrato de litio lithium nitrate
nitrato de lutecio lutetium nitrate
nitrato de manganeso manganese nitrate
nitrato de mercurio mercury nitrate
nitrato de metilo methyl nitrate
nitrato de miconazol miconazole nitrate
nitrato de neodimio neodymium nitrate
nitrato de níquel nickel nitrate
nitrato de paladio palladium nitrate
nitrato de pentilo pentyl nitrate
nitrato de peroxibenzoilo peroxybenzoyl
    nitrate
nitrato de plata silver nitrate

nitrato de plomo lead nitrate
nitrato de propanol propanol nitrate
nitrato de propilo propyl nitrate
nitrato de rodio rhodium nitrate
nitrato de rubidio rubidium nitrate
nitrato de talio thallium nitrate
nitrato de telurio tellurium nitrate
nitrato de terbio terbium nitrate
nitrato de titanio titanium nitrate
nitrato de torio thorium nitrate
nitrato de uranilo uranyl nitrate
nitrato de uranio uranium nitrate
nitrato de urea urea nitrate
nitrato de zirconilo zirconyl nitrate
nitrato de zirconio zirconium nitrate
nitrato fenilmercúrico phenylmercuric
    nitrate
nitrato férrico ferric nitrate
nitrato ferroso ferrous nitrate
nitrato magnésico magnesium nitrate
nitrato manganoso manganous nitrate
nitrato mercúrico mercuric nitrate
nitrato mercurioso mercurous nitrate
nitrato niqueloso nickelous nitrate
nitrato paladioso palladous nitrate
nitrato potásico potassium nitrate
nitrato potásico de cobalto cobalt
    potassium nitrate
nitrato samárico samaric nitrate
nitrato sódico sodium nitrate
nitrato talioso thallous nitrate
nitreno nitrene
nitridación nitridation
nitrido nitrido
nitrificación nitrification
nitrilasa nitrilase
nitrilo nitrile
nitrilo de bencilo benzyl nitrile
nitrilo fórmico formic nitrile
nitrilo graso fatty nitrile
nitrilotriacetato sódico sodium
    nitrilotriacetate
nitrilotriacetonitrilo nitrilotriacetonitrile
nitrito nitrite
nitrito amónico ammonium nitrite
nitrito amónico de cinc zinc ammonium
    nitrite
nitrito bárico barium nitrite
nitrito cálcico calcium nitrite

nitrito cúprico cupric nitrite
nitrito de alquilo alkyl nitrite
nitrito de amilo amyl nitrite
nitrito de butilo butyl nitrite
nitrito de cobre copper nitrite
nitrito de estroncio strontium nitrite
nitrito de etilo ethyl nitrite
nitrito de isoamilo isoamyl nitrite
nitrito de metilo methyl nitrite
nitrito de plata silver nitrite
nitrito de plomo lead nitrite
nitrito de propilo propyl nitrite
nitrito magnésico magnesium nitrite
nitrito potásico potassium nitrite
nitrito potásico cobáltico cobaltic
    potassium nitrite
nitrito sódico sodium nitrite
nitro nitro
nitro de sosa soda niter
nitroacetanilida nitroacetanilide
nitroalmidón nitrostarch
nitroamina nitroamine
nitroamino nitroamino
nitroanilina nitroaniline
nitroanisol nitroanisole
nitroantraceno nitroanthracene
nitroantraquinona nitroanthraquinone
nitroaromático nitroaromatic
nitrobarita nitrobarite
nitrobenceno nitrobenzene
nitrobencilo nitrobenzyl
nitrobenzaldehído nitrobenzaldehyde
nitrobenzamida nitrobenzamide
nitrobenzoilo nitrobenzoyl
nitrobenzonitrilo nitrobenzonitrile
nitrobenzotrifluoruro
    nitrobenzotrifluoride
nitrobromoformo nitrobromoform
nitrobutanol nitrobutanol
nitrocelulosa nitrocellulose
nitroclorobenceno nitrochlorobenzene
nitrocloroformo nitrochloroform
nitrocobalamina nitrocobalamin
nitrocresol nitrocresol
nitrodifenilamina nitrodiphenylamine
nitrodifenilo nitrodiphenyl
nitroestireno nitrostyrene
nitroetano nitroethane
nitroetilpropanodiol

nitroethylpropanediol
nitrofenetol nitrophenetole
nitrofenida nitrophenide
nitrofenilhidrazina nitrophenylhydrazine
nitrofenol nitrophenol
nitrofenolato sódico sodium
    nitrophenolate
nitroferricianuro sódico sodium
    nitroferricyanide
nitrofosfato nitrophosphate
nitrofurano nitrofuran
nitrofurantoína nitrofurantoin
nitrogenado nitrogenated
nitrogenasa nitrogenase
nitrógeno nitrogen
nitrógeno amoniacal ammonia nitrogen
nitrógeno líquido liquid nitrogen
nitroglicerina nitroglycerin
nitroguanidina nitroguanidine
nitrol nitrol
nitromagnesita nitromagnesite
nitromanita nitromannite
nitromanitol nitromannitol
nitromersol nitromersol
nitrometano nitromethane
nitrometilpropanodiol
    nitromethylpropanediol
nitrometoxianilina nitromethoxyaniline
nitrón nitron
nitrona nitrone
nitronaftaleno nitronaphthalene
nitronaftilamina nitronaphthylamine
nitronaftol nitronaphthol
nitronio nitronium
nitroparafina nitroparaffin
nitropropano nitropropane
nitroquinolina nitroquinoline
nitrosilo nitrosyl
nitrosita de estireno styrene nitrosite
nitroso nitroso
nitrosoamina nitrosoamine
nitrosobenceno nitrosobenzene
nitrosodimetilamina
    nitrosodimethylamine
nitrosodimetilanilina
    nitrosodimethylaniline
nitrosoetano nitrosoethane
nitrosofenol nitrosophenol
nitrosoguanidina nitrosoguanidine

nitrosometilurea nitrosomethylurea
nitrosonaftilamina nitrosonaphthylamine
nitrosonaftol nitrosonaphthol
nitrosonio nitrosonium
nitrosonitrato de rutenio ruthenium
    nitrosonitrate
nitrosooxi nitrosooxy
nitrosotolueno nitrosotoluene
nitrosulfatiazol nitrosulfathiazole
nitrotiofeno nitrothiophene
nitrotolueno nitrotoluene
nitrotoluidina nitrotoluidine
nitrotriclorometano
    nitrotrichloromethane
nitrotrifluorometilclorobenceno
    nitrotrifluoromethyl chlorobenzene
nitrourea nitrourea
nitroxileno nitroxylene
nitroxilo nitroxyl
nitruro nitride
nitruro alumínico aluminum nitride
nitruro cálcico calcium nitride
nitruro de berilio beryllium nitride
nitruro de boro boron nitride
nitruro de carbono carbon nitride
nitruro de cinc zinc nitride
nitruro de cobalto cobalt nitride
nitruro de fosforilo phosphoryl nitride
nitruro de fósforo phosphorus nitride
nitruro de hafnio hafnium nitride
nitruro de litio lithium nitride
nitruro de plata silver nitride
nitruro de selenio selenium nitride
nitruro de silicio silicon nitride
nitruro de tántalo tantalum nitride
nitruro de titanio titanium nitride
nitruro de vanadio vanadium nitride
nitruro de zirconio zirconium nitride
nitruro magnésico magnesium nitride
nitruro sódico sodium nitride
nivel atómico atomic level
nivel cuántico quantum level
nivel de energía energy level
nivel de energía atómica atomic energy
    level
nivel de energía electrónica electron
    energy level
nivel de energía molecular molecular
    energy level

nivel de energía nuclear nuclear energy
    level
nivel de Fermi Fermi level
nivel de saturación saturation level
nivel molecular molecular level
nivel superior higher level
nivel vibracional vibrational level
nivel virtual virtual level
niveles de Landau Landau levels
nivenita nivenite
no acuoso nonaqueous
no combustible noncombustible
no destructivo nondestructive
no electrólito nonelectrolyte
no estequiométrico nonstoichiometric
no ferroso nonferrous
no ideal nonideal
no inflamable nonflammable
no iónico nonionic
no ionizante nonionizing
no metal nonmetal
no metálico nonmetallic
no polar nonpolar
no saturación unsaturation
no saturado unsaturated
nobelio nobelium
noble noble
nocivo noxious
nódulo nodule
nombre aprobado approved name
nombre genérico generic name
nombre sistemático systematic name
nombre trivial trivial name
nomenclatura nomenclature
nomenclatura conjuntiva conjuctive
    nomenclature
nomenclatura química chemical
    nomenclature
nomenclatura radicofuncional
    radicofunctional nomenclature
non non
nona nona
nonacíclico nonacyclic
nonacosano nonacosane
nonacosanol nonacosanol
nonadecano nonadecane
nonadecanol nonadecanol
nonadecanona nonadecanone
nonahidrato nonahydrate

nonalactona nonalactone
nonanal nonanal
nonano nonane
nonanoato nonanoate
nonanoato de butilo butyl nonanoate
nonanoato de etilo ethyl nonanoate
nonanoato de nonilo nonyl nonanoate
nonanodiol nonanediol
nonanoilo nonanoyl
nonanol nonanol
nonanona nonanone
nonanonitrilo nonanenitrile
noneno nonene
nonilamina nonylamine
nonilbenceno nonylbenzene
nonilcarbinol nonylcarbinol
nonileno nonylene
nonilfenol nonylphenol
nonilo nonyl
noniltriclorosilano nonyltrichlorosilane
nonino nonyne
nopineno nopinene
nor nor
noradrenalina noradrenaline
norbornadieno norbornadiene
norborneno norbornene
norbornenometanol norbornenemethanol
norepinefrina norepinephrine
norgestrel norgestrel
norleucina norleucine
normal normal
normalidad normality
normas de almacenamiento storage
    guidelines
nornicotina nornicotine
noscapina noscapine
noselita noselite
notación notation
notación de Dyson Dyson notation
notación de Frankland Frankland
    notation
notación de Wiswesser Wiswesser
    notation
notación lineal de Wiswesser Wiswesser
    line notation
novobiocina novobiocin
novobiocina sódica sodium novobiocin
novolaca novolak
nube electrónica electron cloud

nube iónica ion cloud
nucleación nucleation
nucleado nucleate
nuclear nuclear
nucleasa nuclease
nucleido nuclide
nucleido radiactivo radioactive nuclide
nucleidos de Wigner Wigner nuclides
núcleo nucleus
núcleo aromático aromatic nucleus
núcleo atómico atomic nucleus
núcleo compuesto compound nucleus
núcleo de recristalización recrystallization nucleus
núcleo estable stable nucleus
núcleo excitado excited nucleus
núcleo impar-impar odd-odd nucleus
núcleo impar-par odd-even nucleus
núcleo inestable unstable nucleus
núcleo par-impar even-odd nucleus
núcleo par-par even-even nucleus
núcleo pesado heavy nucleus
núcleo primario primary nucleus
núcleo radiactivo radioactive nucleus
nucleofílico nucleophilic
nucleófilo nucleophile
nucleogénesis nucleogenesis
nucleón nucleon
nucleónico nucleonic
nucleoproteína nucleoprotein
núcleos equivalentes equivalent nuclei
núcleos espejos mirror nuclei
nucleósido nucleoside
nucleótido nucleotide
numeración bicíclica bicyclo numbering
número atómico atomic number
número atómico efectivo effective atomic number
número cuántico quantum number
número cuántico acimutal azimuthal quantum number
número cuántico de espín spin quantum number
número cuántico magnético magnetic quantum number
número cuántico orbital orbital quantum number
número cuántico principal principal

quantum number
número de acetona acetone number
número de ácido acid number
número de Avogadro Avogadro number
número de base base number
número de Bellier Bellier number
número de cetano cetane number
número de cobre copper number
número de combinación combining number
número de coordinación coordination number
número de Drew Drew number
número de entalpía de reacción reaction enthalpy number
número de éster ester number
número de Hehner Hehner number
número de Loschmidt Loschmidt number
número de masa mass number
número de masa atómica atomic mass number
número de neutralización neutralization number
número de neutrones neutron number
número de oro gold number
número de oxidación oxidation number
número de peróxido peroxide number
número de Polenske Polenske number
número de Prandtl Prandtl number
número de precipitación precipitation number
número de saponificación saponification number
número de transferencia transference number
número de transporte transport number
número de valencia valence number
número de yodo iodine number
número isotópico isotopic number
número molecular molecular number
número nuclear nuclear number
número protónico proton number
números mágicos magic numbers
nutrición nutrition
nutriente nutrient
nutrificación nutrification

O

ocimeno ocimene
oclusión occlusion
ocre ocher
ocre rojo red ocher
octa octa
octaacetato de sucrosa sucrose
    octaacetate
octabenzona octabenzone
octacloronaftaleno
    octachloronaphthalene
octacloruro de silicio silicon
    octachloride
octacosano octacosane
octadeca octadeca
octadecanal octadecanal
octadecano octadecane
octadecanoato de butilo butyl
    octadecanoate
octadecanol octadecanol
octadeceno octadecene
octadecenol octadecenol
octadecilo octadecyl
octadeciltriclorosilano
    octadecyltrichlorosilane
octadieno octadiene
octafluorobuteno octafluorobutene
octafluorociclobutano
    octafluorocyclobutane
octafluoropropano octafluoropropane
octahedral octahedral
octahedro octahedron
octahidrato octahydrate
octahidrato bárico barium octahydrate
octametiltrisiloxano
    octamethyltrisiloxane
octanal octanal
octano octane
octanoato octanoate
octanoato de cinc zinc octanoate
octanoato de etilo ethyl octanoate
octanoilo octanoyl
octanol octanol
octanona octanone

octavalente octavalent
octeno octene
octeto octet
octeto electrónico electron octet
octilamina octylamine
octilcarbinol octyl carbinol
octileno octylene
octilfenilo octylphenyl
octilfenol octylphenol
octilmercaptano octyl mercaptan
octilo octyl
octilsulfato sódico sodium octyl sulfate
octiltriclorosilano octyltrichlorosilane
octino octyne
octoato cálcico calcium octoate
octoato de cinc zinc octoate
octoato de etilo ethyl octoate
octoato de hierro iron octoate
octoato de manganeso manganese
    octoate
octoato de plomo lead octoate
octoato estannoso stannous octoate
octoato férrico ferric octoate
octoato ferroso ferrous octoate
octometileno octomethylene
octóxido de triuranio triuranium
    octoxide
odorante odorant
odorífero odoriferous
odorimetría odorimetry
odorómetro odorometer
oico oic
oide oid
okonita okonite
oleamida oleamide
oleato oleate
oleato alumínico aluminum oleate
oleato amónico ammonium oleate
oleato bárico barium oleate
oleato cálcico calcium oleate
oleato cobaltoso cobaltous oleate
oleato de bismuto bismuth oleate
oleato de butilo butyl oleate

oleato de butoxietilo butoxyethyl oleate
oleato de cinc zinc oleate
oleato de cobalto cobalt oleate
oleato de cobre copper oleate
oleato de diglicol diglycol oleate
oleato de etilo ethyl oleate
oleato de isocetilo isocetyl oleate
oleato de manganeso manganese oleate
oleato de metilo methyl oleate
oleato de plomo lead oleate
oleato estannoso stannous oleate
oleato fenilmercúrico phenylmercuric
    oleate
oleato férrico ferric oleate
oleato magnésico magnesium oleate
oleato mercúrico mercuric oleate
oleato potásico potassium oleate
oleato sódico sodium oleate
olefina olefin
oleilo oleyl
oleína olein
oleo oleo
oleoilo oleoyl
oleoilsarcosina oleoylsarcosine
oleorresina oleoresin
oligo oligo
oligoelemento trace element
oligómero oligomer
oligonita oligonite
oligonucleótido oligonucleotide
oligopéptido oligopeptide
oligosacárido oligosaccharide
olivenita olivenite
olivetol olivetol
olivina olivine
olor odor
olor de advertencia warning odor
omega omega
ona one
oncógeno oncogen
onda wave
onio onium
opacidad opacity
opaco opaque
ópalo opal
operación operation
opiato opiate
opio opium
opioide opioid

ópticamente optically
ópticamente activo optically active
órbita orbit
órbita atómica atomic orbit
órbita de Bohr Bohr orbit
órbita electrónica electron orbit
orbital orbital
orbital antienlazante antibonding orbital
orbital atómico atomic orbital
orbital de antienlazado antibonding
    orbital
orbital de enlazado bonding orbital
orbital degenerado degenerate orbital
orbital delta delta orbital
orbital electrónico electron orbital
orbital enlazante bonding orbital
orbital exterior outer orbital
orbital fronterizo frontier orbital
orbital híbrido hybrid orbital
orbital molecular molecular orbital
orbital molecular de máxima energía
    ocupado highest occupied molecular
    orbital
orbital molecular de mínima energía
    ocupado lowest occupied molecular
    orbital
orbital molecular desocupado
    unoccupied molecular orbital
orbital molecular ocupado occupied
    molecular orbital
orbitales solapados overlapping orbitals
orcina orcin
orden order
órden de Irving-Williams Irving-
    Williams order
orden de reacción reaction order
orgánico organic
organoarcilla organoclay
organoborano organoborane
organogel organogel
organometálico organometallic
organometaloide organometalloid
organosilano organosilane
organosilíceo organosilicon
organosol organosol
orientación superficial surface
    orientation
ornitina ornithine
oro gold

oro blanco white gold
oro de mosaico mosaic gold
oro musivo mosaic gold
oropimente orpiment
ortamina orthamine
ortita orthite
orto ortho
ortoarseniato de cinc zinc orthoarsenate
ortoclasa orthoclase
ortofosfato orthophosphate
ortofosfato alumínico aluminum orthophosphate
ortofosfato bárico barium orthophosphate
ortofosfato cálcico calcium orthophosphate
ortofosfato de cinc zinc orthophosphate
ortofosfato de etilo ethyl orthophosphate
ortofosfato de litio lithium orthophosphate
ortofosfato de plata silver orthophosphate
ortofosfato de plomo lead orthophosphate
ortofosfato de urea-amonio urea-ammonium orthophosphate
ortofosfato de zirconio zirconium orthophosphate
ortofosfato dicálcico dicalcium orthophosphate
ortofosfato magnésico magnesium orthophosphate
ortofosfato manganoso manganous orthophosphate
ortofosfato potásico potassium orthophosphate
ortofosfato sódico sodium orthophosphate
ortofosfato tricálcico tricalcium orthophosphate
ortofosfato trimercúrico trimercuric orthophosphate
ortofosfato trimercurioso trimercurous orthophosphate
ortofosfato tripotásico tripotassium orthophosphate
ortofosfato trisódico trisodium orthophosphate
ortohidrógeno orthohydrogen

ortonitrógeno orthonitrogen
ortoperyodato bárico barium orthoperiodate
ortopropionato de etilo ethyl orthopropionate
ortosilicato orthosilicate
ortosilicato de etilo ethyl orthosilicate
ortosilicato sódico sodium orthosilicate
ortotungstato cálcico calcium orthotungstate
ortovanadato sódico sodium orthovanadate
osamina osamine
osazona osazone
oscilación oscillation
oscilación atómica atomic oscillation
oscilador armónico harmonic oscillator
oscilador electrónico electron oscillator
osmiato osmate
osmiato potásico potassium osmate
ósmica osmics
osmio osmium
osmiridio osmiridium
osmoceno osmocene
osmol osmole
osmolalidad osmolality
osmolaridad osmolarity
osmometría osmometry
osmómetro osmometer
osmómetro de presión de vapor vapor-pressure osmometer
ósmosis osmosis
ósmosis inversa reverse osmosis
osmoso osmous
osmótico osmotic
osono osone
osotriazol osotriazole
ovaleno ovalene
ovicida ovicide
oxa oxa
oxalato oxalate
oxalato ácido potásico potassium acid oxalate
oxalato alumínico aluminum oxalate
oxalato amónico ammonium oxalate
oxalato amónico de titanio titanium ammonium oxalate
oxalato amónico férrico ferric ammonium oxalate

oxalato antimonioso antimonous oxalate
oxalato bárico barium oxalate
oxalato cálcico calcium oxalate
oxalato ceroso cerous oxalate
oxalato cobaltoso cobaltous oxalate
oxalato cromoso chromous oxalate
oxalato cúprico cupric oxalate
oxalato de alilo allyl oxalate
oxalato de anilina aniline oxalate
oxalato de berilio beryllium oxalate
oxalato de bismuto bismuth oxalate
oxalato de cadmio cadmium oxalate
oxalato de cerio cerium oxalate
oxalato de cesio cesium oxalate
oxalato de cinc zinc oxalate
oxalato de cobre copper oxalate
oxalato de dibutilo dibutyl oxalate
oxalato de dietilo diethyl oxalate
oxalato de erbio erbium oxalate
oxalato de escandio scandium oxalate
oxalato de estroncio strontium oxalate
oxalato de etilo ethyl oxalate
oxalato de europio europium oxalate
oxalato de gadolinio gadolinium oxalate
oxalato de hierro iron oxalate
oxalato de hierro amónico ammonium
    iron oxalate
oxalato de holmio holmium oxalate
oxalato de iterbio ytterbium oxalate
oxalato de itrio yttrium oxalate
oxalato de lantano lanthanum oxalate
oxalato de litio lithium oxalate
oxalato de manganeso manganese
    oxalate
oxalato de neodimio neodymium oxalate
oxalato de niobio niobium oxalate
oxalato de níquel nickel oxalate
oxalato de plata silver oxalate
oxalato de plomo lead oxalate
oxalato de praseodimio praseodymium
    oxalate
oxalato de titanio titanium oxalate
oxalato de torio thorium oxalate
oxalato de tulio thulium oxalate
oxalato de urea urea oxalate
oxalato estannoso stannous oxalate
oxalato férrico ferric oxalate
oxalato ferroso ferrous oxalate
oxalato magnésico magnesium oxalate

oxalato mercúrico mercuric oxalate
oxalato mercurioso mercurous oxalate
oxalato niqueloso nickelous oxalate
oxalato potásico potassium oxalate
oxalato potásico crómico chromium
    potassium oxalate
oxalato potásico de antimonio antimony
    potassium oxalate
oxalato samárico samaric oxalate
oxalato sódico sodium oxalate
oxalato sódico férrico ferric sodium
    oxalate
oxalato titanoso titanous oxalate
oxalilo oxalyl
oxalilurea oxalylurea
oxalo oxalo
oxamato amónico ammonium oxamate
oxamida oxamide
oxamido oxamido
oxamoilo oxamoyl
oxantrol oxanthrol
oxatilo oxatyl
oxazol oxazole
óxdo niqueloso nickelous oxide
oxetano oxetane
oxetilo oxethyl
oxetona oxetone
oxi oxy
oxiácido oxyacid
oxiácido de azufre sulfur oxyacid
oxiácido de fósforo phosphorus oxyacid
oxiácido de selenio selenium oxyacid
oxiamida oxyamide
oxianión oxyanion
oxibromuro de bismuto bismuth
    oxybromide
oxibromuro de carbono carbon
    oxybromide
oxibromuro de fósforo phosphorus
    oxybromide
oxibromuro de zirconio zirconium
    oxybromide
oxicarbonato de bismuto bismuth
    oxycarbonate
oxicarbonilo oxycarbonyl
oxicianógeno oxycyanogen
oxicianuro de carbono carbon
    oxycyanide
oxicianuro mercúrico mercuric

oxycyanide

oxicloruro antimonioso antimonous oxychloride

oxicloruro arsenioso arsenous oxychloride

oxicloruro cálcico calcium oxychloride

oxicloruro crómico chromic oxychloride

oxicloruro de antimonio antimony oxychloride

oxicloruro de azufre sulfur oxychloride

oxicloruro de bismuto bismuth oxychloride

oxicloruro de carbono carbon oxychloride

oxicloruro de cinc zinc oxychloride

oxicloruro de cobre copper oxychloride

oxicloruro de cromo chromium oxychloride

oxicloruro de fósforo phosphorus oxychloride

oxicloruro de rutenio ruthenium oxychloride

oxicloruro de selenio selenium oxychloride

oxicloruro de silicio silicon oxychloride

oxicloruro de telurio tellurium oxychloride

oxicloruro de tungsteno tungsten oxychloride

oxicloruro de uranio uranium oxychloride

oxicloruro de vanadio vanadium oxychloride

oxicloruro de zirconio zirconium oxychloride

oxicloruro sulfúrico sulfuric oxychloride

oxicloruro sulfuroso sulfurous oxychloride

oxicorte flame cutting

oxidación oxidation

oxidación anódica anodic oxidation

oxidación de Jones Jones oxidation

oxidación de Oppenauer Oppenauer oxidation

oxidación de Pfitzner-Moffatt Pfitzner-Moffatt oxidation

oxidación de Riley Riley oxidation

oxidación de Sarett Sarett oxidation

oxidación electrolítica electrolytic

oxidation

oxidación electroquímica electrochemical oxidation

oxidación por aire air oxidation

oxidación-reducción oxidation-reduction

oxidante oxidant

oxidante fotoquímico photochemical oxidant

oxidar oxidize

oxidasa oxidase

oxidasa de ácido ascórbico ascorbic acid oxidase

oxidasa de glucosa glucose oxidase

oxidasa de xantina xanthine oxidase

oxidativo oxidative

oxidicloruro de vanadio vanadium oxydichloride

oxidimetría oxidimetry

oxidipropionitrilo oxydipropionitrile

óxido oxide

óxido acético acetic oxide

óxido ácido acid oxide

óxido alcalinotérreo alkaline-earth oxide

óxido alumínico aluminum oxide

óxido alumínico hidratado hydrated aluminum oxide

óxido anfótero amphoteric oxide

óxido antimonioso antimonous oxide

óxido argéntico argentic oxide

óxido arsenioso arsenous oxide

óxido áurico auric oxide

óxido bárico barium oxide

óxido básico basic oxide

óxido bórico boric oxide

óxido cálcico calcium oxide

óxido cérico ceric oxide

óxido cobáltico cobaltic oxide

óxido cobaltoso cobaltous oxide

óxido colbatosocobáltico cobaltouscobaltic oxide

óxido crómico chromic oxide

óxido cromoso chromous oxide

óxido cúprico cupric oxide

óxido cuproso cuprous oxide

óxido de acetilo acetyl oxide

óxido de alquilo alkyl oxide

óxido de antimonio antimony oxide

óxido de arsénico arsenic oxide

óxido de azufre sulfur oxide

óxido de benzofenona benzophenone
   oxide
óxido de benzoilo benzoyl oxide
óxido de berilio beryllium oxide
óxido de bifenileno biphenylene oxide
óxido de bismuto bismuth oxide
óxido de boro boron oxide
óxido de butileno butylene oxide
óxido de cacodilo cacodyl oxide
óxido de cadmio cadmium oxide
óxido de carbono carbon oxide
óxido de cerio cerium oxide
óxido de ciclohexeno cyclohexene oxide
óxido de cinc zinc oxide
óxido de cloro chlorine oxide
óxido de cobalto cobalt oxide
óxido de cromo chromium oxide
óxido de deuterio deuterium oxide
óxido de dibutilestaño dibutyltin oxide
óxido de dicloro dichlorine oxide
óxido de dicloroetilo dichloroethyl oxide
óxido de didimio didymium oxide
oxido de dietilo diethyl oxide
óxido de difenilo diphenyl oxide
óxido de difenilo clorometilado
   chloromethylated diphenyl oxide
óxido de dinitrógeno dinitrogen oxide
óxido de disprosio dysprosium oxide
óxido de divinilo divinyl oxide
óxido de erbio erbium oxide
óxido de escandio scandium oxide
óxido de estaño tin oxide
óxido de estireno styrene oxide
óxido de estroncio strontium oxide
óxido de etileno ethylene oxide
óxido de etilo ethyl oxide
óxido de europio europium oxide
óxido de fenoxipropilcno
   phenoxypropylene oxide
óxido de flúor fluorine oxide
óxido de fósforo phosphorus oxide
óxido de gadolinio gadolinium oxide
óxido de galio gallium oxide
óxido de germanio germanium oxide
óxido de hafnio hafnium oxide
óxido de hexaclorodifenilo
   hexachlorodiphenyl oxide
óxido de hidrógeno hydrogen oxide
óxido de hierro iron oxide

óxido de holmio holmium oxide
óxido de indio indium oxide
óxido de iridio iridium oxide
óxido de iterbio ytterbium oxide
óxido de itrio yttrium oxide
óxido de lantano lanthanum oxide
óxido de litio lithium oxide
óxido de lutecio lutetium oxide
óxido de manganeso manganese oxide
óxido de mercurio mercury oxide
óxido de mercurio rojo red mercury
   oxide
óxido de mesitilo mesityl oxide
óxido de metileno methylene oxide
óxido de metilo methyl oxide
óxido de molibdeno molybdenum oxide
óxido de neodimio neodymium oxide
óxido de niobio niobium oxide
óxido de níquel nickel oxide
óxido de nitrógeno nitrogen oxide
óxido de octileno octylene oxide
óxido de oro gold oxide
óxido de osmio osmium oxide
óxido de paladio palladium oxide
óxido de pentametileno pentamethylene
   oxide
óxido de pentilo pentyl oxide
óxido de piridina pyridine oxide
óxido de plata silver oxide
óxido de platino platinum oxide
óxido de plomo lead oxide
óxido de plutonio plutonium oxide
óxido de polietileno polyethylene oxide
óxido de polifenileno polyphenylene
   oxide
óxido de polipropileno polypropylene
   oxide
óxido de praseodimio praseodymium
   oxide
óxido de propileno propylene oxide
óxido de renio rhenium oxide
óxido de rodio rhodium oxide
óxido de rubidio rubidium oxide
óxido de rutenio ruthenium oxide
óxido de samario samarium oxide
óxido de selenio selenium oxide
óxido de silicio silicon oxide
óxido de succinilo succinyl oxide
óxido de talio thallium oxide

óxido de tántalo tantalum oxide
óxido de telurio tellurium oxide
óxido de terbio terbium oxide
óxido de titanio titanium oxide
óxido de torio thorium oxide
óxido de tributilestaño tributyltin oxide
óxido de triclorobutileno
    trichlorobutylene oxide
óxido de trifluoroamina trifluoroamine
    oxide
óxido de trimetilamina trimethylamine
    oxide
óxido de trimetileno trimethylene oxide
óxido de tulio thulium oxide
óxido de tungsteno tungsten oxide
óxido de uranilo uranyl oxide
óxido de uranio uranium oxide
óxido de vanadio vanadium oxide
óxido de vinilo vinyl oxide
óxido de yodo iodine oxide
óxido de zirconio zirconium oxide
óxido estánnico stannic oxide
óxido estannoso stannous oxide
óxido etilarsenioso ethylarsenious oxide
óxido férrico ferric oxide
óxido férrico gamma gamma ferric
    oxide
óxido ferromagnético ferromagnetic
    oxide
óxido ferroso ferrous oxide
óxido ferrosoférrico ferrosoferric oxide
óxido fosfórico phosphoric oxide
óxido gálico gallic oxide
óxido galoso gallous oxide
óxido grafítico graphitic oxide
óxido magnésico magnesium oxide
óxido mangánico manganic oxide
óxido manganoso manganous oxide
óxido mercúrico mercuric oxide
óxido mercurioso mercurous oxide
óxido negro de cobre copper oxide black
óxido niquélico nickelic oxide
óxido nítrico nitric oxide
óxido nitroso nitrous oxide
óxido platínico platinic oxide
óxido plumboso plumbous oxide
óxido potásico potassium oxide
óxido rojo red oxide
óxido selenioso selenious oxide

óxido sódico sodium oxide
óxido talioso thallous oxide
óxido trioctilfosfínico trioctylphosphinic
    oxide
óxido túngstico tungstic oxide
óxido uránico uranic oxide
óxido uranoso uranous oxide
oxidorreductasa oxidoreductase
oxifluoruro de carbono carbon
    oxyfluoride
oxifluoruro de cromo chromium
    oxyfluoride
oxifluoruro potásico de niobio niobium
    potassium oxyfluoride
oxifosforano oxyphosphorane
oxigenación oxygenation
oxigenado oxygenated
oxigenar oxygenate
oxígeno oxygen
oxígeno consumido oxygen consumed
oxígeno disuelto dissolved oxygen
oxígeno líquido liquid oxygen
oxígeno pesado heavy oxygen
oxihidrato de bismuto bismuth
    oxyhydrate
oxihidrógeno oxyhydrogen
oxiluminiscencia oxyluminescence
oxima oxime
oxima de acetofenona acetophenone
    oxime
oxima de alcanfor camphor oxime
oxima de benzamida benzamide oxime
oxima de benzoína benzoin oxime
oximetileno oxymethylene
oximeturea oxymethurea
oximido oximido
oxina oxine
oxinitrato de bismuto bismuth oxynitrate
oxiquinolina oxyquinoline
oxiquinona oxyquinone
oxirano oxirane
oxireno oxirene
oxisulfuro antimonioso antimonous
    oxysulfide
oxisulfuro de carbono carbon oxysulfide
oxisulfuro de talio thallium oxysulfide
oxitetraciclina oxytetracycline
oxitoxina oxytoxin
oxitricloruro de vanadio vanadium

oxytrichloride

oxo oxo

oxoácido oxo acid

oxobromuro de bismuto bismuth oxobromide

oxocloruro de bismuto bismuth oxochloride

oxofluoruro de bismuto bismuth oxofluoride

oxoproceso oxo process

oxorreacción oxo reaction

oxosilano oxosilane

oxoyoduro de bismuto bismuth oxoiodide

ozocerita ozocerite, earth wax

ozonación ozonation

ozonidación ozonidation

ozónido ozonide

ozonización ozonization

ozonizador ozonator

ozono ozone

ozonólisis ozonolysis

ozoquerita ozokerite

P

pacnolita pachnolite
pagodita pagodite
paladato palladate
paládico palladic
paladio palladium
paladioso palladous
palamina pallamine
palas pallas
palconato sódico sodium palconate
paligorskita paligorskite
pálmico palmic
palmierita palmierite
palmitato palmitate
palmitato alumínico aluminum palmitate
palmitato amónico ammonium palmitate
palmitato cálcico calcium palmitate
palmitato de cetilo cetyl palmitate
palmitato de cinc zinc palmitate
palmitato de etilo ethyl palmitate
palmitato de isooctilo isooctyl palmitate
palmitato de litio lithium palmitate
palmitato de metilo methyl palmitate
palmitato de plata silver palmitate
palmitato de plomo lead palmitate
palmitato magnésico magnesium
    palmitate
palmitato sódico sodium palmitate
palmitina palmitin
palmitoilo palmitoyl
palmitonitrilo palmitonitrile
palo de campeche logwood
pan de oro gold leaf
panclastita panclastite
pancreatina pancreatin
pancromático panchromatic
pandermita pandermite
pantacromático pantachromatic
pantalla screen
pantenol panthenol
pantetina pantethine
pantocaína pantocaine
pantolactona pantolactone
pantomorfismo pantomorphism

pantotenato cálcico calcium pantothenate
pantotenol pantothenol
papaína papain
papaverina papaverine
papel paper
papel de absorción absorption paper
papel de aluminio aluminum foil
papel de filtro filter paper
papel de filtro plegado fluted filter paper
papel de prueba test paper
papel de tornasol litmus paper
papel indicador indicator paper
par ácido-base acid-base pair
par de electrones electron pair
par donador-aceptor donor-acceptor
    pair
par iónico ion pair
para para
parabeno paraben
parabituminoso parabituminous
paracaseína paracasein
paracetaldehído paracetaldehyde
paracetamol paracetamol
paracianógeno paracyanogen
paraconina paraconine
paradiazina paradiazine
parafina paraffin
parafina clorada chlorinated paraffin
paraformaldehído paraformaldehyde
paragonita paragonite
parahelio parahelium
parahidrógeno parahydrogen
paralaje parallax
paraldehído paraldehyde
paraldol paraldol
paralelosterismo parallelosterism
paralisol paralysol
paramagnético paramagnetic
paramagnetismo paramagnetism
paramagnetismo molecular molecular
    paramagnetism
paramagnetismo nuclear nuclear
    paramagnetism

parámetro parameter
parámetros de Racah Racah parameters
paramomicina paromomycin
paramorfismo paramorphism
paramorfo paramorph
paranitranilina paranitraniline
paraperyodato sódico sodium
    paraperiodate
paraquat paraquat
pararrosanilina pararosaniline
parasiticida parasiticide
parásito parasite
paratión parathion
parawolframato parawolframate
paraxantina paraxanthine
paraxileno paraxylene
parcial partial
parileno parylene
parisita parisite
paroxazina paroxazine
partenina parthenine
partición partition
partícula particle
partícula alfa alpha particle
partícula atómica atomic particle
partícula beta beta particle
partícula cargada charged particle
partícula de alta energía high-energy
    particle
partícula de baja energía low-energy
    particle
partícula de desintegración decay
    particle
partícula elemental elementary particle
partícula fundamental fundamental
    particle
partícula nuclear nuclear particle
partícula relativista relativistic particle
partícula sigma sigma particle
partícula subatómica subatomic particle
partícula subnuclear subnuclear particle
partícula tau tau particle
partícula virtual virtual particle
partícula xi xi particle
particulado particulate
partinio partinium
parvolina parvoline
parvulina parvuline
pasivación passivation

pasivación electroquímica
    electrochemical passivation
pasivación química chemical passivation
pasivador passivator
pasividad passivity
pasividad química chemical passivity
pasivo passive
paso step
paso de reacción reaction step
paso determinante de velocidad rate-
    determining step
pasta paste
pasteurización pasteurization
patentabilidad patentability
patogénico pathogenic
patoquímica pathochemistry
patrón standard
patrón de color color standard
patrón de difracción diffraction pattern
patrón de laboratorio laboratory
    standard
patrón de Laue Laue pattern
patrón de radiactividad radioactivity
    standard
patrón químico chemical standard
patrón termoquímico thermochemical
    standard
patronita patronite
paucina paucine
pebulato pebulate
pecblenda pitchblende
pectato pectate
péctico pectic
pectina pectin
pectinasa pectinase
pectización pectization
pectolita pectolite
pectosa pectose
pedernal flint
pedesis pedesis
pegmatita pegmatite
pelargonato de bencilo benzyl
    pelargonate
pelargonato de butilo butyl pelargonate
pelargonato de etilo ethyl pelargonate
pelargonato de heptilo heptyl
    pelargonate
pelargonato de isoamilo isoamyl
    pelargonate

pelargonato de propilo propyl pelargonate
pelargonilo pelargonyl
pelargonina pelargonin
pelargonitrilo pelargonitrile
peletierina pelletierine
película film
película anódica anodic film
película de acetato acetate film
película de Langmuir-Blodgett Langmuir-Blodgett film
película de plástico plastic film
película fina thin film
película gruesa thick film
película interfacial interfacial film
película metálica metal film
película molecular molecular film
película monomolecular monomolecular film
película poliéster polyester film
película unimolecular unimolecular film
peligro de radiación radiation danger
peligroso dangerous, hazardous
pelotina pellotine
peltre pewter
penetración penetration
penetrante penetrant
penicilamina penicillamine
penicilanasa penicillanase
penicilina penicillin
penicilina potásica potassium penicillin
penta penta
pentaamino pentaamino
pentabásico pentabasic
pentaborano pentaborane
pentaborato amónico ammonium pentaborate
pentaborato sódico sodium pentaborate
pentabromo pentabromo
pentabromobenceno pentabromobenzene
pentabromuro pentabromide
pentabromuro de fósforo phosphorus pentabromide
pentabromuro de tántalo tantalum pentabromide
pentabromuro de yodo iodine pentabromide
pentacarbonilo pentacarbonyl
pentacarbonilo de hierro iron pentacarbonyl
pentacarboxílico pentacarboxylic
pentaceno pentacene
pentacetato pentacetate
pentacíclico pentacyclic
pentacloro pentachloro
pentacloroanilina pentachloroaniline
pentaclorobenceno pentachlorobenzene
pentacloroetano pentachloroethane
pentaclorofenato sódico sodium pentachlorophenate
pentaclorofenol pentachlorophenol
pentacloronaftaleno pentachloronaphthalene
pentacloronitrobenceno pentachloronitrobenzene
pentaclorotiofenol pentachlorothiophenol
pentacloruro pentachloride
pentacloruro de antimonio antimony pentachloride
pentacloruro de azufre sulfur pentachloride
pentacloruro de fósforo phosphorus pentachloride
pentacloruro de molibdeno molybdenum pentachloride
pentacloruro de niobio niobium pentachloride
pentacloruro de renio rhenium pentachloride
pentacloruro de tántalo tantalum pentachloride
pentacloruro de tungsteno tungsten pentachloride
pentacosano pentacosane
pentada pentad
pentadecano pentadecane
pentadecanoato de metilo methyl pentadecanoate
pentadecanol pentadecanol
pentadecanolida pentadecanolide
pentadecanona pentadecanone
pentadecenilo pentadecenyl
pentadecilo pentadecyl
pentadieno pentadiene
pentaeritritol pentaerythritol
pentaetilbenceno pentaethylbenzene
pentafluoruro pentafluoride
pentafluoruro de antimonio antimony

pentafluoride

pentafluoruro de arsénico arsenic
pentafluoride

pentafluoruro de bismuto bismuth
pentafluoride

pentafluoruro de bromo bromine
pentafluoride

pentafluoruro de fósforo phosphorus
pentafluoride

pentafluoruro de oro gold pentafluoride

pentafluoruro de tántalo tantalum
pentafluoride

pentafluoruro de yodo iodine
pentafluoride

pentagaloilo pentagalloyl

pentaglicerina pentaglycerin

pentaglicol pentaglycol

pentaglucosa pentaglucose

pentahidrato pentahydrate

pentahidrato bárico barium
pentahydrate

pentahidro pentahydro

pentahidroxi pentahydroxy

pentahidroxiciclohexano
pentahydroxycyclohexane

pentalina pentalin

pentametilbenceno pentamethylbenzene

pentametilenamina
pentamethyleneamine

pentametilendiamina
pentamethylenediamine

pentametilenglicol pentamethylene
glycol

pentametileno pentamethylene

pentametilfenol pentamethylphenol

pentametilo pentamethyl

pentanal pentanal

pentano pentane

pentanodiol pentanediol

pentanodiona pentanedione

pentanol pentanol

pentanona pentanone

pentanotiol pentanethiol

pentaóxido pentaoxide

pentaprismo pentaprismo

pentaseleniuro de fósforo phosphorus
pentaselenide

pentasulfuro pentasulfide

pentasulfuro de antimonio antimony

pentasulfide

pentasulfuro de arsénico arsenic
pentasulfide

pentasulfuro de fósforo phosphorus
pentasulfide

pentasulfuro de plomo lead pentasulfide

pentasulfuro de vanadio vanadium
pentasulfide

pentasulfuro potásico potassium
pentasulfide

pentatriacontano pentatriacontane

pentavalente pentavalent

pentayodo pentaiodo

pentazdieno pentazdiene

pentazocina pentazocine

pentazolilo pentazolyl

pentel pentel

pentenilo pentenyl

penteno pentene

pentenol pentenol

pentilamina pentylamine

pentilbenceno pentylbenzene

pentilbenzoato pentylbenzoate

pentilcetona pentyl ketone

pentilcinamaldehído
pentylcinnamaldehyde

pentileno pentylene

pentilfenilcetona pentyl phenyl ketone

pentilfenol pentylphenol

pentilideno pentylidene

pentilidino pentylidyne

pentilo pentyl

pentiloxi pentyloxy

pentiloxifenol pentyloxyphenol

pentiloxihidrato pentyloxyhydrate

pentilsulfato bárico barium pentylsulfate

pentilurea pentylurea

pentiluretano pentylurethane

pentino pentyne

pentita pentite

pentitol pentitol

pentlandita pentlandite

pentobarbital pentobarbital

pentobarbital sódico sodium
pentobarbital

pentol pentol

pentolita pentolite

pentosa pentose

pentosana pentosan

pentósido pentoside
pentóxido pentoxide
pentóxido de antimonio antimony
    pentoxide
pentóxido de arsénico arsenic pentoxide
pentóxido de bismuto bismuth pentoxide
pentóxido de fósforo phosphorus
    pentoxide
pentóxido de niobio niobium pentoxide
pentóxido de nitrógeno nitrogen
    pentoxide
pentóxido de tántalo tantalum pentoxide
pentóxido de tungsteno tungsten
    pentoxide
pentóxido de vanadio vanadium
    pentoxide
pentóxido de yodo iodine pentoxide
pentoxifenol pentoxyphenol
pentrita pentrite
pepsina pepsin
pepsinógeno pepsinogen
peptidasa peptidase
péptido peptide
peptización peptization
peptona peptone
per per
peracetato de butilo butyl peracetate
peracidez peracidity
perácido peracid
peralcohol peralcohol
perbenzoato de butilo butyl perbenzoate
perborato perborate
perborato cálcico calcium perborate
perborato de cinc zinc perborate
perborato magnésico magnesium
    perborate
perborato sódico sodium perborate
perbromato perbromate
perbromo perbromo
perbromuro fosfórico phosphoric
    perbromide
percarbamida percarbamide
percarbonato percarbonate
percarbonato potásico potassium
    percarbonate
percarbonato sódico sodium
    percarbonate
percarburo percarbide
perclorato perchlorate

perclorato amónico ammonium
    perchlorate
perclorato bárico barium perchlorate
perclorato cálcico calcium perchlorate
perclorato cobaltoso cobaltous
    perchlorate
perclorato de cesio cesium perchlorate
perclorato de cinc zinc perchlorate
perclorato de estroncio strontium
    perchlorate
perclorato de etilo ethyl perchlorate
perclorato de hidrazina hydrazine
    perchlorate
perclorato de hierro iron perchlorate
perclorato de litio lithium perchlorate
perclorato de níquel nickel perchlorate
perclorato de nitronio nitronium
    perchlorate
perclorato de nitrosilo nitrosyl
    perchlorate
perclorato de plata silver perchlorate
perclorato de plomo lead perchlorate
perclorato de rubidio rubidium
    perchlorate
perclorato férrico ferric perchlorate
perclorato ferroso ferrous perchlorate
perclorato magnésico magnesium
    perchlorate
perclorato potásico potassium
    perchlorate
perclorato sódico sodium perchlorate
perclorilo perchloryl
percloro perchloro
perclorobenceno perchlorobenzene
perclorociclopentadieno
    perchlorocyclopentadiene
percloroetano perchloroethane
percloroéter perchloroether
percloroetileno perchloroethylene
perclorometano perchloromethane
perclorometilmercaptano
    perchloromethyl mercaptan
perclorometilo perchloromethyl
percloropropileno perchloropropylene
percloruro perchloride
percloruro de estaño tin perchloride
percloruro férrico ferric perchloride
percloruro fosfórico phosphoric
    perchloride

percolación percolation
percolador percolator
percolar percolate
percristalización percrystallization
percromato perchromate
percromato amónico ammonium
  perchromate
perdestilación perdistillation
perfenazina perphenazine
perfluoro perfluoro
perfluorobuteno perfluorobutene
perfluorocarburo perfluorocarbon
perfluorociclobutano
  perfluorocyclobutane
perfluoroetileno perfluoroethylene
perfluoropropano perfluoropropane
perfluoropropeno perfluoropropene
perfume perfume
perfusión perfusion
pergenol pergenol
perhidrato perhydrate
perhidro perhydro
perhidrol perhydrol
perhidronaftaleno perhydronaphthalene
peri peri
pericíclico pericyclic
periciclo pericyclo
pericinético perkinetic
periclasa periclase
periclina pericline
perileno perylene
perimidina perimidine
perimidinilo perimidinyl
perimorfo perimorph
periodicidad periodicity
periodo period
periodo de identidad identity period
periodo de inducción induction period
peritéctico peritectic
peritectoide peritectoid
perjudicial hazardous
perlas de vidrio glass beads
perlita perlite
permaleación permalloy
permanencia de olor odor permanence
permanganato permanganate
permanganato amónico ammonium
  permanganate
permanganato bárico barium

permanganate
permanganato cálcico calcium
  permanganate
permanganato de bismuto bismuth
  permanganate
permanganato de cadmio cadmium
  permanganate
permanganato de cesio cesium
  permanganate
permanganato de cinc zinc
  permanganate
permanganato de plata silver
  permanganate
permanganato magnésico magnesium
  permanganate
permanganato potásico potassium
  permanganate
permanganato sódico sodium
  permanganate
permanganilo permanganyl
permeabilidad permeability
permeabilidad efectiva effective
  permeability
permeable permeable
permeametría permeametry
permetrina permethrin
permisible allowable
permitido allowed
permutación permutation
permutita permutite
perovskita perovskite
peroxi peroxy
peroxiacetato de butilo butyl
  peroxyacetate
peroxibenzoato de butilo butyl
  peroxybenzoate
peroxicromo peroxychromium
peroxidación peroxidation
peroxidasa peroxidase
peroxidisulfato potásico potassium
  peroxydisulfate
peroxidisulfato sódico sodium
  peroxydisulfate
peróxido peroxide
peróxido bárico barium peroxide
peróxido cálcico calcium peroxide
peróxido de acetilo acetyl peroxide
peróxido de ácido succínico succinic
  acid peroxide

peróxido de benzoilo benzoyl peroxide
peróxido de bismuto bismuth peroxide
peróxido de butilo butyl peroxide
peróxido de carbamida carbamide
    peroxide
peróxido de carbonato sódico sodium
    carbonate peroxide
peróxido de cesio cesium peroxide
peróxido de ciclohexanona
    cyclohexanone peroxide
peróxido de cinc zinc peroxide
peróxido de clorobenzoilo chlorobenzoyl
    peroxide
peróxido de decanoilo decanoyl peroxide
peróxido de dibenzoilo dibenzoyl
    peroxide
peroxido de dietilo diethyl peroxide
peróxido de estaño tin peroxide
peróxido de estroncio strontium
    peroxide
peróxido de etilo ethyl peroxide
peróxido de hidrógeno hydrogen
    peroxide
peróxido de hidrógeno sódico sodium
    hydrogen peroxide
peróxido de lauroilo lauroyl peroxide
peróxido de litio lithium peroxide
peróxido de manganeso manganese
    peroxide
peróxido de mercurio mercury peroxide
peróxido de miristoilo myristoyl
    peroxide
peróxido de níquel nickel peroxide
peróxido de nitrógeno nitrogen peroxide
peróxido de octilo octyl peroxide
peroxido de pelargonilo pelargonyl
    peroxide
peróxido de pirofosfato sódico sodium
    pyrophosphate peroxide
peróxido de plata silver peroxide
peróxido de plomo lead peroxide
peróxido de propionilo propionyl
    peroxide
peróxido de rubidio rubidium peroxide
peróxido de titanio titanium peroxide
peróxido de urea urea peroxide
peróxido magnésico magnesium
    peroxide
peróxido potásico potassium peroxide

peróxido sódico sodium peroxide
peroxidol peroxydol
peroxiisobutirato de butilo butyl
    peroxyisobutyrate
peroxo peroxo
peroxoborato amónico ammonium
    peroxoborate
peroxoborato de cinc zinc peroxoborate
peroxoborato potásico potassium
    peroxoborate
peroxodicarbonato sódico sodium
    peroxodicarbonate
peroxodisulfato amónico ammonium
    peroxodisulfate
peroxodisulfato bárico barium
    peroxodisulfate
peroxohidrato de carbonato sódico
    sodium carbonate peroxohydrate
peroxosulfato potásico potassium
    peroxosulfate
perrenato perrhenate
persorción persorption
persulfato persulfate
persulfato amónico ammonium
    persulfate
persulfato potásico potassium persulfate
persulfato sódico sodium persulfate
persulfuro persulfide
persulfuro de antimonio antimony
    persulfide
persulfuro de fósforo phosphorus
    persulfide
persulfuro de hidrógeno hydrogen
    persulfide
pertio perthio
pertiocarbonato perthiocarbonate
pervaporación pervaporation
peryodato periodate
peryodato bárico barium periodate
peryodato de cesio cesium periodate
peryodato potásico potassium periodate
peryodato sódico sodium periodate
pesado heavy
pesaje doble double weighing
peso weight
peso atómico atomic weight
peso atómico-gramo gram-atomic weight
peso de combinación combining weight
peso de fórmula formula weight

peso equivalente equivalent weight

peso equivalente-gramo gram-equivalent weight

peso específico specific weight

peso isotópico isotopic weight

peso molar molar weight

peso molecular molecular weight, formula weight

peso molecular-gramo gram-molecular weight

peso molecular medio average molecular weight

peso relativo relative weight

pesticida pesticide

petalita petalite

petrolato petrolatum, petroleum jelly

petróleo petroleum, oil

petróleo bruto crude petroleum, crude oil

petroquímico petrochemical

petroxolina petroxolin

petzita petzite

pez pitch

pH pH

phi phi

pi pi

piceno picene

pico peak

pico de absorción absorption peak

pico de Bragg Bragg peak

picocurie picocurie

picolilamina picolylamine

picolilo picolyl

picolina picoline

picotita picotite

picramato sódico sodium picramate

picramida picramide

picrato picrate

picrato amónico ammonium picrate

picrato de plata silver picrate

picrato de plomo lead picrate

picrato de torio thorium picrate

picrato potásico potassium picrate

picrilo picryl

picro picro

picromerita picromerite

picrotoxina picrotoxin

piedmontita piedmontite

piedra de azufre brimstone

piedra de toque touchstone

piedra preciosa precious stone, gemstone, gem

piezocristalización piezocrystallization

piezoelectricidad piezoelectricity

piezoquímica piezochemistry

pigmento pigment

pigmento de cadmio cadmium pigment

pigmento de cromo chrome pigment

pigmento fluorescente fluorescent pigment

pigmento mineral mineral pigment

pigmento orgánico organic pigment

pila cell

pila de Bunsen Bunsen cell

pila de cadmio cadmium cell

pila de Castner-Kellner Castner-Kellner cell

pila de Clark Clark cell

pila de combustible fuel cell

pila de Daniell Daniell cell

pila de Edison Edison cell

pila de gas gas cell

pila de Leclanché Leclanché cell

pila de mercurio mercury cell

pila de Weston Weston cell

pila galvánica galvanic cell

pila patrón standard cell

pila seca dry cell

pila secundaria secondary cell

pila voltaica voltaic cell

pilocarpina pilocarpine

pimaricina pimaricin

pimelita pimelite

pinacoide pinacoid

pinacol pinacol

pinacolona pinacolone

pinacona pinacone

pinanilo pinanyl

pinano pinane

pineno pinene

pinocarveol pinocarveol

pintura paint

pipecolina pipecoline

piperalina piperaline

piperazina piperazine

piperazina de hidroxietilo hydroxyethyl piperazine

piperidilo piperidyl

piperidina piperidine
piperidinio piperidinium
piperidino piperidino
piperidinoetanol piperidinoethanol
piperileno piperylene
piperilo piperyl
piperina piperine
piperitol piperitol
piperitona piperitone
piperonal piperonal
piperonilideno piperonylidene
piperonilo piperonyl
piperoniloilo piperonyloyl
pipeta pipet
pipeta automática automatic pipet
pipeta capilar capillary pipet
pipeta graduada graduated pipet
pipeta volumétrica volumetric pipet
piranilo pyranyl
pirano pyran
piranona pyranone
piranosa pyranose
piranosida pyranoside
pirantreno pyranthrene
pirazina pyrazine
pirazol pyrazole
pirazolidina pyrazolidine
pirazolilo pyrazolyl
pirazolina pyrazoline
pirazolona pyrazolone
pirenilo pyrenyl
pireno pyrene
piretrina pyrethrin
piretro pyrethrum
piretroide pyrethroid
piretrolona pyrethrolone
piridazina pyridazine
piridilamina pyridylamine
piridilcarbinol pyridylcarbinol
piridilo pyridyl
piridina pyridine
piridinilo pyridinyl
piridinio pyridinium
piridona pyridone
piridoxal pyridoxal
piridoxina pyridoxine
pirimidina pyrimidine
pirimidinilo pyrimidinyl
pirimitato pyrimithate

pirita pyrite
pirita de hierro iron pyrite
piritiamina pyrithiamine
piritiona de cinc zinc pyrithione
piro pyro
piroantimoniato sódico sodium pyroantimonate
piroborato potásico potassium pyroborate
piroborato sódico sodium pyroborate
pirocatecol pyrocatechol
pirocelulosa pyrocellulose
pirocloro pyrochlore
pirocroíta pyrochroite
pirofilita pyrophyllite
pirofórico pyrophoric
pirofosfato pyrophosphate
pirofosfato ácido sódico sodium acid pyrophosphate
pirofosfato bárico barium pyrophosphate
pirofosfato cálcico calcium pyrophosphate
pirofosfato de cinc zinc pyrophosphate
pirofosfato de etilo ethyl pyrophosphate
pirofosfato de hierro iron pyrophosphate
pirofosfato de hierro sódico sodium iron pyrophosphate
pirofosfato de plomo lead pyrophosphate
pirofosfato de tiamina thiamine pyrophosphate
pirofosfato de zirconio zirconium pyrophosphate
pirofosfato estannoso stannous pyrophosphate
pirofosfato férrico ferric pyrophosphate
pirofosfato magnésico magnesium pyrophosphate
pirofosfato manganoso manganous pyrophosphate
pirofosfato potásico potassium pyrophosphate
pirofosfato sódico sodium pyrophosphate
pirofosfato tetrapotásico tetrapotassium pyrophosphate
pirofosfato tetrasódico tetrasodium pyrophosphate
pirofosforilo pyrophosphoryl
pirogalato pyrogallate
pirogalol pyrogallol

pirolán pyrolan
pirolignita de hierro iron pyrolignite
pirólisis pyrolysis
pirolusita pyrolusite
pirometría pyrometry
pirómetro pyrometer
piromorfita pyromorphite
pirona pyrone
pirorracemato sódico sodium
    pyroracemate
pirosina pyrosin
pirosulfato pyrosulfate
pirosulfato potásico potassium
    pyrosulfate
pirosulfato sódico sodium pyrosulfate
pirosulfito pyrosulfite
pirosulfito potásico potassium
    pyrosulfite
pirosulfito sódico sodium pyrosulfite
pirosulfurilo pyrosulfuryl
pirotecnia pyrotechnics
pirovanadato sódico sodium
    pyrovanadate
piroxeno pyroxene
piroxilina pyroxylin
pirrilo pyrryl
pirrocolina pyrrocoline
pirroilo pyrroyl
pirrol pyrrole
pirrolidina pyrrolidine
pirrolidinilo pyrrolidinyl
pirrolidona pyrrolidone
pirrolilo pyrrolyl
pirrolina pyrroline
pirrona pyrrone
pirrotita pyrrhotite
piruvaldehído pyruvaldehyde
piruvato de etilo ethyl pyruvate
piruvato de metilo methyl pyruvate
piruvato sódico sodium pyruvate
piruvonitrilo pyruvonitrile
pivaldehído pivaldehyde
pizarra slate
placa plate
placa perforada perforated plate
placa teorética theoretical plate
plano atómico atomic plane
plano de polarización plane of
    polarization

plano de simetría plane of symmetry
planta plant
planta piloto pilot plant
plasma plasma
plásmido plasmid
plasmina plasmin
plasmógeno plasmogen
plasticidad plasticity
plasticizador plasticizer
plástico plastic
plástico alílico allyl plastic
plástico bituminoso bituminous plastic
plástico celular cellular plastic
plástico celulósico cellulosic plastic
plástico de acrilato acrylate plastic
plástico de estireno styrene plastic
plástico de halocarburo halocarbon
    plastic
plástico reforzado reinforced plastic
plastisol plastisol
plastoquinona plastoquinone
plata silver
plata esterlina sterling silver
plateado silver-plating
platinado platinization
platinicloruro sódico sodium
    platinichloride
platínico platinic
platino platinum
platino-litio platinum-lithium
platinocianuro platinocyanide
platinocloruro platinochloride
platinocloruro sódico sodium
    platinochloride
platinoso platinous
plato de evaporación evaporation dish
plazolita plazolite
pleocroismo pleochroism
plomado lead-coating
plomo lead
plomo azul blue lead
plomo blanco white lead
plomo negro black lead
plomo rojo red lead
plomo telúrico telluric lead
plumbago plumbago
plumbato plumbate
plumbato cálcico calcium plumbate
plumbato sódico sodium plumbate

plúmbico plumbic
plumbito sódico sodium plumbite
plumboso plumbous
plutonio plutonium
poder de absorción absorptive power
poder de resolución resolving power
poder rotatorio rotatory power
polar polar
polaridad polarity
polaridad química chemical polarity
polarimetría polarimetry
polarímetro polarimeter
polarizabilidad polarizability
polarizabilidad molecular molecular
    polarizability
polarizable polarizable
polarización polarization
polarización anódica anodic polarization
polarización atómica atomic polarization
polarización catódica cathodic
    polarization
polarización circular circular
    polarization
polarización electrolítica electrolytic
    polarization
polarización electroquímica
    electrochemical polarization
polarización nuclear nuclear polarization
polarizado polarized
polarizado en un plano plane-polarized
polarizador polarizer
polarizante polarizing
polarizar polarize
polarografía polarography
polarograma polarogram
poli poly
poliacetal polyacetal
poliacetaldehído polyacetaldehyde
poliacetileno polyacetylene
poliacrilamida polyacrylamide
poliacrilato polyacrylate
poliacrilonitrilo polyacrylonitrile
polialcano polyalkane
polialcohol polyalcohol
polialómero polyallomer
polialqueno polyalkene
polialquilideno polyalkylidene
poliamida polyamide
poliamina polyamine

poliaminotriazol polyaminotriazole
poliatómico polyatomic
polibásico polybasic
polibencimidazol polybenzimidazole
polibutadieno polybutadiene
polibuteno polybutene
polibutileno polybutylene
policarbonato polycarbonate
policarboxílico polycarboxylic
policíclico polycyclic
policloral polychloral
policloro polychlor
policloropreno polychloroprene
policlorotrifluoroetano
    polychlorotrifluoroethane
policondensación polycondensation
policromático polychromatic
polidimetilsiloxano polydimethylsiloxane
polidispersión polydispersion
polielectrólito polyelectrolyte
polieno polyene
poliespiro polyspiro
poliéster polyester
poliéster de sucrosa sucrose polyester
poliestireno polystyrene
polietenoide polyethenoid
poliéter polyether
poliéter clorado chlorinated polyether
polieterglicól polyether glycol
polietilenglicol polyethylene glycol
polietilenimina polyethyleneimine
polietileno polyethylene
polifásico polyphase
polifenilo polyphenyl
poliformación polyforming
poliformaldehído polyformaldehyde
polifosfato polyphosphate
polifosfato amónico ammonium
    polyphosphate
polifosfato sódico sodium polyphosphate
polifuncional polyfunctional
polígeno polygen
poliglicerol polyglycerol
poliglicol polyglycol
polihalita polyhalite
polihexafluoropropeno
    polyhexafluoropropene
polihidrato polyhydrate
polihídrico polyhydric

poliimida polyimide
poliinsaturado polyunsaturated
poliisobuteno polyisobutene
poliisobutileno polyisobutylene
poliisocianurato polyisocyanurate
poliisopreno polyisoprene
polimérico polymeric
polimerismo polymerism
polimerización polymerization
polimerización aniónica anionic
    polymerization
polimerización catalítica catalytic
    polymerization
polimerización catiónica cationic
    polymerization
polimerización de vinilo vinyl
    polymerization
polimerización de Ziegler-Natta
    Ziegler-Natta polymerization
polimerización en emulsión emulsion
    polymerization
polimerización iónica ionic
    polymerization
polimerizar polymerize
polímero polymer
polímero acrílico acrylic polymer
polímero alternante alternating polymer
polímero amorfo amorphous polymer
polímero atáctico atactic polymer
polímero con flúor fluorine polymer
polímero con injerto graft polymer
polímero de ácido acrílico acrylic acid
    polymer
polímero de acrilamida acrylamide
    polymer
polímero de acrilonitrilo acrylonitrile
    polymer
polímero de acroleína acrolein polymer
polímero de adición addition polymer
polímero de aldehído aldehyde polymer
polímero de alilo allyl polymer
polímero de cadena chain polymer
polímero de clorotrifluoroetileno
    chlorotrifluoroethylene polymer
polímero de cloruro de vinilideno
    vinylidene chloride polymer
polímero de condensación condensation
    polymer
polímero de coordinación coordination

polymer
polímero de estereobloques stereoblock
    polymer
polímero de estireno styrene polymer
polímero de eteno ethene polymer
polímero de fluorocarburo fluorocarbon
    polymer
polímero de isopreno isoprene polymer
polímero de piridina pyridine polymer
polímero de propeno propene polymer
polímero de tetrahidrofurano
    tetrahydrofuran polymer
polímero de vinilo vinyl polymer
polímero en escalera ladder polymer
polímero epóxido epoxy polymer
polímero estereoespecífico stereospecific
    polymer
polímero estereorregular stereoregular
    polymer
polímero inorgánico inorganic polymer
polímero lineal linear polymer
polímero nitroso nitroso polymer
polímero regular regular polymer
polímero táctico tactic polymer
polimetafosfato potásico potassium
    polymetaphosphate
polimetanal polymethanal
polimetilbenceno polymethylbenzene
polimetilenglicol polymethylene glycol
polimetilpenteno polymethylpentene
polimezcla polyblend
polimixina polymyxin
polimorfismo polymorphism
polinuclear polynuclear
polinuclídico polynuclidic
poliol polyol
poliolefina polyolefin
poliolefina clorada chlorinated
    polyolefin
polioxadiazol polyoxadiazole
polioxamida polyoxamide
polioxi polyoxy
polioxietileno polyoxyethylene
polioxietilenoxipropileno
    polyoxyethyleneoxypropylene
polioximetileno polyoxymethylene
polioxipropilendiamina
    polyoxypropylenediamine
polipéptido polypeptide

polipirrolidina polypyrrolidine
poliprenol polyprenol
polipropilenbenceno
    polypropylenebenzene
polipropilenglicol polypropylene glycol
polipropileno polypropylene
polirrotaxama polyrotaxme
polisacárido polysaccharide
polisilicato cálcico calcium polysilicate
polisiloxano polysiloxane
polisorbato polysorbate
polisulfona polysulfone
polisulfuro polysulfide
polisulfuro amónico ammonium
    polysulfide
polisulfuro potásico potassium
    polysulfide
polisulfuro sódico sodium polysulfide
politeno polythene
politetrafluoroeteno
    polytetrafluoroethene
politetrafluoroetileno
    polytetrafluoroethylene
politiacilo polythiazyl
politiadazol polythiadazole
politionato polythionate
poliuretano polyurethane
polivalencia polyvalence
polivalente polyvalent
polivinilacetal polyvinyl acetal
polivinilbutiral polyvinyl butyral
polivinilcarbazol polyvinyl carbazole
polivinilideno polyvinylidene
polivinilo polyvinyl
polivinilpirrolidina polyvinylpyrrolidine
polixematileno polyhexamethylene
poliyoduro polyiodide
polonio polonium
polucita pollucite
polvo powder, dust
polvo azul blue powder
polvo de blanqueo bleaching powder
polvo de cinc zinc dust
polvo de hierro iron powder
polvo de hornear baking powder
polvo negro black powder
pólvora gunpowder
pólvora de algodón guncotton
pólvora sin humo smokeless powder

pómez pumice
porcelana porcelain
porcentaje percentage
porfina porphin
porfirina porphyrin
porfirinógeno porphyrinogen
poro pore
porométrico porometric
porosidad porosity
porosímetro porosimeter
poroso porous
portador carrier
portador de catalizador catalyst carrier
portador de oxígeno oxygen carrier
portadora de hidrógeno hydrogen
    carrier
posición adjacente adjacent position
posición alfa alpha position
posición beta beta position
positivo positive
positrón positron
positronio positronium
posprecipitación postprecipitation
potasa potash
potasa cáustica caustic potash
potasán potasan
potásico potassic
potasio potassium
potencial potential
potencial atómico atomic potential
potencial crítico critical potential
potencial de contacto contact potential
potencial de deposición deposition
    potential
potencial de electrodo electrode
    potential
potencial de equilibrio equilibrium
    potential
potencial de ionización ionization
    potential, ionizing potential
potencial de Morse Morse potential
potencial de Nernst Nernst potential
potencial de oxidación oxidation
    potential
potencial de oxidación-reducción
    oxidation-reduction potential
potencial de polarización polarization
    potential
potencial de redox redox potential

potencial de reducción reduction
  potential
potencial de sedimentación
  sedimentation potential
potencial electrocinético electrokinetic
  potential
potencial electrolítico electrolytic
  potential
potencial electronegativo electronegative
  potential
potencial electropositivo electropositive
  potential
potencial electroquímico electrochemical
  potential
potencial noble noble potential
potencial químico chemical potential
potencial termodinámico
  thermodynamic potential
potencial zeta zeta potential
potenciometría redox redox
  potentiometry
potenciométrico potentiometric
praseodimio praseodymium
precauciones de almacenamiento
  storage precautions
precipitabilidad precipitability
precipitable precipitable
precipitación precipitation
precipitación ácida acid precipitation
precipitación fraccionaria fractional
  precipitation
precipitación isoeléctrica isoelectric
  precipitation
precipitación rítmica rhythmic
  precipitation
precipitado precipitate, deposit
precipitado amarillo yellow precipitate
precipitado blanco white precipitate
precipitado coherente coherent
  precipitate
precipitado rojo red precipitate
precipitador precipitator
precipitador electrostático electrostatic
  precipitator
precipitante precipitant
precursor precursor
predisociación predissociation
prednisolona prednisolone
prednisona prednisone

preferencial preferential
pregananodiol pregnanediol
pregnenodiona pregnenedione
pregnenolona pregnenolone
prehnita prehnite
prenilo prenyl
preparación preparation
prepolímero prepolymer
preservativo preservative
presión pressure
presión ambiente ambient pressure
presión atmosférica atmospheric
  pressure
presión barométrica barometric pressure
presión constante constant pressure
presión crítica critical pressure
presión de disociación dissociation
  pressure
presión de disolución solution pressure
presión de gas gas pressure
presión de vapor vapor pressure
presión normal normal pressure,
  standard pressure
presión osmótica osmotic pressure
presión parcial partial pressure
presión superficial surface pressure
primario primary
primera ley de termodinámica first law
  of thermodynamics
principal principal
principio amargo bitter principle
principio de correspondencia
  correspondence principle
principio de distribución distribution
  principle
principio de exclusión exclusion
  principle
principio de exclusión de Pauli Pauli
  exclusion principle
principio de exclusión de Pauli-Fermi
  Pauli-Fermi exclusion principle
principio de Franck-Condon Franck-
  Condon principle
principio de incertidumbre uncertainty
  principle
**principio de incertidumbre de
  Heisenberg** Heisenberg uncertainty
  principle
principio de Le Chatelier Le Chatelier

principle
principio de Pauli Pauli principle
prisma prism
prisma de Nicol Nicol prism
pristano pristane
probeta test tube
procedimento de Pregl Pregl procedure
procedimiento procedure
procedimiento de Darzens Darzens
  procedure
procedimiento de Wilzbach Wilzbach
  procedure
procesamiento processing
procesamiento químico chemical
  processing
proceso process
proceso adiabático adiabatic process
proceso cíclico cyclic process
proceso de Acker Acker process
proceso de Acheson Acheson process
proceso de amalgamación amalgamation
  process
proceso de Barff Barff process
proceso de Bayer Bayer process
proceso de Benfield Benfield process
proceso de Bergius Bergius process
proceso de Bessemer Bessemer process
proceso de Betterton-Kroll Betterton-
  Kroll process
proceso de Betts Betts process
proceso de Birkeland-Eyde Birkeland-
  Eyde process
proceso de Borcher Borcher process
proceso de Bosch Bosch process
proceso de Brewster Brewster process
proceso de Brin Brin process
proceso de Castner Castner process
proceso de cianamida cyanamide
  process
proceso de cianuro cyanide process
proceso de Clark Clark process
proceso de Claude Claude process
proceso de Coahran Coahran process
proceso de contacto contact process
proceso de conversión conversion
  process
proceso de copelación cupellation
  process
proceso de Creighton Creighton process

proceso de cumeno cumene process
proceso de Chance-Claus Chance-Claus
  process
proceso de Deacon Deacon process
proceso de Downs Downs process
proceso de Edeleanu Edeleanu process
proceso de Fament Fament process
proceso de Farrar Farrar process
proceso de fenolato phenolate process
proceso de Fischer-Tropsch Fischer-
  Tropsch process
proceso de Frasch Frasch process
proceso de Goldschmidt Goldschmidt
  process
proceso de Haber Haber process
proceso de Hall Hall process
proceso de Hargreaves Hargreaves
  process
proceso de Houdriflow Houdriflow
  process
proceso de Houdry Houdry process
proceso de Hunter Hunter process
proceso de Keyes Keyes process
proceso de Kroll Kroll process
proceso de Lebedev Lebedev process
proceso de Linde Linde process
proceso de Lyovac Lyovac process
proceso de Marathon-Howard
  Marathon-Howard process
proceso de Mond Mond process
proceso de oxidación oxidation process
proceso de Parkes Parkes process
proceso de Pattinson Pattinson process
proceso de Raschig Raschig process
proceso de recombinación
  recombination process
proceso de Reich Reich process
proceso de Reppe Reppe process
proceso de Roesler Roesler process
proceso de Rosenstein Rosenstein
  process
proceso de Siemens Siemens process
proceso de Siemens-Halske Siemens-
  Halske process
proceso de Simons Simons process
proceso de Sperry Sperry process
proceso de Steffen Steffen process
proceso de Stengel Stengel proccss
proceso de Thermofor Thermofor

process
proceso de Toth Toth process
proceso de Trona Trona process
proceso de Twitchell Twitchell process
proceso de Wacker Wacker process
proceso de Weldon Weldon process
proceso de Wohlwill Wohlwill process
proceso de Wulff Wulff process
proceso de Ziegler Ziegler process
proceso de Ziervogel Ziervogel process
proceso electrolítico electrolytic process
proceso electroquímico electrochemical
    process
proceso elemental elementary process
proceso en cascada cascade process
proceso estocástico stochastic process
proceso fotoquímico photochemical
    process
proceso irreversible irreversible process
proceso isentrópico isentropic process
proceso isotérmico isothermal process
proceso nuclear nuclear process
proceso químico chemical process
proceso radioquímico radiochemical
    process
proceso reversible reversible process
proceso termatómico thermatomic
    process
producción production
producción quimionuclear
    chemonuclear production
producto product
producto de desintegración decay
    product
producto de desintegración radiactiva
    radioactive decay product
producto de fisión fission product
producto de fusión fusion product
producto de reacción reaction product
producto de solubilidad solubility
    product
producto de volatilidad volatility
    product
producto iónico ionic product
producto químico chemical product,
    chemical
producto químico agrícola agricultural
    chemical
producto químico alelopático

allelopathic chemical
producto químico fino fine chemical
producto químico nocivo noxious
    chemical
producto químico peligroso hazardous
    chemical
producto químico pesado heavy
    chemical
producto químico seco dry chemical
producto radiactivo radioactive product
producto secundario by-product
productor de neutrones neutron
    producer
progesterona progesterone
proguanil proguanil
prolactina prolactin
prolamina prolamine
prolilo prolyl
prolina proline
promazina promazine
prometazina promethazine
prometio promethium
promotor promoter
promotor de catalizador catalyst
    promotor
promotor de reacción reaction promotor
propadieno propadiene
propagación propagation
propanal propanal
propanamida propanamide
propanil propanil
propano propane
propanodiamina propanediamine
propanodiol propanediol
propanoilo propanoyl
propanol propanol
propanolamina propanolamine
propanolpiridina propanolpyridine
propanona propanone
propanonitrilo propanenitrile
propanosultona propane sultone
propanotiol propanethiol
propargilo propargyl
propargita propargite
propelente propellant
propenal propenal
propenilamina propenylamine
propenilanisol propenylanisole
propenileno propenylene

propenilguaetol propenyl guaethol
propenilideno propenylidene
propenilo propenyl
propeniltiourea propenylthiourea
propeno propene
propenol propenol
propenonitrilo propenenitrile
propenotiol propenethiol
propidedad periódica periodic property
propiedad property
propiedad coligativa colligative property
propiedad física physical property
propiedad química chemical property
propiedades atómicas atomic properties
propiedades de grupo group properties
propiedades intensivas intensive properties
propiedades seudocríticas pseudocritical properties
propiedades seudorreducidas pseudoreduced properties
propiedades tóxicas toxic properties
propilacetona propylacetone
propilamina propylamine
propilanilina propylaniline
propilbenceno propylbenzene
propilendiamina propylenediamine
propilenglicol propylene glycol
propilenimina propyleneimine
propileno propylene
propilhidroxilamina propylhydroxylamine
propilideno propylidene
propilmercaptano propyl mercaptan
propilo propyl
propilparabeno propylparaben
propilpiperidina propylpiperidine
propilpiridina propylpyridine
propiltriclorosilano propyltrichlorosilane
propilurea propylurea
propinal propynal
propinilo propynyl
propino propyne
propinol propynol
propiofenona propiophenone
propiolactona propiolactone
propiolato de etilo ethyl propiolate
propioloilo propioloyl
propiona propione

propionaldehído propionaldehyde
propionato propionate
propionato bárico barium propionate
propionato cálcico calcium propionate
propionato cobaltoso cobaltous propionate
propionato de amilo amyl propionate
propionato de bencilo benzyl propionate
propionato de bismuto bismuth propionate
propionato de butilo butyl propionate
propionato de cadmio cadmium propionate
propionato de celulosa cellulose propionate
propionato de cinc zinc propionate
propionato de etilo ethyl propionate
propionato de geranilo geranyl propionate
propionato de glicol glycol propionate
propionato de isoamilo isoamyl propionate
propionato de isobutilo isobutyl propionate
propionato de linalilo linalyl propionate
propionato de metilo methyl propionate
propionato de pentilo pentyl propionate
propionato de plomo lead propionate
propionato de propilo propyl propionate
propionato de vinilo vinyl propionate
propionato fenilmercúrico phenylmercuric propionate
propionato magnésico magnesium propionate
propionato sódico sodium propionate
propionilbenceno propionylbenzene
propionilo propionyl
propionitrilo propionitrile
propiono propiono
proporción proportion
proporción constante constant proportion
proporción fija fixed proportion
proporcional proportional
proporciones definidas definite proportions
proporciones equivalentes equivalent proportions
propoxi propoxy

propóxido alumínico aluminum
   propoxide
propoxifeno propoxyphene
propoxipropanol propoxypropanol
propranonol propranonol
proquiral prochiral
prostaglandina prostaglandin
protactinio protactinium
protamina protamine
proteasa protease
protección ambiental environmental
   protection
protección catódica cathodic protection
protección contra fuego fire protection
protección contra humedad humidity
   protection
protección sacrificial sacrificial
   protection
protector químico chemical protector
proteína protein
proteína hidrolizada hydrolized protein
proteólisis proteolysis
protilo protyl
protio protium
proto proto
protocloruro de estaño tin protochloride
protocloruro de hierro iron
   protochloride
protofílico protophilic
protogénico protogenic
protólisis protolysis
protón proton
protón nuclear nuclear proton
protón retardado delayed proton
protonación protonation
protoplasma protoplasm
protosulfato protosulfate
protosulfuro de estaño tin protosulfide
protosulfuro de hierro iron protosulfide
prototrópico prototropic
protóxido bárico barium protoxide
protóxido de estaño tin protoxide
protóxido de manganeso manganese
   protoxide
protóxido de níquel nickel protoxide
protóxido de plomo lead protoxide
protrombina prothrombin
provitamina provitamin
proyección projection

proyección de Fischer Fischer projection
proyección de Haworth Haworth
   projection
proyección de Newman Newman
   projection
proyección en cabrilla sawhorse
   projection
prueba test, proof
prueba con llama flame test
prueba cualitativa qualitative test
prueba cuantitativa quantitative test
prueba de acroleína acrolein test
prueba de Almen Almen test
prueba de anillo ring test
prueba de anillo pardo brown-ring test
prueba de azufre sulfur test
prueba de Babcock Babcock test
prueba de banco bench test
prueba de Barfoed Barfoed test
prueba de Beilstein Beilstein test
prueba de bencidina benzidine test
prueba de Benedict Benedict test
prueba de Bettendorf Bettendorf test
prueba de biuret biuret test
prueba de Boettger Boettger test
prueba de Brown-Boverti Brown-
   Boverti test
prueba de color de Rosenheim
   Rosenheim color test
prueba de dureza de Brinell Brinell
   hardness test
prueba de Fehling Fehling test
prueba de Foulger Foulger test
prueba de Hinsberg Hinsberg test
prueba de Kjeldahl Kjeldahl test
prueba de Lassaigne Lassaigne test
prueba de Marsh Marsh test
prueba de Millon Millon test
prueba de Molisch Molisch test
prueba de perla bead test
prueba de perla con bórax borax-bead
   test
prueba de Reinsch Reinsch test
prueba de Riegler Riegler test
prueba de Schiff Schiff test
prueba de solubilidad solubility test
prueba de Tauber Tauber test
prueba de Tollens Tollens test
prueba de yodo iodine test

prueba física physical test

prueba no destructiva nondestructive test

prueba ordinal ordinal test

prueba química chemical test

prusiato amarillo de potasa yellow prussiate of potash

prusiato amarillo de sosa yellow prussiate of soda

prusiato potásico potassium prussiate

prusiato sódico sodium prussiate

pseudo pseudo

psi psi

psilocibina psilocybin

psilocina psilocin

psilomelano psilomelane

pteridina pteridine

pterina pterin

ptialina ptyalin

ptomaína ptomaine

puente bridge

puente hidrídico hydridic bridge

puente protónico protonic bridge

puente salino salt bridge

pulegona pulegone

pulpa pulp

pulpa de madera wood pulp

pulpa de sosa soda pulp

pulpa de sulfito sulfite pulp

pululanasa pullulanase

pulverización pulverization, comminution

pulverización catódica cathodic sputtering, sputtering

punta peak

punto crítico critical point

punto cuádruple quadruple point

punto de anilina aniline point

punto de autoignición autoignition point

punto de condensación al vacío vacuum condensing point

punto de congelación freezing point, ice point

punto de Curie Curie point

punto de descomposición decomposition point

punto de ebullición boiling point

punto de ebullición absoluto absolute boiling point

punto de ebullición inicial initial boiling point

punto de equilibrio equilibrium point

punto de equivalencia equivalence point

punto de fluidez pour point

punto de flujo pour point

punto de fusión melting point, fusion point

punto de fusión mixto mixed melting point

punto de ignición ignition point

punto de inflamabilidad flash point

punto de inversión inversion point

punto de rocío dew point

punto de saturación saturation point

punto de solidificación setting point

punto de sublimación sublimation point

punto de transición transition point

punto de turbidez turbidity point

punto de vapor steam point

punto eutéctico eutectic point

punto final end point

punto invariante invariant point

punto isobéstico isobestic point

punto isoeléctrico isoelectric point

punto lambda lambda point

punto peritéctico peritectic point

punto seco dry point

punto triple triple point

pureza purity, fineness

pureza colorimétrica colorimetric purity

pureza óptica optical purity

pureza química chemical purity

purificación purification

purificación de agua water purification

purina purine

puro pure

puromicina puromycin

púrpura de bromocresol bromocresol purple

púrpura de cresol cresol purple

púrpura de Tiro Tyrian purple

purpurina purpurin

putrefacción putrefaction

putrescina putrescine

Q

quark quark
quebracho quebracho
quebradizo brittle
queiramidina cheiramidine
quelación chelation
quelato chelate
quelato de titanio titanium chelate
queleritrina chelerythrine
quemado burnt
quemadura química chemical burn
queratina keratin
queratinasa keratinase
quercitol quercitol
quercitrina quercitrin
queroseno kerosene
quevkinita chevkinite
quianita kyanite
quilate karat, carat
química chemistry
química ambiental environmental chemistry
química analítica analytical chemistry
química aplicada applied chemistry
química bioinorgánica bioinorganic chemistry
química biomimética biomimetic chemistry
química clínica clinical chemistry
química computacional computational chemistry
química cuántica quantum chemistry
química de estado sólido solid-state chemistry
química de radiación radiation chemistry
química física physical chemistry
química forense forensic chemistry
química industrial industrial chemistry
química inorgánica inorganic chemistry
química legal legal chemistry
química medicinal medicinal chemistry
química nuclear nuclear chemistry
química orgánica organic chemistry

química superficial surface chemistry
química supramolecular supramolecular chemistry
químicamente inducido chemically induced
químicamente puro chemically pure
químico chemist
quimiodinámica chemodynamics
quimiólisis chemolysis
quimioluminiscencia chemiluminescence
quimiometría chemometrics
quimionuclear chemonuclear
quimiosíntesis chemosynthesis
quimiósmosis chemosmosis
quimiosmótico chemosmotic
quimiosorción chemosorption
quimióstato chemostat
quimiotaxis chemotaxis
quimiotaxonomía chemotaxonomy
quimioterapia chemotherapy
quimiotropismo chemotropism
quimisorción chemisorption
quimosina chymosin
quimotripsina chymotrypsin
quimotripsinógeno chymotrypsinogen
quina cinchona bark
quinacrina quinacrine
quinaftol quinaphthol
quinaldina quinaldine
quinamina quinamine
quinazolina quinazoline
quinhidrona quinhydrone
quinidina quinidine
quinina quinine
quininona quininone
quinitol quinitol
quinizarina quinizarin
quinógeno quinogen
quinoidina chinoidine
quinoilo quinoyl
quinol quinol
quinolilo quinolyl
quinolina quinoline

quinolinio quinolinium
quinolinol quinolinol
quinolinona quinolinone
quinolizina quinolizine
quinolona quinolone
quinona quinone
quinonadioxima quinonedioxime
quinonilo quinonyl
quinosol quinosol
quinoxalina quinoxaline
quinoxalinilo quinoxalinyl
quinquevalente quinquevalent

quinteto quintet
quintoceno quintozene
quinuclidina quinuclidine
quinurenina kynurenine
quiral chiral
quiralidad chirality, handedness
quiroóptico chirooptical
quitapintura paint remover
quitina chitin
quitinasa chitinase
quitosana chitosan

R

racema raceme
racemación racemation
racemasa racemase
racemato racemate
racémico racemic
racemización racemization
racemización parcial partial racemization
rad rad
radiación radiation
radiación actínica actinic radiation
radiación beta beta radiation
radiación dañina harmful radiation
radiación de fondo background radiation
radiación de fuga leakage radiation
radiación de ionización ionizing radiation
radiación de resonancia resonance radiation
radiación dipolar dipole radiation
radiación electromagnética electromagnetic radiation
radiación fluorescente fluorescent radiation
radiación gamma gamma radiation
radiación infrarroja infrared radiation
radiación nuclear nuclear radiation
radiación primaria primary radiation
radiación secundaria secondary radiation
radiación ultravioleta ultraviolet radiation
radiactividad radioactivity
radiactividad artificial artificial radioactivity
radiactivo radioactive
radiado radiated
radiador radiator
radiador alfa alpha radiator
radiancia radiance
radiante radiant, radiating
radiar radiate
radiativo radiative

radical radical
radical de ácido acid radical
radical electronegativo electronegative radical
radical enmascarado masked radical
radical libre free radical
radical negativo negative radical
radical orgánico organic radical
radical positivo positive radical
radio radium, radius
radio atómico atomic radius
radio covalente covalent radius
radio de Bohr Bohr radius
radio de electrón clásico classical electron radius
radio de van der Waals van der Waals radius
radio hidrodinámico hydrodynamic radius
radio iónico ionic radius
radio nuclear nuclear radius
radioactinio radioactinium
radioanálisis radioassay
radiobario radiobarium
radiobiología radiobiology
radiocarbono radiocarbon
radiocesio radiocesium
radiocristalografía radiocrystallography
radiocromatografía radiochromatography
radioelemento radioelement
radiogénico radiogenic
radiografía radiography
radioinmunoanálisis radioimmunoassay
radioisótopo radioisotope
radiólisis radiolysis
radiolita radiolite
radiología radiology
radiológico radiological
radioluminiscencia radioluminescence
radiometalografía radiometallography
radiometría radiometry
radiométrico radiometric

radiómetro radiometer
radiomimético radiomimetic
radión radion
radionitrógeno radionitrogen
radionucleido radionuclide
radioplomo radiolead
radioquímica radiochemistry
radioquímico radiochemical
radiosodio radiosodium
radiotelurio radiotellurium
radiotor radiothor
radiotorio radiothorium
radioyodo radioiodine
radón radon
rafinato raffinate
rafinosa raffinose
ralstonita ralstonite
ramificación branching
ramificado branched
ramio ramie
ramnitol rhamnitol
ramnosa rhamnose
rancio rancid, sour
rankinita rankinite
ranksita ranksite
rápido fast
rarefacción rarefaction
rarefactor getter
rareficar rarefy
rasorita rasorite
rauwolfia rauwolfia
rayo ray
rayo beta beta ray
rayo delta delta ray
rayón rayon
rayos alfa alpha rays
rayos canales canal rays
rayos catódicos cathodic rays
rayos de Becquerel Becquerel rays
rayos gamma gamma rays
rayos moleculares molecular rays
razón ratio
razón de abundancia abundance ratio
razón de carbono carbon ratio
razón de carga a masa charge-mass
  ratio
razón de conversión conversion ratio
razón de dilución dilution ratio
razón de equilibrio equilibrium ratio

razón de masa a carga mass-to-charge
  ratio
razón de mezcla de gases gas admixture
  ratio
razón de neutralización neutralization
  ratio
razón de radios radius ratio
razón giromagnética gyromagnetic ratio
razón isotópica isotopic ratio
reacción reaction
reacción ácida acid reaction
reacción alcalina alkaline reaction
reacción analítica analytical reaction
reacción balanceada balanced reaction
reacción bimolecular bimolecular
  reaction
reacción caótica chaotic reaction
reacción compleja complex reaction
reacción con llama flame reaction
reacción cualitativa qualitative reaction
reacción cuantitativa quantitative
  reaction
reacción de Acree-Rosenheim Acree-
  Rosenheim reaction
reacción de Adamkiewicz Adamkiewicz
  reaction
reacción de adición addition reaction
reacción de Adkins-Peterson Adkins-
  Peterson reaction
reacción de Allan-Robinson Allan-
  Robinson reaction
reacción de anillo ring reaction
reacción de Bart Bart reaction
reacción de Barton Barton reaction
reacción de Baudisch Baudisch reaction
reacción de Béchamp Béchamp reaction
reacción de Beckmann Beckmann
  reaction
reacción de Belousov-Zhabotinskii
  Belousov-Zhabotinskii reaction
reacción de Benary Benary reaction
reacción de Berthelot Berthelot reaction
reacción de Betti Betti reaction
reacción de Biginelli Biginelli reaction
reacción de biuret biuret reaction
reacción de Blaise Blaise reaction
reacción de Blanc Blanc reaction
reacción de Bodroux Bodroux reaction
reacción de Bohn-Schmidt Bohn-

Schmidt reaction

reacción de Bradsher Bradsher reaction

reacción de Bucherer Bucherer reaction

reacción de Bucherer-Bergs Bucherer-Bergs reaction

reacción de cadena ramificada branched-chain reaction

reacción de Cannizzaro Cannizzaro reaction

reacción de Carroll Carroll reaction

reacción de ciclización de Nazarov Nazarov cyclization reaction

reacción de Claisen Claisen reaction

reacción de condensación condensation reaction

reacción de condensación de Mannich Mannich condensation reaction

reacción de Conrad-Limpach Conrad-Limpach reaction

reacción de Criegee Criegee reaction

reacción de Curtius Curtius reaction

reacción de Chichibabin Chichibabin reaction

reacción de Chugaev Chugaev reaction

reacción de Dakin Dakin reaction

reacción de Darzens Darzens reaction

reacción de De Mayo De Mayo reaction

reacción de Delepine Delepine reaction

reacción de desplazamiento displacement reaction

reacción de Dieckmann Dieckmann reaction

reacción de Diels-Alder Diels-Alder reaction

reacción de Dische Dische reaction

reacción de disociación dissociation reaction

reacción de Doebner Doebner reaction

reacción de Doebner-Miller Doebner-Miller reaction

reacción de Dragendorff Dragendorff reaction

reacción de Duff Duff reaction

reacción de Dutt-Wormall Dutt-Wormall reaction

reacción de Ehrlich-Sachs Ehrlich-Sachs reaction

reacción de Einhorn-Brunner Einhorn-Brunner reaction

reacción de Elbs Elbs reaction

reacción de eliminación elimination reaction

reacción de eliminación de Cope Cope elimination reaction

reacción de Eltekoff Eltekoff reaction

reacción de Emmert Emmert reaction

reacción de enaminas de Stork Stork enamine reaction

reacción de Étard Étard reaction

reacción de Fenton Fenton reaction

reacción de Ferrario Ferrario reaction

reacción de Feulgen Feulgen reaction

reacción de Finkelstein Finkelstein reaction

reacción de Fischer-Tropsch Fischer-Tropsch reaction

reacción de Fittig Fittig reaction

reacción de Flood Flood reaction

reacción de Forster Forster reaction

reacción de fragmentación de Grob Grob fragmentation reaction

reacción de Franchimont Franchimont reaction

reacción de Frankland Frankland reaction

reacción de Frankland-Duppa Frankland-Duppa reaction

reacción de Friedel-Crafts Friedel-Crafts reaction

reacción de Fujimoto-Belleau Fujimoto-Belleau reaction

reacción de Fujiwara Fujiwara reaction

reacción de Gabriel Gabriel reaction

reacción de Gattermann Gattermann reaction

reacción de Gattermann-Koch Gattermann-Koch reaction

reacción de Goldschmidt Goldschmidt reaction

reacción de Gomberg Gomberg reaction

reacción de Gomberg-Bachmann Gomberg-Bachmann reaction

reacción de Gould-Jacobs Gould-Jacobs reaction

reacción de Graebe-Ullman Graebe-Ullman reaction

reacción de Griess Griess reaction

reacción de Grignard Grignard reaction

reacción de Guerbet Guerbet reaction

reacción de haces cruzadas crossed-beam reaction

reacción de Haller-Bauer Haller-Bauer reaction

reacción de Hammick Hammick reaction

reacción de Hell-Volhard-Zelinsky Hell-Volhard-Zelinsky reaction

reacción de Henkel Henkel reaction

reacción de Henry Henry reaction

reacción de Herz Herz reaction

reacción de Hilbert-Johnson Hilbert-Johnson reaction

reacción de Hill Hill reaction

reacción de Hinsberg Hinsberg reaction

reacción de Hofmann Hofmann reaction

reacción de Hofmann-Sand Hofmann-Sand reaction

reacción de Hooker Hooker reaction

reacción de Hopkins-Cole Hopkins-Cole reaction

reacción de Houben-Hoesch Houben-Hoesch reaction

reacción de Hunsdiecker Hunsdiecker reaction

reacción de indofenina indophenine reaction

reacción de inserción insertion reaction

reacción de intercambio exchange reaction

reacción de Janovsky Janovsky reaction

reacción de Japp-Klingemann Japp-Klingemann reaction

reacción de Kendall-Mattox Kendall-Mattox reaction

reacción de Knoevenagel Knoevenagel reaction

reacción de Kochi Kochi reaction

reacción de Kolbe Kolbe reaction

reacción de Komarowsky Komarowsky reaction

reacción de Kucherov Kucherov reaction

reacción de Kuhn-Winterstein Kuhn-Winterstein reaction

reacción de Lehmsted-Tanasescu Lehmsted-Tanasescu reaction

reacción de Leuckart Leuckart reaction

reacción de Lieben Lieben reaction

reacción de Liebermann Liebermann reaction

reacción de Maillard Maillard reaction

reacción de Malaprade Malaprade reaction

reacción de Mannich Mannich reaction

reacción de McFayden-Stevens McFayden-Stevens reaction

reacción de Mendius Mendius reaction

reacción de Menschutkin Menschutkin reaction

reacción de Meyer Meyer reaction

reacción de Michael Michael reaction

reacción de Michaelis-Arbuzov Michaelis-Arbuzov reaction

reacción de Mignonac Mignonac reaction

reacción de Mitsunobu Mitsunobu reaction

reacción de Nef Nef reaction

reacción de Nencki Nencki reaction

reacción de Nierenstein Nierenstein reaction

reacción de orden cero zero-order reaction

reacción de oxidación oxidation reaction

reacción de oxidación-reducción oxidation-reduction reaction

reacción de ozónidos de Harries Harries ozonide reaction

reacción de Passerini Passerini reaction

reacción de Pasteur Pasteur reaction

reacción de Paterno-Buchi Paterno-Buchi reaction

reacción de Pellizzari Pellizzari reaction

reacción de Perkin Perkin reaction

reacción de Perkow Perkow reaction

reacción de Peterson Peterson reaction

reacción de Pfitzinger Pfitzinger reaction

reacción de Pictet-Hubert Pictet-Hubert reaction

reacción de Pinner Pinner reaction

reacción de Piria Piria reaction

reacción de Polonovski Polonovski reaction

reacción de Pomeranz-Fritsch Pomeranz-Fritsch reaction

reacción de Ponzio Ponzio reaction
reacción de Prevost Prevost reaction
reacción de primer orden first-order reaction
reacción de Prins Prins reaction
reacción de Pschorr Pschorr reaction
reacción de Quelet Quelet reaction
reacción de Ramberg-Backlund Ramberg-Backlund reaction
reacción de redox redox reaction
reacción de Reed Reed reaction
reacción de Reformatsky Reformatsky reaction
reacción de Reimer Reimer reaction
reacción de Reimer-Tiemann Reimer-Tiemann reaction
reacción de Reissert Reissert reaction
reacción de reordenamiento rearrangement reaction
reacción de Reverdin Reverdin reaction
reacción de Ritter Ritter reaction
reacción de Robinson Robinson reaction
reacción de Rosenmund Rosenmund reaction
reacción de Rothemund Rothemund reaction
reacción de Sachse Sachse reaction
reacción de Sandmeyer Sandmeyer reaction
reacción de Schiemann Schiemann reaction
reacción de Schmidt Schmidt reaction
reacción de Scholl Scholl reaction
reacción de Schorigin Schorigin reaction
reacción de Schotten-Baumann Schotten-Baumann reaction
reacción de segundo orden second-order reaction
reacción de Semmler-Wolff Semmler-Wolff reaction
reacción de Serini Serini reaction
reacción de Sharpless Sharpless reaction
reacción de Simmons-Smith Simmons-Smith reaction
reacción de Simonini Simonini reaction
reacción de Sommelet Sommelet reaction
reacción de Staudinger Staudinger reaction

reacción de Stobbe Stobbe reaction
reacción de Strecker Strecker reaction
reacción de substitución substitution reaction
reacción de Sullivan Sullivan reaction
reacción de Swarts Swarts reaction
reacción de telomerización telomerization reaction
reacción de tercer orden third-order reaction
reacción de Thiele Thiele reaction
reacción de Thorpe Thorpe reaction
reacción de Tishchenko Tishchenko reaction
reacción de Tscherniac-Einhorn Tscherniac-Einhorn reaction
reacción de Ullman Ullman reaction
reacción de Varrentrapp Varrentrapp reaction
reacción de Vilsmeier-Haack Vilsmeier-Haack reaction
reacción de von Braun von Braun reaction
reacción de von Richter von Richter reaction
reacción de Wackenroder Wackenroder reaction
reacción de Wacker Wacker reaction
reacción de Wagner-Jauregg Wagner-Jauregg reaction
reacción de Wharton Wharton reaction
reacción de Whiting Whiting reaction
reacción de Willgerodt Willgerodt reaction
reacción de Williamson Williamson reaction
reacción de Wittig Wittig reaction
reacción de Wohl Wohl reaction
reacción de Wohl-Ziegler Wohl-Ziegler reaction
reacción de Wolff-Kishner Wolff-Kishner reaction
reacción de Wolffenstein-Boters Wolffenstein-Boters reaction
reacción de Wurtz Wurtz reaction
reacción de Wurtz-Fittig Wurtz-Fittig reaction
reacción de Ziesel Ziesel reaction
reacción de Zimmermann Zimmermann

reaction

reacción de Zincke-Suhl Zincke-Suhl
reaction

reacción electrocíclica electrocyclic
reaction

reacción electroorgánica electroorganic
reaction

reacción en cadena chain reaction

reacción en cadena nuclear nuclear
chain reaction

reacción endotérmica endothermic
reaction

reacción espontánea spontaneous
reaction

reacción estereoespecífica stereospecific
reaction

reacción exotérmica exothermic reaction

reacción explosiva explosive reaction

reacción fotonuclear photonuclear
reaction

reacción fotoquímica photochemical
reaction

reacción heterogénea heterogeneous
reaction

reacción heterolítica heterolytic reaction

reacción homogénea homogeneous
reaction

reacción homolítica homolytic reaction

reacción incompleta incomplete reaction

reacción inducida induced reaction

reacción inducida por neutrones
neutron-induced reaction

reacción interna internal reaction

reacción intramolecular intramolecular
reaction

reacción irreversible irreversible
reaction

reacción isotópica isotopic reaction

reacción lenta slow reaction

reacción metatética metathetical reaction

reacción neutra neutral reaction

reacción nuclear nuclear reaction

reacción nuclear directa direct nuclear
reaction

reacción obscura dark reaction

reacción oscilante oscillating reaction

reacción oxidante oxidizing reaction

reacción pericíclica pericyclic reaction

reacción primaria primary reaction

reacción química chemical reaction

reacción química lenta slow chemical
reaction

reacción química rápida fast chemical
reaction

reacción química reversible reversible
chemical reaction

reacción rápida fast reaction

reacción reductora reducing reaction

reacción reversible reversible reaction

reacción secundaria secondary reaction,
side reaction

reacción sigmatrópica sigmatropic
reaction

reacción simultánea simultaneous
reaction

reacción subatómica subatomic reaction

reacción superficial surface reaction

reacción termonuclear thermonuclear
reaction

reacción topoquímica topochemical
reaction

reacción unimolecular unimolecular
reaction

reacciones de Baeyer-Villiger Baeyer-
Villiger reactions

reactante reactant

reactivación reactivation

reactividad reactivity

reactividad química chemical reactivity

reactivo reagent

reactivo catiónico cationic reagent

reactivo de Abel Abel reagent

reactivo de Agulhon Agulhon reagent

reactivo de Barfoed Barfoed reagent

reactivo de Bessonoff Bessonoff reagent

reactivo de Bettendorf Bettendorf
reagent

reactivo de Bolton Bolton reagent

reactivo de Brady Brady reagent

reactivo de Carnot Carnot reagent

reactivo de Caro Caro reagent

reactivo de Cleland Cleland reagent

reactivo de Collman Collman reagent

reactivo de Denigè Denigè reagent

reactivo de Dobbin Dobbin reagent

reactivo de Dragendorff Dragendorff
reagent

reactivo de Erdmann Erdmann reagent

reactivo de Fehling Fehling reagent
reactivo de Fenton Fenton reagent
reactivo de Fischer Fischer reagent
reactivo de Fraude Fraude reagent
reactivo de Froehde Froehde reagent
reactivo de Gerard Gerard reagent
reactivo de Girard Girard reagent
reactivo de Griess Griess reagent
reactivo de Grignard Grignard reagent
reactivo de Günzberg Günzberg reagent
reactivo de Hanus Hanus reagent
reactivo de Huber Huber reagent
reactivo de Hübl Hübl reagent
reactivo de Ivanov Ivanov reagent
reactivo de Jacquemart Jacquemart
   reagent
reactivo de Jones Jones reageant
reactivo de Karl Fischer Karl Fischer
   reagent
reactivo de Klein Klein reagent
reactivo de Mandelin Mandelin reagent
reactivo de Marme Marme reagent
reactivo de Mayer Mayer reagent
reactivo de Millon Millon reagent
reactivo de Nessler Nessler reagent
reactivo de Normant Normant reagent
reactivo de Nylander Nylander reagent
reactivo de Obermayer Obermayer
   reagent
reactivo de Petroff Petroff reagent
reactivo de Scheibler Scheibler reagent
reactivo de Schiff Schiff reagent
reactivo de Schweitzer Schweitzer
   reagent
reactivo de Simmons-Smith Simmons-
   Smith reagent
reactivo de Stead Stead reagent
reactivo de Tollens Tollens reagent
reactivo de Twitchell Twitchell reagent
reactivo de Vilsmeier Vilsmeier reagent
reactivo de Wagner Wagner reagent
reactivo de Zerewitinoff Zerewitinoff
   reagent
reactivo electrofílico electrophilic
   reagent
reactivo polimérico polymeric reagent
reactor reactor
reactor de producción production
   reactor

reactor de torio thorium reactor
reactor de uranio uranium reactor
reactor engendrador breeder
reactor nuclear nuclear reactor
reactor térmico thermal reactor
recalescencia recalescence
receptor receptor, receiver
receptor de Bruehl Bruehl receiver
reciclado recycling
recipiente de Dewar Dewar flask
recocido annealing
recolección electrónica electron
   collection
reconstitución reconstitution
recorrido libre free path
recorrido libre medio mean free path
recristalización recrystallization
recristalizar recrystallize
rectificación rectification
rectivo de Halphen Halphen reagent
rectorita rectorite
recuperación recovery
red lattice
red centrada en el cuerpo body-centered
   lattice
red centrada en las caras face-centered
   lattice
red cristalina crystal lattice
red molecular molecular lattice
redestilación redistillation
redingita reddingite
redisposición rearrangement
redox redox
reducción reduction
reducción de Béchamp Béchamp
   reduction
reduccion de Bouveault-Blanc
   Bouveault-Blanc reduction
reducción de Clemmensen Clemmensen
   reduction
reducción de punto de congelación
   freezing-point reduction
reducción de punto de ebullición
   boiling-point reduction
reducción de Rosenmund Rosenmund
   reduction
reducción de Sabatier-Senderens
   Sabatier-Senderens reduction
reducción de Wolff-Kishner Wolff-

Kishner reduction
reducción diferencial differential
    reduction
reducción electrolítica electrolytic
    reduction
reducción electroquímica
    electrochemical reduction
reducción química chemical reduction
reducido reduced
reducir reduce
reductasa reductase
reductona reductone
reductor reducer, reductor
reductor de Jones Jones reductor
reemplazo replacement
reemplazo doble double replacement
referencia reference
refinación refining
refinación de petróleo petroleum
    refining
refinación por solvente solvent refining
refinado refined
refinamiento refinement
refinamiento por zonas zone refining
refinar refine
refinería refinery
reflujo reflux, refluxing
reflujo total total reflux
reformación reforming
reformación al vapor steam reforming
reformación catalítica catalytic
    reforming
reformación térmica thermal reforming
refracción refraction
refracción atómica atomic refraction
refracción doble double refraction
refractario refractory
refractividad refractivity
refractividad específica specific
    refractivity
refractivo refractive
refractómetro refractometer
refrigeración refrigeration
refrigerante refrigerant, coolant
regelación regelation
regeneración regeneration
regenerativo regenerative
región de Geiger-Müller Geiger-Müller
    region

región espectral spectral region
regioselectividad regioselectivity
registrador de pH pH recorder
regla de Abbegg Abbegg rule
regla de Alder-Rickert Alder-Rickert
    rule
regla de Aston Aston rule
regla de Badger Badger rule
regla de Barlow Barlow rule
regla de Blanc Blanc rule
regla de Bragg Bragg rule
regla de Bredt Bredt rule
regla de Compton Compton rule
regla de Cram Cram rule
regla de Dieterici Dieterici rule
regla de dilución dilution rule
regla de Dühring Dühring rule
regla de Erlenmeyer Erlenmeyer rule
regla de exclusión mutua mutual
    exclusion rule
regla de fases phase rule
regla de fases de Gibbs Gibbs phase rule
regla de Fries Fries rule
regla de Geiger-Nutall Geiger-Nutall
    rule
regla de Gibbs Gibbs rule
regla de Guldberg Guldberg rule
regla de Hofmann Hofmann rule
regla de Hückel Hückel rule
regla de Hume-Rothery Hume-Rothery
    rule
regla de intervalos de Landé Landé
    interval rule
regla de Konowaloff Konowaloff rule
regla de Kopp Kopp rule
regla de Kundt Kundt rule
regla de lactonas de Hudson Hudson
    lactone rule
regla de Leduc Leduc rule
regla de Magnus Magnus rule
regla de Markovnikoff Markovnikoff
    rule
regla de Markovnikov Markovnikov
    rule
regla de Nernst-Thomson Nernst-
    Thomson rule
regla de Ostwald Ostwald rule
regla de Paneth Paneth rule
regla de Pauli Pauli rule

regla de Perrin Perrin rule
regla de selección de Laporte Laporte selection rule
regla de Stuffer Stuffer rule
regla de Teller-Redlich Teller-Redlich rule
regla de Welter Welter rule
reglas de Alder-Stein Alder-Stein rules
reglas de Baldwin Baldwin rules
reglas de clorinación de Hass Hass chlorination rules
reglas de Fajans Fajans rules
reglas de Hammick-Illingworth Hammick-Illingworth rules
reglas de Hund Hund rules
reglas de isorrotación de Hudson Hudson isorotation rules
reglas de Pascal Pascal rules
reglas de secuencia sequence rules
reglas de selección selection rules
reglas de Woodward-Hoffmann Woodward-Hoffmann rules
regulador regulator
regulador de aire air regulator
regulador de gas gas regulator
regulador de temperatura temperature regulator
rehidratación rehydration
reinita reinite
rejalgar realgar
rejilla grating, grid
rejilla de difracción diffraction grating
rejilla escalonada echelon grating
relación de Keesom Keesom relationship
relación diagonal diagonal relationship
relajación relaxation
relajación molecular molecular relaxation
relatividad relativity
relativista relativistic
relativo relative
reloj radiactivo radioactive clock
relleno filler
rem rem
renato rhenate
renaturación renaturation
rendimiento yield
rendimiento cuántico quantum yield
rendimiento de fisión fission yield

rendimiento de fluorescencia fluorescence yield
rendimiento fotoquímico photochemical yield
renina rennin, renin
renio rhenium
reología rheology
reómetro rheometer
reordenamiento rearrangement
reordenamiento alílico allylic rearrangement
reordenamiento de Amadori Amadori rearrangement
reordenamiento de Beckmann Beckmann rearrangement
reordenamiento de Ciamician-Dennstedt Ciamician-Dennstedt rearrangement
reordenamiento de Claisen Claisen rearrangement
reordenamiento de Cope Cope rearrangement
reordenamiento de Curtius Curtius rearrangement
reordenamiento de Chapman Chapman rearrangement
reordenamiento de Demjanov Demjanov rearrangement
reordenamiento de dienona-fenol dienone-phenol rearrangement
reordenamiento de éteres de Wittig Wittig ether rearrangement
reordenamiento de Favorskii Favorskii rearrangement
reordenamiento de Fischer-Hepp Fischer-Hepp rearrangement
reordenamiento de Fries Fries rearrangement
reordenamiento de Gabriel-Colman Gabriel-Colman rearrangement
reordenamiento de Hayasi Hayashi rearrangement
reordenamiento de Hofmann-Martius Hofmann-Martius rearrangement
reordenamiento de Jacobsen Jacobsen rearrangement
reordenamiento de Ladenburg Ladenburg rearrangement
reordenamiento de Lossen Lossen rearrangement

reordenamiento de McLafferty
McLafferty rearrangement
reordenamiento de Meisenheimer
Meisenheimer rearrangement
reordenamiento de Neber Neber
rearrangement
reordenamiento de Perkin Perkin
rearrangement
reordenamiento de pinacol pinacol
rearrangement
reordenamiento de Pummerer
Pummerer rearrangement
reordenamiento de retropinacol
retropinacol rearrangement
reordenamiento de Rowe Rowe
rearrangement
reordenamiento de Smiles Smiles
rearrangement
reordenamiento de Sommelet-Hauser
Sommelet-Hauser rearrangement
reordenamiento de Stevens Stevens
rearrangement
reordenamiento de Stieglitz Stieglitz
rearrangement
reordenamiento de Tafel Tafel
rearrangement
reordenamiento de Tiemann Tiemann
rearrangement
reordenamiento de Wagner-Meerwein
Wagner-Meerwein rearrangement
reordenamiento de Wallach Wallach
rearrangement
reordenamiento de Wessely-Moser
Wessely-Moser rearrangement
reordenamiento de Westphalen-Lettre
Westphalen-Lettre rearrangement
reordenamiento de Wittig Wittig
rearrangement
reordenamiento de Wolff Wolff
rearrangement
reordenamiento molecular molecular
rearrangement
repelencia repellency
repelente repellent
repercolación repercolation
repetible repeatable
reprocesamiento reprocessing
reproducibilidad reproducibility
reproducible reproducible

reptación reptation
repulsión repulsion
repulsión molecular molecular repulsion
repulsión nuclear nuclear repulsion
requisitos ambientales environmental
requirements
resbenzofenona resbenzophenone
reseno resene
reserpina reserpine
reserva reserve
residual residual
residuo residue
residuo de destilación distillation residue
residuos anódicos anodic mud, anodic
slime
resina resin, rosin
resina acetálica acetal resin
resina acrílica acrylic resin
resina alílica allyl resin
resina alquídica alkyd resin
resina amarilla yellow resin
resina aniónica anionic resin
resina celulósica cellulosic resin
resina de acetato de vinilo vinyl acetate
resin
resina de acrilato acrylate resin
resina de adición addition resin
resina de cloruro de vinilo vinyl
chloride resin
resina de contacto contact resin
resina de copaiba copaiba resin
resina de estireno-acrilonitrilo styrene-
acrylonitrile resin
resina de fenol-formaldehído phenol-
formaldehyde resin
resina de fenol-furfural phenol-furfural
resin
resina de fluorocarburo fluorocarbon
resin
resina de halocarburo halocarbon resin
resina de hidrocarburos hydrocarbon
resin
resina de intercambio iónico ion-
exchange resin
resina de ionómero ionomer resin
resina de isocianato isocyanate resin
resina de Manila Manila resin
resina de melamina melamine resin
resina de poliamida polyamide resin

resina de poliamina-metileno polyamine-methylene resin
resina de polietileno polyethylene resin
resina de poliimida polyimide resin
resina de polimetacrilato polymethacrylate resin
resina de polioxialquileno polyoxyalkylene resin
resina de politerpeno polyterpene resin
resina de politetrafluoroetileno polytetrafluoroethylene resin
resina de poliuretano polyurethane resin
resina de polivinilideno polyvinylidene resin
resina de polivinilo polyvinyl resin
resina de resorcina-formaldehído resorcinol-formaldehyde resin
resina de triazona triazone resin
resina de urea-formaldehído urea-formaldehyde resin
resina de uretano urethane resin
resina de vinilacetal vinyl acetal resin
resina de viniléter vinyl ether resin
resina de vinilideno vinylidene resin
resina de vinilo vinyl resin
resina en polvo powdered resin
resina epóxica epoxy resin
resina etilénica ethylene resin
resina fenólica phenolic resin
resina fenoxi phenoxy resin
resina olefínica olefin resin
resina poliéster polyester resin
resina poliéter polyether resin
resina sintética synthetic resin
resina soluble en agua water-soluble resin
resina termoplástica thermoplastic resin
resinamina resinamine
resinato resinate
resinato alumínico aluminum resinate
resinato cálcico calcium resinate
resinato cobaltoso cobaltous resinate
resinato de cinc zinc resinate
resinato de cobalto cobalt resinate
resinato de cobre copper resinate
resinato de estaño tin resinate
resinato de hierro iron resinate
resinato de manganeso manganese resinate

resinato de plomo lead resinate
resinato férrico ferric resinate
resinato sódico sodium resinate
resinificación resinification
resinografía resinography
resinoide resinoid
resinol resinol
resinoso resinous
resistencia al fuego fire resistance
resistente al ácido acid resistant
resistente al fuego fire resistant
resita resite
resitol resitol
resocianina resocyanin
resol resol
resolución resolution
resonancia resonance
resonancia atómica atomic resonance
resonancia cuadripolar nuclear nuclear quadripole resonance
resonancia cuádruple quadruple resonance
resonancia de espín spin resonance
resonancia de espín electrónico electron-spin resonance
resonancia de Fermi Fermi resonance
resonancia electrónica electron resonance
resonancia magnética magnetic resonance
resonancia magnética electrónica electron magnetic resonance
resonancia magnética nuclear nuclear magnetic resonance
resonancia nuclear nuclear resonance
resonancia paramagnética paramagnetic resonance
resonancia paramagnética electrónica electron paramagnetic resonance
resonante resonant
resorcilo resorcyl
resorcina resorcin
resorcinol resorcinol
restitución restitution
resultante resultant
retardador retarder
retardante de llama flame-retarding
retardo retardation
retardo iónico ion retardation

retención retention
reteno retene
reticulado reticulated
retículo grating, lattice
retinal retinal
retineno retinene
retinita retinite
retinol retinol
retorta retort
retro retro
retroalimentación feedback
retrodispersión backscatter
retrodonación back donation
retroextracción back-extraction, stripping
retrogradación retrogradation
retrorcina retrorsine
retrosíntesis retrosynthesis
retrotitulación back titration
retrovaloración back titration
revelador developer
reversible reversible
reversión reversion
revertosa revertose
revestido coated
revestimiento coating, cladding
revestimiento anódico anodic coating
revestimiento básico basic lining
revestimiento catódico cathodic coating
revestimiento de oro gold coating
revestimiento protector protective coating
revivir revive
rho rho
riboflavina riboflavin
ribofuranosiladenina ribofuranosyladenine
ribonucleasa ribonuclease
ribonucleoproteína ribonucleoprotein
ribosa ribose
ribosida riboside
ribosilo ribosyl
ribosoma ribosome
ribulosa ribulose
ricina ricin
ricinina ricinine
ricinoleato ricinoleate
ricinoleato amónico ammonium ricinoleate

ricinoleato cálcico calcium ricinoleate
ricinoleato de butilo butyl ricinoleate
ricinoleato de cadmio cadmium ricinoleate
ricinoleato de cinc zinc ricinoleate
ricinoleato de litio lithium ricinoleate
ricinoleato de metilo methyl ricinoleate
ricinoleato potásico potassium ricinoleate
ricinoleato sódico sodium ricinoleate
ricinoleína ricinolein
rickardita rickardite
richmondita richmondite
riesgo aceptable acceptable risk
rigoleno rhigolene
ripidolita ripidolite
ristocetina ristocetin
roca carbonatada carbonate rock
roca de carbonato carbonate rock
roca de fosfato phosphate rock
rodalina rhodalline
rodamina rhodamine
rodanato rhodanate
rodanato sódico sodium rhodanate
rodanida rhodanide
rodanina rhodanine
rodanometría rhodanometry
rodenticida rodenticide
rodinilo rhodinyl
rodinol rhodinol, rodinol
rodio rhodium
rodita rhodite
rodizita rhodizite
rodocrosita rhodochrosite
rodolita rhodolite
rodonita rhodonite
rodopsina rhodopsin
rodoxantina rhodoxanthin
roentgen roentgen, röntgen
roentgenograma roentgenogram
rojo congo congo red
rojo de alizarina alizarin red
rojo de antimonio antimony red
rojo de bromofenol bromophenol red
rojo de cadmio cadmium red
rojo de clorofenol chlorophenol red
rojo de cresol cresol red
rojo de cromo chrome red
rojo de índigo indigo red

rojo de la India   Indian red
rojo de metilo   methyl red
rojo de óxido férrico   ferric oxide red
rojo de paranitranilina   paranitraniline red
rojo de París   Paris red
rojo de Persia   Persian red
rojo de quinolina   quinoline red
rojo de rutenio   ruthenium red
rojo de toluidina   toluidine red
rojo fenol   phenol red
rojo neutro   neutral red
rojo para   para red
rojo turco   Turkey red
rojo veneciano   Venetian red
rómbico   rhombic
rombohedral   rhombohedral
rombohedro   rhombohedron
rompedor de emulsión   emulsion breaker
rompimiento de anillo   ring breakage
rompimiento de emulsión   emulsion breaking
rosa de Bengala   Rose Bengal
rosanilina   rosaniline
roscoelita   roscoelite
roselita   roselite
rosinol   rosinol
rotación   rotation
rotación específica   specific rotation
rotación interna restringida   restricted internal rotation
rotación molecular   molecular rotation
rotación óptica   optical rotation
rotacional   rotational
rotámero   rotamer
rotámetro   rotameter
rotanio   rhotanium
rotatorio   rotatory
rotaversión   rotaversion
rotenona   rotenone
rotulación   labeling
rotulado   labeled
rowlandita   rowlandite
rubeno   rubene
ruberita   ruberite
rubí   ruby
rubí espinela   ruby spinel
rubiceno   rubicene
rubidina   rubidine
rubidio   rubidium
rubreno   rubrene
rutefordita   rutherfordite
rutenato   ruthenate
ruténico   ruthenic
rutenio   ruthenium
rutenioso   ruthenious
rutenoceno   ruthenocene
rutilo   rutile
rutina   rutin
rydberg   rydberg

S

sabineno sabinene
sabinol sabinol
sabor taste, flavor
sacarasa saccharase
sacarato saccharate
sacarato ácido potásico potassium acid saccharate
sacarato de estroncio strontium saccharate
sacárico saccharic
sacárido saccharide
sacarificación saccharification
sacarificar saccharify
sacarímetro saccharimeter
sacarina saccharin
sacarosa saccharose, sucrose
saccarina sódica sodium saccharin
safranina safranine
safrol safrole
sahlita sahlite
sal salt
sal ácida acid salt
sal amarilla yellow salt
sal amoníaco sal ammoniac
sal amónica ammonium salt
sal amónica cuaternaria quaternary ammonium salt
sal amortiguadora buffer salt
sal azul blue salt
sal básica basic salt
sal biliar bile salt
sal binaria binary salt
sal compleja complex salt
sal común common salt
sal cuaternaria quaternary salt
sal de anilina aniline salt
sal de antimonio antimony salt
sal de diazonio diazonium salt
sal de Engel Engel salt
sal de Erdmann Erdmann salt
sal de estaño tin salt
sal de Étard Étard salt
sal de Fischer Fischer salt

sal de fosfonio phosphonium salt
sal de Fremy Fremy salt
sal de Glauber Glauber salt
sal de Graham Graham salt
sal de Lemery salt of Lemery
sal de Macquer Macquer salt
sal de Maddrell Maddrell salt
sal de Magnus Magnus salt
sal de mesa table salt
sal de Mohr Mohr salt
sal de oro gold salt
sal de Pelilgot Peligot salt
sal de plata silver salt
sal de Reinecke Reinecke salt
sal de roca rock salt
sal de Rochelle Rochelle salt
sal de Schäffer Schäffer salt
sal de Seidlitz Seidlitz salt
sal de tártaro salt of tartar
sal de Vauquelin Vauquelin salt
sal de Wurster Wurster salt
sal de Zeise Zeise salt
sal débil weak salt
sal doble double salt
sal esencial essential salt
sal fundida molten salt, fused salt
sal mixta mixed salt
sal neutra neutral salt
sal normal normal salt
sal orgánica organic salt
sal simple simple salt
sal sosa sal soda
sal verde green salt
sala limpia clean room
salamida salamide
sales de Epsom Epsom salts
sales de tierras raras rare-earth salts
salicilado salicylated
salicilal salicylal
salicilaldehído salicylaldehyde
salicilamida salicylamide
salicilanilida salicylanilide
salicilato salicylate

salicilato alumínico aluminum salicylate
salicilato amónico ammonium salicylate
salicilato bárico barium salicylate
salicilato de amilo amyl salicylate
salicilato de atropina atropine salicylate
salicilato de bencilo benzyl salicylate
salicilato de bornilo bornyl salicylate
salicilato de cafeína caffeine salicylate
salicilato de cesio cesium salicylate
salicilato de cinc zinc salicylate
salicilato de codeína codeine salicylate
salicilato de estroncio strontium
    salicylate
salicilato de etilo ethyl salicylate
salicilato de feniletilo phenylethyl
    salicylate
salicilato de fenilo phenyl salicylate
salicilato de isoamilo isoamyl salicylate
salicilato de isobornilo isobornyl
    salicylate
salicilato de isobutilo isobutyl salicylate
salicilato de litio lithium salicylate
salicilato de mentilo menthyl salicylate
salicilato de metilo methyl salicylate
salicilato de octilfenilo octylphenyl
    salicylate
salicilato de pentilo pentyl salicylate
salicilato de plata silver salicylate
salicilato de plomo lead salicylate
salicilato fenilmercúrico phenylmercuric
    salicylate
salicilato magnésico magnesium
    salicylate
salicilato mercúrico mercuric salicylate
salicilato potásico potassium salicylate
salicilato sódico sodium salicylate
salicilideno salicylidene
salicilo salicyl
saliciloilo salicyloyl
salicilonitrilo salicylonitrile
salicina salicin
sálico salic
salicoilo salicoyl
salida de gas gas outlet
saligenina saligenin
saligenol saligenol
salímetro salimeter
salinidad salinity
salino saline

salinómetro salinometer
salitre saltpeter
salmina salmine
salmuera brine
salol salol
salto cuántico quantum jump
samárico samaric
samario samarium
samaroso samarous
samarskita samarskite
sandaraca sandarac
sangrar bleed
sanguinaria sanguinaria
santalilo santalyl
santalol santalol
saponificación saponification
saponina saponin
saponita saponite
sarcolactato sarcolactate
sarcolisina sarcolysin
sarcolita sarcolite
sarcosina sarcosine
sarcosinato sódico sodium sarcosinate
sardinianita sardinianite
sardónica sardonyx
sarina sarin
saturable saturable
saturación saturation
saturación con agua water saturation
saturado saturated
saturador saturator
saturante saturating
saturar saturate
saxitoxina saxitoxin
scheelita scheelite
schonita schönite
sebacato de butilo butyl sebacate
sebacato de dimetilo dimethyl sebacate
sebacato de dioctilo dioctyl sebacate
sebacoilo sebacoyl
sebaconitrilo sebaconitrile
sebo tallow
sec sec
secado drying
secado instantáneo flash drying
secado por solvente solvent drying
secador dryer
sección transversal cross-section
sección transversal de fisión fission

cross-section
sección transversal gamma gamma
 cross-section
sección transversal nuclear nuclear
 cross-section
seco dry
secobarbital secobarbital
secobarbital sódico sodium secobarbital
secuencia isoeléctrica isoelectronic
 sequence
secuestrante sequestering
secuestro sequestration
secundario secondary
seda silk
sedante sedative
sedimentación sedimentation
sedimento sediment, sludge
sedimento químico chemical sediment
sedimentos de aceite foots
segregación segregation
segunda ley de termodinámica second
 law of thermodynamics
segundo second
seguridad safety
selectividad selectivity
selectividad de catalizador catalyst
 selectivity
selectividad de filtro filter selectivity
selectivo selective
selena selena
seleniato selenate
seleniato amónico ammonium selenate
seleniato bárico barium selenate
seleniato cálcico calcium selenate
seleniato cúprico cupric selenate
seleniato de cadmio cadmium selenate
seleniato de plomo lead selenate
seleniato sódico sodium selenate
selénico selenic
selenilo selenyl
seleninilo seleninyl
selenino selenino
selenio selenium
selenioso selenious
selenita selenite
selenito amónico ammonium selenite
selenito de cobalto cobalt selenite
selenito sódico sodium selenite
seleniuro selenide

seleniuro arsenioso arsenous selenide
seleniuro bárico barium selenide
seleniuro cuproso cuprous selenide
seleniuro de arsénico arsenic selenide
seleniuro de bismuto bismuth selenide
seleniuro de cadmio cadmium selenide
seleniuro de cinc zinc selenide
seleniuro de dietilo diethyl selenide
seleniuro de dimetilo dimethyl selenide
seleniuro de hidrógeno hydrogen
 selenide
seleniuro de mercurio mercury selenide
seleniuro de plata silver selenide
seleniuro de plomo lead selenide
seleniuro de tetrafósforo
 tetraphosphorus selenide
seleniuro ferroso ferrous selenide
seleno seleno
selenocianato selenocyanate
selenofurano selenofuran
selenol selenol
selenomio selenomium
selenona selenone
selenonilo selenonyl
selenonio selenonium
selenono selenono
selenourea selenourea
selenóxido selenoxide
sellado sealed
sellador sealant
sellaíta sellaite
semi semi
semicarbazida semicarbazide
semicarbazido semicarbazido
semicarbazino semicarbazino
semicarbazona semicarbazone
semicarbazona de acetona acetone
 semicarbazone
semicarbazono semicarbazono
semicelda half-cell
semiconductor semiconductor
semidina semidine
semiempírico semiempirical
semilla seed
semilla de algodón cottonseed
semimetal semimetal
semimetálico semimetallic
semimicroanálisis semimicroanalysis
semimicroquímica semimicrochemistry

semipermeable semipermeable
semipila half-cell
semiprecioso semiprecious
semisintético semisynthetic
semisólido semisolid
semivalencia semivalence
senarmontita senarmontite
sensibilizador sensitizer
sensible sensitive
sensible a la luz light-sensitive
sensible a la temperatura temperature-sensitive
sensible a radiación radiation-sensitive
sensor sensor
sensor de humedad humidity sensor
separación separation
separación centrífuga centrifugal separation
separación de aminas de Hofmann Hofmann amine separation
separación de isótopos isotope separation
separación electrolítica electrolytic separation
separación electromagnética electromagnetic separation
separación magnética magnetic separation
separador separator
separador de isótopos isotope separator
sepia sepia
sepiolita sepiolite
septifeno septiphene
serendipismo serendipity
serie series
serie actínida actinide series
serie alifática aliphatic series
serie aromática aromatic series
serie colateral collateral series
serie de actinio actinium series
serie de actividad activity series
serie de alquenos alkene series
serie de Balmer Balmer series
serie de benceno benzene series
serie de Brackett Brackett series
serie de desintegración decay series
serie de desintegración de uranio uranium decay series
serie de desintegración radiactiva

radioactive decay series
serie de desplazamiento displacement series
serie de hidrocarburos hydrocarbon series
serie de Hofmeister Hofmeister series
serie de líneas series of lines
serie de Lyman Lyman series
serie de Paschen Paschen series
serie de redox redox series
serie de torio thorium series
serie de uranio uranium series
serie de uranio-radio uranium-radium series
serie de Volta Volta series
serie electromotriz electromotive series
serie electroquímica electrochemical series
serie espectral spectral series, spectrum series
serie espectroquímica spectrochemical series
serie eutrópica eutropic series
serie fundamental fundamental series
serie galvánica galvanic series
serie homogénea homogeneous series
serie homóloga homologous series
serie lantánida lanthanide series
serie liotrópica lyotropic series
serie perilógica perilogic series
serie periódica periodic series
serie principal principal series
serie química chemical series
serie radiactiva radioactive series
serina serine
serotonina serotonin
serpentina serpentine
sesamina sesamin
sesamol sesamol
sesamolina sesamolin
sesona sesone
sesqui sesqui
sesquibromuro férrico ferric sesquibromide
sesquicarbonato sesquicarbonate
sesquicarbonato sódico sodium sesquicarbonate
sesquicloruro sesquichloride
sesquicloruro de cromo chromium

sesquichloride

sesquicloruro de etilaluminio
ethylaluminum sesquichloride

sesquicloruro de rutenio ruthenium
sesquichloride

sesquicloruro de talio thallium
sesquichloride

sesquicloruro talioso thallous
sesquichloride

sesquióxido sesquioxide

sesquióxido de galio gallium sesquioxide

sesquióxido de iridio iridium sesquioxide

sesquióxido de lantano lanthanum
sesquioxide

sesquióxido de manganeso manganese
sesquioxide

sesquióxido de molibdeno molybdenum
sesquioxide

sesquióxido de níquel nickel sesquioxide

sesquióxido de plomo lead sesquioxide

sesquióxido de rutenio ruthenium
sesquioxide

sesquióxido de vanadio vanadium
sesquioxide

sesquisal sesquisalt

sesquisilicato sódico sodium
sesquisilicate

sesquisosa sesquisoda

sesquisulfato de titanio titanium
sesquisulfate

sesquisulfuro de fósforo phosphorus
sesquisulfide

sesquiterpeno sesquiterpene

seudo pseudo

seudoácido pseudo acid

seudoaromático pseudoaromatic

seudoasimétrico pseudoasymmetric

seudobase pseudo base

seudocumeno pseudocumene

seudocumidina pseudocumidine

seudocumilo pseudocumyl

seudocumol pseudocumol

seudohalógeno pseudohalogen

seudoisomerismo pseudoisomerism

seudoisótopo pseudoisotope

seudomórfico pseudomorphic

sexivalente sexivalent

sexteto sextet

siderita siderite

siderotilato siderotilate

sidnona sydnone

siembra seeding

siena sienna

sievert sievert

sigma sigma

sigmatrópico sigmatropic

siicato de zirconio zirconium silicate

silano silane

silanodiilo silanediyl

silanol silanol

silanotriilo silanetriyl

silatrano silatrane

silazano silazane

silicano silicane

silicato silicate

silicato alumínico aluminum silicate

silicato bárico barium silicate

silicato cálcico calcium silicate

silicato cobaltoso cobaltous silicate

silicato de berilio beryllium silicate

silicato de cinc zinc silicate

silicato de cobre copper silicate

silicato de etilbutilo ethylbutyl silicate

silicato de etilo ethyl silicate

silicato de litio lithium silicate

silicato de manganeso manganese
silicate

silicato de plomo lead silicate

silicato de sosa silicate of soda

silicato magnésico magnesium silicate

silicato manganoso manganous silicate

silicato potásico potassium silicate

silicato sódico sodium silicate

silicato tricálcico tricalcium silicate

sílice silica

sílice fundida fused silica

sílice hidratada hydrated silica

silíceo siliceous

silícico silicic

silicileno silicylene

silicilo silicyl

silicio silicon

silicio-cobre silicon-copper

siliciuro silicide

siliciuro bárico barium silicide

siliciuro cálcico calcium silicide

siliciuro crómico chromic silicide

siliciuro de cobalto cobalt silicide

siliciuro de niobio niobium silicide
siliciuro de rutenio ruthenium silicide
siliciuro de tungsteno tungsten silicide
siliciuro de zirconio zirconium silicide
siliciuro magnésico magnesium silicide
silico silico
silicoaluminato sódico sodium silicoaluminate
silicobromoformo silicobromoform
silicocloroformo silicochloroform
silicoetano silicoethane
silicofluoruro silicofluoride
silicofluoruro alumínico sódico sodium aluminum silicofluoride
silicofluoruro bárico barium silicofluoride
silicofluoruro cálcico calcium silicofluoride
silicofluoruro de plomo lead silicofluoride
silicofluoruro magnésico magnesium silicofluoride
silicofluoruro potásico potassium silicofluoride
silicofluoruro sódico sodium silicofluoride
silicoheptano silicoheptane
silicol silicol
silicomanganeso silicomanganese
silicometano silicomethane
silicomolibdato sódico sodium silicomolybdate
silicona silicone
siliconato de metilo sódico sodium methyl siliconate
silicotungstato sódico sodium silicotungstate
silicoyodoformo silicoiodoform
sililación silylation
silileno silylene
sililo sylyl
silimanita sillimanite
siloxano siloxane
siloxi siloxy
silumina silumin
silvanita sylvanite
silvano sylvan
silvestreno silvestrene
silvex silvex

silvina sylvine
silvinita sylvinite
silviquímico silvichemical
silvita sylvite
simazina simazine
símbolo symbol
símbolo atómico atomic symbol
símbolo químico chemical symbol
simetría symmetry
simetría de cristal crystal symmetry
simetría interna internal symmetry
simetría orbital orbital symmetry
simétrico symmetrical
sin solvente solventless
sindiotáctico syndiotactic
sinéresis syneresis
sinergismo synergism
sinergista synergist
singulete singlet
sinigrina sinigrin
sintáctico syntactic
sínter sinter
sinterización sintering
sinterizar sinterize
síntesis synthesis
síntesis alicíclica de Perkin Perkin alicyclic synthesis
síntesis asimétrica asymmetric synthesis
síntesis de aldehídos de Bouveault Bouveault aldehyde synthesis
síntesis de aldehídos de Gattermann Gattermann aldehyde synthesis
síntesis de aldehídos de Kröhnke Kröhnke aldehyde synthesis
síntesis de Arndt-Eistert Arndt-Eistert synthesis
síntesis de Bogert-Cook Bogert-Cook synthesis
síntesis de Campbell Campbell synthesis
síntesis de cetenas de Schmidlin Schmidlin ketene synthesis
síntesis de ciclopropanos de Kishner Kishner cyclopropane synthesis
síntesis de Darzens Darzens synthesis
síntesis de Erlenmeyer Erlenmeyer synthesis
síntesis de ésteres malónicos malonic ester synthesis
síntesis de éteres de Williamson

Williamson ether synthesis

síntesis de Favorskii-Babayan
Favorskii-Babayan synthesis

síntesis de Feist-Benary Feist-Benary synthesis

síntesis de fenilhidrazinas de Fischer
Fischer phenylhydrazine synthesis

síntesis de Fischer-Tropsch Fischer-Tropsch synthesis

síntesis de Fittig Fittig synthesis

síntesis de Frankland Frankland synthesis

síntesis de Freund Freund synthesis

síntesis de Friedlander Friedlander synthesis

síntesis de Gabriel Gabriel synthesis

síntesis de Gastaldi Gastaldi synthesis

síntesis de Gattermann Gattermann synthesis

síntesis de Gattermann-Koch
Gattermann-Koch synthesis

síntesis de Gogte Gogte synthesis

síntesis de Graebe-Ullman Graebe-Ullman synthesis

síntesis de Grignard Grignard synthesis

síntesis de Grove Grove synthesis

síntesis de Grundmann Grundmann synthesis

síntesis de Haworth Haworth synthesis

síntesis de Heumann-Pfleger Heumann-Pfleger synthesis

síntesis de Hinsberg Hinsberg synthesis

síntesis de Hoch-Campbell Hoch-Campbell synthesis

síntesis de Hoesch Hoesch synthesis

síntesis de Houben-Fischer Houben-Fischer synthesis

síntesis de Houben-Hoesch Houben-Hoesch synthesis

síntesis de indoles de Fischer Fischer indole synthesis

síntesis de indoles de Nenitzescu
Nenitzescu indole synthesis

síntesis de indoles de Reissert Reissert indole synthesis

síntesis de isonitrilos de Hofmann
Hofmann isonitrile synthesis

síntesis de Kiliani-Fischer Kiliani-Fischer synthesis

síntesis de Kolbe Kolbe synthesis

síntesis de Letts Letts synthesis

síntesis de Madelung Madelung synthesis

síntesis de Martinet Martinet synthesis

síntesis de Meyer Meyer synthesis

síntesis de Nef Nef synthesis

síntesis de oxazoles de Fischer Fischer oxazole synthesis

síntesis de Paal-Knorr Paal-Knorr synthesis

síntesis de Patterson Patterson synthesis

síntesis de Pelouze Pelouze synthesis

síntesis de Piloty-Robinson Piloty-Robinson synthesis

síntesis de Pinner Pinner synthesis

síntesis de pirazinas de Staedel-Rugheimer Staedel-Rugheimer pyrazine synthesis

síntesis de pirazoles de Knorr Knorr pyrazole synthesis

síntesis de pirazoles de Pechmann
Pechmann pyrazole synthesis

síntesis de piridinas de Chichibabin
Chichibabin pyridine synthesis

síntesis de piridinas de Hantzsch
Hantzsch pyridine synthesis

síntesis de pirroles de Hantach Hantach pyrrole synthesis

síntesis de pirroles de Knorr Knorr pyrrole synthesis

síntesis de polipéptidos de Fischer
Fischer polypeptide synthesis

síntesis de purinas de Traube Traube purine synthesis

síntesis de quinazolinas de Niementowski Niementowski quinazoline synthesis

síntesis de quinolinas de Camps Camps quinoline synthesis

síntesis de quinolinas de Combes
Combes quinoline synthesis

síntesis de quinolinas de Knorr Knorr quinoline synthesis

síntesis de quinolinas de Niementowski
Niementowski quinoline synthesis

síntesis de quinolinas de Riehm Riehm quinoline synthesis

síntesis de Riemschneider

Riemschneider synthesis

síntesis de Rosenmund-von Braun Rosenmund-von Braun synthesis

síntesis de Sandmeyer Sandmeyer synthesis

síntesis de Skraup Skraup synthesis

síntesis de Stolle Stolle synthesis

síntesis de Strecker Strecker synthesis

síntesis de Urech Urech synthesis

síntesis de von Richter von Richter synthesis

síntesis de Widman-Stoermer Widman-Stoermer synthesis

síntesis de Williamson Williamson synthesis

síntesis de Wöhler Wöhler synthesis

síntesis de Wurtz Wurtz synthesis

síntesis estereoespecífica stereospecific synthesis

síntesis orgánica organic synthesis

síntesis química chemical synthesis

sintético synthetic

sintetizar synthesize

sintol synthol

sirosingopina syrosingopine

sisal sisal

sistema absoluto absolute system

sistema binario binary system

sistema coloidal colloidal system

sistema conjugado conjugated system

sistema cristalográfico crystal system

sistema cuantificado quantized system

sistema cuaternario quaternary system

sistema cúbico cubic system

sistema de anillo ring system

sistema de Claude Claude system

sistema de Ginebra Geneva system

sistema de Hermann-Mauguin Hermann-Mauguin system

sistema de redox redox system

sistema de Schoenflies Schoenflies system

sistema de Stock Stock system

sistema degenerado degenerate system

sistema dinámico dynamic system

sistema disipativo dissipative system

sistema disperso dispersed system

sistema espiro spiro system

sistema estable stable system

sistema eutéctico eutectic system

sistema extintor de fuego fire-extinguisher system

sistema heterogéneo heterogeneous system

sistema hexagonal hexagonal system

sistema homogéneo homogeneous system

sistema internacional international system

sistema monodisperso monodisperse system

sistema ortorrómbico orthorhombic system

sistema periódico periodic system

sistema peritéctico peritectic system

sistema polifuncional polyfunctional system

sistema regular regular system

sistema rómbico rhombic system

sistema rombohedral rhombohedral system

sistema ternario ternary system

sistema tetragonal tetragonal system

sistema triclínico triclinic system

sistema trigonal trigonal system

sistémico systemic

sitosterol sitosterol

sklodowskita sklodowskite

skutterudita skutterudite

smog smog

sobrenadante supernatant

sobrepotencial overpotential

sobretensión overvoltage

soda soda

sodalita sodalite

sodamida sodamide

sodilo sodyl

sodio sodium

sodioacetato sódico sodium sodioacetate

soja soybean

sol sol

solación solation

solana solan

solanina solanine

solato solate

soldadura welding

soldadura con hidrógeno atómico atomic hydrogen welding

soldadura fuerte brazing

soldante solder
solidificación solidification, setting
solidificar solidify
sólido solid
sólidos suspendidos suspended solids
solidus solidus
solubilidad solubility
solubilidad de equilibrio equilibrium solubility
solubilidad de gas gas solubility
solubilizar solubilize
soluble soluble
soluble en agua water soluble
soluble en grasa fat soluble
solución solution
solución ácida acid solution
solución acuosa aqueous solution
solución alcalina alkaline solution
solución amortiguadora buffer solution
solución anisotónica anisotonic solution
solución anódica anodic solution
solución básica basic solution
solución centinormal centinormal solution
solución coloidal colloidal solution
solución concentrada concentrated solution
solución de Benedict Benedict solution
solución de Bouchardat Bouchardat solution
solución de Brodie Brodie solution
solución de Carl Carl solution
solución de Channing Channing solution
solución de Dakin Dakin solution
solución de Eder Eder solution
solución de Fehling Fehling solution
solución de Feiser Feiser solution
solución de Fischer Fischer solution
solución de Folin Folin solution
solución de Gabbet Gabbet solution
solución de Hanus Hanus solution
solución de Hübl Hübl solution
solución de Lugol Lugol solution
solución de nutrientes nutrient solution
solución de Pavy Pavy solution
solución de Pearson Pearson solution
solución de reactivo reagent solution
solución de Ringer Ringer solution
solución de Rohrbach Rohrbach solution

solución de sal de Pasteur Pasteur salt solution
solución de Schiff Schiff solution
solución de titulación titrant
solución de valoración titrant
solución de Volhard Volhard solution
solución de Wagner Wagner solution
solución de Wijs Wijs solution
solución decimolar decimolar solution
solución decinormal decinormal solution, tenth-normal solution
solución diluida dilute solution
solución electrolítica electrolytic solution
solución física physical solution
solución hipertónica hypertonic solution
solución hipotónica hypotonic solution
solución ideal ideal solution
solución iónica ionic solution
solución isotónica isotonic solution
solución molal molal solution
solución molar molar solution
solución molecular molecular solution
solución molecular-gramo gram-molecular solution
solución neutra neutral solution
solución no acuosa nonaqueous solution
solución no saturada unsaturated solution
solución normal normal solution, standard solution
solución perfecta perfect solution
solución química chemical solution
solución saturada saturated solution
solución sólida solid solution
solución supersaturada supersaturated solution
solución volumétrica volumetric solution
soluto solute
solutropo solutrope
solvatación solvation
solvato solvate
solvatocromismo solvatochromism
solvente solvent
solvente ácido acid solvent
solvente acuoso aqueous solvent
solvente aprótico aprotic solvent
solvente básico basic solvent
solvente clorado chlorinated solvent
solvente de ionización ionizing solvent

solvente de Stoddard Stoddard solvent
solvente físico physical solvent
solvente latente latent solvent
solvente no acuosa nonaqueous solvent
solvente no ionizante nonionizing
    solvent
solvente no polar nonpolar solvent
solvente orgánico organic solvent
solvente polar polar solvent
solvente químico chemical solvent
solvente volátil volatile solvent
solvólisis solvolysis
solvolítico solvolytic
solvus solvus
sonda probe
sonda electrónica electron probe
sonólisis sonolysis
sonoquímica sonochemistry
sorbato cálcico calcium sorbate
sorbato de butilo butyl sorbate
sorbato potásico potassium sorbate
sorbato sódico sodium sorbate
sorbente sorbent
sórbido sorbide
sorbitán sorbitan
sorbitol sorbitol
sorbosa sorbose
sorción sorption
sosa soda
sosa cáustica caustic soda
sosa de lavar washing soda
sosoloide sosoloid
sub sub
subacetato subacetate
subacetato de plomo lead subacetate
subatómica subatomics
subatómico subatomic
subcapa subshell
subcarbonato subcarbonate
subcarbonato cúprico cupric
    subcarbonate
subcarbonato de bismuto bismuth
    subcarbonate
subcarbonato de plomo lead
    subcarbonate
subcarbonato férrico ferric subcarbonate
subcloruro de azufre sulfur subchloride
subcloruro de bismuto bismuth
    subchloride

subenfriado undercooling, supercooling
suberano suberane
suberilo suberyl
suberona suberone
subestructura substructure
subgalato de bismuto bismuth subgallate
sublación sublation
subletal sublethal
sublimación sublimation
sublimador sublimator
sublimador de Hortvet Hortvet
    sublimator
sublimar sublimate
subnitrato subnitrate
subnitrato de bismuto bismuth
    subnitrate
subnivel sublevel
subnuclear subnuclear
subóxido suboxide
subóxido de carbono carbon suboxide
subóxido de plata silver suboxide
subóxido de plomo lead suboxide
subóxido de vanadio vanadium suboxide
subsalicilato de bismuto bismuth
    subsalicylate
substancia substance
substancia aprótica aprotic substance
substancia controlada controlled
    substance
substancia corrosiva corrosive substance
substancia dañina harmful substance
substancia de contacto contact substance
substancia delicuescente deliquescent
    substance
substancia explosiva explosive substance
substancia fotosensible photosensitive
    substance
substancia inerte inert substance
substancia inflamable flammable
    substance
substancia nociva noxious substance
substancia peligrosa hazardous
    substance
substancia química chemical substance
substancia racémica racemic substance
substancia radiomimética radiomimetic
    substance
substancia tóxica toxic substance
substitución substitution

substitución electrofílica electrophilic substitution

substitución nucleofílica nucleophilic substitution

substituir substitute

substituto de azúcar sugar substitute

substrato substrate

subsulfato de paladio palladium subsulfate

subsulfato férrico ferric subsulfate

subsulfito sódico sodium subsulfite

subtilina subtilin

succinaldehído succinaldehyde

succinamilo succinamyl

succinamoilo succinamoyl

succinato succinate

succinato amónico ammonium succinate

succinato bárico barium succinate

succinato cálcico calcium succinate

succinato cobaltoso cobaltous succinate

succinato de bencilo benzyl succinate

succinato de bencilo sódico sodium benzyl succinate

succinato de cadmio cadmium succinate

succinato de cinc zinc succinate

succinato de dietilo diethyl succinate

succinato de etilo ethyl succinate

succinato magnésico magnesium succinate

succinato mercúrico mercuric succinate

succinato sódico sodium succinate

succinilo succinyl

succinimida succinimide

succinimido succinimido

succiniodimida succiniodimide

succinonitrilo succinonitrile

sucrasa sucrase

sucrato sucrate

sucrol sucrol

sucrosa sucrose

suelo soil

suero serum

suero de la leche whey

suero de la verdad truth serum

sujetador de crisol crucible holder

sulfacetamida sulfacetamide

sulfadimidina sulfadimidine

sulfadiazina sulfadiazine

sulfaguanidina sulfaguanidine

sulfaldehído sulfaldehyde

sulfalizarato sódico sodium sulfalizarate

sulfamato sulfamate

sulfamato amónico ammonium sulfamate

sulfamato cálcico calcium sulfamate

sulfamerazina sulfamerazine

sulfametizol sulfamethizole

sulfámico sulfamic

sulfamida sulfamide

sulfamilo sulfamyl

sulfamoilo sulfamoyl

sulfanilamida sulfanilamide

sulfanilato sulfanilate

sulfanilato sódico sodium sulfanilate

sulfano sulfane

sulfantimoniato sódico sodium sulfantimonate

sulfarseniato sulfarsenate

sulfatación sulfation

sulfatasa sulfatase

sulfatiazol sulfathiazole

sulfato sulfate

sulfato ácido de hidroxilamina hydroxylamine acid sulfate

sulfato ácido de metilo acid methyl sulfate

sulfato ácido potásico potassium acid sulfate

sulfato ácido sódico sodium acid sulfate

sulfato alumínico aluminum sulfate

sulfato alumínico de cesio cesium aluminum sulfate

sulfato alumínico potásico potassium aluminum sulfate

sulfato alumínico sódico sodium aluminum sulfate

sulfato amónico ammonium sulfate

sulfato amónico ácido acid ammonium sulfate

sulfato amónico alumínico aluminum ammonium sulfate

sulfato amónico cobaltoso cobaltous ammonium sulfate

sulfato amónico de cobalto cobalt ammonium sulfate

sulfato amónico de cromo chromium ammonium sulfate

sulfato amónico de hierro iron ammonium sulfate

sulfato amónico de níquel nickel ammonium sulfate

sulfato amónico férrico ferric ammonium sulfate

sulfato amónico ferroso ferrous ammonium sulfate

sulfato amónico sódico sodium ammonium sulfate

sulfato antimonioso antimonous sulfate

sulfato áurico auric sulfate

sulfato bárico barium sulfate

sulfato cálcico calcium sulfate

sulfato cérico ceric sulfate

sulfato ceroso cerous sulfate

sulfato cobáltico cobaltic sulfate

sulfato cobaltoso cobaltous sulfate

sulfato crómico chromic sulfate

sulfato cromoso chromous sulfate

sulfato cúprico cupric sulfate

sulfato cuproso cuprous sulfate

sulfato de aluminio amónico ammonium aluminum sulfate

sulfato de anilina aniline sulfate

sulfato de antimonio antimony sulfate

sulfato de atropina atropine sulfate

sulfato de bencidina benzidine sulfate

sulfato de berilio beryllium sulfate

sulfato de bismuto bismuth sulfate

sulfato de brucina brucine sulfate

sulfato de cadmio cadmium sulfate

sulfato de cerio cerium sulfate

sulfato de cesio cesium sulfate

sulfato de cinc zinc sulfate

sulfato de cinc alumínico aluminum zinc sulfate

sulfato de cobalto cobalt sulfate

sulfato de cobre copper sulfate

sulfato de cocaína cocaine sulfate

sulfato de codeína codeine sulfate

sulfato de cromo chromium sulfate

sulfato de cromo amónico ammonium chromium sulfate

sulfato de dextrano sódico sodium dextran sulfate

sulfato de didimio didymium sulfate

sulfato de dietilo diethyl sulfate

sulfato de dihidrazina dihydrazine sulfate

sulfato de dimetilo dimethyl sulfate

sulfato de disprosio dysprosium sulfate

sulfato de efedrina ephedrine sulfate

sulfato de erbio erbium sulfate

sulfato de escandio scandium sulfate

sulfato de estaño tin sulfate

sulfato de estroncio strontium sulfate

sulfato de etilo ethyl sulfate

sulfato de europio europium sulfate

sulfato de fenilo phenyl sulfate

sulfato de fisostigmina physostigmine sulfate

sulfato de gadolinio gadolinium sulfate

sulfato de galio gallium sulfate

sulfato de guanilurea guanylurea sulfate

sulfato de hafnio hafnium sulfate

sulfato de hidrazina hydrazine sulfate

sulfato de hidrógeno hydrogen sulfate

sulfato de hidrógeno amónico ammonium hydrogen sulfate

sulfato de hidrógeno potásico potassium hydrogen sulfate

sulfato de hidrógeno sódico sodium hydrogen sulfate

sulfato de hidroxilamina hydroxylamine sulfate

sulfato de hidroxiquinolina hydroxyquinoline sulfate

sulfato de hierro iron sulfate

sulfato de hierro amónico ammonium iron sulfate

sulfato de indio indium sulfate

sulfato de iterbio ytterbium sulfate

sulfato de itrio yttrium sulfate

sulfato de kanamicina kanamycin sulfate

sulfato de lantano lanthanum sulfate

sulfato de litio lithium sulfate

sulfato de lutecio lutetium sulfate

sulfato de manganeso manganese sulfate

sulfato de manganeso amónico ammonium manganese sulfate

sulfato de metilo methyl sulfate

sulfato de neodimio neodymium sulfate

sulfato de níquel nickel sulfate

sulfato de níquel amónico ammonium nickel sulfate

sulfato de paladio palladium sulfate

sulfato de paramomicina paromomycin sulfate

sulfato de pentilo pentyl sulfate

sulfato de plata silver sulfate
sulfato de platino platinum sulfate
sulfato de plomo lead sulfate
sulfato de praseodimio praseodymium
　　sulfate
sulfato de quinina quinine sulfate
sulfato de radio radium sulfate
sulfato de rodio rhodium sulfate
sulfato de rubidio rubidium sulfate
sulfato de rubidio alumínico aluminum
　　rubidium sulfate
sulfato de talio thallium sulfate
sulfato de talio alumínico aluminum
　　thallium sulfate
sulfato de telurio tellurium sulfate
sulfato de terbio terbium sulfate
sulfato de titanilo titanyl sulfate
sulfato de titanio titanium sulfate
sulfato de torio thorium sulfate
sulfato de uranilo uranyl sulfate
sulfato de uranio uranium sulfate
sulfato de vanadilo vanadyl sulfate
sulfato de vanadio vanadium sulfate
sulfato de yodo iodine sulfate
sulfato de zirconilo zirconyl sulfate
sulfato de zirconio zirconium sulfate
sulfato de zirconio sódico sodium
　　zirconium sulfate
sulfato estánnico stannic sulfate
sulfato estannoso stannous sulfate
sulfato férrico ferric sulfate
sulfato ferroso ferrous sulfate
sulfato magnésico magnesium sulfate
sulfato magnésico amónico ammonium
　　magnesium sulfate
sulfato manganoso manganous sulfate
sulfato mercúrico mercuric sulfate
sulfato mercurioso mercurous sulfate
sulfato niqueloso nickelous sulfate
sulfato platínico platinic sulfate
sulfato potásico potassium sulfate
sulfato potásico alumínico aluminum
　　potassium sulfate
sulfato potásico berílico beryllium
　　potassium sulfate
sulfato potásico crómico chromium
　　potassium sulfate
sulfato potásico de cobalto cobalt
　　potassium sulfate

sulfato potásico de manganeso
　　manganese potassium sulfate
sulfato potásico de níquel nickel
　　potassium sulfate
sulfato samárico samaric sulfate
sulfato samaroso samarous sulfate
sulfato sódico sodium sulfate
sulfato sódico alumínico aluminum
　　sodium sulfate
sulfato sódico amónico ammonium
　　sodium sulfate
sulfato sódico anhidro anhydrous
　　sodium sulfate
sulfato talioso thallous sulfate
sulfato titánico titanic sulfate
sulfato titanoso titanous sulfate
sulfato vanádico vanadic sulfate
sulfazida sulfazide
sulfenamida sulfenamide
sulfénico sulfenic
sulfeno sulfeno
sulfenona sulfenone
sulfhidrato cálcico calcium sulfhydrate
sulfhidrato de etilo ethyl sulfhydrate
sulfhidrato de glicol glycol sulfhydrate
sulfhidrato sódico sodium sulfhydrate
sulfhidrilo sulfhydryl
sulfima sulfime
sulfimida sulfimide
sulfimido benzoico benzoic sulfimide
sulfina sulfine
sulfinato sulfinate
sulfinilimida sulfinylimide
sulfinilo sulfinyl
sulfino sulfino
sulfito sulfite
sulfito ácido potásico potassium acid
　　sulfite
sulfito ácido sódico sodium acid sulfite
sulfito amónico ammonium sulfite
sulfito bárico barium sulfite
sulfito cálcico calcium sulfite
sulfito cúprico cupric sulfite
sulfito cuproso cuprous sulfite
sulfito de bismuto bismuth sulfite
sulfito de cesio cesium sulfite
sulfito de cinc zinc sulfite
sulfito de dietilo diethyl sulfite
sulfito de escandio scandium sulfite

sulfito de estroncio strontium sulfite
sulfito de hidrógeno cálcico calcium hydrogen sulfite
sulfito de hidrógeno sódico sodium hydrogen sulfite
sulfito de hierro iron sulfite
sulfito de litio lithium sulfite
sulfito de manganeso manganese sulfite
sulfito de plata silver sulfite
sulfito de plomo lead sulfite
sulfito de rubidio rubidium sulfite
sulfito magnésico magnesium sulfite
sulfito manganoso manganous sulfite
sulfito potásico potassium sulfite
sulfito sódico sodium sulfite
sulfo sulfo
sulfoamino sulfoamino
sulfobromoftaleína sódica sodium sulfobromophthalein
sulfocarbanilida sulfocarbanilide
sulfocarbimida sulfocarbimide
sulfocarbolato sódico sodium sulfocarbolate
sulfocarbolida sulfocarbolide
sulfocarbonato potásico potassium sulfocarbonate
sulfocarbonato sódico sodium sulfocarbonate
sulfocianato sulfocyanate
sulfocianato cálcico calcium sulfocyanate
sulfocianato mercúrico mercuric sulfocyanate
sulfocianato potásico potassium sulfocyanate
sulfocianato sódico sodium sulfocyanate
sulfocianuro sulfocyanide
sulfocianuro bárico barium sulfocyanide
sulfocianuro de plomo lead sulfocyanide
sulfocianuro potásico potassium sulfocyanide
sulfocianuro sódico sodium sulfocyanide
sulfohidrato sulfohydrate
sulfolano sulfolane
sulfona sulfone
sulfonación sulfonation
sulfonamida sulfonamide
sulfonato sulfonate
sulfonato de alizarina alizarin sulfonate
sulfonato de alquilbenceno alkylbenzene sulfonate

sulfonato sódico sodium sulfonate
sulfonildifenol sulfonyldiphenol
sulfonilo sulfonyl
sulfonio sulfonium
sulfonóxido sulfinoxide
sulfopropionitrilo sódico sodium sulfopropionitrile
sulforricinoleato sódico sodium sulforicinoleate
sulfoseleniuro sulfoselenide
sulfóxido sulfoxide
sulfóxido de dimetilo dimethyl sulfoxide
sulfóxido de talio thallium sulfoxide
sulfoxilato sulfoxylate
sulfoxilato de cinc zinc sulfoxylate
sulfoxilato sódico sodium sulfoxylate
sulfuración sulfuration
sulfurado sulfurated
sulfúrico sulfuric
sulfurilo sulfuryl
sulfuro sulfide
sulfuro alumínico aluminum sulfide
sulfuro amónico ammonium sulfide
sulfuro antimónico antimonic sulfide
sulfuro antimonioso antimonous sulfide
sulfuro arsenioso arsenous sulfide
sulfuro áurico auric sulfide
sulfuro bárico barium sulfide
sulfuro cálcico calcium sulfide
sulfuro cérico ceric sulfide
sulfuro cobáltico cobaltic sulfide
sulfuro cobaltoso cobaltous sulfide
sulfuro cúprico cupric sulfide
sulfuro cuproso cuprous sulfide
sulfuro de alilo allyl sulfide
sulfuro de amilo amyl sulfide
sulfuro de antimonio antimony sulfide
sulfuro de arsénico arsenic sulfide
sulfuro de bencilo benzyl sulfide
sulfuro de benzofenona benzophenone sulfide
sulfuro de benzoilo benzoyl sulfide
sulfuro de bismuto bismuth sulfide
sulfuro de boro boron sulfide
sulfuro de butilo butyl sulfide
sulfuro de cadmio cadmium sulfide
sulfuro de carbonilo carbonyl sulfide
sulfuro de caucho rubber sulfide

sulfuro de cianógeno cyanogen sulfide
sulfuro de cinc zinc sulfide
sulfuro de cobre copper sulfide
sulfuro de dialilo diallyl sulfide
sulfuro de diamilo diamyl sulfide
sulfuro de dibromodietilo dibromodiethyl sulfide
sulfuro de dibutilestaño dibutyltin sulfide
sulfuro de diclorodietilo dichlorodiethyl sulfide
sulfuro de didecilo didecyl sulfide
sulfuro de diectilo diectyl sulfide
sulfuro de diestearilo distearyl sulfide
sulfuro de dietilo diethyl sulfide
sulfuro de dimetilo dimethyl sulfide
sulfuro de dimiristilo dimyristyl sulfide
sulfuro de dipentilo dipentyl sulfide
sulfuro de dipropilo dipropyl sulfide
sulfuro de divinilo divinyl sulfide
sulfuro de estaño tin sulfide
sulfuro de estroncio strontium sulfide
sulfuro de etilo ethyl sulfide
sulfuro de fósforo phosphorus sulfide
sulfuro de galio gallium sulfide
sulfuro de germanio germanium sulfide
sulfuro de hidrógeno hydrogen sulfide
sulfuro de hidrógeno amónico ammonium hydrogen sulfide
sulfuro de hidrógeno sódico sodium hydrogen sulfide
sulfuro de hierro iron sulfide
sulfuro de indio indium sulfide
sulfuro de itrio yttrium sulfide
sulfuro de lantano lanthanum sulfide
sulfuro de litio lithium sulfide
sulfuro de manganeso manganese sulfide
sulfuro de metilo methyl sulfide
sulfuro de molibdeno molybdenum sulfide
sulfuro de neodimio neodymium sulfide
sulfuro de nitrógeno nitrogen sulfide
sulfuro de oro gold sulfide
sulfuro de plata silver sulfide
sulfuro de plomo lead sulfide
sulfuro de renio rhenium sulfide
sulfuro de rodio rhodium sulfide
sulfuro de rubidio rubidium sulfide
sulfuro de rutenio ruthenium sulfide

sulfuro de selenio selenium sulfide
sulfuro de silicio silicon sulfide
sulfuro de talio thallium sulfide
sulfuro de telurio tellurium sulfide
sulfuro de uranilo uranyl sulfide
sulfuro de vanadio vanadium sulfide
sulfuro de viniletoxietilo vinylethoxyethyl sulfide
sulfuro de vinilo vinyl sulfide
sulfuro estánnico stannic sulfide
sulfuro estannoso stannous sulfide
sulfuro férrico ferric sulfide
sulfuro ferroso ferrous sulfide
sulfuro fosfórico phosphoric sulfide
sulfuro magnésico magnesium sulfide
sulfuro manganoso manganous sulfide
sulfuro mercúrico mercuric sulfide
sulfuro mercurioso mercurous sulfide
sulfuro niquélico nickelic sulfide
sulfuro plumboso plumbous sulfide
sulfuro potásico potassium sulfide
sulfuro samárico samaric sulfide
sulfuro sódico sodium sulfide
sulfuro talioso thallous sulfide
sulfuro vanádico vanadic sulfide
sulfuroso sulfurous
sultama sultam
sultona sultone
sulvanita sulvanite
superácido superacid
superactínido superactinide
superaleación superalloy
supercalentamiento superheating
supercarbonato supercarbonate
superconductividad superconductivity
superconductor superconductor
supercrítico supercritical
superficie surface
superficie de Fermi Fermi surface
superficie molar molar surface
superfluidez superfluidity
superfluido superfluid
superfosfato superphosphate
superfosfato amoniacal ammoniated superphosphate
superfosfato cálcico calcium superphosphate
superfosfato triple triple superphosphate
supernormal supernormal

superóxido superoxide
superóxido bárico barium superoxide
superóxido cálcico calcium superoxide
superóxido de plomo lead superoxide
superóxido potásico potassium
     superoxide
superóxido sódico sodium superoxide
superpalita superpalite
superpesado superheavy
superplasticidad superplasticity
superpolímero superpolymer
supersaturación supersaturation
supersaturado supersaturated
supersolubilidad supersolubility

supra supra
surfactante surfactant
surfusión surfusion
surrosión surrosion
susceptibilidad atómica atomic
     susceptibility
susceptibilidad magnética magnetic
     susceptibility
susceptibilidad molar molar
     susceptibility
suspensión suspension
suspensión acuosa slurry
suspensión coloidal colloidal suspension
suspensoide suspensoid

T

tabaco tobacco
tabla periódica periodic table
tabún tabun
taconita taconite
tacticidad tacticity
tactosol tactosol
taenita taenite
talato thallate
talato de manganeso manganese tallate
talato de plomo lead tallate
talato férrico ferric tallate
talco talc
tálico thallic
talidomida thalidomide
talio thallium
talioso thallous
talosa talose
tamaño de partícula particle size
tamiz screen
tamizado screening
tampón buffer
tanasa tannase
tanato cálcico calcium tannate
tanato de bismuto bismuth tannate
tanato férrico ferric tannate
tanilo tannyl
tanino tannin
tanque solar solar pond
tantalato tantalate
tantalato ferroso ferrous tantalate
tantalio tantalum
tantalita tantalite
tántalo tantalum
tantaloso tantalous
tantirón tantiron
tapa lid
tapioca tapioca
tapolita tapiolite
tapón stopper
tapón con llave stopper with tap
tapón de vidrio glass stopper
taquiol tachiol
tara tare

tártaro emético tartar emetic
tartaroilo tartaroyl
tartrato tartrate
tartrato ácido amónico ammonium acid tartrate
tartrato ácido potásico potassium acid tartrate
tartrato ácido sódico sodium acid tartrate
tartrato alumínico aluminum tartrate
tartrato amónico ammonium tartrate
tartrato amónico ácido acid ammonium tartrate
tartrato bárico barium tartrate
tartrato cálcico calcium tartrate
tartrato cobaltoso cobaltous tartrate
tartrato crómico chromic tartrate
tartrato cúprico cupric tartrate
tartrato de cadmio cadmium tartrate
tartrato de cinc zinc tartrate
tartrato de dibutilo dibutyl tartrate
tartrato de dietilo diethyl tartrate
tartrato de estroncio strontium tartrate
tartrato de hidrazina hydrazine tartrate
tartrato de hidrógeno amónico ammonium hydrogen tartrate
tartrato de hidrógeno potásico potassium hydrogen tartrate
tartrato de hidrógeno sódico sodium hydrogen tartrate
tartrato de hierro amónico ammonium iron tartrate
tartrato de plata silver tartrate
tartrato de plomo lead tartrate
tartrato de rubidio rubidium tartrate
tartrato estannoso stannous tartrate
tartrato ferroso ferrous tartrate
tartrato magnésico magnesium tartrate
tartrato manganoso manganous tartrate
tartrato mercurioso mercurous tartrate
tartrato potásico potassium tartrate
tartrato potásico alumínico aluminum potassium tartrate

tartrato potásico de antimonio antimony potassium tartrate
tartrato potásico de bismuto bismuth potassium tartrate
tartrato potásico sódico sodium potassium tartrate
tartrato sódico sodium tartrate
tartrato sódico de antimonio antimony sodium tartrate
tartrato sódico potásico potassium sodium tartrate
tartrazina tartrazine
tartronoilo tartronoyl
tau tau
taurilo tauryl
taurina taurine
tautomerasa tautomerase
tautomería tautomerism
tautomérico tautomeric
tautomerismo de valencia valence tautomerism
tautómero tautomer
tautourea tautourea
taxol taxol
tebaína thebaine
tebenidina thebenidine
tecnecio technetium
técnica de datación dating technique
técnica de Paneth Paneth technique
técnico technical
tecnología química chemical technology
tectita tektite
tefroíta tephroite
teína theine
tela de alambre wire cloth
telodrina telodrin
telomerización telomerization
teluluro de hidrógeno hydrogen telluride
teluluro de indio indium telluride
telurato tellurate
telurato amónico ammonium tellurate
telurato de hidrógeno hydrogen tellurate
telurato sódico sodium tellurate
telurilo telluryl
telurinilo tellurinyl
telurio tellurium
telurita tellurite
telurita potásica potassium tellurite
telurita sódica sodium tellurite

telurocetona telluroketone
teluronio telluronium
teluroso tellurous
telururo telluride
telururo de bismuto bismuth telluride
telururo de cadmio cadmium telluride
telururo de cinc zinc telluride
telururo de dimetilo dimethyl telluride
telururo de germanio germanium telluride
telururo de mercurio mercury telluride
telururo de plata silver telluride
telururo de plomo lead telluride
temperatura temperature
temperatura absoluta absolute temperature
temperatura ambiente ambient temperature
temperatura ambiente room temperature
temperatura característica de Debye Debye characteristic temperature
temperatura constante constant temperature
temperatura correspondiente corresponding temperature
temperatura crítica critical temperature
temperatura de cero absoluto absolute zero temperature
temperatura de condensación condensation temperature
temperatura de Curie Curie temperature
temperatura de Flory Flory temperature
temperatura de ignición ignition temperature
temperatura de inversión inversion temperature
temperatura de Néel Néel temperature
temperatura de sala room temperature
temperatura de solución crítica critical solution temperature
temperatura de transición transition temperature
temperatura de transición de vidrio glass-transition temperature
temperatura efectiva effective temperature
temperatura eutéctica eutectic temperature

temperatura normal standard temperature, normal temperature
temperatura peritéctica peritectic temperature
temperatura y presión normales standard temperature and pressure
templa mash
templar temper
tenazas tongs
tenazas de crisol crucible tongs
tenildiamina thenyldiamine
tenilo thenyl
tennantita tennantite
tenorita tenorite
tensión acuosa aqueous tension
tensión de electrodo electrode voltage
tensión de vapor vapor tension
tensión superficial surface tension
teobromina theobromine
teofilina theophylline
teorema de Bloch Bloch theorem
teorema de Kramers Kramers theorem
teorético theoretical
teoría atómica atomic theory
teoría atómica de Dalton Dalton atomic theory
teoría cinética kinetic theory
teoría cuántica quantum theory
teoría cuántica relativista relativistic quantum theory
teoría de Avogadro Avogadro theory
teoría de Bohr Bohr theory
teoría de Bohr-Sommerfeld Bohr-Sommerfeld theory
teoría de Brönsted Brönsted theory
teoría de Brönsted-Lowry Brönsted-Lowry theory
teoría de campo cristalino crystal-field theory
teoría de campo de ligandos ligand-field theory
teoría de Debye-Hückel Debye-Hückel theory
teoría de deformación strain theory
teoría de deformación de Baeyer Baeyer strain theory
teoría de Langmuir Langmuir theory
teoría de Lewis Lewis theory
teoría de Lowry-Brönsted Lowry-Brønsted theory
teoría de Lucas Lucas theory
teoría de McMillan-Mayer McMillan-Mayer theory
teoría de octeto octet theory
teoría de valencia valence theory
teoría de van't Hoff van't Hoff theory
teoría electrónica de Lewis Lewis electron theory
teoría nuclear nuclear theory
teoría orbital orbital theory
ter ter
teratógeno teratogen
terbia terbia
terbio terbium
tercera ley de termodinámica third law of thermodynamics
terciario tertiary
terebeno terebene
tereftalaldehído terephthalaldehyde
tereftalato de dimetilo dimethyl terephthalate
tereftalonitrilo terephthalonitrile
terfenilo terphenyl
termal thermal, thermic
termalizar thermalize
térmico thermic, thermal
terminación de cadena chain termination
terminología terminology
termita thermite
termo thermo
termoaislamiento heat insulation
termoanálisis thermoanalysis
termobalanza thermobalance
termocatalítico thermocatalytic
termocroico thermocroic
termodifusión thermodiffusion
termodinámica thermodynamics
termodinámica química chemical thermodynamics
termodinámico thermodynamic
termoelectricidad thermoelectricity
termoeléctrico thermoelectric
termoestable thermostable
termofisión thermofission
termofusión thermofusion
termogénico thermogenic
termógrafo thermograph
termólisis thermolysis

termolítico thermolytic
termoluminiscencia thermoluminescence
termoluminiscente thermoluminescent
termometamorfismo
thermometamorphism
termométrico thermometric
termómetro thermometer
termómetro armado armored
thermometer
termómetro de ampolleta húmeda wet-
bulb thermometer
termómetro de Beckmann Beckmann
thermometer
termómetro de Einchluss Einchluss
thermometer
termómetro de gas gas thermometer
termómetro de pentano pentane
thermometer
termonatrita thermonatrite
termoneutralidad thermoneutrality
termonuclear thermonuclear
termopar thermocouple
termoplástico thermoplastic
termoquímica thermochemistry
termoquímico thermochemical
termorregulador thermoregulator
termostático thermostatic
termostato thermostat
termotrópico thermotropic
ternario ternary
teroría de Werner Werner theory
teróxido teroxide
terpadieno terpadiene
terpeno terpene
terpenoide terpenoid
terpenol terpenol
terpilenol terpilenol
terpineno terpinene
terpineol terpineol
terpinoleno terpinolene
terpolímero terpolymer
tervalente tervalent
testosterona testosterone
tetina thetine
tetra tetra
tetraacetato de pentaeritritol
pentaerythritol tetraacetate
tetraacetato de plomo lead tetraacetate
tetraacetilacetonato de zirconio

zirconium tetraacetylacetonate
tetrabásico tetrabasic
tetrabencilo tetrabenzyl
tetraborano tetraborane
tetraborato de cinc zinc tetraborate
tetraborato de litio lithium tetraborate
tetraborato de mercurio mercury
tetraborate
tetraborato potásico potassium
tetraborate
tetraborato sódico sodium tetraborate
tetrabromo tetrabromo
tetrabromoaurato sódico sodium
tetrabromoaurate
tetrabromobenceno tetrabromobenzene
tetrabromoetano tetrabromoethane
tetrabromoetileno tetrabromoethylene
tetrabromofluoresceína
tetrabromofluorescein
tetrabromometano tetrabromomethane
tetrabromosilano tetrabromosilane
tetrabromuro de acetileno acetylene
tetrabromide
tetrabromuro de carbono carbon
tetrabromide
tetrabromuro de estaño tin tetrabromide
tetrabromuro de etileno ethylene
tetrabromide
tetrabromuro de germanio germanium
tetrabromide
tetrabromuro de iridio iridium
tetrabromide
tetrabromuro de selenio selenium
tetrabromide
tetrabromuro de silicio silicon
tetrabromide
tetrabromuro de telurio tellurium
tetrabromide
tetrabromuro de uranio uranium
tetrabromide
tetrabutilestaño tetrabutyltin
tetracaína tetracaine
tetracarbonilo de cobalto cobalt
tetracarbonyl
tetracarbonilo de níquel nickel
tetracarbonyl
tetracarboxibutano tetracarboxybutane
tetraceno tetrazene
tetracetona tetraketone

tetraciano tetracyano
tetracianoaurato potásico potassium tetracyanoaurate
tetracianoetileno tetracyanoethylene
tetracianoquinonadimetano tetracyanoquinonedimethane
tetraciclina tetracycline
tetraciclona tetracyclone
tetracloro tetrachloro
tetracloroaluminato sódico sodium tetrachloroaluminate
tetracloroanilina tetrachloroaniline
tetracloroaurato sódico sodium tetrachloroaurate
tetraclorobenceno tetrachlorobenzene
tetraclorodifeniletano tetrachlorodiphenylethane
tetracloroetano tetrachloroethane
tetracloroetileno tetrachloroethylene
tetraclorofenato sódico sodium tetrachlorophenate
tetraclorofenol tetrachlorophenol
tetraclorometano tetrachloromethane
tetracloronaftaleno tetrachloronaphthalene
tetracloropaladato potásico potassium tetrachloropalladate
tetracloropaladato sódico sodium tetrachloropalladate
tetracloroplatinato sódico sodium tetrachloroplatinate
tetraclorosalicilanilida tetrachlorosalicylanilide
tetraclorosilano tetrachlorosilane
tetracloruro tetrachloride
tetracloruro de azufre sulfur tetrachloride
tetracloruro de carbono carbon tetrachloride
tetracloruro de estaño tin tetrachloride
tetracloruro de etileno ethylene tetrachloride
tetracloruro de germanio germanium tetrachloride
tetracloruro de iridio iridium tetrachloride
tetracloruro de manganeso manganese tetrachloride
tetracloruro de osmio osmium tetrachloride

tetracloruro de platino platinum tetrachloride
tetracloruro de plomo lead tetrachloride
tetracloruro de silicio silicon tetrachloride
tetracloruro de titanio titanium tetrachloride
tetracloruro de torio thorium tetrachloride
tetracloruro de uranio uranium tetrachloride
tetracloruro de vanadio vanadium tetrachloride
tetracloruro de zirconio zirconium tetrachloride
tetracosano tetracosane
tetracosilo tetracosyl
tetrada tetrad
tetradecano tetradecane
tetradecanol tetradecanol
tetradeceno tetradecene
tetradecilamina tetradecylamine
tetradecileno tetradecylene
tetradecilo tetradecyl
tetradecilsulfato sódico sodium tetradecyl sulfate
tetradeciltiol tetradecyl thiol
tetradifón tetradifon
tetraédrico tetrahedronal
tetraedro tetrahedron
tetraetilbenceno tetraethylbenzene
tetraetilenglicol tetraethylene glycol
tetraetilenopentamina tetraethylenepentamine
tetraetilestaño tetraethyltin
tetraetilgermanio tetraethylgermanium
tetraetilo tetraethyl
tetraetilplomo tetraethyllead
tetraetoxipropano tetraethoxypropane
tetrafenilborato sódico sodium tetraphenylborate
tetrafenilbutadieno tetraphenylbutadiene
tetrafenilciclopentadienona tetraphenylcyclopentadienone
tetrafenileno tetraphenylene
tetrafenilestaño tetraphenyltin
tetrafeniletano tetraphenylethane
tetrafenilmetano tetraphenylmethane

tetrafenilo tetraphenyl
tetrafenilsilano tetraphenylsilane
tetrafenilurea tetraphenylurea
tetrafluoroborato tetrafluoroborate
tetrafluoroborato potásico potassium
    tetrafluoroborate
tetrafluorodicloroetano
    tetrafluorodichloroethane
tetrafluoroetileno tetrafluoroethylene
tetrafluorohidrazina
    tetrafluorohydrazine
tetrafluorometano tetrafluoromethane
tetrafluorosilano tetrafluorosilane
tetrafluoruro de azufre sulfur
    tetrafluoride
tetrafluoruro de carbono carbon
    tetrafluoride
tetrafluoruro de germanio germanium
    tetrafluoride
tetrafluoruro de manganeso manganese
    tetrafluoride
tetrafluoruro de selenio selenium
    tetrafluoride
tetrafluoruro de silicio silicon
    tetrafluoride
tetrafluoruro de uranio uranium
    tetrafluoride
tetrafluoruro de xenón xenon
    tetrafluoride
tetrafluoruro de zirconio zirconium
    tetrafluoride
tetrafosfato de hexaetilo hexaethyl
    tetraphosphate
tetrafosfato sódico sodium
    tetraphosphate
tetraftalato de polietileno polyethylene
    tetraphthalate
tetragaloilo tetragalloyl
tetragonal tetragonal
tetrahidrato tetrahydrate
tetrahidrato de perborato sódico
    sodium perborate tetrahydrate
tetrahidro tetrahydro
tetrahidroaluminato de litio lithium
    tetrahydroaluminate
tetrahidrobenceno tetrahydrobenzene
tetrahidroborato potásico potassium
    tetrahydroborate
tetrahidroborato sódico sodium

    tetrahydroborate
tetrahidrocannibol tetrahydrocannibol
tetrahidrofenol tetrahydrophenol
tetrahidrofurano tetrahydrofuran
tetrahidrofuranodimetanol
    tetrahydrofurandimethanol
tetrahidrofurfurilamina
    tetrahydrofurfurylamine
tetrahidronaftaleno
    tetrahydronaphthalene
tetrahidropiranometanol
    tetrahydropyranmethanol
tetrahidropiridina tetrahydropyridine
tetrahidrotiofeno tetrahydrothiophene
tetrahidroxi tetrahydroxy
tetrahidroxibenceno
    tetrahydroxybenzene
tetrahidroxibutano tetrahydroxybutane
tetrahidruro de germanio germanium
    tetrahydride
tetrahidruro de silicio silicon
    tetrahydride
tetrakis tetrakis
tetralina tetralin
tetralita tetralite
tetralona tetralone
tetrámero tetramer
tetrametilbenceno tetramethylbenzene
tetrametilbencidina tetramethylbenzidine
tetrametilciclobutanodiol
    tetramethylcyclobutanediol
tetrametildiaminobenzhidrol
    tetramethyldiaminobenzhydrol
tetrametildiaminobenzofenona
    tetramethyldiaminobenzophenone
tetrametildiaminodifenilsulfona
    tetramethyldiaminodiphenylsulfone
tetrametilendiamina
    tetramethylenediamine
tetrametileno tetramethylene
tetrametilestaño tetramethyltin
tetrametiletilendiamina
    tetramethylethylenediamine
tetrametilmetano tetramethylmethane
tetrametilo tetramethyl
tetrametilplomo tetramethyllead
tetrametilsilano tetramethylsilane
tetrametilurea tetramethylurea
tetrametoxipropano

tetramethoxypropane

tetramina tetramine

tetramolecular tetramolecular

tetramorfismo tetramorphism

tetranitrato tetranitrate

tetranitrato de pentaeritritol
pentaerythritol tetranitrate

tetranitro tetranitro

tetranitroanilina tetranitroaniline

tetranitrobifenilo tetranitrobiphenyl

tetranitrofenol tetranitrophenol

tetranitrometano tetranitromethane

tetraóxido tetraoxide

tetrapropileno tetrapropylene

tetrasilano tetrasilane

tetrasulfuro bárico barium tetrasulfide

tetrasulfuro de arsénico arsenic
tetrasulfide

tetrasulfuro de cromo chromium
tetrasulfide

tetrasulfuro de molibdeno molybdenum
tetrasulfide

tetrasulfuro sódico sodium tetrasulfide

tetratioarseniato tetrathioarsenate

tetratioarsenito tetrathioarsenite

tetrationato tetrathionate

tetravalente tetravalent

tetrayodo tetraiodo

tetrayodoaurato potásico potassium
tetraiodoaurate

tetrayoduro de carbono carbon
tetraiodide

tetrayoduro de estaño tin tetraiodide

tetrayoduro de germanio germanium
tetraiodide

tetrayoduro de iridio iridium tetraiodide

tetrayoduro de silicio silicon tetraiodide

tetrazol tetrazole

tetrazolilo tetrazolyl

tetrazolio tetrazolium

tetrazona tetrazone

tetrel tetrel

tetrilo tetryl

tetritol tetritol

tetrol tetrol

tetrona tetrone

tetrosa tetrose

tetroxalato potásico potassium
tetroxalate

tetróxido tetroxide

tetróxido de bismuto bismuth tetraoxide

tetróxido de cesio cesium tetraoxide

tetróxido de dinitrógeno dinitrogen
tetraoxide

tetróxido de manganeso manganese
tetraoxide

tetróxido de nitrógeno nitrogen
tetraoxide

tetróxido de osmio osmium tetraoxide

tetróxido de plomo lead tetraoxide

tetróxido de rutenio ruthenium
tetraoxide

tetróxido de vanadio vanadium
tetraoxide

texafirina texaphyrin

texilo thexyl

textura texture

thalenita thalenite

theta theta

thomsonita thomsonite

thortveitita thortveitite

tia thia

tiabendazol thiabendazole

tiacilo thiazyl

tial thial

tiamina thiamine

tiantreno thianthrene

tiazina thiazine

tiazol thiazole

tiazolilo thiazolyl

tiempo de generación generation time

tiempo de reacción reaction time

tiempo de relajación relaxation time

tiempo de retención retention time

tiempo de transición transition time

tiempo libre medio mean free time

tienilo thienyl

tierra earth

tierra alcalina alkaline earth

tierra de batán fuller's earth

tierra de diatomeas diatomaceous earth

tierra de pipas pipe clay

tierra de sombra umber

tierra rara rare earth

timerosal thimerosal

timidina thymidine

timilo thymyl

timina thymine

timol thymol
timolftaleína thymolphthalein
timolol timolol
timolsulfonaftaleína
    thymolsulfonephthalein
tincal tincal
tinta ink
tinte dye, colorant
tintómetro tintometer
tintura tincture, stain
tintura de yodo iodine tincture
tio thio
tioacetal thioacetal
tioacetaldehído thioacetaldehyde
tioacetamida thioacetamide
tioacetanilida thioacetanilide
tioacetato amónico ammonium
    thioacetate
tioácido thioacid
tioalcohol thioalcohol
tioaldehído thioaldehyde
tioanhídrido thioanhydride
tioantimoniato sódico sodium
    thioantimonate
tioarseniato thioarsenate
tioarseniato de arsénico arsenic
    thioarsenate
tioato thioate
tioaurato sódico sodium thioaurate
tiobenzaldehído thiobenzaldehyde
tiobenzamida thiobenzamide
tiobenzofenona thiobenzophenone
tiocarbamoilo thiocarbamoyl
tiocarbanilida thiocarbanilide
tiocarbonato thiocarbonate
tiocarbonato celulósico cellulosic
    thiocarbonate
tiocarbonato sódico sodium
    thiocarbonate
tiocarbonilo thiocarbonyl
tiocarboxi thiocarboxy
tiocetona thioketone
tiocianato thiocyanate
tiocianato alumínico aluminum
    thiocyanate
tiocianato amónico ammonium
    thiocyanate
tiocianato bárico barium thiocyanate
tiocianato cálcico calcium thiocyanate

tiocianato cuproso cuprous thiocyanate
tiocianato de arsénico arsenic
    thiocyanate
tiocianato de bencilo benzyl thiocyanate
tiocianato de cinc zinc thiocyanate
tiocianato de etilo ethyl thiocyanate
tiocianato de guanidina guanidine
    thiocyanate
tiocianato de hierro iron thiocyanate
tiocianato de litio lithium thiocyanate
tiocianato de nonilo nonyl thiocyanate
tiocianato de pentilo pentyl thiocyanate
tiocianato de plata silver thiocyanate
tiocianato de plomo lead thiocyanate
tiocianato férrico ferric thiocyanate
tiocianato ferroso ferrous thiocyanate
tiocianato magnésico magnesium
    thiocyanate
tiocianato mercúrico mercuric
    thiocyanate
tiocianato potásico potassium
    thiocyanate
tiocianato sódico sodium thiocyanate
tiocianiato de butilo butyl thiocyanate
tiocianoacetato de isobornilo isobornyl
    thiocyanoacetate
tiocianógeno thiocyanogen
tiocianuro thiocyanide
tiocloruro de fósforo phosphorus
    thiochloride
tiocromo thiochrome
tiodifenol thiodiphenol
tiodiglicol thiodiglycol
tiodipropionitrilo thiodipropionitrile
tioestannato amónico ammonium
    thiostanate
tioéster thioester
tioetano de etilo ethyl thioethane
tioetanol de etilo ethyl thioethanol
tioéter thioether
tioéter alílico allyl thioether
tiofano thiophane
tiofenaldehído thiophenealdehyde
tiofenato de cinc zinc thiophenate
tiofenilo thiophenyl
tiofeno thiophene
tiofenol thiophenol
tioflavina thioflavine
tiofosforilo thiophosphoryl

tiofosgeno thiophosgene
tiofurano thiofuran
tioglicerol thioglycerol
tioglicolato de isooctilo isooctyl thioglycolate
tioglicolato sódico sodium thioglycolate
tiohidantoína thiohydantoin
tiohidroxi thiohydroxy
tiol thiol
tiolato thiolate
tiomolibdato thiomolybdate
tionato thionate
tionilo thionyl
tiopental sódico thiopental sodium
tioridazina thioridazine
tiosemicarbazida thiosemicarbazide
tiosorbitol thiosorbitol
tiosulfato thiosulfate
tiosulfato amónico ammonium thiosulfate
tiosulfato bárico barium thiosulfate
tiosulfato de estroncio strontium thiosulfate
tiosulfato de hierro iron thiosulfate
tiosulfato de plata silver thiosulfate
tiosulfato de plata sódico sodium silver thiosulfate
tiosulfato de plomo lead thiosulfate
tiosulfato de plomo sódico sodium lead thiosulfate
tiosulfato magnésico magnesium thiosulfate
tiosulfato potásico potassium thiosulfate
tiosulfato sódico sodium thiosulfate
tiosulfato sódico de plata silver sodium thiosulfate
tiotungstato thiotungstate
tiouracil thiouracil
tiourea thiourea
tiourea de etileno ethylene thiourea
tioxanteno thioxanthene
tioxilenol thioxylenol
tioxo thioxo
típico typical
tipo type
tiram thiram
tiramina tyramine
tirocidina tyrocidine
tironina thyronine

tirosilo tyrosyl
tirosina tyrosine
tirosinasa tyrosinase
tirotricina thyrothricin
tiroxina thyroxine
tisonita tysonite
titanato titanate
titanato bárico barium titanate
titanato cálcico calcium titanate
titanato de butilo butyl titanate
titanato de estroncio strontium titanate
titanato de litio lithium titanate
titanato de níquel nickel titanate
titanato de plomo lead titanate
titanato de tetrabutilo tetrabutyl titanate
titanato de tetraetilhexilo tetraethylhexyl titanate
titanato de tetraisopropilo tetraisopropyl titanate
titanato ferroso ferrous titanate
titanato potásico potassium titanate
titanato sódico sodium titanate
titania titania
titánico titanic
titanilo titanyl
titanio titanium
titanoso titanous
titulación titration
titulación ácido-base acid-base titration
titulación amperométrica amperometric titration
titulación automática automatic titration
titulación complejométrica complexometric titration
titulación con permanganato permanganate titration
titulación conductiométrica conductometric titration
titulación de Mohr Mohr titration
titulación de Volhard Volhard titration
titulación diferencial differential titration
titulación electrométrica electrometric titration
titulación entálpica enthalpy titration
titulación fotométrica photometric titration
titulación magnetométrica magnetometric titration

titulación oscilométrica oscillometric titration

titulación potenciométrica potentiometric titration

titulación radiométrica radiometric titration

titulación redox redox titration

titulación térmica thermal titration

titulación termométrica thermometric titration

título titer

tixotropía thixotropy

tocoferol tocopherol

tocofersolán tocophersolan

tolano tolan

tolidina tolidine

tolilaldehído tolylaldehyde

tolildietanolamina tolyldiethanolamine

tolilendiamina tolylenediamine

tolileno tolylene

tolilideno tolylidene

tolilo tolyl

tolnaftato tolnaftate

tolualdehído tolualdehyde

toluato sódico sodium toluate

toluendiamina toluenediamine

toluendiisocianato toluenediisocyanate

toluenilo toluenyl

tolueno toluene

toluenosulfonato sódico sodium toluenesulfonate

toluensulfamina toluenesulfamine

toluensulfanilida toluenesulfanilide

toluensulfonamida toluenesulfonamide

toluensulfonilo toluenesulfonyl

toluentiol toluenethiol

toluidina toluidine

toluidino toluidino

toluilendiamina toluylenediamine

toluileno toluylene

toluilo toluyl

toluoilo toluoyl

toluquinona toluquinone

toma de gas gas outlet

tomatina tomatine

tonca tonka

topacio topaz

topoquímica topochemistry

topotáctico topotactic

toria thoria

toriado thoriated

torianita thorianite

torina thorin

torio thorium

toriode toroid

torita thorite

tornasol litmus

torón thoron

torr torr

torre tower

torre de absorción absorption tower

torre de destilación distillation tower

torre de Gay-Lussac Gay-Lussac tower

torre empaquetada packed tower

torta amarilla yellow cake

torta de sal salt cake

tosilación tosylation

tosilo tosyl

toxafeno toxaphene

toxicidad toxicity

tóxico toxic

toxicología toxicology

toxina toxin

trans trans

transactínido transactinide

transalquilación transalkylation

transaminación transamination

transaminasa transaminase

transductor transducer

transesterificación transesterification

transferasa transferase

transferecia térmica heat transfer

transferencia de cadena chain transfer

transferencia de carga charge transfer

transferencia de fase phase transfer

transferencia electrónica electron transfer

transferencia radiativa radiative transfer

transformación transformation

transformación atómica atomic transformation

transformación de Curtius Curtius transformation

transformación de Fourier Fourier transform

transformación de Hofmann Hofmann transformation

transformación nuclear nuclear

transformation

transformación radiactiva radioactive
transformation

transformación termonuclear
thermonuclear transformation

transición transition

transición cuántica quantum transition

transición de fase phase transition

transición de Fermi Fermi transition

transición electrónica electronic
transition

transición gamma gamma transition

transición isomérica isomeric transition

transición permitida allowed transition

transición prohibida forbidden transition

transicional transitional

transmitancia transmittance

transmutación transmutation

transmutación atómica atomic
transmutation

transoide transoid

transparencia transparency

transporte transport

transporte activo active transport

transposición transposition,
rearrangement

transuránico transuranic

travertino travertine

trayectoria electrónica electron
trajectory

trayectoria libre molecular molecular
free path

traza trace

trazador tracer

trazador isotópico isotopic tracer

trazador radiactivo radioactive tracer

trementina turpentine

trementina de madera wood turpentine

tremolita tremolite

trenardita threnardite

treo threo

treonina threonine

treosa threose

tretamina tretamine

tri tri

triaceno triazene

triacetato triacetate

triacetato alumínico aluminum triacetate

triacetato de celulosa cellulose triacetate

triacetato de glicerilo glyceryl triacetate

triacetilgalato de etilo ethyl
triacetylgallate

triacetina triacetin

triacetonamina triacetoneamine

triacontano triacontane

triacontanol triacontanol

triacontilo triacontyl

tríada triad

trialilamina triallylamine

trialilo triallyl

trialquilsilanol trialkylsilanol

triamcinolona triamcinolone

triamilamina triamylamine

triamilbenceno triamylbenzene

triamina triamine

triaminobenceno triaminobenzene

triaminotolueno triaminotoluene

triángulo de crisol crucible triangle

triatómico triatomic

triazano triazane

triazina triazine

triazinilo triazinyl

triazol triazole

tribásico tribasic

tribencilamina tribenzylamine

tribencilo tribenzyl

triboluminiscencia triboluminescence

tribromo tribromo

tribromoacetaldehído
tribromoacetaldehyde

tribromoanilina tribromoaniline

tribromobenceno tribromobenzene

tribromoetanal tribromoethanal

tribromoetano tribromoethane

tribromoetanol tribromoethanol

tribromofenol tribromophenol

tribromometano tribromomethane

tribromopropano tribromopropane

tribromosalicilanilida
tribromosalicylanilide

tribromuro tribromide

tribromuro de alilo allyl tribromide

tribromuro de antimonio antimony
tribromide

tribromuro de arsénico arsenic
tribromide

tribromuro de bismuto bismuth
tribromide

tribromuro de boro boron tribromide
tribromuro de fósforo phosphorus tribromide
tribromuro de iridio iridium tribromide
tribromuro de nitrógeno nitrogen tribromide
tribromuro de oro gold tribromide
tribromuro férrico ferric tribromide
tributilaluminio tributylaluminum
tributilamina tributylamine
tributilborano tributylborane
tributilestaño tributyltin
tributilo tributyl
tricarbimida tricarbimide
tricetona triketone
triciano tricyano
tricianuro de oro gold tricyanide
tricíclico tricyclic
triciclodecano tricyclodecane
triclorfón trichlorfon
tricloro trichloro
tricloroacetaldehído trichloroacetaldehyde
tricloroacetamida trichloroacetamide
tricloroacetato sódico sodium trichloroacetate
tricloroanisol trichloroanisole
triclorobenceno trichlorobenzene
tricloroborazol trichloroborazole
triclorobromometano trichlorobromomethane
tricloroetanal trichloroethanal
tricloroetano trichloroethane
tricloroetanol trichloroethanol
tricloroeteno trichloroethene
tricloroetileno trichloroethylene
triclorofenato potásico potassium trichlorophenate
triclorofenato sódico sodium trichlorophenate
triclorofenol trichlorophenol
triclorofluorometano trichlorofluoromethane
triclorometano trichloromethane
triclorometilo trichloromethyl
tricloronaftaleno trichloronaphthalene
tricloronitrosometano trichloronitrosomethane
tricloropropano trichloropropane

triclorosilano trichlorosilane
triclorosilano de ciclohexilo cyclohexyl trichlorosilane
triclorotolueno trichlorotoluene
triclorotrifluoroacetona trichlorotrifluoroacetone
triclorotrifluoroetano trichlorotrifluoroethane
tricloruro trichloride
tricloruro benzoico benzoic trichloride
tricloruro de antimonio antimony trichloride
tricloruro de arsénico arsenic trichloride
tricloruro de bencenilo benzenyl trichloride
tricloruro de bismuto bismuth trichloride
tricloruro de boro boron trichloride
tricloruro de butilina butylin trichloride
tricloruro de carbono carbon trichloride
tricloruro de cromo chromium trichloride
tricloruro de etileno ethylene trichloride
tricloruro de formilo formyl trichloride
tricloruro de fósforo phosphorus trichloride
tricloruro de galio gallium trichloride
tricloruro de hierro iron trichloride
tricloruro de indio indium trichloride
tricloruro de iridio iridium trichloride
tricloruro de nitrógeno nitrogen trichloride
tricloruro de oro gold trichloride
tricloruro de renio rhenium trichloride
tricloruro de rodio rhodium trichloride
tricloruro de rutenio ruthenium trichloride
tricloruro de titanio titanium trichloride
tricloruro de tolueno toluene trichloride
tricloruro de trimesoilo trimesoyl trichloride
tricloruro de vanadio vanadium trichloride
tricloruro de vinilo vinyl trichloride
tricloruro de yodo iodine trichloride
tricloruro férrico ferric trichloride
tricosano tricosane
tricosanol tricosanol
tricresilo tricresyl

tridecano tridecane
tridecanoato de metilo methyl tridecanoate
tridecanol tridecanol
trideceno tridecene
tridecilbenceno tridecylbenzene
tridecilo tridecyl
tridimita tridymite
tridodecilamina tridodecyl amine
triestearato de glicerilo glyceryl tristearate
triestearato de glicerol glycerol tristearate
trietanolamina triethanolamine
trietilaluminio triethylaluminum
trietilamina triethylamine
trietilbenceno triethylbenzene
trietilborano triethylborane
trietilendiamina triethylenediamine
trietilenfosforamida triethylenephosphoramide
trietilenglicol triethylene glycol
trietilenmelamina triethylenemelamine
trietileno triethylene
trietilentetramina triethylenetetramine
trietilentriamina triethylenetriamine
trietilmetano triethylmethane
trietilo triethyl
trietilortoformiato triethylorthoformate
trietoxi triethoxy
trietoxihexano triethoxyhexane
trietoximetano triethoxymethane
trietoximetoxipropano triethoxymethoxypropane
trifásico triphasic
trifenilantimonio triphenylantimony
trifenilbenceno triphenylbenzene
trifenilboro triphenylboron
trifenilcarbinol triphenylcarbinol
trifenileno triphenylene
trifenilestaño triphenyltin
trifenilfosfina triphenylphosphine
trifenilfósforo triphenylphosphorus
trifenilguanidina triphenylguanidine
trifenilmetilo triphenylmethyl
trifenilo triphenyl
trifenol triphenol
triflato triflate
trifluoro trifluoro

trifluoroacetato de talio thallium trifluoroacetate
trifluoroacetilacetonato de cobre copper trifluoroacetylacetonate
trifluorobromometano trifluorobromomethane
trifluoroclorometano trifluorochloromethane
trifluoroestireno trifluorostyrene
trifluorometilo trifluoromethyl
trifluoronitrosometano trifluoronitrosomethane
trifluorotricloroetano trifluorotrichloroethane
trifluoroyodometano trifluoroiodomethane
trifluoruro de antimonio antimony trifluoride
trifluoruro de arsénico arsenic trifluoride
trifluoruro de boro boron trifluoride
trifluoruro de bromo bromine trifluoride
trifluoruro de cloro chlorine trifluoride
trifluoruro de cobalto cobalt trifluoride
trifluoruro de cromo chromium trifluoride
trifluoruro de nitrógeno nitrogen trifluoride
trifluoruro de rutenio ruthenium trifluoride
trifluoruro de tolueno toluene trifluoride
trifosfato bárico barium triphosphate
trifosfato de adenosina adenosine triphosphate
trifosfato pentasódico pentasodium triphosphate
trifosfato sódico sodium triphosphate
trifosgeno triphosgene
trigaloilo trigalloyl
triglicérido triglyceride
triglicerol triglycerol
triglicina triglycine
triglime triglyme
trigol trigol
trigonal trigonal
trigonelina trigonelline
trihexilo trihexyl
trihidrato trihydrate
trihidrato de alúmina alumina trihydrate

trihidrato de bismuto bismuth trihydrate
trihídrico trihydric
trihidroxi trihydroxy
trihidroxiantraquinona
    trihydroxyanthraquinone
trihidroxibenceno trihydroxybenzene
trihidróxido de bismuto bismuth
    trihydroxide
trihidroxipropano trihydroxypropane
triisobutilaluminio triisobutylaluminum
triisobutileno triisobutylene
triisopropanolamina triisopropanolamine
trilaurilamina trilaurylamine
trilaurilfosfito trilauryl phosphite
trilaurilo trilauryl
trilauriltritiofosfito trilauryl
    trithiophosphite
trímero trimer
trimetadiona trimethadione
trimetálico trimetallic
trimetano trimethano
trimetilaluminio trimethylaluminum
trimetilamina trimethylamine
trimetilanilina trimethylaniline
trimetilbenceno trimethylbenzene
trimetilbutano trimethylbutane
trimetilciclododecatrieno
    trimethylcyclododecatriene
trimetilciclohexano
    trimethylcyclohexane
trimetilciclohexanol
    trimethylcyclohexanol
trimetilclorosilano trimethylchlorosilane
trimetilenglicol trimethylene glycol
trimetileno trimethylene
trimetilestaño trimethyltin
trimetilhexano trimethylhexane
trimetilhexanol trimethylhexanol
trimetilmetano trimethylmethane
trimetilnonanona trimethylnonanone
trimetilo trimethyl
trimetiloletano trimethylolethane
trimetilolpropano trimethylolpropane
trimetilpentano trimethylpentane
trimetilpentanodiol trimethylpentanediol
trimetilpenteno trimethylpentene
trimetilpiridina trimethylpyridine
trimetilsililo trimethylsilyl
trimetoxi trimethoxy

trimetoxiboroxina trimethoxyboroxine
trimetoximetano trimethoxymethane
trimolecular trimolecular
trimorfismo trimorphism
trinitrato trinitrate
trinitrato de bismuto bismuth trinitrate
trinitrato de celulosa cellulose trinitrate
trinitrato de fenol phenol trinitrate
trinitrato de glicerilo glyceryl trinitrate
trinitrato de glicerol glycerol trinitrate
trinitro trinitro
trinitroanilina trinitroaniline
trinitroanisol trinitroanisole
trinitrobenceno trinitrobenzene
trinitrofenol trinitrophenol
trinitrofenolato amónico ammonium
    trinitrophenolate
trinitroglicerina trinitroglycerin
trinitrometano trinitromethane
trinitronaftaleno trinitronaphthalene
trinitrorresorcina trinitroresorcinol
trinitrorresorcinato de plomo lead
    trinitroresorcinate
trinitrotolueno trinitrotoluene
trinor trinor
triol triol
triona trione
triosa triose
trioxa trioxa
trioxano trioxane
trioxicloruro de renio rhenium
    trioxychloride
trióxido trioxide
trióxido de antimonio antimony trioxide
trióxido de arsénico arsenic trioxide
trióxido de azufre sulfur trioxide
trióxido de bismuto bismuth trioxide
trióxido de cesio cesium trioxide
trióxido de cromo chromium trioxide
trióxido de hierro iron trioxide
trióxido de iridio iridium trioxide
trióxido de lantano lanthanum trioxide
trióxido de manganeso manganese
    trioxide
trióxido de molibdeno molybdenum
    trioxide
trióxido de nitrógeno nitrogen trioxide
trióxido de oro gold trioxide
trióxido de samario samarium trioxide

trióxido de telurio   tellurium trioxide
trióxido de titanio   titanium trioxide
trióxido de tungsteno   tungsten trioxide
trióxido de uranio   uranium trioxide
trióxido de vanadio   vanadium trioxide
trióxido férrico   ferric trioxide
trioxígeno   trioxygen
trioxima   trioxime
trioximetileno   trioxymethylene
trioxina   trioxin
tripalmitato de glicerilo   glyceryl tripalmitate
tripalmitina   tripalmitin
triplete   triplet
triplita   triplite
trípode   tripod
trípoli   tripoli
tripolifosfato potásico   potassium tripolyphosphate
tripolifosfato sódico   sodium tripolyphosphate
tripolita   tripolite
tripropilaluminio   tripropylaluminum
tripropilamina   tripropylamine
tripropilenglicol   tripropylene glycol
tripropileno   tripropylene
tripropilo   tripropyl
tripsina   trypsin
tripsinógeno   trypsinogen
triptano   triptane
tripticeno   triptycene
triptófano   tryptophan
triptofila   tryptophyl
triquitas   whiskers
tris   tris
trishidroxifenilpropano   trishydroxyphenylpropane
trisilano   trisilane
trisilicato magnésico   magnesium trisilicate
trisulfato de antimonio   antimony trisulfate
trisulfato férrico   ferric trisulfate
trisulfuro   trisulfide
trisulfuro bárico   barium trisulfide
trisulfuro de antimonio   antimony trisulfide
trisulfuro de arsénico   arsenic trisulfide
trisulfuro de bismuto   bismuth trisulfide

trisulfuro de boro   boron trisulfide
trisulfuro de cesio   cesium trisulfide
trisulfuro de fósforo   phosphorus trisulfide
trisulfuro de molibdeno   molybdenum trisulfide
trisulfuro de oro   gold trisulfide
trisulfuro de vanadio   vanadium trisulfide
triteluNuro de bismuto   bismuth tritelluride
triterpeno   triterpene
tritilo   trityl
tritio   tritium
tritioacetaldehído   trithioacetaldehyde
tritiocarbonato   trithiocarbonate
tritiocarbonato sódico   sodium trithiocarbonate
tritionato bárico   barium trithionate
trititanato sódico   sodium trititanate
tritón   triton
triturar   triturate
trivalente   trivalent
triyodo   triiodo
triyodometano   triiodomethane
triyodotironina   triiodothyronine
triyoduro de antimonio   antimony triiodide
triyoduro de bismuto   bismuth triiodide
triyoduro de formilo   formyl triiodide
triyoduro de fósforo   phosphorus triiodide
triyoduro de nitrógeno   nitrogen triiodide
triyoduro de oro   gold triiodide
triyoduro potásico   potassium triiodide
trombina   thrombin
trona   trona
tropanol   tropanol
tropilidina   tropilidine
tropilio   tropylium
tropina   tropine
tropoilo   tropoyl
tropolona   tropolone
tubería de vidrio   glass tubing
tubo con embudo   funnel tube
tubo de absorción   absorption tube
tubo de absorción de Babo   Babo absorption tube
tubo de Arndt   Arndt tube

tubo de Bowen Bowen tube
tubo de Carius Carius tube
tubo de combustión combustion tube
tubo de condensación fraccionaria
    fractional condensation tube
tubo de Coolidge Coolidge tube
tubo de Crookes Crookes tube
tubo de destilación fraccionaria
    fractional distillation tube
tubo de ensayo test tube
tubo de Faraday Faraday tube
tubo de fermentación fermentation tube
tubo de filtro filter tube
tubo de Fleming Fleming tube
tubo de fusión fusion tube
tubo de LaBel LaBel tube
tubo de plástico plastic pipe
tubo de rayos X X-ray tube
tubo de vidrio glass tube
tubo en U U-tube
tubo espectral spectral tube
tubo secante drying tube
tubos de Nessler Nessler tubes
tulia thulia
tulio thulium
tungstato tungstate
tungstato amónico ammonium tungstate
tungstato cálcico calcium tungstate
tungstato cobaltoso cobaltous tungstate
tungstato de bismuto bismuth tungstate

tungstato de cadmio cadmium tungstate
tungstato de cobalto cobalt tungstate
tungstato de cobre copper tungstate
tungstato de litio lithium tungstate
tungstato de manganeso manganese
    tungstate
tungstato de plomo lead tungstate
tungstato ferroso ferrous tungstate
tungstato magnésico magnesium
    tungstate
tungstato potásico potassium tungstate
tungstato sódico sodium tungstate
tungsteno tungsten
túngstico tungstic
tungstita tungstite
tungstofosfato tungstophosphate
tungstofosfato sódico sodium
    tungstophosphate
tungstosilicato tungstosilicate
tungstosilicato sódico sodium
    tungstosilicate
tunsgstato bárico barium tungstate
turba peat
turbidez turbidity
turbidimetría turbidimetry
turbio cloudy
turmalina tourmaline
turquesa turquoise
tuyeno thujene
tuyona thujone

U

ubiquinona ubiquinone
ulexina ulexine
ulexita ulexite
ulmina ulmin
ultra ultra
ultracentrífuga ultracentrifuge
ultrafiltración ultrafiltration
ultramicroscópico ultramicroscopic
ultramicroscopio ultramicroscope
ultrapuro ultrapure
ultrarrápido ultrarapid
ultrasensible ultrasensitive
ultrasónica ultrasonics
ultrasónico ultrasonic
ultravioleta ultraviolet
umangita umangite
umbeliferona umbelliferone
umbelulona umbellulone
umbral threshold
umbral de detección threshold of
    detection
umbral de fisión fission threshold
unario unary
undeca undeca
undecalactona undecalactone
undecanal undecanal
undecano undecane
undecanol undecanol
undecanona undecanone
undecenilo undecenyl
undeceno undecene
undecenol undecenol
undecilamina undecylamine
undecilenato cálcico calcium
    undecylenate
undecilenato de cinc zinc undecylenate
undecilenato potásico potassium
    undecylenate
undecilenato sódico sodium undecylenate
undecileno undecylene
undecilo undecyl
ungüento ointment
uni uni

uniaxial uniaxial
unidad unit
unidad absoluta absolute unit
unidad angstrom angstrom unit
unidad atómica atomic unit
unidad básica base unit
unidad cuántica quantum unit
unidad de masa atómica atomic mass
    unit
unidad de peso atómico atomic weight
    unit
unidad de radiación radiation unit
unidad de radio radium unit
unidad de sacarosa saccharose unit
unidad derivada derived unit
unidad fundamental fundamental unit
unidad internacional international unit
unidad repetitiva repeating unit
unimolecular unimolecular
unión union
unión espiro spiro union
univalente univalent
ur ur
uracil uracil
uralita uralite
uramido uramido
uramina uramine
uramino uramino
uranato uranate
uranato sódico sodium uranate
urania urania
uránico uranic
uranilo uranyl
uranina uranine
uraninita uraninite
uranio uranium
uranio agotado depleted uranium
uranio enriquecido enriched uranium
uranita uranite
uranoceno uranocene
uranocircita uranocircite
uranofano uranophane
uranosferita uranospherite

uranoso uranous
uranotalita uranothallite
urato magnésico magnesium urate
urea urea
urea de etileno ethylene urea
ureasa urease
ureido ureido

uretano urethane
uridina uridine
urilón urylon
uronio uronium
urotropina urotropin
ursina ursin
uvarovita uwarowite

V

vacante vacancy
vacío vacuum
vainillilideno vanillylidene
vainillilo vanillyl
vainillina vanillin
vainilloilo vanilloyl
valacidina valacidin
valencia valence, valency
valencia principal principal valency
valentinita valentinite
valeral valeral
valeraldehído valeraldehyde
valeramida valeramide
valerato valerate
valerato amónico ammonium valerate
valerato cálcico calcium valerate
valerato ceroso cerous valerate
valerato de atropina atropine valerate
valerato de butilo butyl valerate
valerato de cadmio cadmium valerate
valerato de etilo ethyl valerate
valerato de fenilo phenyl valerate
valerato de isoamilo isoamyl valerate
valerato de isobutilo isobutyl valerate
valerato de mentol menthol valerate
valerato de pentilo pentyl valerate
valerilo valeryl
valerolactona valerolactone
valilo valyl
valina valine
valor calorífico calorific value
valor de acetilo acetyl value
valor de ácido acid value
valor de amortiguador buffer value
valor de bario barium value
valor de peróxido peroxide value
valor de pH pH value
valor de saponificación saponification
    value
valor de yodo iodine value
valor delta delta value
valor sigma sigma value
valor tau tau value

valor umbral threshold value
valor umbral límite threshold limit value
valoración titration
valoración ácido-base acid-base titration
valoración amperométrica
    amperometric titration
valoración automática automatic
    titration
valoración complejométrica
    complexometric titration
valoración con permanganato
    permanganate titration
valoración conductométrica
    conductometric titration
valoración de Mohr Mohr titration
valoración de Volhard Volhard titration
valoración diferencial differential
    titration
valoración electrométrica electrometric
    titration
valoración entálpica enthalpy titration
valoración fotométrica photometric
    titration
valoración magnetométrica
    magnetometric titration
valoración oscilométrica oscillometric
    titration
valoración potenciométrica
    potentiometric titration
valoración radiométrica radiometric
    titration
valoración redox redox titration
valoración térmica thermal titration
valoración termométrica thermometric
    titration
valproato sódico sodium valproate
válvula de admisión de gas gas-
    admittance valve
válvula de entrada de gas gas-inlet valve
vanadato vanadate
vanadato amónico ammonium vanadate
vanadato de itrio yttrium vanadate
vanadato de litio lithium vanadate

vanadato de plata   silver vanadate
vanadato de plomo   lead vanadate
vanadato férrico   ferric vanadate
vanadato sódico   sodium vanadate
vanadílico   vanadylic
vanadilo   vanadyl
vanadiloso   vanadylous
vanadinita   vanadinite
vanadio   vanadium
vanadioso   vanadous
vanadita   vanadite
vanadofosfato sódico   sodium vanadophosphate
vanadol   vanadol
vanilato de etilo   ethyl vanillate
vanilina de etilo   ethyl vanillin
vapor   vapor
vapor de agua   steam
vapor saturado   saturated vapor
vapor venenoso   poison vapor
vaporización   vaporization
vaporizador de Lundergardh   Lundergardh vaporizer
vaporizar   vaporize
varilla agitadora   stirring rod
variscita   variscite
vaso de laboratorio   beaker
vasopresina   vasopressin
vauquelinita   vauquelinite
vecinal   vicinal
vehículo   vehicle
velocidad   velocity, rate
velocidad de combustión   combustion rate
velocidad de desintegración   decay rate
velocidad de electrón   electron velocity
velocidad de fisión   fission rate
velocidad de formación   rate of formation
velocidad de migración   migration velocity
velocidad de propagación   propagation rate
velocidad de reacción   reaction rate, reaction velocity
velocidad de reacción específica   specific reaction rate
velocidad de sedimentación   sedimentation rate

velocidad mínima de ionización   minimum ionizing speed
velocidad molecular   molecular velocity
veneno   poison
veneno catalítico   catalytic poison
veneno de serpiente   snake venom
venenoso   poisonous
venturi   venturi
veratraldehído   veratraldehyde
veratrilideno   veratrylidene
veratrilo   veratryl
veratroilo   veratroyl
veratrol   veratrole
verbenona   verbenone
verde brillante   brilliant green
verde de benzaldehído   benzaldehyde green
verde de bromocresol   bromocresol green
verde de cobalto   cobalt green
verde de cobre   copper green
verde de cromo   chrome green
verde de naftol   naphthol green
verde de París   Paris green
verde esmeralda   emerald green
verde imperial   imperial green
verde malaquita   malachite green
verde Scheele   Scheele green
verde ultramarino   ultramarine green
verde Victoria   Victoria green
vermiculita   vermiculite
vernolepina   vernolepin
verxita   verxite
vetivona   vetivone
vía   pathway
vibración   vibration
vibración atómica   atomic vibration
vibración molecular   molecular vibration
vida media   mean life
vidrio   glass
vidrio cerámico   glass-ceramic
vidrio de agua   water glass
vidrio de bórax   borax glass
vidrio de borosilicato   borosilicate glass
vidrio de fosfato   phosphate glass
vidrio de plomo   lead glass
vidrio de seguridad   safety glass
vidrio de uranio   uranium glass
vidrio esmerilado   ground glass

vidrio líquido liquid glass
vidrio metálico metal glass
vidrio óptico optical glass
vidrio protector protective glass
vidrio rojo red glass
vidrio soluble soluble glass
villiaumita villiaumite
vinagre vinegar
vinilacetileno vinylacetylene
vinilacetonitrilo vinylacetonitrile
vinilación vinylation
vinilamina vinylamine
vinilbenceno vinylbenzene
vinilbutiléter vinyl butyl ether
vinilcarbazol vinylcarbazole
vinilcetona vinyl ketone
vinilciclohexeno vinylcyclohexene
vinileno vinylene
vinilestireno vinylstyrene
viniléter vinyl ether
viniletileno vinylethylene
viniletiléter vinyl ethyl ether
viniletilhexiléter vinylethylhexyl ether
viniletilpiridina vinylethylpyridine
vinilideno vinylidene
vinilimina vinylimine
vinilisobutiléter vinyl isobutyl ether
vinilmetilcetona vinyl methyl ketone
vinilmetiléter vinyl methyl ether
vinilo vinyl
vinílogo vinylog
vinilpiridina vinylpyridine
vinilpirrolidona vinylpyrrolidone
viniltolueno vinyltoluene
viniltriclorosilano vinyltrichlorosilane
vino wine
vioformo vioform
violantrona violanthrone
violeta de antraceno anthracene violet
violeta de cobalto cobalt violet
violeta de metilo methyl violet
violeta de París Paris violet
violeta neutra neutral violet
viomicina viomycin
virus virus
viscocidad superficial surface viscocity
viscoplástico viscoplastic
viscosa viscose
viscosidad viscosity

viscosidad intrínseca intrinsic viscosity
viscosimetría viscosimetry
viscosímetro viscometer, viscosimeter
viscoso viscid
visualizador analógico analog display
vitamina vitamin
vitelina vitellin
vítreo vitreous
vitrificación vitrification
vitriolo vitriol
vitriolo azul blue vitriol
vitriolo blanco white vitriol
vitriolo de cinc zinc vitriol
vitriolo verde green vitriol
vivianita vivianite
volátil volatile
volatilidad volatility
volatilización volatilization
volatilizar volatilize
volframio wolfram
voltaíta voltaite
voltzina voltzite
volumen volume
volumen atómico atomic volume
volumen crítico critical volume
volumen de combinación combining volume
volumen de retención retention volume
volumen específico specific volume
volumen incompresible incompressible volume
volumen molal molal volume
volumen molar molar volume, mole volume
volumen molar parcial partial molar volume
volumen molecular molecular volume
volumen molecular-gramo gram-molecular volume
volumen nuclear nuclear volume
volumenómetro volumenometer
volumetría titration
volumétrico volumetric
volúmetro volumeter
vulcanita vulcanite
vulcanización vulcanization
vulcanizar vulcanize
vulpinita vulpinite

W

wagnerita wagnerite
warfarina warfarin
warfarina sódica sodium warfarin
wavellita wavellite
whisky whiskey
whitneyita whitneyite
willemita willemite
witherita witherite
wohlerita wöhlerite
wolframato wolframate
wolframato de plomo lead wolframate

wolframato magnésico magnesium
 wolframate
wolframato potásico potassium
 wolframate
wolframato sódico sodium wolframate
wolframio wolfram
wolframita wolframite
wollastonita wollastonite
wulfenita wulfenite
wurtzita wurtzite

X

xantano xanthan
xantato xanthate
xantato de almidón starch xanthate
xantato de amilo amyl xanthate
xantato de celulosa cellulose xanthate
xantato de pentilo pentyl xanthate
xantato de propilo propyl xanthate
xantato potásico potassium xanthate
xantato sódico sodium xanthate
xantenilo xanthenyl
xanteno xanthene
xantenol xanthenol
xantilio xanthylium
xantilo xanthyl
xantina xanthine
xanto xantho
xantocromo xanthochromium
xantofila xanthophyll
xantogenato sódico sodium xanthogenate
xantona xanthone
xantopterina xanthopterin
xantosiderita xanthosiderite
xantosina xanthosine
xantoxileno xanthoxylene
xenato xenate
xenilamina xenylamine

xenilo xenyl
xenobiótico xenobiotic
xenol xenol
xenolita xenolite
xenón xenon
xenotima xenotime
xerogel xerogel
xerografía xerography
xi xi
xilano xylan
xileno xylene
xileno de almizcle musk xylene
xilenodiol xylenediol
xilenol xylenol
xilenosulfonato sódico sodium
    xylenesulfonate
xilideno xylidene
xilidina xylidine
xilileno xylylene
xililo xylyl
xilita xylite
xilitol xylitol
xilocaína xylocaine
xiloilo xyloyl
xilol xylol
xilosa xylose

Y

yeso gypsum
yeso de París plaster of Paris
yodado iodized
yodato iodate
yodato amónico ammonium iodate
yodato bárico barium iodate
yodato cálcico calcium iodate
yodato de cadmio cadmium iodate
yodato de cesio cesium iodate
yodato de cinc zinc iodate
yodato de litio lithium iodate
yodato de plata silver iodate
yodato de plomo lead iodate
yodato de rubidio rubidium iodate
yodato magnésico magnesium iodate
yodato potásico potassium iodate
yodato sódico sodium iodate
yodeosina iodeosin
yodilo iodyl
yodimetría iodimetry
yodipamida iodipamide
yodisán iodisan
yodita iodite
yodo iodine
yodoacetato de etilo ethyl iodoacetate
yodoalcano iodoalkane
yodoanilina iodoaniline
yodobehenato cálcico calcium iodobehenate
yodobenceno iodobenzene
yodoetano iodoethane
yodoetileno iodoethylene
yodoformo iodoform
yodóforo iodophor
yodohipurato sódico sodium iodohippurate
yodometano iodomethane
yodometría iodometry
yodométrico iodometric
yodonio iodonium
yodopropano iodopropane
yodosilo iodosyl
yodoso iodoso

yodosobenceno iodosobenzene
yodosuccinimida iodosuccinimide
yodoxi iodoxy
yodoxilo iodoxyl
yoduro iodide
yoduro alumínico aluminum iodide
yoduro amónico ammonium iodide
yoduro antimonioso antimonous iodide
yoduro arsenioso arsenous iodide
yoduro áurico auric iodide
yoduro bárico barium iodide
yoduro bárico mercúrico mercuric barium iodide
yoduro cálcico calcium iodide
yoduro ceroso cerous iodide
yoduro cobaltoso cobaltous iodide
yoduro crómico chromic iodide
yoduro cúprico cupric iodide
yoduro cuproso cuprous iodide
yoduro de acetilo acetyl iodide
yoduro de alilo allyl iodide
yoduro de antimonio antimony iodide
yoduro de arsénico arsenic iodide
yoduro de azufre sulfur iodide
yoduro de bencilo benzyl iodide
yoduro de benzoilo benzoyl iodide
yoduro de berilio beryllium iodide
yoduro de bismuto bismuth iodide
yoduro de bromo bromine iodide
yoduro de butilo butyl iodide
yoduro de cadmio cadmium iodide
yoduro de cesio cesium iodide
yoduro de cianógeno cyanogen iodide
yoduro de cinc zinc iodide
yoduro de cinc potásico potassium zinc iodide
yoduro de cobalto cobalt iodide
yoduro de cobre copper iodide
yoduro de estaño tin iodide
yoduro de estroncio strontium iodide
yoduro de etilo ethyl iodide
yoduro de fosfonio phosphonium iodide
yoduro de galio gallium iodide

yoduro de hexadecilo hexadecyl iodide
yoduro de hidrógeno hydrogen iodide
yoduro de hierro iron iodide
yoduro de isopropilo isopropyl iodide
yoduro de litio lithium iodide
yoduro de manganeso manganese iodide
yoduro de metileno methylene iodide
yoduro de metilmagnesio
    methylmagnesium iodide
yoduro de metilo methyl iodide
yoduro de neodimio neodymium iodide
yoduro de níquel nickel iodide
yoduro de nitrógeno nitrogen iodide
yoduro de octilo octyl iodide
yoduro de oro gold iodide
yoduro de paladio palladium iodide
yoduro de pentilo pentyl iodide
yoduro de plata silver iodide
yoduro de platino platinum iodide
yoduro de plomo lead iodide
yoduro de propilo propyl iodide
yoduro de rubidio rubidium iodide
yoduro de silicio silicon iodide
yoduro de talio thallium iodide
yoduro de telurio tellurium iodide
yoduro de timol thymol iodide
yoduro de titanio titanium iodide
yoduro de uranio uranium iodide
yoduro de vinilo vinyl iodide

yoduro de zirconio zirconium iodide
yoduro estánnico stannic iodide
yoduro estannoso stannous iodide
yoduro férrico ferric iodide
yoduro ferroso ferrous iodide
yoduro gálico gallic iodide
yoduro magnésico magnesium iodide
yoduro manganoso manganous iodide
yoduro mercúrico mercuric iodide
yoduro mercúrico bárico barium
    mercuric iodide
yoduro mercúrico de plata silver
    mercury iodide
yoduro mercúrico potásico potassium
    mercuric iodide
yoduro mercurioso mercurous iodide
yoduro niqueloso nickelous iodide
yoduro paladioso palladous iodide
yoduro platinoso platinous iodide
yoduro potásico potassium iodide
yoduro potásico de bismuto bismuth
    potassium iodide
yoduro potásico de cadmio cadmium
    potassium iodide
yoduro potásico de cinc zinc potassium
    iodide
yoduro sódico sodium iodide
yoduro talioso thallous iodide
yohimbina yohimbine

Z

zafiro sapphire
zahína sorghum
zaratita zaratite
zeaxantina zeaxanthin
zeína zein
zeolita zeolite
zeolita en jaula cage zeolite
zimasa zymase
zimohexasa zymohexase
zimólisis zymolysis
zimoquímica zymochemistry
zimosa zymose
zimosis zymosis
zimurgia zymurgy
zinc zinc
zingibereno zingiberene
zinnwaldita zinnwaldite
zippeíta zippeite
ziram ziram
zircón zircon
zirconato zirconate

zirconato bárico barium zirconate
zirconato cálcico calcium zirconate
zirconato de estroncio strontium
    zirconate
zirconato de litio lithium zirconate
zirconato de tetrabutilo tetrabutyl
    zirconate
zirconato de tetraisopropilo
    tetraisopropyl zirconate
zirconato magnésico magnesium
    zirconate
zirconia zirconia
zircónico zirconic
zirconilo zirconyl
zirconio zirconium
zirlita zirlite
zoaleno zoalene
zoisita zoisite
zooquímica zoochemistry
zwitterion zwitterion